Lecture Notes in Computer Science 9076

Commenced Publication in 1973
Founding and Former Series Editors:
Gerhard Goos, Juris Hartmanis, and Jan van Leeuwen

More information about this series at http://www.springer.com/series/7407

Rahul Jain · Sanjay Jain
Frank Stephan (Eds.)

Theory and Applications of Models of Computation

12th Annual Conference, TAMC 2015
Singapore, May 18–20, 2015
Proceedings

 Springer

Editors
Rahul Jain
National University of Singapore
Singapore
Singapore

Frank Stephan
National University of Singapore
Singapore
Singapore

Sanjay Jain
National University of Singapore
Singapore
Singapore

ISSN 0302-9743   ISSN 1611-3349 (electronic)
Lecture Notes in Computer Science
ISBN 978-3-319-17141-8   ISBN 978-3-319-17142-5 (eBook)
DOI 10.1007/978-3-319-17142-5

Library of Congress Control Number: 2015937497

Springer Cham Heidelberg New York Dordrecht London

Printed on acid-free paper

Springer International Publishing AG Switzerland is part of Springer Science+Business Media
(www.springer.com)

Preface

This volume contains the papers presented at TAMC 2015: Theory and Applications of Models of Computation held during May 18–20, 2015 in Singapore.

There were 78 submissions. Each submission was reviewed by 2–5 (on the average by 3.1) reviewers. The Program Committee decided to accept 35 papers. We would like to thank the Program Committee members and their sub reviewers for their hard work in putting together this program. The program also included 3 invited talks by Lance Fortnow, Miklos Santha, and Alexandra Shlapentokh.

The conference series TAMC started in the year 2004 and has been held annually since then. The previous conferences are the following: Beijing China 2004, Kunming China 2005, Beijing China 2006, Shanghai China 2007, Xian China 2008, Changsha China 2009, Prague Czech Republic 2010, Tokyo Japan 2011, Beijing China 2012, Hong Kong China 2013, and Chennai India 2014.

We thank the Steering Committee for their timely advice on various matters. The Steering Committee comprises Manindra Agrawal, Jin-Yi Cai, Barry Cooper, John Hopcroft, Angsheng Li, and Zhiyong Liu.

We would also like to thank the local team here in Singapore:

(1) Siew Foong Ho (Local Arrangements Chair) and Pei Pei Teo for Local Organization.

(2) Yong Ngee Kee and Philip Lim for Registration website.

(3) Several members of the finance and administrative staff of the School of Computing, National University of Singapore, for taking care of various matters.

We kindly acknowledge the financial support from the School of Computing and the Centre for Quantum Technologies, National University of Singapore.

We greatly appreciate the help offered by Easychair in handling all matters related to the conference, right from receiving the submissions to the creation of the proceedings. Easychair really makes the life of the PC Chairs and the PC very easy! We are also grateful to Springer for their continuous support and for publishing of the proceedings of the conference series TAMC.

February 2015

Sanjay Jain
Rahul Jain
Frank Stephan

Organization

Program Committee

Ajith Abraham	Machine Intelligence Research Laboratories, Auburn, Washington, USA
Anthony Bonato	Ryerson University, Canada
Yijia Chen	Shanghai Jiao Tong University, China
Rodney G. Downey	Victoria University of Wellington, New Zealand
Henning Fernau	Universität Trier, Germany
Dimitris Fotakis	National Technical University of Athens, Greece
T.V. Gopal	Anna University Chennai, India
Rahul Jain	National University of Singapore, Singapore
Steffen Lempp	University of Wisconsin - Madison, USA
Jiamou Liu	Auckland University of Technology, New Zealand
Frédéric Magniez	Université Paris Diderot, France
Klaus Meer	Brandenburgische Technische Universität Cottbus-Senftenberg, Germany
Mia Minnes	University of California, San Diego, USA
Philippe Moser	National University of Ireland, Maynooth, Ireland
Mitsunori Ogihara	University of Miami, USA
Yota Otachi	Japan Advanced Institute of Science and Technology, Japan
Yicheng Pan	Institute of Software, Chinese Academy of Sciences, China
Pan Peng	Technische Universität Dortmund, Germany
Anil Seth	Indian Institute of Technology Kanpur, India
Frank Stephan	National University of Singapore, Singapore
Chaitanya Swamy	University of Waterloo, Canada
Hongan Wang	Institute of Software, Chinese Academy of Sciences, China
Wei Wang	Sun Yat-Sen University, Guangzhou, China
Guohua Wu	Nanyang Technological University, Singapore
Sun Xiaoming	Institute of Computing Technology, Chinese Academy of Sciences, China
Yitong Yin	Nanjing University, China
Mingsheng Ying	University of Technology Sydney, Australia
Neal Young	University of California, Riverside, USA
Thomas Zeugmann	Hokkaido University, Sapporo, Japan
Shengyu Zhang	Chinese University of Hong Kong, Hong Kong
Conghua Zhou	Jiangsu University, China

Additional Reviewers

Ambainis, Andris
Baartse, Martijn
Bazgan, Cristina
Bei, Xiaohui
Bouyer, Patricia
Bury, Marc
Chan, Timothy M.
Chang, Hong
Chrobak, Marek
Cooper, S. Barry
Damaschke, Peter
Dell, Holger
Dittmann, Christoph
Elberfeld, Michael
Fellows, Michael
Fomin, Fedor
Fournier, Jacques
Gadouleau, Maximilien
Har-Peled, Sariel
Homan, Christopher
Hu, Guangda
Huang, Xiuzhen
Jaggi, Sidharth
Jain, Sanjay
Janssen, Jeannette
Jeffery, Stacey
Kalimullin, Iskander
Kaplan, Marc
Kiyomi, Masashi

Konrad, Christian
Korman, Matias
Kosub, Sven
Krzyzak, Adam
Lauriere, Mathieu
Lemire, Daniel
Lescanne, Pierre
Leupold, Peter
Limaye, Nutan
Lin, Chengyu
Liu, Xingwu
M.S., Ramanujan
Maehara, Takanori
Markham, Damian
Mengel, Stefan
Mestre, Julian
Milis, Ioannis
Mouawad, Amer
Munteanu, Alexander
Nandakumar, Satyadev
Papakonstantinou, Periklis
Pappa, Anna
Penninger, Rainer
Poirier, Antoine
Prouff, Emmanuel
R., Venkatesh
Raman, Rajeev
Rao, Michael
Ravi, S.S.

Rawitz, Dror
Reimann, Jan
Reinhardt, Klaus
Schwiegelshohn, Chris
Schwoon, Stefan
Schöpp, Ulrich
Seki, Shinnosuke
Serre, Olivier
Shen, Yuping
Srinivasan, Venkatesh
Staiger, Ludwig
Strehler, Martin
Toda, Takahisa
Tsang, Hing Yin
Tzoumas, Vasileios
Wang, Haitao
Wang, Yaoyu
Wei, Zhaohui
Wright, John
Xiao, Tao
Youseff, Arman
Zhang, Chihao
Zhang, Jialin
Zhang, Peng
Zhang, Yi
Zhou, Gelin
Zhou, Hong
Zissimopoulos, Vassilis

Contents

Invited Papers

Hilbert's Tenth Problem for Subrings of $\mathbb{Q}$ and Number Fields (Extended Abstract)

Alexandra Shlapentokh[✉]

Department of Mathematics, East Carolina University,
Greenville, NC 27858, USA
shlapentokha@ecu.edu
http://myweb.ecu.edu/shlapentokha

1   Some History

In 1900 David Hilbert presented a list of questions at an international meeting of Mathematicians in Paris. The tenth problem on the list asked the following question (rephrased here in modern terms): given an arbitrary polynomial equation in several variables over $\mathbb{Z}$, is there a uniform algorithm to determine whether such an equation has solutions in $\mathbb{Z}$? This question, became known as Hilbert's Tenth Problem, and has been answered negatively in the work of M. Davis, H. Putnam, J. Robinson and Yu. Matijasevich. (See [3,4,12]).

Since the time when this result was obtained, similar questions have been raised for other fields and rings. In other words, if R is a computable ring, then, given an arbitrary polynomial equation in several variables over R, is there a uniform algorithm to determine whether such an equation has solutions in R? Arguably, the most prominent open questions in the area are the questions of decidability of an analog of Hilbert's Tenth Problem (HTP) for $R = \mathbb{Q}$ and R equal to the ring of integers of an arbitrary number field. In this talk we survey what is known about the status of Hilbert's Tenth Problem over subrings of number fields, including $\mathbb{Q}$ to date.

We start our survey with a discussion of HTP for the field of rational numbers. The first attempts to show that HTP was undecidable over $\mathbb{Q}$ involved efforts to construct an existential (Diophantine) definition of $\mathbb{Z}$ over $\mathbb{Q}$ in the language of rings. It is not hard to show that in general any ring of characteristic 0 where $\mathbb{Z}$ is existentially definable as a subset has an undecidable HTP.

Unfortunately, the Diophantine definition plan for $\mathbb{Q}$ quickly ran into problems. In 1992 Barry Mazur formulated a series of conjectures which were to play an important role in the development of the subject (see [13–15]). One of these conjectures implied that $\mathbb{Z}$ has not Diophantine (existential) definition over $\mathbb{Q}$. Quite a few years later Jochen Königsmann showed in [11] that a strong version of Bombieri-Lang conjecture also implies that there is no Diophantine definition of $\mathbb{Z}$ over $\mathbb{Q}$.

A. Shlapentokh—The author has been partially supported by the NSF grant DMS-1161456.

© Springer International Publishing Switzerland 2015
R. Jain et al. (Eds.): TAMC 2015, LNCS 9076, pp. 3–9, 2015.
DOI: 10.1007/978-3-319-17142-5_1

Since the plan to construct a Diophantine definition of $\mathbb{Z}$ over $\mathbb{Q}$ ran into substantial difficulties, alternative ways were considered for showing that HTP had no solution over $\mathbb{Q}$. One of the alternative methods required construction of a Diophantine model of $\mathbb{Z}$, i.e. a map $\phi : \mathbb{Z} \longrightarrow \mathbb{Q}^k$ such that the graphs of addition and multiplication over $\mathbb{Z}$ are Diophantine. Such a map, if it exists, maps Diophantine sets to Diophantine sets and one can then show that $\mathbb{Q}$ like $\mathbb{Z}$ has undecidable Diophantine sets implying that HTP is undecidable over $\mathbb{Q}$.

An old plan for building a Diophantine model of $\mathbb{Z}$ over $\mathbb{Q}$ involved elliptic curves of rank one, using multiples of a point of infinite order to model integers. Under such a map the graph of addition is clearly Diophantine, but it is not clear what happens to the graph of multiplication. Unfortunately, it turns out that the situation with Diophantine models is not any better than with Diophantine definitions. A theorem of Cornelissen and Zahidi (see [2]) showed that multiplication of indices of elliptic curve points is probably not existentially definable.

Theorem 1. *If Mazur's conjecture on topology of rational points holds, then there is no Diophantine model of $\mathbb{Z}$ over $\mathbb{Q}$.*

At this point two main research directions emerged trying to approach HTP for $\mathbb{Q}$: investigation of big rings and reduction of the number of universal quantifiers in the first-order definitions of $\mathbb{Z}$ over $\mathbb{Q}$. In this talk we will concentrate on subrings of $\mathbb{Q}$.

2   Big Rings

We start with a definition of the rings in question whose first appearance on the scene in [27] and [26] dates back to 1994.

Definition 2 (A Ring in Between). Let $\mathscr{S}$ be a set of primes of $\mathbb{Q}$. Let $O_{\mathbb{Q},\mathscr{S}}$ be the following subring of $\mathbb{Q}$.

$$\left\{ \frac{m}{n} : m, n \in \mathbb{Z}, n \neq 0, n \text{ is divisible by primes of } \mathscr{S} \text{ only} \right\}$$

If $\mathscr{S} = \emptyset$, then $O_{\mathbb{Q},\mathscr{S}} = \mathbb{Z}$. If $\mathscr{S}$ contains all the primes of $\mathbb{Q}$, then $O_{\mathbb{Q},\mathscr{S}} = \mathbb{Q}$. If $\mathscr{S}$ is finite, we call the ring *small*. If $\mathscr{S}$ is infinite, we call the ring *big*.

Some of these rings have other (canonical) names: the small rings are also called rings of $\mathscr{S}$-integers, and when $\mathscr{S}$ contains all but finitely many primes, the rings are called semi-local subrings of $\mathbb{Q}$. To measure the "size" of big rings we use the natural density of prime sets defined below.

Definition 3 (Natural Density). If $\mathscr{A}$ is a set of primes, then the natural density of $\mathscr{A}$ is equal to the limit below (if it exists):

$$\lim_{X \to \infty} \frac{\#\{p \in A, p \leq X\}}{\#\{p \leq X\}}$$

It turned out that we already knew everything we needed to know about small rings from the work of Julia Robinson (see [24]). In particular from her work on the first-order definability of integers over $\mathbb{Q}$ one can deduce the following theorem and corollaries.

Theorem 4 (Julia Robinson). *For every p, the ring $R_p = \{x \in \mathbb{Q} | x = \frac{m}{n}, m, n \in \mathbb{Z}, n > 0, p \nmid n\}$ has a Diophantine definition over $\mathbb{Q}$.*

This theorem of Julia Robinson plays a role in many other results. In particular, we get the following corollary.

Corollary 5. *1. $\mathbb{Z}$ has a Diophantine definition over any small subring of $\mathbb{Q}$.*
2. HTP is unsolvable over all small subrings of $\mathbb{Q}$.

Over big rings the questions turned out to be far more difficult. In 2003 Bjorn Poonen in [21] proved the first result on Diophantine undecidability (unsolvability of HTP) over a big subring of $\mathbb{Q}$.

Theorem 6. *There exist recursive sets of primes $\mathcal{T}_1$ and $\mathcal{T}_2$, both of natural density zero and with an empty intersection, such that for any set $\mathcal{S}$ of primes containing $\mathcal{T}_1$ and avoiding $\mathcal{T}_2$, the following hold:*

– $\mathbb{Z}$ has a Diophantine model over $O_{\mathbb{Q},\mathcal{S}}$.
– Hilbert's Tenth Problem is undecidable over $O_{\mathbb{Q},\mathcal{S}}$.

Poonen used elliptic curves to prove his result but the model he constructed was very different from the one envisioned by the old elliptic curve plan we described earlier. Poonen modeled integers *by approximation*. The construction of the model does start with an elliptic curve of rank one

Poonen's method was further extended by Eisenträger and Everest in [7], by Perlega in [18] and finally by Eisenträger, Everest and the author in [8]. The theorem proved in [8] provides a "covering" of $\mathbb{Q}$ by big rings where HTP is undecidable.

Theorem 7 (Eisenträger, Everest, Shlapentokh). *For any finite set of positive computable real numbers (i.e. real numbers which can be approximated by a computable sequence of rational numbers) $r_1, \ldots, r_k$ such that $r_1 + \ldots + r_k = 1$ we can partition the set of all (rational) primes into set $\mathcal{P}_1, \ldots, \mathcal{P}_k$ such that the natural density of each $\mathcal{P}_i$ is r_i, each ring $O_{\mathbb{Q},\mathcal{P}_i}$ has a Diophantine model of $\mathbb{Z}$ and therefore HTP is undecidable over each $O_{\mathbb{Q},\mathcal{P}_i}$.*

The author also constructed a model of $\mathbb{Z}$ using Diophantine equivalence classes (a class model of $\mathbb{Z}$) over a big ring using the old idea of trying to make multiplication of indices Diophantine in [29].

2.1 The Other End of the Spectrum

In this section we would like to describe some work (still in progress) which approached HTP over big rings from the other end of the spectrum, i.e. from the point of view of $\mathbb{Q}$.

We convert the question of decidability of HTP of a recursive ring R into a question of the Turing degree of a subset of $\mathbb{Z}_{>0}$. To that end let $\{p_i(\bar{x})\}$ be an effective enumeration of all polynomials over R and let $\mathrm{HTP}(R)$ denote the set of indices corresponding to polynomials having a root in R. Now given DPRM result we have that $\mathrm{HTP}(\mathbb{Z}) \equiv_T H$, where H is the halting set. Further, any recursive or computably presentable ring R with a Diophantine model of $\mathbb{Z}$ has $\mathrm{HTP}(R) \equiv_T H$.

At the same time, by results of Richard Friedberg [10] and Albert Muchnik [17] we know that there are Turing degrees containing undecidable r.e. sets not as hard as H, i.e. H is not Turing equivalent to these sets. What if $\mathrm{HTP}(\mathbb{Q})$ is one of these sets? If this were the case, there would be neither an algorithm to solve HTP over $\mathbb{Q}$ nor a Diophantine model of $\mathbb{Z}$ over $\mathbb{Q}$. So if $\mathrm{HTP}(\mathbb{Q}) \not\equiv_T \mathrm{HTP}(\mathbb{Z})$ it makes sense to see if there are big subrings R of $\mathbb{Q}$, "infinitely" far away from $\mathbb{Q}$ with $\mathrm{HTP}(R) \equiv_T \mathrm{HTP}(\mathbb{Q})$.

In the project still in progress (see [9]) Kirsten Eisenträger, Russell Miller, Jennifer Park, and the author using an effective form of Theorem 4 have constructed families of computably presentable subrings R of $\mathbb{Q}$ with $\mathrm{HTP}(R) \equiv_T \mathrm{HTP}(\mathbb{Q})$. The constructed rings consist of rational numbers where an infinite set of primes is allowed to divide the denominator, but the complement of this set of primes, that is the set of primes that are not allowed to divide the denominator is also infinite. Priority method was used to make the set of inverted primes c.e. (and thus the rings computably presentable). Further, the set of primes which can occur as divisors of the denominators of elements in the ring can be arranged to have the lower natural density equal to 0. So we are truly looking at a ring "in the middle", "infinitely far away" from both $\mathbb{Z}$ and $\mathbb{Q}$. These rings also have the property that the set of inverted primes, i.e. primes allowed to divide the denominators is computable from $\mathrm{HTP}(\mathbb{Q})$. So if $\mathrm{HTP}(\mathbb{Q})$ is decidable, these prime sets are also decidable and the rings in question are computable subrings of $\mathbb{Q}$ (not just computably presentable).

The co-authors have also obtained an analog of Theorem 7, though a weaker one. More specifically, for any positive integer k, one can partition the set of all prime numbers into k sets $\mathscr{S}_1, \ldots, \mathscr{S}_K$, each of lower density 0, and construct rings $R_1, \ldots, R_k$ where the primes allowed to divide the denominators are precisely $\mathscr{S}_1, \ldots, \mathscr{S}_K$ respectively and such that $\mathrm{HTP}(R_i) \equiv_T \mathrm{HTP}(\mathbb{Q})$. Unfortunately, these rings are not necessarily computably presentable and we can only say that each $\mathscr{S}_i$ is Turing reducible to $\mathrm{HTP}(\mathbb{Q})$, so that again if $\mathrm{HTP}(\mathbb{Q})$ is decidable, these prime sets are also decidable and the rings in question are computable subrings of $\mathbb{Q}$.

Now if we combine the results above with results constructing big rings with HTP equivalent to the halting problem, then one can conclude that if HTP over $\mathbb{Z}$ is different from HTP over $\mathbb{Q}$, in particular if $\mathrm{HTP}(\mathbb{Q})$ is decidable, then we have an extremely strange picture of tightly intermingled recursive rings inside $\mathbb{Q}$ with different levels of difficulty for HTP. Such a picture seems unlikely, though of course we cannot rule it out without a proof.

3 Number Fields

The state of knowledge concerning the rings of integers and HTP is summarized in the theorem below.

Theorem 8. $\mathbb{Z}$ *is Diophantine and HTP is unsolvable over the rings of integers of the following fields:*

- *Extensions of degree 4 of $\mathbb{Q}$ (except for a totally complex extension without a degree-two subfield), totally real number fields and their extensions of degree 2. (See [5, 6].) Note that these fields include all abelian extensions.*
- *Number fields with exactly one pair of non-real embeddings (See [19, 25].)*
- *Any number field K such that there exists an elliptic curve E of positive rank defined over $\mathbb{Q}$ with $[E(K) : E(\mathbb{Q})] < \infty$. (See [20, 22, 28].)*
- *Any number field K such that there exists an elliptic curve of rank 1 over K and an abelian variety over $\mathbb{Q}$ keeping its rank over K. (See [1].)*

All the gaps in the results are "almost" filled by a theorem proved by B. Mazur and K. Rubin (see [16]).

Theorem 9. *Suppose K/L is a cyclic extension of prime degree of number fields. If the Shafarevich-Tate Conjecture is true for L, then there is an elliptic curve E over L with $rank(E(L)) = rank(E(K)) = 1$. (Here $E(K)$ denote the points on the elliptic curve with coordinates in K and $E(L)$ is defined analogously with respect to L.)*

From this theorem we can obtain the following corollary.

Corollary 10. *If Shafarevich-Tate Conjecture is true for all number fields, then $\mathbb{Z}$ has a Diophantine definition over the ring of integers of any number field and therefore HTP is undecidable over the ring of integers of any number field.*

4 Big Rings Inside Number Fields

As over $\mathbb{Q}$ we can define big and small subrings of number fields. If a number field has an elliptic curve of rank 1 (which would follow if Shafarevich-Tate Conjecture is true), then there are generalizations of Poonen's construction and the complementary subring construction. (See [8, 23, 29].) At the same time "the other end of spectrum" results generalize unconditionally.

References

1. Cornelissen, G., Thanases, P., Zahidi, K., Zahidi, K.: Division-ample sets and diophantine problem for rings of integers. J. de Théorie des Nombres Bordeaux **17**, 727–735 (2005)
2. Cornelissen, G., Zahidi, K.: Topology of diophantine sets: remarks on Mazur's conjectures. In: Denef, J., Lipshitz, L., Pheidas, T., Van Geel, J. (eds.) Hilbert's Tenth Problem: Relations with Arithmetic and Algebraic Geometry. Contemporary Mathematics, pp. 253–260. American Mathematical Society, Rhods Island (2000)
3. Davis, M.: Hilbert's tenth problem is unsolvable. Am. Math. Monthly **80**, 233–269 (1973)
4. Davis, M., Matiyasevich, Y., Robinson, J.: Hilbert's tenth problem. Diophantine equations: Positive aspects of a negative solution. In: Proceedings of Symposia in Pure Mathematics, vol. 28, pp. 323–378. American Mathematical Society (1976)
5. Denef, J.: Hilbert's tenth problem for quadratic rings. Proc. Am. Math. Soc. **48**, 214–220 (1975)
6. Denef, J., Lipshitz, L.: Diophantine sets over some rings of algebraic integers. J. London Math. Soc. **18**(2), 385–391 (1978)
7. Eisenträger, K., Everest, G.: Descent on elliptic curves and Hilbert's tenth problem. Proc. Am. Math. Soc. **137**(6), 1951–1959 (2009)
8. Eisenträger, K., Everest, G., Shlapentokh, A.: Hilbert's tenth problem and Mazur's conjectures in complementary subrings of number fields. Math. Res. Lett. **18**(6), 1141–1162 (2011)
9. Eisenträger, K., Miller, R., Park, J., Shlapentokh, A.: Easy as $\mathbb{Q}$. (work in progress)
10. Friedberg, R.M.: Two recursively enumerable sets of incomparable degrees of unsolvability. Proc. National Acad. Sci. USA **43**, 236–238 (1957)
11. Jochen K.: Defining $\mathbb{Z}$ in $\mathbb{Q}$ Annals of Mathematics to appear
12. Matiyasevich, Y.V.: Hilbert's tenth problem. Foundations of Computing Series. MIT Press, Cambridge (1993). Translated from the 1993 Russian original by the author, With a foreword by Martin Davis
13. Mazur, B.: The topology of rational points. Exp. Math. **1**(1), 35–45 (1992)
14. Mazur, B.: Questions of decidability and undecidability in number theory. J. Symb.Logic **59**(2), 353–371 (1994)
15. Mazur, B.: Galois representations in arithmetic algebraic geometry. In: Scholl, A.J., Taylor, R.L. (eds.) Open problems regarding rational points on curves and varieties. Cambridge University Press, Cambridge (1998)
16. Mazur, B., Rubin, K.: Ranks of twists of elliptic curves and Hilbert's Tenth Problem. Inventiones Mathematicae **181**, 541–575 (2010)
17. Muchnik, A.A.: On the separability of recursively enumerable sets. Doklady Akademii Nauk SSSR (N.S.), vol. 109,pp. 29–32 (1956)
18. Perlega, S.: Additional results to a theorem of Eisenträger and Everest. Archiv der Mathematik (Basel) **97**(2), 141–149 (2011)
19. Pheidas, T.: Hilbert's tenth problem for a class of rings of algebraic integers. Proc. Am. Math. Soc. **104**(2), 611–620 (1988)
20. Bjorn, P.: Elliptic curves whose rank does not grow and Hilbert's Tenth Problem over the rings of integers. Private Communication
21. Poonen, B.: Using elliptic curves of rank one towards the undecidability of Hilbert's tenth problem over rings of algebraic integers. In: Fieker, C., Kohel, D.R. (eds.) ANTS 2002. LNCS, vol. 2369, pp. 33–42. Springer, Heidelberg (2002)

22. Poonen, B.: Hilbert's tenth problem and Mazur's conjecture for large subrings of ℚ. J Am. Math. Soc. **16**(4), 981–990 (2003)
23. Poonen, B., Shlapentokh, A.: Diophantine definability of infinite discrete non-archimedean sets and diophantine models for large subrings of number fields. Journal für die Reine und Angewandte Mathematik **27–48**, 2005 (2005)
24. Robinson, J.: Definability and decision problems in arithmetic. J. Symb. Logic **14**, 98–114 (1949)
25. Shlapentokh, A.: Extension of Hilbert's tenth problem to some algebraic number fields. Commun. Pure Appl. Math. **XLII**, 939–962 (1989)
26. Shlapentokh, A.: Diophantine classes of holomorphy rings of global fields. J. Algebra **169**(1), 139–175 (1994)
27. Shlapentokh, A.: Diophantine equivalence andcountable rings. J. Symb. Logic **59**, 1068–1095 (1994)
28. Shlapentokh, A.: Elliptic curves retaining their rank in finite extensions and Hilbert's tenth problem for rings of algebraic numbers. Trans. Am. Math. Soc. **360**(7), 3541–3555 (2008)
29. Shlapentokh, A.: Elliptic curve points and Diophantine models of ℤ in large subrings of number fields. Int. J. Number Theor. **8**(6), 1335–1365 (2012)

Nondeterministic Separations

Lance Fortnow[✉]

Georgia Institute of Technology, Atlanta, USA
fortnow@cc.gatech.edu

Abstract. We survey recent research on the power of nondeterministic computation and how to use nondeterminism to get new separations of complexity classes. Results include separating NEXP from NP with limited advice, a new proof of the nondeterministic time hierarchy and a surprising relativized world where NP is as powerful as NEXP infinitely often.

1 Results

In this talk we focus on new results by the speaker about the power of nondeterminism which sits at the heart of the famous P versus NP problem. The results in this paper first appeared in works by Buhrman, Fortnow and Santhanam [1–3]

Theorem 1. *For any constant c,* NEXP $\not\subseteq$ NP/n^c.

Eric Allender asked whether even Theorem 1 (NEXP $\not\subseteq$ NP/n^c) can be strengthened to a lower bound that works on almost all input lengths, rather than on infinitely many. Direct diagonalizations tend to work on almost all input lengths–our separation is indirect, and technique does not give this stronger property. We give a new relativized world showing that relativizing techniques cannot get the stronger separation even without the advice.

Theorem 2. *There exists a relativized world such that* NEXP $\subseteq$ i.o.NP.

Cook [4] first showed a nondeterministic time hierarchy, given in its strongest form by Seiferas, Fischer and Meyer [5] and simplified by Žàk [6]. We give yet a new proof that gives a far more compact diagonalization.

Theorem 3. *If* t_1 *and* t_2 *are time-constructable functions such that*

- $t_1(n) = o(t_2(n))$, *and*
- $n \leqslant t_1(n) \leqslant n^c$ *for some constant c*

then NTIME($t_2(n)$) $\not\subseteq$ NTIME($t_1(n)$).

Corollary 1. *For any reals* $1 \leqslant r < s$, NTIME(n^s) $\not\subseteq$ NTIME(n^r).

We can use the techniques of this new proof to get a time hierarchy with advice.

© Springer International Publishing Switzerland 2015
R. Jain et al. (Eds.): TAMC 2015, LNCS 9076, pp. 10–17, 2015.
DOI: 10.1007/978-3-319-17142-5_2

Theorem 4. *Let $d \geqslant 1$ be any constant, and let t be a time-constructible time bound such that $t = o(n^d)$. Then* $\mathsf{NTIME}(n^d) \not\subseteq \mathsf{NTIME}(t)/n^{1/d}$.

Theorem 4 improves on known results handling advice in two respects. First, the amount of advice in the lower bound can be as high as $n^{\Omega(1)}$, in contrast to earlier results in which it was limited to be $O(\log(n))$. Second, the hierarchy is provably tight in terms of the time bounds, while earlier results handling advice could only separate $\mathsf{NTIME}(n^d)$ from $\mathsf{NTIME}(n^c)$ with advice, where $c < d$.

We are able to use Theorem 4 to derive a new circuit lower bound for NP, improving a 30-year old result of Kannan [7].

Corollary 2. *Let $k > 1$ be any constant.* NP *does not have* NP-*uniform non-deterministic circuits of size $O(n^k)$.*

2   Proof of Theorem 1

We first need the following lemma, a slightly stronger version of a result in Homer and Mocas [8] about lower bounds for deterministic exponential time against advice. The proof we give is folklore.

Lemma 1. *For any constant d,* $\mathsf{EXP} \not\subseteq$ i.o.$\mathsf{DTIME}(2^{n^d})/n^d$.

Proof. The proof is by diagonalization. We define a diagonalizing language L which is not in i.o.$\mathsf{DTIME}(2^{n^d})/n^d$ by defining a machine M which runs in exponential time and decides L.

M operates as follows on input x of length n. It enumerates advice taking machines $M_1, M_2 \ldots M_{\log(n)}$ each running in time at most 2^{n^d} and taking advice of length n^d. It then enumerates all $\log(n)2^{n^d}$ truth tables computed by these machines when every possible string of length n^d is given as advice. It then computes a truth table of an n-bit function f which is distinct from all the truth tables enumerated so far–this can be done in exponential time by a simple pruning strategy. Finally it outputs $f(x)$.

Now we are ready to prove our lower bound for NEXP.

Proof. We will show that either $\mathsf{NEXP} \not\subseteq \mathsf{NP}/poly$ or $\mathsf{NEXP} \not\subseteq \mathsf{NE}/n^c$. From this, the result follows.

Assume, to the contrary, that both these inclusions hold, i.e., $\mathsf{NEXP} \subseteq \mathsf{NP}/poly$ and $\mathsf{NEXP} \subseteq \mathsf{NE}/n^c$. We will derive a contradiction. Let L be a complete language for NE with respect to linear-time reductions. Since $\mathsf{NEXP} \subseteq \mathsf{NP}/poly$, we get that $L \in \mathsf{NTIME}(n^k)/n^k$ for some constant k. Since L is complete for NE with respect to linear-time reductions, we get that $\mathsf{NE} \subseteq \mathsf{NTIME}(n^k)/O(n^k)$.

By translation, we get that $\mathsf{NE}/n^c \subseteq \mathsf{NTIME}(n^{kc})/O(n^{kc})$. To see this, let L' be a language in NE/n^c, and let M' be an advice-taking NE machine accepting L' with advice length n^c. Define a language $L'' \in \mathsf{NE}$ as follows: a string $< x, a >$ is in L'' iff M' accepts x with advice a. Since M' is an NE machine, it follows that

$L'' \in$ NE. Thus, by assumption $L'' \in$ NTIME$(m^k)/O(m^k)$, where m is the input length for L''. Let M'' be an advice-taking machine solving L'' using resources as stated. Now we can solve L' in NTIME$(n^{kc})/O(n^{kc})$ as follows. The advice-taking machine M we construct for solving L' interprets its advice as consisting of two parts: the first part is an advice string a of length n^c, where n is the input size, and the second part is an advice string b of length $O((n + n^c)^k) = O(n^{kc})$. M simulates M'' on input $< x, a >$ with advice string b, where x is the input for L'. M accepts iff M'' accepts. M operates within time $O(n^{kc})$ (since it simulates an $O(n^k)$ time machine on an input of length $O(n^c)$), uses advice of length $O(n^{kc})$, and decides L' correctly, by definition of L'' and the assumption on M''.

Thus, we have NEXP $\subseteq$ NE/n^c and NE/$n^c \subseteq$ NTIME$(n^{kc})/O(n^{kc})$, which together imply NEXP $\subseteq$ NTIME$(n^{kc})/O(n^{kc})$. But since EXP $\subseteq$ NEXP and NTIME$(n^{kc})/O(n^{kc}) \subseteq$ DTIME$(2^{n^{kc}})/O(n^{kc})$ we get EXP $\subseteq$ DTIME$(2^{n^{kc}})/O(n^{kc})$, which is a contradiction to Lemma 1.

3   Proof of Theorem 2

We show the surprising relativized world where NEXP is infinitely often contained in NP.

Proof. Let M_i be a standard enumeration of non-deterministic relativized Turing machines that runs in time at most 2^{n^i}. Since these machines are paddable, for any A and any $L \in$ NEXP^A there will some i such that $L = L(M_i^A)$. We will create A such that for every i there are an infinite number of n such that for all x of length n,

$$x \in L(M_i^A) \Leftrightarrow \text{there exists a } y \text{ with } |y| = 2|x|^i \text{ and } (i, x, y) \in A$$

which immediately implies Theorem 2.

Start with $A = \varnothing$. We construct A in stages (i, j) chosen in any order that cover all possible (i, j).

Stage (i, j): Pick n such that n is larger than any frozen string as well as the n chosen in any previous stage.

Set all strings x of length n to be unmarked.

Repeat the following as long as there is an unmarked x of length n such that $M_i^A(x)$ accepts: Fix an accepting path of $M_i^A(x)$ and freeze every string queried along that path. Mark x. Pick a y, $|y| = 2|x|^i$ such that (i, x, y) is not frozen and let $A = A \cup \{(i, x, y)\}$.

We can always find such a y since we have 2^{2n^i} possible (i, x, y) and at this point since we have frozen at most 2^{n^i} strings for at most 2^n possible x's for a total of $2^{n^i} 2^n < 2^{2n^i}$ frozen strings.

By adding every (i, x, y) that is non frozen in the proof above one can get an even stronger oracle.

Corollary 3. *There exists a relativized world such that* NEXP $\subseteq$ i.o.RP.

4 New Proof of Nondeterministic Time Hierarchy

Here we give an alternate proof of Theorem 3.

Proof (Proof of Theorem 3). Let $M_1, M_2, \ldots$ be an enumeration of multitape nondeterministic machines that run in time $t_1(n)$.

Define a nondeterministic Turing machine M that on input $1^i 01^m 0w$ does as follows:

- If $|w| < t_1(i+m+2)$ accept if both $M_i(1^i 01^m 0w0)$ and $M_i(1^i 01^m 0w1)$ accept.
- If $|w| \geqslant t_1(i+m+2)$ accept if $M_i(1^i 01^m 0)$ rejects on the path specified by the bits of w.

Since we can universally simulate $t(n)$-time nondeterministic multitape Turing machines on an $O(t(n))$-time 2-tape nondeterministic Turing machine, $L(M) \in$ $\mathsf{NTIME}(O(t_1(n+1))) \subseteq \mathsf{NTIME}(t_2(n))$. Note $(n+1)^c = O(n^c)$ for any c.

Suppose $\mathsf{NTIME}(t_2(n)) \subseteq \mathsf{NTIME}(t_1(n))$. Pick a c such that $t_1(n) \ll n^c$. By assumption there is a language $L \in \mathsf{NTIME}(t_1(n))$ such that $L(M) = L$. Fix i such that $L = L(M_i)$. Then $z \in L(M_i) \Leftrightarrow z \in L(M)$ for all $z = 1^i 01^{n_0} 0w$ for $w \leqslant t_1(i + n_0 + 2)$.

By induction we have $M_i(1^i 01^{n_0} 0)$ accepts if $M_i(1^i 01^{n_0} 0w)$ accepts for all $w \leqslant t_1(i + n_0 + 2)$. So $M_i(1^i 01^{n_0} 0)$ accepts if and only if $M_i(1^i 01^{n_0} 0)$ rejects on every computation path, contradicting the definition of nondeterministic time.

5 Proof of Theorem 4

Theorem 4 follows immediately from the following result.

Theorem 5. *Fix any constant $d > 1$. Let t_1 and t_2 be time-constructible functions such that $t_2 = O(n^d)$ and $t_1(n+1) = o(t_2(n))$. Then there is a language in $\mathsf{NTIME}(t_2)$ which is not in $\mathsf{NTIME}(t_1)/t_2^{-1}(n)$.*

We need a new notion of "cumulative advice", defined as follows. Given a time function $t : \mathbb{N} \to \mathbb{N}$ and an advice function $a : \mathbb{N} \to \mathbb{N}$, a language L is said to be in $\mathsf{NTIME}(t)/_c a$ if there is an advice-taking non-deterministic machine M such that, for each n, there is a string b_n of length at most $a(n)$ for which M, given $< n, b_n >$ on its advice tape, halts in time $t(n)$ and accepts an input x of length at most n iff $x \in L$.

The notion of cumulative advice is defined here for non-deterministic time but it extends naturally to any complexity measure.

Informally, an advice string given as cumulative advice helps to decide all inputs of length *at most* a given length, while the traditional notion of advice only applies to inputs which are all of the same length. If a language L is in $\mathsf{NTIME}(t)/a$ and a is a non-decreasing function, then it is obvious that L is $\mathsf{NTIME}(t)/_c na$, since cumulative advice for length n can be formed simply by concatenating all advice strings of length at most n. However, it is far from clear whether advice of length a can be simulated with cumulative advice $o(na)$, when a is polynomially bounded.

We will first prove a hierarchy theorem for non-deterministic polynomial time against sub-linear cumulative advice, and then show how to strengthen this to a hierarchy theorem for non-deterministic polynomial time against sub-linear advice. Note that though the notion of cumulative advice plays an important role in our proof, it does not appear in our main theorem - the main theorem holds for the traditional notion of advice.

Lemma 2. *Fix any constant $d > 1$. Let t_1 and t_2 be time-constructible functions such that $t_2 = O(n^d)$, $t_1(n + 1) = o(t_2(n))$. Then there is a language $L \in$ NTIME(t_2) which is not in NTIME$(t_1)/_c t_2^{-1}(n)$.*

Note that the statement of Lemma 2 is identical to that of Theorem 5, except that the lower bound is against cumulative advice.

Proof. First fix a function $f : \mathbb{N} \to \mathbb{N}$ such that $f(n)$ is computable in time $O(n)$, and for each constant k, there are only finitely many triples (n_1, n_2, n_3) of integers such that $n_1 \leqslant n_2 \leqslant n_3 \leqslant n_1^k$ such that $f(n_1), f(n_2), f(n_3)$ are all distinct, and also such that each positive integer has infinitely many pre-images under f. We will use the function $f(n) = i$ if $2^{2^{2^m}} \leqslant n < 2^{2^{2^{m+1}}}$, where i is the unique number such that $bin(m)$ is of the form $1^k 0 bin(i)$ for some $k \geqslant 0$. Here $bin(j)$ denotes the binary representation of the number j.

Intuitively, f selects which cumulative advice-taking non-deterministic Turing machine we attempt to diagonalize against at a given input length n. The properties of f ensure that the same machine is being diagonalized against for a long enough stretch of inputs, and that it is easy to compute for any given input length which machine we're diagonalizing against. Let $M_1, M_2, M_3 \ldots$ be an efficiently computable enumeration of all cumulative advice-taking 2-tape non-deterministic Turing machines. We define a non-deterministic machine M without advice which operates as follows.

On input x, M first computes $n = |x|$, $i = f(n)$ and the number $t_2(n)$, the last of which it uses as a clock for its computation. It then computes the largest m such that $2^{2^{2^m}} \leqslant n < 2^{2^{2^{m+1}}}$. Set $A = 2^{2^{2^m}}$. If $n > t_2(A)$, M simply rejects. Otherwise M decomposes x as yz, where $|y| = A$. If $n < t_2(A)$, M simulates M_i on input $x0$ with advice $< t_2(A), y >$ on the advice tape[1]. If M_i halts within the allotted time, M next simulates M_i on input $x1$ with advice $< t_2(A), y >$ on the advice tape. If this simulation halts as well within the allotted time, M accepts iff both simulations (i.e., of M_i on $x0$ and M_i on $x1$) accept. In every other case, M rejects.

If $n = t_2(A)$, M simulates M_i on y with guess sequence z (i.e., z is treated as an encoding of all the non-deterministic choices of M_i), and with advice $< n, y >$ on the advice tape. It accepts iff the simulation halts and rejects. Note that the simulation on such an input length n is completely deterministic.

---

[1] We assume that if M_i needs only $r < |y|$ bits of advice, then only the first r bits of y are used.

By definition of M, $L(M) \in \mathsf{NTIME}(t_2)$. We claim $L(M) \notin \mathsf{NTIME}(t_1)/_c t_2^{-1}(n)$. The proof of this claim is by contradiction. Suppose, to the contrary, that there is a cumulative advice-taking non-deterministic Turing machine deciding $L(M)$ in time $O(t_1)$ with $t_2^{-1}(n)$ bits of advice. By the tape reduction theorem for non-deterministic time, there is a 2-tape advice-taking non-deterministic machine M_i which decides $L(M)$ in time $O(t_1)$ with $t_2^{-1}(n)$ bits of advice.

Let $g : \mathbb{N} \to \mathbb{N}$ be a function such that the simulation of t steps of a machine M_i is performed within $g(i)t$ steps of M. Choose A a power of a power of a power of 2 large enough so that $f(A) = i$ and $2g(i)t_1(n' + 1) + 100n' < t_2(n')$ for all $n' \geqslant A$. By choice of f and since $t_1(n + 1) = o(t_2(n))$, such an A exists. Now, for all n such that $A \leqslant n < t_2(A)$, the simulations of M_i by M halt within the allotted time, since all the extra computations (of $n, i, t_2(n)$ and the decomposition) can be performed in time $< 100n$. Note also that the simulations at length $n = t_2(A)$ complete succesfully since $t_2(n) - n \geqslant t_1(n)$.

By assumption, there is a sequence of advice strings $\{b_m\}$ such that for each m, for each x of length at most m, M_i accepts x with advice $< m, b_m >$ iff $x \in L(M)$, and $|b_m| \leqslant t_2^{-1}(m)$. Let y be any string of length A such that $b_{t_2(A)}$ is a prefix of y. By the assumption on size of advice strings, such a string y exists.

Now we have that M accepts on y iff M_i accepts on both $y0$ and $y1$ with $< t_2(A), y >$ on the advice tape. Continuing inductively, we have that M accepts y iff M_i accepts on all strings of the form yz, $|z| \leqslant t_2(A) - A$ with $< t_2(A), y >$ on the advice tape. Now we take advantage of the behavior of M on strings of length $t_2(A)$. M accepts on a string yz, $|z| = t_2(A) - A$ iff z is not a sequence of non-deterministic choices leading to acceptance of M_i on y with $< t_2(A), y >$ on the advice tape. Hence, if M_i with $< t_2(A), y >$ on the advice tape agrees with M on all strings of the form $yz, |yz| = t_2(A)$, we have that M accepts y iff M_i rejects y with $< t_2(A), y >$ on the advice tape, which contradicts the assumption that M on y agrees with M_i on y with $< t_2(A), y >$ on the advice tape.

Lemma 3. *Let L be any language, and let $L' = \{0^k 1 x | x \in L, k \geqslant 0\}$. For any non-decreasing advice function $a : \mathbb{N} \to \mathbb{N}$, and for any non-decreasing time function $t : \mathbb{N} \to \mathbb{N}$ which is $\Omega(n)$, we have that $L \in \mathsf{NTIME}(t(n+1))/_c a(n+1)$ iff $L' \in \mathsf{NTIME}(t(n))/a(n)$.*

Proof. We define $L' = \{0^k 1 x | x \in L, k \geqslant 0\}$. We first show the forward implication, and then the reverse one.

Suppose $L \in \mathsf{NTIME}(t(n+1))/_c a(n+1)$, for some time function t and cumulative advice function a. Let M be an advice-taking non-deterministic Turing machine which always halts in time $t(n + 1)$ on inputs of length n and decides L correctly with $a(n+1)$ bits of cumulative advice. For each input length m, let b_m be a correct advice string of length at most $a(m+1)$ for M at length m, i.e., for all x of length at most m, M accepts x given advice $< m, b_m >$ iff $x \in L$. We define an advice-taking non-deterministic Turing machine M' which always halts in time $t(n)$ on inputs of length n and decides L correctly with at most $a(n)$ bits of advice.

Given an input x', M' operates as follows. M' first computes the unique string x such that $0^k 1 x = x'$, for some $k \geqslant 0$. This computation can be done easily in linear time. M' then interprets its advice string c_n as the cumulative advice b_{n-1} for M at length $n-1$, and simulates M on x with advice $< n-1, c_n >$. It accepts iff M accepts. M' always halts in time $O(t(n))$ since the string x' is of length at most $n-1$ and since M always halts in time $t(m+1)$ on inputs of length m. The correctness of M' follows from the fact that M is a correct advice-taking machine deciding L with cumulative advice.

For the reverse implication, suppose $L' \in \mathsf{NTIME}(t(n))/a(n)$. Let M' be an advice-taking non-deterministic machine which always halts in time $t(\cdot)$ and accepts L' with at most $a(n)$ bits of advice. We define an advice-taking machine M halting in time $t(n+1)$ and accepting L with at most $a(n+1)$ bits of cumulative advice as follows.

Say M is given a string x on its input tape, and $< m, b_m >$ on its advice tape, with $m \geqslant |x|$. Note that we can assume wlog that $m \geqslant |x|$, since otherwise M is allowed to behave arbitrarily. M forms the string $x' = 0^{m-|x|} 1 x$ and then simulates M' on input x' with advice b_m. Namely, it interprets its advice string as advice for M' at length $m+1$. The time taken for the simulation is $O(t(m+1))$ since t is at least linear, and the advice is of length at most $a(m+1)$. The correctness of M follows from the correctness of M'.

Proof of Theorem 5. Applying Lemma 2 to the time functions $t_1(n+1), t_2(n+1)$ and the cumulative advice function $t_2^{-1}(n+1)$, we have that there is a language L which is in $\mathsf{NTIME}(t_2(n+1))$ but not in $\mathsf{NTIME}(t_1(n+1))/t_2^{-1}(n+1)$. Using Lemma 3 with $t = t_2$ and $a = 0$, we have that $L' \in \mathsf{NTIME}(t_2)$. Using Lemma 3 with $t = t_1$ and $a = t_2^{-1}$, we have that $L' \notin \mathsf{NTIME}(t_1)/t_2^{-1}(n)$. Thus L' satisfies the required conditions.

We note that the polynomial upper bound on t_2 in Theorem 5 is in fact redundant. It helps to simplify the choice of f in the proof, but in fact for any time-constructible t_2 an appropriate f can be chosen to make the proof go through.

References

1. Buhrman, H., Fortnow, L., Santhanam, R.: Unconditional lower bounds against advice. In: Albers, S., Marchetti-Spaccamela, A., Matias, Y., Nikoletseas, S., Thomas, W. (eds.) ICALP 2009, Part I. LNCS, vol. 5555, pp. 195–209. Springer, Heidelberg (2009)
2. Fortnow, L., Santhanam, R.: Robust simulations and significant separations. In: Aceto, L., Henzinger, M., Sgall, J. (eds.) ICALP 2011, Part I. LNCS, vol. 6755, pp. 569–580. Springer, Heidelberg (2011)
3. Fortnow, L., Santhanam, R.: Hierarchies against sublinear advice. Technical report TR14-171, Electronic Colloquium on Computational Complexity (2014)
4. Cook, S.: A hierarchy for nondeterministic time complexity. J. Comput. Syst. Sci. **7**(4), 343–353 (1973)
5. Seiferas, J., Fischer, M., Meyer, A.: Separating nondeterministic time complexity classes. J. ACM **25**(1), 146–167 (1978)

6. Žàk, S.: A turing machine time hierarchy. Theor. Comput. Sci. **26**(3), 327–333 (1983)
7. Kannan, R.: Circuit-size lower bounds and non-reducibility to sparse sets. Inf. Control **55**, 40–56 (1982)
8. Homer, S., Mocas, S.: Nonuniform lower bounds for exponential time classes. In: Hájek, Petr, Wiedermann, Jiří (eds.) MFCS 1995. LNCS, vol. 969. Springer, Heidelberg (1995)

Quantum and Randomized Query Complexities
(Extended Abstract)

Miklos Santha[1,2](✉)

[1] CNRS–LIAFA, Université Paris Diderot, 75205 Paris, France
miklos.santha@gmail.com
[2] Centre for Quantum Technologies, National University of Singapore,
Singapore 117543, Singapore

Deterministic query complexity is a simplified model of computation where the resource measured is only the number of questions to the input to get information about individual input bits, while all other operations are for free. In the randomized model the queries can be chosen probabilistically, and in the quantum model they can be in superposition. While we have made significant progress in understanding all three models, numerous important questions (some of them over 40 years old) remain still unsolved.

Quantum query complexity has been very useful for studying the power of quantum computation. Important quantum algorithms, in particular the search algorithm of Grover and the period finding subroutine of Shor's factoring algorithm, can be formulated in this model. Yet, the model is still simple enough that one can often hope to prove tight lower bounds. Recently there have been very exciting developments in quantum query complexity. In particular, in a series of works, Reichardt has shown that the general adversary bound of Høyer, Lee and Špalek, formerly just a lower bound technique, was also an upper bound up to constant factors. This characterization clearly opened a new way for designing quantum query algorithms. Indeed, the dual adversary bound can be written as a relatively simple SDP, and therefore any feasible solution yields an upper bound for quantum query complexity. Nonetheless, in practice this approach can be quite difficult to implement since the minimization form of the SDP has exponentially many constraints. Even for simple functions it can be challenging to explicit a feasible solution, and even more to find one with a good objective value.

To surmount this problem, Belovs has introduced the beautiful model of learning graphs, which can be viewed as the minimization form of the general adversary bound with an additional structure imposed on the form of the solution. Learning graphs have a simple, combinatorial description. The vertices of a learning graph correspond to the sets of input variables known to the algorithm. The transitions of the algorithm are governed by a unit flow injected to

Research supported by the European Commission IST STREP project Quantum Algorithms (QALGO) 600700, the French ANR Blanc program under contract ANR-12-BS02-005 (RDAM project), the Singapore Ministry of Education and the National Research Foundation, also through the Tier 3 Grant "Random numbers from quantum processes".

R. Jain et al. (Eds.): TAMC 2015, LNCS 9076, pp. 18–19, 2015.
DOI: 10.1007/978-3-319-17142-5_3

the empty set, and every sink of the flow must contain a positive certificate for the function. The flow constraint makes learning graphs easy to reason about by ensuring that the SDP constraints are automatically satisfied, leaving one only to worry about optimizing the objective value. Learning graphs come in several flavor. In the non-adaptive version they are only sensitive to the certificate structure of the function to be computed, implying that various problems with the same certificate structure have the same learning graph complexity. For example, finding a triangle in a graph has the same certificate structure as the problem of finding one whose edge labels satisfy some specific property in a labeled graph.

In this survey talk we demonstrate the pertinence of learning graphs by their impact in two distinct research directions. Firstly, they gave significant impetus to the design of new quantum query algorithms for various problems, in particular for finding constant size subgraphs, associativity testing, and k-distinctness. Secondly, they inspired new concepts and techniques in the theory of quantum walks. We also illustrate the strong connection between learning graphs and quantum walks by the parallel between their classical analogies. Indeed, if quantum walks are often based on classical random walks in graphs, learning graphs are closely related to electric networks. The complexity of a learning graph is in fact defined in function of the overall weight and the effective resistance of the corresponding electric network. This relation actually makes possible a quadratic simulation of learning graphs by random query algorithms. While such a simulation could possibly exist for generic query algorithms in the case of total functions, proving such a relationship seems currently quite elusive.

Unlike in the case of quantum query complexity, we are not aware of any useful equivalent characterization of the randomized query complexity. We know several techniques to lower bound the bounded-error randomized query complexity, such as the approximate polynomial degree, the block sensitivity, the randomized certificate complexity and the classical adversary bound. All these techniques were recently subsumed by the public-coin partition bound of Jain, Lee and Vishnoi, which was also proven to be within a quadratic factor of the randomized query complexity. They raised the question whether this bound could actually be asymptotically equivalent to it. We describe the recent result of Racicot–Desloges, Kothari and Santha which answers the question by the negative. Indeed, they present a function whose randomized query complexity is asymptotically higher than its deterministic subcube complexity, which in turn is always lower bounded by the partition bound. This result shows that an equivalent characterization of the randomized query complexity should necessarily overcome the randomized subcube barrier which is not the case of the currently available techniques.

Recursion Theory
and Mathematical Logic

Algorithmically Random Functions and Effective Capacities

Douglas Cenzer[✉] and Christopher P. Porter

Department of Mathematics, University of Florida, Gainesville 32611, USA
cenzer@ufl.edu, cp@cpporter.com

Abstract. We continue the investigation of algorithmically random functions and closed sets, and in particular the connection with the notion of capacity. We study notions of random continuous functions given in terms of a family of computable measures called symmetric Bernoulli measures. We isolate one particular class of random functions that we refer to as random online functions F, where the value of $y(n)$ for $y = F(x)$ may be computed from the values of $x(0), \ldots, x(n)$. We show that random online functions are neither onto nor one-to-one. We give a necessary condition on the members of the ranges of random online functions in terms of initial segment complexity and the associated computable capacity. Lastly, we introduce the notion of Martin-Löf random online *partial* function on 2^ω and give a family of online partial random functions the ranges of which are precisely the random closed sets introduced in [2].

Keywords: Algorithmic randomness · Computability theory · Random closed sets · Random continuous functions · Capacity.

1 Introduction

In a series of recent papers [2–4,7], Barmpalias, Brodhead, Cenzer et al. have developed the notion of algorithmic randomness for closed sets and continuous functions on 2^ω as part of the broad program of algorithmic randomness. The study of random closed sets was furthered by Axon [1], Diamondstone and Kjos-Hanssen [8], and others. Cenzer et al. [7] studied the relationship between notions of random closed sets with respect to different computable probability measures and effective capacities.

Here we look more closely at the relationship between random continuous functions and effective capacity. First, we generalize the notion of random continuous function from [4] to a wider class of computable measures that we call symmetric Bernoulli measures. Then we study properties of the effective capacities associated to the classes of functions that are random with respect to various symmetric Bernoulli measures. We isolate one such class of functions, which we refer to as random online continuous functions. We study the reals in the range of a random online continuous function, as well as the average values of random online continuous functions.

© Springer International Publishing Switzerland 2015
R. Jain et al. (Eds.): TAMC 2015, LNCS 9076, pp. 23–37, 2015.
DOI: 10.1007/978-3-319-17142-5_4

It turns out that a number of effective capacities cannot be generated by a class of functions that are random with respect to a symmetric Bernoulli measure. We identify a class of measures on the space of functions that yield random online partial continuous functions and prove that a wide class of effective capacities can be generated by such functions, including the effective capacity that is associated to the original definition of algorithmically random closed set from [2].

Algorthmic randomness for closed sets was defined in [2] starting from a natural computable measure on the space $\mathcal{C}(2^\omega)$ of closed subsets of 2^ω and using the notion of Martin-Löf randomness given by Martin-Löf tests. It was shown that Δ_2^0 random closed sets exist but there are no random Π_1^0 closed sets. It is shown that any random closed set is perfect, has measure 0, and has box dimension $\log_2 \frac{4}{3}$. A random closed set has no n-c.e. elements.

Algorithmic randomness for continuous functions on 2^ω was defined in [4] by defining a representation of such functions in 3^ω and using the uniform measure on 3^ω to induce a measure on the space $\mathcal{F}(2^\omega)$ of continuous functions. It was shown that random Δ_2^0 continuous functions exist, but no computable function can be random and no random function can map a computable real to a computable real. The image of a random continuous function is always a perfect set and hence uncountable. For any $y \in 2^\omega$, there exists a random continuous function F with y in the image of F. Thus the image of a random continuous function need not be a random closed set. The set of zeros of a random continuous function is a random closed set (if nonempty).

The connection between measure and capacity for the space $\mathcal{C}(2^\omega)$ was investigated in [7]. For any computable measure μ^* on $\mathcal{C}(2^\omega)$, a computable capacity may be defined by letting $T(Q)$ be the μ^*-measure of the family of closed sets K which have nonempty intersection with Q for each $Q \in \mathcal{C}(2^\omega)$. An effective version of the Choquet's theorem was obtained by showing that every computable capacity may be obtained from a computable measure in this way. Conditions were given on a measure ν^* on $\mathcal{C}(2^\omega)$ that characterize when the capacity of all ν^*-random closed sets equals zero. For certain computable measures, effectively closed sets with positive capacity and with Lebesgue measure zero are constructed. For computable measures, a real q is upper semi-computable if and only if there is an effectively closed set with capacity q.

The problem of characterizing the possible members of random closed sets was studied by Diamondstone and Kjos-Hanssen in [8]. They gave an alternative presentation for random closed sets and showed a strong connection between the effective Hausdorff dimension of a real x and the membership of x in a random closed set.

The outline of the paper is as follows. In Sect. 2, we provide the requisite background. In Sect. 3 we define symmetric Bernoulli measures on the space of continuous functions on 2^ω and prove basic facts about the domains and ranges of functions that are random with respect to such measures. We study the connection between random functions and effective capacities on the space of closed subsets of 2^ω in Sect. 4. Next, we introduce and study the notion of a random online function in Sect. 5. Lastly, in Sect. 6, we define random online partial functions and establish a correspondence between the ranges of such functions and various families of random closed sets.

The authors would like to thank Laurent Bienvenu and the anonymous referees for helpful comments on an earlier draft of this paper.

2 Background

Some definitions are needed. For a finite string $\sigma \in \{0,1\}^n$, let $|\sigma| = n$ denote the length of n. For two strings σ, τ, say that τ *extends* σ and write $\sigma \prec \tau$ if $|\sigma| \leq |\tau|$ and $\sigma(i) = \tau(i)$ for $i < |\sigma|$. For $x \in 2^\omega$, $\sigma \prec x$ means that $\sigma(i) = x(i)$ for $i < |\sigma|$. Let $\sigma^\frown \tau$ denote the concatenation of σ and τ and let $\sigma^\frown i$ denote $\sigma^\frown(i)$ for $i = 0, 1$. Let $x\lceil n = (x(0), \ldots, x(n-1))$. The empty string will be denoted ϵ. Two reals x and y may be coded together into $z = x \oplus y$, where $z(2n) = x(n)$ and $z(2n+1) = y(n)$ for all n. For a finite string σ, let $[\![\sigma]\!]$ denote $\{x \in 2^\omega : \sigma \prec x\}$. We shall refer to $[\![\sigma]\!]$ as the *interval* determined by σ. Each such interval is a clopen set and the clopen sets are just finite unions of intervals. Now a nonempty closed set P may be identified with a tree $T_P \subseteq \{0,1\}^*$ where $T_P = \{\sigma : P \cap [\![\sigma]\!] \neq \emptyset\}$. Note that T_P has no dead ends. That is, if $\sigma \in T_P$, then either $\sigma^\frown 0 \in T_P$ or $\sigma^\frown 1 \in T_P$ (or both). For an arbitrary tree $T \subseteq \{0,1\}^*$, let $[T]$ denote the set of infinite paths through T. It is well-known that $P \subseteq 2^\omega$ is a closed set if and only if $P = [T]$ for some tree T. P is a Π_1^0 class, or an effectively closed set, if $P = [T]$ for some computable tree T.

A measure ν on 2^ω is *computable* if there is a computable function $\hat{\nu} : 2^{<\omega} \times \omega \to \mathbb{Q}_2$ (where $\mathbb{Q}_2 = \{\frac{m}{2^n} : n, m \in \omega\}$) such that $|\nu([\![\sigma]\!]) - \hat{\nu}(\sigma, i)| \leq 2^{-i}$ for every $\sigma \in 2^{<\omega}$ and $i \in \omega$. A computable measure on 3^ω is similarly defined.

Martin-Löf [10] observed that stochastic properties could be viewed as special kinds of effectively presented measure zero sets and defined a random real as one that avoids these measure 0 sets. More precisely, a real $x \in 2^\omega$ is Martin-Löf random if for every effective sequence $S_1, S_2, \ldots$ of c.e. open sets with $\mu(S_n) \leq 2^{-n}$, $x \notin \bigcap_n S_n$ (where μ is the uniform measure on 2^ω). This can be straightforwardly extended to any computable measure ν on 2^ω or 3^ω by replacing the condition $\mu(S_n) \leq 2^{-n}$ with $\nu(S_n) \leq 2^{-n}$.

Given a measure μ on 3^ω, we define a measure μ^* on the space $\mathcal{C}(2^\omega)$ of closed subsets of 2^ω as follows. Given a closed set $Q \subseteq 2^\omega$, let $T = T_Q$ be the tree without dead ends such that $Q = [T]$. Let $\sigma_0, \sigma_1, \ldots$ enumerate the elements of T in order, first by length and then lexicographically. We then define the *(canonical) code* $x = x_Q = x_T$ of Q by recursion such that for each n, $x(n) = 2$ if both $\sigma_n^\frown 0$ and $\sigma_n^\frown 1$ are in T, $x(n) = 1$ if $\sigma_n^\frown 0 \notin T$ and $\sigma_n^\frown 1 \in T$, and $x(n) = 0$ if $\sigma_n^\frown 0 \in T$ and $\sigma_n^\frown 1 \notin T$. We then define μ^* by setting

$$\mu^*(\mathcal{X}) = \mu(\{x_Q : Q \in \mathcal{X}\}) \tag{1}$$

for any $\mathcal{X} \subseteq \mathcal{C}(2^\omega)$. For the uniform measure, this means that given $\sigma \in T_Q$, there is probability $\frac{1}{3}$ that both $\sigma^\frown 0 \in T_Q$ and $\sigma^\frown 1 \in T_Q$ and, for $i = 0, 1$, there is probability $\frac{1}{3}$ that only $\sigma^\frown i \in T_Q$. Brodhead, Cenzer, and Dashti [2] defined a closed set $Q \subseteq 2^\omega$ to be (Martin-Löf) random if x_Q is (Martin-Löf) random.

We will sometimes refer to the random closed sets given by the uniform measure on 3^ω as the *standard random closed sets*.

Given a continuous function F on 2^ω, observe that for any $\sigma \in 2^{<\omega}$ there is some $n \in \omega$ and $\tau \in 2^{<\omega}$ of length n such that for all $x \in [\![\sigma]\!]$, $F(x){\restriction}n = \tau$.

Let $\mathcal{F}(2^\omega)$ denote the collection of all continuous functions $F : 2^\omega \to 2^\omega$. Each $F \in \mathcal{F}(2^\omega)$ may be represented by a function $f : 2^{<\omega} \setminus \{\epsilon\} \to \{0,1,2\}$, defined inductively as follows. Suppose we have defined $f(\sigma{\restriction}i) = e_i$ for $i = 1, \ldots, n$ and every σ of length n. Then given some σ of length $n+1$, where $f(\sigma{\restriction}i) = e_i$ for $i = 1, \ldots, n$, let $\rho = (n_1, \ldots, n_k)$ be the result of deleting all 2 s from $(e_1, \ldots, e_n)$. If for all $x \in [\![\sigma]\!]$, $F(x){\restriction}(k+1) = \rho^\frown j$ for some $j \in \{0,1\}$, then we may set $e_{n+1} = j$, although we may set $e_{n+1} = 2$. If there is no such j, we must set $e_{n+1} = 2$. It is helpful to think of the 2's as delaying the output of F along initial segments of some $x \in 2^\omega$. For each $F \in \mathcal{F}(2^\omega)$, there are infinitely many functions that represent F, and $f : 2^{<\omega} \setminus \{\epsilon\} \to \{0,1,2\}$ defines a (possibly partial) $F \in \mathcal{F}(2^\omega)$. Each representing function $f : 2^{<\omega} \setminus \{\epsilon\} \to \{0,1,2\}$ can be straightforwardly coded as some $z \in 3^\omega$. We can thus define a measure μ^{**} on $\mathcal{F}(2^\omega)$ induced by the uniform measure on 3^ω. As with the case of computable measures on $\mathcal{C}(2^\omega)$, every computable measure ν on 3^ω induces a computable measure ν^{**} on $\mathcal{F}(2^\omega)$. Brodhead, Cenzer, and Remmel [6] defined $F \in \mathcal{F}(2^\omega)$ to be Martin-Löf random if F is represented by a representing function coded by a Martin-Löf random $z \in 3^\omega$. We will sometimes refer to the random continuous functions given by the uniform measure on 3^ω as the *standard random continuous functions*.

Next we consider the notion of a capacity.

Definition 1. *A* capacity *on $\mathcal{C}(2^\omega)$ is a function $T : \mathcal{C}(2^\omega) \to [0,1]$ with $T(\emptyset) = 0$ such that*

1. T *is monotone increasing, that is, $Q_1 \subseteq Q_2$ implies $T(Q_1) \leq T(Q_2)$.*
2. T *has the* alternating of infinite order *property, that is, for $n \geq 2$ and any $Q_1, \ldots, Q_n \in \mathcal{C}$*

$$T\left(\bigcap_{i=1}^n Q_i\right) \leq \sum\left\{(-1)^{|I|+1} T\left(\bigcup_{i \in I} Q_i\right) : \emptyset \neq I \subseteq \{1,2,\ldots,n\}\right\}.$$

3. *If $Q = \bigcap_n Q_n$ and $Q_{n+1} \subseteq Q_n$ for all n, then $T(Q) = \lim_{n\to\infty} T(Q_n)$.*

We will also assume, unless otherwise specified, that $T(2^\omega) = 1$. We will say that a capacity T is computable if it is computable on the family of clopen sets, that is, if there is a computable function F from the Boolean algebra $\mathcal{B}$ of clopen sets into $[0,1]$ such that $F(B) = T(B)$ for any $B \in \mathcal{B}$.

Given a measure μ^* on the space $\mathcal{C}(2^\omega)$ of closed sets, define

$$T_\mu(Q) = \mu^*(\{\mathcal{X} \in \mathcal{C}(2^\omega) : \mathcal{X} \cap Q \neq \emptyset\}),$$

That is, $T_\mu(Q)$ is the probability that a randomly chosen closed set meets Q. The following effective version of the Choquet Capacity Theorem was shown in [7].

Theorem 1 ([7])

1. *For any computable probability measure μ on $\mathcal{C}(2^\omega)$, $\mathcal{T}_\mu$ is a computable capacity.*
2. *For any computable capacity $\mathcal{T}$ on $\mathcal{C}(2^\omega)$, there is a computable measure μ on the space of closed sets such that $\mathcal{T} = \mathcal{T}_\mu$.*

For a given computable capacity $\mathcal{T}$, if μ^* is a computable measure on $\mathcal{C}(2^\omega)$ such that $\mathcal{T} = \mathcal{T}_\mu$, we will refer to μ^*-random closed sets as the random closed sets associated to $\mathcal{T}$ and $\mathcal{T}$ as the capacity associated to the μ^*-random closed sets.

3 Symmetric Bernoulli Measures on $\mathcal{F}(2^\omega)$

In this section, we consider continuous functions that are random with respect to some measure from a specific class of computable measures on 3^ω.

Definition 2. *Let μ be a measure on 3^ω.*

(i) μ is a Bernoulli *measure if there are $p_0, p_1, p_2 \in [0,1]$ such that $p_0 + p_1 + p_2 = 1$ and $\mu(\sigma^\frown i) = p_i \cdot \mu(\sigma)$ for each $i \in \{0,1,2\}$.*
(ii) μ is a symmetric Bernoulli *measure if μ is a Bernoulli measure and there is some $r \in [0,1/2]$ such that $r = p_0 = p_1$ (so that $p_2 = 1 - 2r$).*

The symmetric Bernoulli measure with parameter $r \in [0,1/2]$ will be denoted μ_r. Note that μ_r is computable if and only if r is a computable real number.

We are interested in the behavior of the μ_r^{**}-random continuous functions on 2^ω. Note that in the case that $r = 1/3$, μ_r is the uniform measure on 3^ω and the μ_r^{**}-random continuous functions are the standard random continuous functions discussed in the previous section. In fact, the results in this section generalize certain results from [3] concerning $\mu_{1/3}^{**}$-random continuous functions.

First, it was shown in [3] that every $\mu_{1/3}^{**}$-random continuous function is total. However, if we allow the parameter r to vary, which results in a change of the probability of the occurrence of delays (i.e., the occurrence of 2s), the situation becomes slightly more interesting. Specifically, if μ_r is such that the probability of delay is greater than or equal to $1/2$, then not every μ_r^{**}-random function will be total.

The following lemma will be needed.

Lemma 1. *Let μ_r be a symmetric Bernoulli measure on 3^ω, let $A \subseteq \{0,1,2\}$, and let $p = \sum_{i \in A} p_i$, where $p_0 = p_1 = r$ and $p_2 = 1 - 2r$. Then the μ^{**}-measure q of the functions $F \in \mathcal{F}(2^\omega)$ such that there exists $x \in 2^\omega$ with $f(x\lceil n) \in A$ for all n (where f is the function representing F) equals 0 if $p \leq 1/2$ and equals $\frac{2p-1}{p^2}$ if $p > 1/2$.*

Proof. It follows from the compactness of 2^ω that there exists x such that $f(x\lceil n) \in A$ for all $n > 0$ if and only if for every n, there exists $\sigma \in \{0,1\}^n$ such that $f(\sigma\lceil m) \in A$ for all $0 < m < n$. Let q_n be the probability that such

$\sigma \in \{0,1\}^n$ exists. Then $q_0 = 1$, $q_{n+1} \le q_n$ for all n, and $q = \lim_{n \to \infty} q_n$. Considering the cases of $f(i)$ for $i \in \{0,1\}$, we calculate that

$$q_{n+1} = 2pq_n - p^2 q_n^2.$$

Taking the limit of both sides, we see that $q = 2pq - p^2 q^2$, so that either $q = 0$ or $q = \frac{2p-1}{p^2}$. In the case that $p < 1/2$, the latter is negative. Thus $q = 0$ if $p \le 1/2$.

For the other case, note first that $2pq_n - p^2 q_n^2 = 1 - (1 - pq_n)^2$, so that $q_n \ge x$ implies that $2pq_n - p^2 q_n^2 \ge 2px - p^2 x^2$. Let $s = \frac{2p-1}{p^2}$. We now show by induction that $q_n \ge s$ for all n. Initially we have $q_0 = 1 \ge s$. Now assuming that $q_n \ge s$, it follows that

$$q_{n+1} = 2pq_n - p^2 q_n^2 \ge 2ps - p^2 s^2 = s(2p - p^2 s) = s(2p - (2p - 1)) = s.$$

Now suppose that $p > 1/2$, so that $s = \frac{2p-1}{p^2} > 0$. Since the sequence $(q_n)_{n \in \omega}$ is decreasing and $q_n \ge s$ for all n, it follows that the limit $q = \lim_n q_n \ge s$ and hence $q = s$.

Proposition 1. *Let μ_r be a symmetric Bernoulli measure on 3^ω for some $r \in [0, 1/2]$. Then the μ_r^{**}-measure of the collection of partial continuous functions on 2^ω is 0 if $r \ge 1/4$ and is 1 if $r < 1/4$.*

Proof. First note that the measure must be either 0 or 1 in either case. This is because a function F is total if and only if the restrictions of F to both $[\![0]\!]$ and $[\![1]\!]$ are total, so that if p is the measure of the set of total functions, then $p = p^2$. Next observe that the function represented by $f : 2^{<\omega} \to \{0,1,2\}$ is partial if and only if there exists $x \in 2^\omega$ and n such that $f(x \restriction m) = 2$ for all $m \ge n$. It is enough to compute the probability q that there exists x such that $f(x \restriction m) = 2$ for all $m > 0$.

Let $A = \{2\}$, so that $f(\sigma) \in A$ with probability $p = 1 - 2r$ for each $\sigma \in 2^{<\omega} \setminus \{\epsilon\}$. Then by Lemma 1, the μ^{**}-measure of functions F such that there exists $x \in 2^\omega$ with $f(x \restriction n) \in A$ for all $n > 0$ equals 0 if $r \ge 1/4$ and equals $\frac{2p-1}{p^2} = \frac{1-4r}{(1-2r)^2}$ if $r < 1/4$. Since for $r < 1/4$, there are positive μ^{**}-measure many functions F for which such an x exists, it follows that the collection of partial functions has μ^{**}-measure 1.

Next, it was also shown in [3] that the probability that the range of a random continuous function includes a fixed $y \in 2^\omega$ is equal to 3/4. This was obtained by computing, for each $\sigma \in 2^{<\omega}$ of length n, the probability p_n that the range of a random continuous function has non-empty intersection with $[\![\sigma]\!]$ and then proving that $\lim_{n \to \infty} p_n = 3/4$. We consider the analogous result in the general case of a symmetric Bernoulli measure.

Theorem 2. *Let μ_r be a symmetric Bernoulli measure on 3^ω for some $r \in (0, 1/2]$ and let $y \in 2^\omega$. Then the μ_r^{**}-measure of the collection of continuous functions F such that $y \in \mathrm{ran}(F)$ is equal to*

$$\frac{1 - 2r}{(1 - r)^2}.$$

Proof. By symmetry of the measure μ_r, it suffices to show that μ_r^{**}-measure of the collection of continuous functions F such that $0^\infty \in \text{ran}(F)$ is equal to $\frac{1-2r}{(1-r)^2}$. Let $A = \{0, 2\}$, so that $f(\sigma) \in A$ with probability $p = 1 - r$ for $\sigma \in 2^{<\omega} \setminus \{\epsilon\}$. Then by Lemma 1, the μ^{**}-measure of functions F such that there exists $x \in 2^\omega$ with $f(x{\restriction}n) \in A$ for all n equals 0 if $r \geq 1/2$ and equals $\frac{2p-1}{p^2} = \frac{1-2r}{(1-r)^2}$ if $r < 1/2$.

Note that even if a function F satisfies $f(x{\restriction}n) \in A$ for every $n > 0$ for some $x \in 2^\omega$, this does not guarantee that $0^\infty \in \text{ran}(F)$, since we may have $f(x{\restriction}n) = 2$ for all but finitely many n. For a given $F \in \mathcal{F}(2^\omega)$, let $\mathcal{C}_F = \{x \in 2^\omega : (\forall n)f(x{\restriction}n) \in A\}$. One can verify that the probability that $0^\infty \in \text{ran}(F)$, given that $\mathcal{C}_F$ is non-empty, is 1 as follows. Suppose that $\mathcal{C}_F$ is non-empty. Then if we consider the left-most path x of $\mathcal{C}_F$, by the law of large numbers, as the occurrence of the label 0 on initial segments of x is $\frac{r}{1-r}$, the limiting frequency of 0s along x is $\frac{r}{1-r}$ with probability 1. Since the μ_r^{**}-measure of the collection of functions F such that $\mathcal{C}_F$ is non-empty is $\frac{1-2r}{(1-r)^2}$, the conclusion follows.

Observe that as r approaches 0, the above probability approaches 1. This means that as the probability of delay approaches 1, we have more chances to hit any given real, and so this probability approaches one. However, for the value $r = 0$, we have a discontinuity, as the resulting measure is concentrated on the function coded by 2^∞, which never outputs any bits but only delays indefinitely on every possible input. Lastly, as r approaches $1/2$, the above probability approaches 0. In fact, this probability only attains the value 0 when $r = 1/2$, that is, when the μ_r^{**}-random functions have no delay. Hereafter, we will refer to $\mu_{1/2}^{**}$-random functions as *random online functions*, which we study in detail in Sect. 5.

4   From Functions to Capacities

The significance of the proof of Theorem 2 is that it reveals a connection between a notion of random continuous function and a notion of effective capacity. In particular, we have the following result.

Theorem 3. *Let ν^{**} be a computable measure on $\mathcal{F}(2^\omega)$ and suppose that every ν^{**}-random function is total. Then the function*

$$T(\mathcal{S}) = \nu^{**}(\{F \in \mathcal{F}(2^\omega) : \text{ran}(F) \cap \mathcal{S} \neq \emptyset\})$$

is a computable capacity on $\mathcal{C}(2^\omega)$.

Proof. First we show that the map taking a ν^{**}-random function to its range induces a computable measure on $\mathcal{C}(2^\omega)$. Let F be a ν^{**}-random function. Since F is a continuous map from a compact space to a Hausdorff space, F is a closed map. By assumption, F is total, and hence $\text{ran}(F) = F(2^\omega)$ is a closed set. Moreover, it is not hard to see that there is a (partial) Turing functional $\Phi : 3^\omega \to 3^\omega$ that, given a real in 3^ω that codes a representing function f of some ν^{**}-random function F, outputs a real that codes the range of F. One can verify that Φ is defined on a subset of 3^ω of ν-measure one. It follows that Φ and ν together induce a computable measure ν_Φ on 3^ω defined by

$$\nu_\Phi(\mathcal{X}) = \nu(\Phi^{-1}(\mathcal{X}))$$

for all measurable $\mathcal{X} \subseteq 3^\omega$ (see [5, Lemma 2.6]). It follows from the preservation of randomness theorem ([5, Theorem 3.2]) that the image of a ν-random real under Φ is a ν_Φ-random real. In addition, by the no randomness ex nihilo principle ([5, Theorem 3.5]), every ν_Φ-random real is the image of a ν-random real under Φ. Thus, it follows that the range of a ν^{**}-random continuous function is a ν_Φ^*-random closed set and every ν_Φ^*-random is in the range of some ν^{**}-random continuous function.

Thus we have

$$T(Q) = \nu^{**}(\{F \in \mathcal{F}(2^\omega) : \mathrm{ran}(F) \cap Q \neq \emptyset\}) = \nu_\Phi^*(\{C \in \mathcal{C}(2^\omega) : C \cap Q \neq \emptyset\})$$

for every $Q \in \mathcal{C}(2^\omega)$. By the Theorem 1, it follows that T is a computable capacity.

In the proof of Theorem 3, we showed that if ν^{**} is a computable measure on $\mathcal{F}(2^\omega)$ such that the ν^{**}-random functions are total, then the ranges of the ν^{**}-random functions yield a notion of random closed sets with respect to some computable measure ν_Φ^* on $\mathcal{C}(2^\omega)$. This raises the following question: Is there a computable measure ν^{**} on $\mathcal{F}(2^\omega)$ such that the ranges of the ν^{**}-random functions are the standard random closed sets?

We will provide a full answer to this question in Sect. 6, but as a first step, we prove the following.

Proposition 2. *Let μ_r be a symmetric Bernoulli measure on 3^ω with $r \in (0, 1/2)$. Then the collection of ranges of the μ_r^{**}-random functions is not the collection of standard random closed sets.*

Proof. Let $r \in (0, 1/2)$. By Theorem 2, the μ_r^{**}-measure of the collection of continuous functions F such that $0^\infty \in \mathrm{ran}(F)$ is equal to $\dfrac{1 - 2r}{(1 - r)^2} > 0$. However, as shown in [2], no standard random closed set contains a computable real, and thus the conclusion follows.

A more significant difference between the collection of ranges of the μ_r^{**}-random functions and the collection of standard random closed sets can be seen by considering the computable capacity associated to each of these two collections. First, let μ be the uniform measure on 3^ω. Then the capacity $T_\mu(Q)$ on $\mathcal{C}(2^\omega)$ associated to the collection of standard random closed sets (see Theorem 1) can be shown to satisfy $T([\![\sigma]\!]) = \left(\frac{2}{3}\right)^n$ for every $n \in \omega$ and every $\sigma \in 2^{<\omega}$ of length n. Thus for $x \in 2^\omega$, $T_\mu(\{x\}) = \lim_{n \to \infty} T_\mu([\![x{\restriction}n]\!]) = 0$.

Now suppose that $r \in (0, 1/2)$. Let $\nu = \mu_r$ and let T_r be the capacity from Theorem 3. Then as we proved $T_r(\{x\}) > 0$ for every $x \in 2^\omega$. Thus, if we want to find a family of random functions such that the ranges of all such functions are the standard random closed sets, then we need the capacity $\mathcal{T}$ associated to this family to satisfy $\mathcal{T}(\{x\}) = 0$ for every $x \in 2^\omega$.

One such candidate is the collection of $\mu^{**}_{1/2}$-random functions, for by Theorem 3, in the case that $r = 1/2$, we have $\mathcal{T}_r(\{x\}) = 0$ for every $x \in 2^\omega$. Is it the case that the ranges of the $\mu^{**}_{1/2}$-random functions are the standard random closed sets? To answer this question, we will look more closely at the $\mu^{**}_{1/2}$-random functions.

5   Random Online Functions

In this section, we study the collection of functions that are random with respect to the measure $\mu^{**}_{1/2}$ induced by the symmetric Bernoulli measure $\mu_{1/2}$ on 3^ω. We will hereafter refer to the $\mu^{**}_{1/2}$-random functions as the *random online functions* due to the absence of 2s in their codes in 3^ω, which means that each bit given as input to such a function immediately (and randomly) yields one bit as output. Given this absence of 2s, we can equivalently define a random online function to be given by a representing function $f : 2^{<\omega}\backslash\{\epsilon\} \to \{0,1\}$. In this case, each online function has precisely one representing function. To see this, let $(\sigma_n)_{n\in\omega}$ be the canonical listing of $2^{<\omega}$ in length-lexicographical order. Then given $x \in 2^\omega$, we define a representing function f_x such that $f_x(\sigma_{n+1}) = x(n)$ for every $n \in \omega$. One can readily verify that the function F_X defined by

$$F_x(y) = f_x(y{\upharpoonright}1)^\frown f_x(y{\upharpoonright}2)^\frown f_x(y{\upharpoonright}3)^\frown \ldots$$

is an online function, and that every online function can be obtained in this way. Thus, a function $F \in \mathcal{F}(2^\omega)$ is a random online function if and only if F has a representing function f coded by a Martin-Löf random $x \in 2^\omega$.

Note that by Proposition 1, every random online function is total. We establish several additional results.

Theorem 4. *No computable real is in the range of a random online function.*

Proof. The proof can be obtained by modifying the proof of Theorem 2.4 from [3], according to which no standard random continuous function is partial.

Corollary 1. *No random online function is onto.*

Theorem 5. *Let F be a random online function and let $x \in 2^\omega$ code the representing function of F. If y is Martin-Löf random with relative to x, then $F^{-1}(\{F(y)\})$ is a standard random closed set.*

Proof (Sketch). We define a map $\Theta : 2^\omega \to 3^\omega$ that maps the join of two reals $x \oplus y \in 2^\omega$ to some $z \in 3^\omega$, where x is the code of the representing function of a random online function and z is a code of the closed set $F^{-1}(\{F(y)\})$. One can verify that Θ induces the uniform measure on 3^ω. Given $y \in \mathsf{MLR}^x$, by van Lambalgen's theorem (see [9, Theorem 6.9.1]) the real $x \oplus y$ is random, and hence by the preservation of randomness theorem, $\Theta(x \oplus y) = z$ is random with respect to the measure induced by Θ, namely the uniform measure on 2^ω, which establishes the theorem.

Corollary 2. *No random online function is one-to-one.*

Proof. Given a random online function F, let let $x \in 2^\omega$ code the representing function of F. Since F is total, F is defined on some y that is Martin-Löf random relative to x. Then by Theorem 5, $F^{-1}(\{F(y)\})$ is a random closed set, which is perfect (as shown in [2]). Thus F is not one-to-one.

By Theorem 2, for a fixed $y \in 2^\omega$, the probability that a random online function will have y in its range is 0. In fact, if for each n we let p_n be the probability that a random online function hits $[\![\sigma]\!]$ for a fixed σ of length n (where F *hits* $[\![\sigma]\!]$ if $\mathrm{ran}(F) \cap [\![\sigma]\!] \neq \emptyset$), by considering the cases of $f(i)$ for $i \in \{0,1\}$, one can show that

(i) $p_1 = 3/4$, and
(ii) $p_{n+1} = p_n(1 - \frac{1}{4}p_n)$.

Moreover, one can verify that $\lim_{n\to\infty} p_n = 0$ for each $n \geq 1$. Hereafter, we will refer to the p_i's as *hitting probabilities*.

Using the notation of the previous section, it follows that $\mathcal{T}_{1/2}(\sigma) = p_n$ for every n and every σ of length n. We can use this fact to determine the computable measure ν on 3^ω with the property that the ν^*-random closed sets are precisely the ranges of random online functions. Following the proof of the effective Choquet capacity theorem from [7] to find the values of ν, the key observation to make is that for each $n \in \omega$ and each $\sigma \in 2^{<\omega}$ of length n,

$$\nu(\sigma 2 \mid \sigma) = 2\left(\frac{p_n}{p_{n-1}}\right) - 1 = 2(1 - \frac{1}{4}p_n) - 1 = 1 - \frac{1}{2}p_n$$

for $n \geq 1$ (where $p_0 = 1$). Here $\nu(\sigma i \mid \sigma)$ is the probability, under ν, that a random function F hits $[\![\sigma i]\!]$ given that F hits $[\![\sigma]\!]$. For each such σ, we thus have $\nu(\sigma 0 \mid \sigma) = \nu(\sigma 1 \mid \sigma) = \frac{1}{4}p_n$. Since $\lim_{n\to\infty} p_n = 0$, $\nu(\sigma 2 \mid \sigma)$ approaches 1 while $\nu(\sigma 0 \mid \sigma)$ and $\nu(\sigma 1 \mid \sigma)$ both approach 0 as we consider longer and longer strings σ. Thus one can prove:

Theorem 6. *For each random online function F, the range of F is not a standard random closed set.*

Proof. Let μ be the uniform measure on 3^ω and let ν be the measure on 3^ω as defined above. Then one can verify that μ/ν is a computable ν-martingale on 3^ω, where $d : 2^{<\omega} \to [0, +\infty)$ is a ν-martingale on 3^ω if

$$\nu(\sigma)d(\sigma) = \nu(\sigma 0)d(\sigma 0) + \nu(\sigma 1)d(\sigma 1) + \nu(\sigma 2)d(\sigma 2).$$

Given a $x \in 3^\omega$, for each $n \geq 0$ we can write

$$\frac{\mu\big(x{\upharpoonright}(n+1)\big)}{\nu\big(x{\upharpoonright}(n+1)\big)} = \frac{\mu\big(x{\upharpoonright}(n+1) \mid x{\upharpoonright}n\big)}{\nu\big(x{\upharpoonright}(n+1) \mid x{\upharpoonright}n\big)} \frac{\mu(x{\upharpoonright}n)}{\nu(x{\upharpoonright}n)},$$

Since $\lim_{n\to\infty} p_n = 0$, for each k, there is some n_k such that $p_{n_k} \leq 2^{-k}$. Then for any σ of length greater than n_k, we have $1 \geq \nu(\sigma 2 \mid \sigma) \geq 1 - 2^{-(k+1)}$ and

$\nu(\sigma 0 \mid \sigma) = \nu(\sigma 1 \mid \sigma) \leq 2^{-(k+2)}$. If $x \in 3^\omega$ is μ-random, then for each $n \geq n_k$ such that $x(n) = 2$, which happens roughly $1/3$ of the time, we have

$$\frac{\mu(x{\upharpoonright}(n+1))}{\nu(x{\upharpoonright}(n+1))} = \frac{1/3}{\nu((x{\upharpoonright}n)^\frown 2 \mid x{\upharpoonright}n)} \frac{\mu(x{\upharpoonright}n)}{\nu(x{\upharpoonright}n)} \geq 1/3 \frac{\mu(x{\upharpoonright}n)}{\nu(x{\upharpoonright}n)},$$

For each $n \geq n_k$ such that $x(n) = 0$ or $x(n) = 1$, which happens roughly $2/3$ of the time, we have for $i = 0, 1$,

$$\frac{\mu(x{\upharpoonright}(n+1))}{\nu(x{\upharpoonright}(n+1))} = \frac{1/3}{\nu((x{\upharpoonright}n)^\frown i \mid x{\upharpoonright}n)} \nu(x{\upharpoonright}n) \geq \frac{1/3}{2^{-(k+2)}} \frac{\mu(x{\upharpoonright}n)}{\nu(x{\upharpoonright}n)} \geq 2^k \frac{\mu(x{\upharpoonright}n)}{\nu(x{\upharpoonright}n)}.$$

One can verify that $\lim_{n\to\infty} \frac{\mu(x{\upharpoonright}n+1)}{\nu(x{\upharpoonright}n+1)} = \infty$ for every μ-random $x \in 3^\omega$. It is well-known that this implies that no such x can be ν-random, and the conclusion follows.

It is reasonable to ask which reals are in the range of some random online function. We give a partial answer to this question by providing a necessary condition for being a member of the range of some random online function. We first prove a more general result, which is an extension of a result in [8], according to which every member of a standard random closed set must have sufficiently high effective Hausdorff dimension. Recall that $K(\sigma)$ is the prefix-free Kolmogorov complexity of σ.

Theorem 7. *Let μ^* be a computable measure on $\mathcal{C}(2^\omega)$ and T_μ the computable capacity associated to μ. If x is a member of some μ^*-random closed set, then there is some c such that*

$$K(x{\upharpoonright}n) \geq -\log T_\mu(\llbracket x{\upharpoonright}n \rrbracket) - c$$

for all n.

Proof. Suppose that x is such that for every c, there is some n such that

$$K(x{\upharpoonright}n) < -\log T_\mu(\llbracket x{\upharpoonright}n \rrbracket) - c.$$

We first define

$$S_i = \{\sigma \in 2^{<\omega} : K(\sigma) < -\log T_\mu(\llbracket \sigma \rrbracket) - i\}.$$

Next, we let $\widehat{S}_i$ consist of those strings in S_i with no proper initial segments in S_i, so that $\llbracket \widehat{S}_i \rrbracket = \llbracket S_i \rrbracket$. Lastly, we define

$$\mathcal{U}_i = \{Q \in \mathcal{C}(2^\omega) : (\exists \sigma \in \widehat{S}_i)[Q \cap \llbracket \sigma \rrbracket \neq \emptyset]\}.$$

Then

$$\mu^*(\mathcal{U}_i) \leq \sum_{\sigma \in \widehat{S}_i} \mu^*(\{Q \in \mathcal{C}(2^\omega) : Q \cap \llbracket \sigma \rrbracket\} \neq \emptyset) = \sum_{\sigma \in \widehat{S}_i} T_\mu(\llbracket \sigma \rrbracket) < \sum_{\sigma \in \widehat{S}_i} 2^{-K(\sigma)-i} \leq 2^{-i},$$

where the last inequality follows from the fact that $\sum_{\sigma \in 2^{<\omega}} 2^{-K(\sigma)} \leq 1$. Thus, $(\mathcal{U}_i)_{i \in \omega}$ forms a μ^*-Martin-Löf test. Now let $Q \in \mathcal{C}(2^\omega)$ be such that $x \in Q$. Then for each i, there is some least n such that $x \restriction n \in \widehat{S}_i$, and thus $Q \in \mathcal{U}_i$. It follows that no $Q \in \mathcal{C}(2^\omega)$ containing x is μ^*-random.

An order function $f : \omega \to \omega$ is a non-decreasing, unbounded function. Recall further that a real $x \in 2^\omega$ is *complex* if there is some computable order function f such that $K(x \restriction n) \geq f(n)$ for every n. Let $(p_n)_{n \in \omega}$ be the collection of hitting probabilities determined by the collection of random online functions. Since $(p_n)_{n \in \omega}$ is a computable, strictly decreasing sequence of rationals that converges to 0, it follows that the function $f(n) = -\log p_n$ is a computable order function.

This observation, combined with Theorem 7, yields:

Corollary 3. *If $x \in 2^\omega$ is in the range of a random online function, then*

$$K(x \restriction n) \geq -\log p_n - c$$

for some $c \in \omega$. In particular, x is complex.

We conjecture that the converse, or some minor variant thereof, holds as well.

6   Random Online Partial Functions

As we have seen, for each symmetric Bernoulli measure μ_r on 2^ω with $r \in (0, 1/2)$, the collection of ranges of the μ_r^{**}-random functions is not the collection of standard random closed sets. The collection of ranges of random online functions was, at first glance, a reasonable candidate for being equal to the collection of standard random closed sets, but this too fails by Theorem 6. Thus, we cannot use symmetric Bernoulli measures to obtain such a class of random functions.

As discussed in Sect. 4, the capacity $\mathcal{T}$ associated to the standard random closed sets satisfies $\mathcal{T}(\{x\}) = 0$ for every $x \in 2^\omega$. Thus, for any collection of random functions the ranges of which are the standard random closed sets, we need the capacity associated with this collection of functions to converge to zero quickly. Note, however, that by Theorem 2, as we increase the possibility of delay in our functions, this actually increases the probability that we hit a given real.

The first step to a solution is to introduce a notion of random online *partial* function. As with the representing functions of continuous functions on 2^ω, we define an online partial function to be given by a $\{0, 1, 2\}$-valued representing function. The values 0 and 1 play the same role as before, but the 2s play a different role. If F is the partial function given by a $\{0, 1, 2\}$-valued representing function f, for each $\sigma \in 2^{<\omega}$ with $f(\sigma) = 2$, we have $F(X)\uparrow$ for every $X \succ \sigma$. That is, instead of causing our function to delay at a given node, a node labelled with a '2' indicates that our function is undefined on all reals extending this node.

Observe that each symmetric Bernoulli measure μ_r on 3^ω yields a notion of random online partial function. However, for certain choices of r, we are not even guaranteed to have any functions with non-empty domain.

Proposition 3. *If μ_r is a computable symmetric Bernoulli measure on 3^ω, then the probability that a μ_r^{**}-random online partial function has non-empty domain is 0 if $r < 1/4$ and is*

$$\frac{4r-1}{4r^2}$$

if $r \geq 1/4$.

Proof. An online partial function F has non-empty domain if and only if there is some $x \in 2^\omega$ such that $f(x{\restriction}n) \neq 2$ for every $n > 0$. Let $A = \{0,1\}$, so that $f(\sigma) \in A$ with probability $p = 2r$ for every $\sigma \in 2^{<\omega} \setminus \{\epsilon\}$. Applying Lemma 1, the μ^{**}-measure of functions F such that there exists $x \in 2^\omega$ with $f(x{\restriction}n) \in A$ for all n equals 0 if $r < 1/4$ and equals $\frac{2p-1}{p^2} = \frac{4r-1}{4r^2}$ if $r \geq 1/4$. $\square$

The final step to obtaining a collection of random functions whose ranges are the standard random closed sets is to consider a wider class of measures, namely, computable, symmetric *generalized* Bernoulli measures on 3^ω. Such a measure is given by a computable sequence of rationals $\boldsymbol{r} = (r_i)_{i \in \omega}$ with $r_i \leq 1/2$ for every i such that for each n and each σ of length n, $\mu_{\boldsymbol{r}}(\sigma 0 \mid \sigma) = \mu(\sigma 1 \mid \sigma) = r_n \cdot \mu(\sigma)$ and $\mu_{\boldsymbol{r}}(\sigma 2 \mid \sigma) = (1 - 2r_n)\mu(\sigma)$. We can now prove the following.

Theorem 8. *Let T be an computable capacity on $\mathcal{C}(2^\omega)$ such that there is a computable sequence of rationals $(p_i)_{i \in \omega}$ satisfying*

(i) for each n, $T(\llbracket \sigma \rrbracket) = p_n$ for every $\sigma \in 2^n$, and
(ii) $\lim_{n \to \infty} p_n = 0$.

Then there is a computable, generalized symmetric Bernoulli measure $\mu_{\boldsymbol{r}}$ on 3^ω such that the ranges of the $\mu_{\boldsymbol{r}}^{**}$-random online partial functions are precisely the random closed sets associated with the capacity T. Moreover, in the case that $\lim_{n \to \infty} \frac{p_{n+1}}{p_n} = p$ for some $p \in [0,1]$, we have $\lim_{n \to \infty} r_n = \frac{p}{2}$.

Proof. To obtain the measure $\mu_{\boldsymbol{r}}$, we suppose we have a collection of $\mu_{\boldsymbol{r}}$-random functions that yield the hitting probabilities $(p_n)_{n \in \omega}$ then follow the proof of Theorem 2 to recover the values of the sequence $(r_i)_{i \in \omega}$.

Without loss of generality, we can consider the probability of hitting $\llbracket 0^n \rrbracket$ for each n. By convention, $p_0 = T(\emptyset) = 1$. For $n \geq 0$, to determine the relationship between p_{n+1} and p_n, we consider the possible initial values $f(0)$ and $f(1)$ of a representing function $f : 2^{<\omega} \setminus \{\epsilon\} \to \{0, 1, 2\}$ corresponding to an arbitrary $F \in \mathcal{F}(2^\omega)$. Due to our new interpretation of 2s, we only have a total of four cases to consider:

Case 1: $f(0) \neq 0$ and $f(1) \neq 0$, then $\mathrm{ran}(F) \cap \llbracket 0^{n+1} \rrbracket = \emptyset$.
Case 2: If $f(0) = f(1) = 0$, which occurs with probability r_{n+1}^2, then $\mathrm{ran}(F) \cap \llbracket 0^{n+1} \rrbracket \neq \emptyset$ with probability $1 - (1 - p_n)^2 = 2p_n - p_n^2$.
Case 3: $f(i) = 0$ and $f(1 - i) = 1$, which occurs with probability $2r_{n+1}^2$, then $\mathrm{ran}(F) \cap \llbracket 0^{n+1} \rrbracket \neq \emptyset$ with probability p_n.
Case 4: $f(i) = 0$ and $f(1 - i) = 2$, which occurs with probability $2r_{n+1}(1 - 2r_{n+1})$, then $\mathrm{ran}(F) \cap \llbracket 0^{n+1} \rrbracket \neq \emptyset$ with probability p_n.

Combining these cases yields

$$p_{n+1} = (2p_n - p_n^2)r_{n+1}^2 + 2p_n r_{n+1}^2 + 2r_{n+1}(1 - 2r_{n+1})p_n,$$

which simplifies to

$$p_{n+1} = 2p_n r_{n+1} - p_n^2 r_{n+1}^2.$$

Solving for r_{n+1} yields

$$r_{n+1} = \frac{p_{n+1}}{p_n(1 + \sqrt{1 - p_{n+1}})}.$$

It follows that the capacity induced by the family of μ_r^{**}-random online partial functions is the capacity T. Now, the map Φ that maps a μ_r^{**}-random online partial function F to its range is still a computable map, as we can effectively determine those basic open neighborhoods $[\![\sigma]\!]$ on which F is undefined. Then if we let ν^* be the computable measure on $\mathcal{C}(2^\omega)$ induced by Φ and μ_r (as in the proof of Theorem 3), then we will have

$$T(Q) = \mu_r^{**}(\{F \in \mathcal{F}(2^\omega) : \operatorname{ran}(F) \cap Q \neq \emptyset\}) = \nu^*(\{C \in \mathcal{C}(2^\omega) : C \cap Q \neq \emptyset\}).$$

for every $Q \in \mathcal{C}(2^\omega)$. Thus, the ranges of the μ_r^{**}-random online partial functions are the random closed sets associated to T.

Lastly, observe that

$$\lim_{n \to \infty} r_n = \lim_{n \to \infty} \frac{p_{n+1}}{p_n(1 + \sqrt{1 - p_{n+1}})} = \left(\lim_{n \to \infty} \frac{p_{n+1}}{p_n}\right)\left(\lim_{n \to \infty} \frac{1}{1 + \sqrt{1 - p_{n+1}}}\right) = \frac{p}{2}.$$

Theorem 9. *Let $r = (r_i)_{i \in \omega}$ be defined by*

$$r_i = \frac{2/3}{1 + \sqrt{1 - \left(\frac{2}{3}\right)^i}}.$$

Then the collection of ranges of the μ_r^{**}-random online partial functions is equal to the collection of the standard random closed sets.

Proof. Let T be the capacity associated to the standard random closed sets. As discussed in Sect. 4, we have $T([\![\sigma]\!]) = \left(\frac{2}{3}\right)^n$ for every $n \in \omega$. Then T satisfies the conditions of Theorem 8. By the proof of Theorem 8, if μ_r is the computable, symmetric generalized Bernoulli measure on 3^ω where

$$r_i = \frac{2/3}{1 + \sqrt{1 - \left(\frac{2}{3}\right)^i}}$$

for every $i \in \omega$, then the ranges of the μ_r^{**}-random online partial functions are precisely the standard random closed sets.

References

1. Axon, L.M.: Algorithmically random closed sets and probability. PhD thesis, University of Notre Dame (2010)
2. Barmpalias, G., Brodhead, P., Cenzer, D., Dashti, S., Weber, R.: Algorithmic randomness of closed sets. J. Logic Comput. **17**, 1041–1062 (2007)
3. Barmpalias, G., Brodhead, P., Cenzer, D., Remmel, J.B., Weber, R.: Algorithmic randomness of continuous functions. Arch. Math. Logic **46**, 533–546 (2008)
4. Barmpalias, G., Cenzer, D., Remmel, J.B., Weber, R.: k-triviality of closed sets and continuous functions. J. Logic Comput. **19**, 3–16 (2009)
5. Bienvenu, L., Porter, C.: Strong reductions in effective randomness. Theoret. Comput. Sci. **459**, 55–68 (2012)
6. Brodhead, P., Cenzer, D., Remmel, J.B.: Random continuous functions. Electron. Notes Theor. Comput. Sci. **167**, 275–287 (2007)
7. Cenzer, D., Brodhead, P., Toska, F., Wyman, S.: Algorithmic randomness and capacity of closed sets. Log. Methods Comput. Sci. **6**, 1–16 (2011)
8. Diamondstone, D., Kjos-Hanssen, B.: Martin-Löf randomness and galton-watson processes. Ann. Pure Appl. Logic **163**, 519–529 (2012)
9. Downey, R., Hirschfeldt, D.: Algorithmic Randomness Complex. Springer, Heidelberg (2011)
10. Martin-Lof, P.: The definition of random sequences. Inf. Control **9**, 602–619 (1966)

Where Join Preservation Fails in the Bounded Turing Degrees of C.E. Sets

Nadine Losert[(✉)]

Department of Mathematics and Computer Science,
Heidelberg University, Heidelberg, Germany
nadine.losert@informatik.uni-heidelberg.de

Abstract. We will look at the question for which bounded Turing reducibilities r and r' such that r is stronger than r' join preservation holds, i.e. for which r and r' every join in the computably enumerable (c.e.) r-degrees is also a join in the c.e. r'-degrees. We will also have a look at the corresponding question for meets. We will consider the class of monotone admissible (uniformly) bounded Turing reducibilities, i.e. the reflexive and transitive Turing reducibilities with use bounded by a function that is contained in a (uniformly computable) family of strictly increasing computable functions. This class contains for example ibT- and cl-reducibility. We will show that join preservation does not hold for cl and any admissible uniformly bounded Turing reducibility. We will show that, on the other hand, for all monotone admissible bounded Turing reducibilities r and r' such that r is stronger than r', meet preservation holds.

1 Introduction

Various notions of reducibilities stronger than Turing reducibility have been studied in computability theory, e.g. the so called classical strong reducibilities: one-one reducibility (1-reducibility), many-one reducibility (m-reducibility), truth-table reducibility (tt-reducibility), and weak truth-table reducibility (wtt-reducibility) (see e.g. Odifreddi [13]). More recently, one has started to look at the so called strongly bounded Turing reducibilities: identity bounded Turing reducibility (ibT-reducibility) and computable Lipschitz reducibility (cl-reducibility) which are defined in terms of Turing functionals where the use is bounded by the identity function and the identity function plus a constant and which were introduced by Soare [14] and Downey, Hirschfeldt, and LaForte [9,10], respectively. cl-reducibility is not only a notion of relative complexity but can also be viewed as a notion of relative randomness and is hence important in the field of algorithmic randomness (see the monograph [8] by Downey and Hirschfeldt for more background). The degree structures of the strongly bounded Turing reducibilities on the c.e. sets have been studied intensively. Barmpalias [5] showed that the partial ordering $(\mathbf{R}_{cl}, \leq)$ of the c.e. cl-degrees has no maximal elements;

I would like to thank my advisor, Klaus Ambos-Spies, for his help and guidance during my work on this paper.

R. Jain et al. (Eds.): TAMC 2015, LNCS 9076, pp. 38–49, 2015.
DOI: 10.1007/978-3-319-17142-5_5

Fan and Lu [12] showed that there are maximal pairs hence the partial orderings of the ibT- and cl-degrees are not upper semilattices, and Barmpalias and Lewis [6] and Day [7] showed that these partial orderings are not dense. Ambos-Spies, Bodewig, Kräling, and Yu [3] embedded the nonmodular lattice N5 into the c.e. ibT- and cl-degrees thereby showing that these partial orderings are not distributive, and Ambos-Spies [1] proved some global results, e.g. showed that the first order theories of the partial orderings of the c.e. ibT- and cl-degrees are undecidable. Recently, Ambos-Spies [2] introduced a more general class of bounded Turing reducibilities, the uniformly bounded Turing reducibilities. A reducibility r is a (uniformly) bounded Turing reducibility ((u)bT-reducibility) if there is a family $\mathcal{F}$ of (uniformly) computable functions such that, for all sets A and B, A is r-reducible to B if and only if A is Turing reducible to B with use bounded by some function f in $\mathcal{F}$. We call a (uniformly) bounded Turing reducibility admissible if it is reflexive and transitive and we call it monotone if it is induced by a family of strictly increasing functions. Examples of monotone admissible ubT-reducibilities are the strongly bounded Turing reducibilities ibT and cl as well as the linearly bounded and the primitive recursively bounded Turing reducibilities. An example of an admissible monotone bT-reducibility which is not uniformly bounded is wtt-reducibility. Here, we will only look at the monotone admissible bT-reducibilities.

If a reducibility r is stronger than a reducibility r', of course, every upper r-bound for some sets A and B is also an upper r'-bound for A and B and the same holds for lower bounds. But this does not necessarily imply that least upper r-bounds (joins) have to be a least upper r'-bounds, too. Again, the same holds for greatest lower bounds (meets). Here, we ask the question for which reducibilities r and r', joins and meets in the c.e. r-degrees are preserved in the c.e. r'-degrees. We say r-r' join (meet) preservation holds if, for all noncomputable c.e. sets A, B, and C such that the r-degree of C is the join (meet) of the r'-degrees of A and B, it holds that the r'-degree of C is the join (meet) of the r'-degrees of A and B, too.

For most of the classical reducibilities mentioned above, the structure of the c.e. degrees is an upper semilattice where the join of the degrees of two sets A and B is induced by the effective disjoint union $A \oplus B$. So, for two such reducibilities where r is stronger than r', of course, r-r' join preservation holds. So, for example, m-tt join preservation, tt-wtt join preservation and wtt-T join preservation hold. For reducibilities r whose degree structures are not an upper semilattice with join induced by the effective disjoint union, the question of r-r' join preservation is less obvious. For the classical strong reducibilities, 1-reducibility is an example of such a reducibility, but, as one can easily show (see Lemma 2 below), 1-m join preservation holds. It easily follows that r-r' join preservation holds for all classical strong reducibilities where r is stronger than r'. For the (uniformly) bounded Turing reducibilities, the question of join preservation is less straightforward. Ambos-Spies, Ding, Fan, and Merkle [4] showed that ibT-cl join preservation holds and Ambos-Spies, Bodewig, Kräling, and Yu (see [1]) showed that cl-wtt join preservation holds, too. This may lead one to conjecture that – just as in case of the classical strong reducibilities – r-r' join preservation holds for any monotone admissible (u)bT-reducibilities where r is stronger than r', too. As we will show here, however, this is not the case. In fact, for $r =$ ibT,cl and for *any* monotone amissible ubT-reducibility r' which is strictly stronger than cl, r-r' join preservation fails (see Theorem 1 below).

We complement our main result by considering meet preservation in the monotone admissible bt-reducibilities, too. There we generalize the result in [4] that ibT-cl meet preservation holds by showing that indeed, r-r' meet preservation holds for all monotone admissible bT-reducibilities r and r' such that r is stronger than r' (see Lemma 5).

So, for the monotone admissible (uniformly) bounded Turing reducibilities, meet preservation holds in general while, in some instances, join preservation fails. For the classical reducibilities, i.e. the strong reducibilities together with Turing reducibility, the converse is true. There join preservation holds in general, whereas, as Downey and Stob [11] showed, wtt-T meet preservation fails.

2   Preliminaries

A reducibility r is *admissible* if it is reflexive and transitive. For two reducibilities r and r', we say that r is *stronger* than r' (denoted by $r \preceq r'$) if, for all sets A and B, from $A \leq_r B$, it follows that $A \leq_{r'} B$, and r is *strictly stronger* than r' ($r \prec r'$) if $r \preceq r'$ and $r \neq r'$.

Definition 1. *For two admissible reducibilities r and r', we say that r-r' join preservation holds (in the c.e. degrees) if, for any noncomputable c.e. sets A, B, and C,*

$$deg_r(A) \vee deg_r(B) = deg_r(C) \Rightarrow deg_{r'}(A) \vee deg_{r'}(B) = deg_{r'}(C)$$

holds. Otherwise, we say that r-r' join preservation fails. Similarly, r-r' meet preservation holds (in the c.e. degrees) if, for any noncomputable c.e. sets A, B, and C,

$$deg_r(A) \wedge deg_r(B) = deg_r(C) \Rightarrow deg_{r'}(A) \wedge deg_{r'}(B) = deg_{r'}(C)$$

holds and r-r' meet preservation fails otherwise.

Let $\{\Phi_e^X : e \geq 0\}$ be a fixed enumeration of all Turing functionals obtained by Gödelization of the oracle Turing machines. Then, we obtain an enumeration $\{\Phi_e^{X,f} : e \geq 0\}$ of all f-bounded Turing functionals by bounding the use of each Φ_e^X on input x by $f(x)$ (by making the computation divergent in case of longer oracle queries). For any pair of sets A and B, A is f-bounded Turing reducible to B (denoted by $A \leq_{f-\mathrm{T}} B$) if and only if there is an e such that $A = \Phi_e^{B,f}$. By letting $f = id$, we obtain an enumeration $\{\hat{\Phi}_e^X\}$ of all identity bounded Turing functionals.

We call a reducibility r a *bounded Turing reducibility* (*bT-reducibility*) if there is a family $\mathcal{F}$ of computable functions such that $A \leq_r B$ if and only if

$A \leq_{f-T} B$ for some function $f \in \mathcal{F}$; in this case we say that r is *induced* by $\mathcal{F}$. If $\mathcal{F}$ is uniformly computable, r is called a *uniformly bounded Turing reducibility* (*ubT-reducibility*). We call a bounded Turing reducibility *monotone* if it is induced by a family $\mathcal{F}$ which consists only of strictly increasing functions. Note that ibT and cl are ubT-reducibilities which are induced by $\mathcal{F}_{ibT} = \{id\}$ and $\mathcal{F}_{cl} = \{id + e : e \geq 0\}$, respectively.

Lemma 1 (Ambos-Spies [2]). *Let r and r' be admissible ubT-reducibilities. Then, $r \preceq r'$ if and only if there are uniformly computable families $\mathcal{F}$ and $\mathcal{F}'$ that induce r and r', respectively, such that $\mathcal{F} \leq^* \mathcal{F}'$, i.e. for every function $f \in \mathcal{F}$, there is a function $f' \in \mathcal{F}'$ such that $f(x) \leq f'(x)$ for almost all $x \in \omega$.*

3   Join Preservation

It is a straightforward observation that r-r' join preservation holds for reducibilities r and r' such that r is stronger than r' and such that the structures of the c.e. r-degrees and of the c.e. r'-degrees form upper semilattices with join induced by the effective disjoint union. We will now observe (by giving an example) that r-r' join preservation may hold even if the structure of the c.e. r-degrees does not form an upper semilattice.

Lemma 2. 1-m *join preservation holds.*

Proof. Given c.e. sets A_0, A_1, and B such that

$$deg_1(A_0) \vee deg_1(A_1) = deg_1(B) \tag{1}$$

holds, we have to show that $deg_m(A_0) \vee deg_m(A_1) = deg_m(B)$ holds, too. As we know that $deg_m(A_0) \vee deg_m(A_1) = deg_m(A_0 \oplus A_1)$, we only have to show that $B =_m A_0 \oplus A_1$. It is obvious that $A_i \leq_1 A_0 \oplus A_1$ via $f_i(x) = 2x + i$ for $i = 0, 1$, so, it follows from (1) that $B \leq_1 A_0 \oplus A_1$, hence $B \leq_m A_0 \oplus A_1$. On the other hand, if we fix g_i such that $A_i \leq_1 B$ via g_i for $i = 0, 1$, it follows that $A_0 \oplus A_1 \leq_m B$ via g where $g(2x + i) = g_i(x)$ for all $x \geq 0$ and for $i = 0, 1$.   □

More examples of reducibilities r and r' where the structure of r does not form an upper semilattice but where r-r' join preservation still holds have been given in the bounded Turing degrees.

Lemma 3 (Ambos-Spies, Ding, Fan, and Merkle [4]; Ambos-Spies [1]). ibT-cl, ibT-wtt, *and* cl-wtt *join preservation hold.*

This result might lead to the assumption that cl-r join preservation holds for all reducibilities r with cl $\preceq r \preceq$ wtt, but this is not the case. We will now show that cl-r join preservation even fails for *all* admissible monotone ubT-reducibilities with cl $\prec r$.

Theorem 1. *Let r be a monotone admissible ubT-reducibility such that* cl $\prec r$. *Then, for $r' = $ ibT, cl, r'-r join-preservation fails.*

Proof. By Lemma 3, ibT-cl join preservation holds. So, it is enough to prove the theorem for $r' = \text{ibT}$. Since, by cl $\prec r$, any upper ibT-bound for two sets A_0 and A_1 is also an upper r-bound for A_0 and A_1, it suffices to construct c.e. sets A_0, A_1, B, and C such that $deg_{\text{ibT}}(A_0) \vee deg_{\text{ibT}}(A_1) = deg_{\text{ibT}}(B)$ and such that $A_0, A_1 \leq_r C$ but $B \not\leq_r C$. Let $\mathcal{F}$ be a uniformly computable admissible family of strictly increasing functions such that r is induced by $\mathcal{F}$. As $\mathcal{F}$ is uniformly computable, we can fix a computable function f such that $f \geq^* h$ for all $h \in \mathcal{F}$. As cl $\prec r$, hence $r \not\leq$ cl, $\mathcal{F} \not\leq^* \{id + e : e \geq 0\}$ holds, so, there is a function $g \in \mathcal{F}$ such that $\{g\} \not\leq^* \{id + e : e \geq 0\}$, i.e. for any $e \geq 0$, $g(x) > x + e$ for infinitely many x. Since g is strictly increasing, this implies that for all $e \geq 0$, $g(x) > x + e$ for all but finitely many x, so, $id + e \leq^* g$ for all $e \geq 0$. So, in order to complete the proof, it suffices to show that the following lemma holds.

Lemma 4. *Let g be a strictly increasing computable function such that $id + e \leq^* g$ for all e and let f be any computable function (in particular, f can be chosen as above). Then, there are c.e. sets A_0, A_1, B and C such that the following hold.*

$$deg_{\text{ibT}}(A_0) \vee deg_{\text{ibT}}(A_1) = deg_{\text{ibT}}(B) \tag{2}$$

$$A_0, A_1 \leq_{g\text{-T}} C \tag{3}$$

$$B \not\leq_{f\text{-T}} C. \tag{4}$$

Proof. We will enumerate c.e. sets A_0, A_1, B, and C such that (2) to (4) hold using a tree argument. The construction will use ideas introduced in the proof that the nondistributive lattice N5 can be embedded into the partial orderings $(\mathbf{R}_{\text{ibT}}, \leq)$ and $(\mathbf{R}_{\text{cl}}, \leq)$ in [3]. Our notation will be the same as in that proof. To guarantee that (3) holds and that B is an upper ibT-bound for A_0 and A_1, we will satisfy the following global *permitting* (or *coding*) requirement for $i = 0, 1$.

$$(x \searrow_{s+1} A_i \Rightarrow \exists y \leq x(y \searrow_{s+1} B)) \,\&\, (x \searrow_{s+1} A_i \Rightarrow \exists y \leq g(x)(y \searrow_{s+1} C)) \tag{5}$$

To guarantee that B is in fact the least upper ibT-bound for A_0 and A_1, i.e. that (2) holds, we will meet the following *join requirements* for $e \geq 0$.

$$\mathcal{Q}_e : A_0 = \hat{\Phi}_{e_1}^{W_{e_0}} \,\&\, A_1 = \hat{\Phi}_{e_2}^{W_{e_0}} \Rightarrow B \leq_{\text{ibT}} W_{e_0} \ (e = \langle e_0, e_1, e_2 \rangle).$$

Finally, we will satisfy condition (4) by meeting the *nonordering requirements*

$$\mathcal{P}_e : B \neq \Phi_e^{C, f}$$

for $e \geq 0$. Before we give the actual construction, we will explain the ideas underlying the strategies for meeting the individual requirements and how to combine them.

As the join requirements $\mathcal{Q}_e$ are conditional requirements whose hypotheses are not decidable, we have to guess on the correctness of the hypotheses.

We define the length of agreement between A_0 and $\hat{\Phi}_{e_1}^{W_{e_0}}$ and between A_1 and $\hat{\Phi}_{e_2}^{W_{e_0}}$ at stage s by letting

$$l(e, s) = \max\{x : \forall y < x(A_{0,s}(y) = \hat{\Phi}_{e_1,s}^{W_{e_0,s}}(y) \;\&\; A_{1,s}(y) = \hat{\Phi}_{e_2,s}^{W_{e_0,s}}(y))\}.$$

Since the $\hat{\Phi}$ are bounded functionals, $\lim_{s\to\infty} l(e, s) \leq \infty$ exists and the following holds.

$$(A_0 = \hat{\Phi}_{e_1}^{W_{e_0}} \;\&\; A_1 = \hat{\Phi}_{e_2}^{W_{e_0}}) \Leftrightarrow \lim_{s\to\infty} l(e, s) = \infty \Leftrightarrow \limsup_{s\to\infty} l(e, s) = \infty. \qquad (6)$$

In the following, we call a join requirement $\mathcal{Q}_e$ *infinitary* if its hypothesis is true (i.e., if $\lim_{s\to\infty} l(e, s) = \infty$) and we call $\mathcal{Q}_e$ *finitary* otherwise. The strategy for meeting the join requirements is the join strategy used by Ambos-Spies, Bodewig, Kräling, and Yu in [3]. For meeting an infinitary join requirement $\mathcal{Q}_e$, we guarantee $B \leq_{\text{ibT}} W_{e_0}$ by permitting (up to some computable subset of B). We work with a computable set $S = \{s_n : n \geq 0\}$ of $\mathcal{Q}_e$-*expansionary stages*, i.e., $s_0 < s_1 < s_2 < \dots$ and $l(e, s_0) < l(e, s_1) < l(e, s_2) < \dots$. We ensure that numbers put into B between stages $s_n + 1$ and $s_{n+1} + 1$ are greater than $s_n + 1$. So, it suffices to guarantee that if a number x enters B at a stage $s + 1$ where $s \in S$ and $x < l(e, s)$ then a number $\leq x$ will be enumerated into W_{e_0} after stage s. This change in W_{e_0} is forced by putting a sufficiently small number into A_0 or A_1. As one can easily check, this is achieved by guaranteeing the following.

$$x \searrow_{s+1} B \;\&\; x < l(e, s) \Rightarrow \exists y < \min(x', l(e, s))(y \searrow_{s+1} A_0 \text{ or } y \searrow_{s+1} A_1) \qquad (7)$$
$$\text{where } x' = \mu z(z > x \;\&\; z \notin W_{e_0,s})$$

For meeting the nonordering requirements $\mathcal{P}_e$, we will use the Friedberg-Muchnik strategy. For a fixed unused number x, we ensure $B(x) \neq \Phi_e^{C,f}(x)$ by waiting for a stage s such that $\Phi_{e,s}^{C_s,f}(x) = 0$. Then, at stage $s + 1$, we put x into B and, in order to preserve the computation $\Phi_{e,s}^{C_s,f}(x)$, we impose a restraint of length $f(x) + 1$ on C, thereby ensuring

$$B(x) = 1 \neq 0 = B_s(x) = \Phi_{e,s}^{C_s,f}(x) = \Phi_e^{C,f}(x). \qquad (8)$$

In the presence of the join requirements and the global permitting requirement, this strategy needs some amendments. To describe the potential conflicts, consider the situation in which we wish to meet requirement $\mathcal{P}_e$ and simultaneously satisfy the global permitting requirement (5) and follow the join strategy (7) for a single infinitary join requirement $\mathcal{Q}_{e'}$ of higher priority.

Now, when we put a number x into B at stage $s + 1$ in order to guarantee (8), then, according to (7), we have to put a number $y < x'$ into A_0 or A_1 at stage $s + 1$ where

$$x' = \mu z(z > x \;\&\; z \notin W_{e'_0,s}).$$

(In our case, we choose to put y into A_1.) If we do so, then, as long as $x \leq y$, this is consistent with the first part of condition (5). But, for the second part of this condition, we have to put a number $z \leq g(y)$ into C. In case that $z \leq f(x)$, however, this will injure the restraint imposed on C in order to preserve the computation $\Phi_{e,s}^{C_s,f}(x)$. In order to overcome this problem, we will make sure that we can find a number y such that $f(x) < y < x'$ where y is not yet in A_1 and the interval $[y, g(y)]$ is not yet completely enumerated into C. (Then putting y into A_1 and some new number z with $y \leq z \leq g(y)$ into C makes the enumeration of x into B compatible with (5) and (7).)

For that matter, we will assign a sufficiently long interval I_n of unused numbers to $\mathcal{P}_e$. I_n will contain finitely many candidates $x_{n,k}$ for a possible attack on $\mathcal{P}_e$ where these numbers are chosen so that $x_{n,k+1} > f(x_{n,k})$ and $g(x_{n,k}) \geq x_{n,k} + k + 2$ for all k. (Note that the latter can be achieved since, by choice of g, $g(y) > y + k + 2$ for all sufficiently large y; also note that $g(x_{n,k}) \geq x_{n,k} + k + 2$ implies $g(y) \geq y + k + 2$ for all $y \geq x_{n,k}$.) We will arrange that, for some k (and some stage s), $(x_{n,k}, x_{n,k+1}] \subseteq W_{e_0',s}$ where $x_{n,k}$ is not in B_s, $x_{n,k+1}$ is not in $A_{1,s}$ and the interval $[x_{n,k+1}, g(x_{n,k+1})]$ is not completely contained in C_s. (Hence, for $x = x_{n,k}$ and $y = x_{n,k+1}$, $y < x'$ whence we can ensure (8) and simultaneously obey (5) and (7) by putting $x_{n,k}$ into B, $x_{n,k+1}$ into A_1, and some unused number from the interval $[x_{n,k+1}, g(x_{n,k+1})]$ into C at stage $s + 1$.) In order to ensure $(x_{n,k}, x_{n,k+1}] \subseteq W_{e_0'}$ for some k, we will successively and in decreasing order put numbers w from I_n into A_0 at stages $s + 1$ where $l(e, s)$ is greater than the endpoint of I_n. This forces $W_{e_0'}$ to respond by enumerating more and more numbers from I_n (or smaller ones). As we will argue, this implies that, at some point s, there will be an interval $(x_{n,k}, \ldots x_{n,k+1}] \subset I_n$ such that the enumeration of the numbers $\geq x_{n,k} + 1$ from I_n into A_0 has forced all the numbers $x_{n,k} + 1, \ldots, x_{n,k+1}$ into $W_{e_0'}$. (In the actual construction, all the numbers actually have to be forced simultaneously into all sets $W_{e_0'}$ attached to the infinitary higher priority join requirements, but we will show that this can be achieved.) So we can use $x_{n,k}$ for an attack on $\mathcal{P}_e$ – provided that $x_{n,k} \notin B_s$, $x_{n,k+1} \notin A_{1,s}$ and $[x_{n,k+1}, g(x_{n,k+1})] \not\subseteq C_s$.

The latter, however, is not trivially true, since to make the enumeration of w into A_0 compatible with (5) simultaneously we have to put a trace $w_B \leq w$ into B and a trace $w_C \leq g(w)$ into C. So whenever we put w into A_0, then, simultaneously we put w into B (which is compatible with (7) since w goes simultaneously into A_0) and a number from the interval $[w, g(w))$ into C. Since we put only numbers $w > x_{n,k}$ into A_0 this procedure also puts only numbers $> x_{n,k}$ into B and no numbers into A_1 hence guarantees $x_{n,k} \notin B_s$ and $x_{n,k+1} \notin A_{1,s}$. To ensure that $[x_{n,k+1}, g(x_{n,k+1})] \not\subseteq C_s$, however, we have to choose the trace $w_C \in [w, g(w))$ to be put into C carefully. Here we let $w_C = w + k' + 1$ for the unique k' such that $w \in (x_{n,k'}, x_{n,k'+1}]$. Note that, by choice of the numbers $x_{n,k'}$ this ensures that $w_C \leq g(w)$. On the other hand, this ensures that $x_{n,k+1} + k + 2$ is not enumerated into C since, for $w \leq x_{n,k+1}$, $w_C \leq w + k + 1$ while, for $w > x_{n,k+1} < x_{n,k+1} + k + 2$, $w_C \geq w + (k + 1) + 1 > x_{n,k+1} + k + 2$.

This completes the discussion of the basic conflicts among the different goals of the construction and how these conflicts can be resolved. We now turn to the actual construction.

We implement the guesses about which of the join requirements are infinitary on the full binary tree $T = \{0,1\}^{<\omega}$. A node α codes a guess about the first n join requirements $\mathcal{Q}_0, \ldots, \mathcal{Q}_{n-1}$ where, for $e < n$, $\alpha(e) = 0$ codes the guess that $\mathcal{Q}_e$ is infinitary and $\alpha(e) = 1$ codes the guess that $\mathcal{Q}_e$ is finitary. So the *true path* $f : \omega \to \{0,1\}$ of the construction is defined by

$$f(e) = \begin{cases} 0 & \text{if } A_0 = \hat{\Phi}_{e_1}^{W_{e_0}} \And A_1 = \hat{\Phi}_{e_2}^{W_{e_0}} \\ 1 & \text{otherwise.} \end{cases}$$

For each node α of length e there is a strategy $\mathcal{P}_\alpha$ for meeting requirement $\mathcal{P}_e$ which is based on the guess α. We will show that the strategy $\mathcal{P}_{f \restriction e}$ on the true path will succeed in meeting $\mathcal{P}_e$.

At any stage s of the construction we have an approximation δ_s of $f \restriction s$, i.e., a guess which of the first s join requirements are infinitary. For the definition of δ_s, first we inductively define α-*stages* for each node α as follows. Each stage $s \geq 0$ is a λ-stage. If s is an α-stage, then we call s α-*expansionary* if $l(|\alpha|, s) > l(|\alpha|, t)$ for all α-stages $t < s$, and we call s an $\alpha 0$-stage if s is α-expansionary and an $\alpha 1$-stage if s is an α-stage but not α-expansionary. Now, for each $s \geq 0$, let $\delta_s \in T$ be the unique α of length s such that s is an α-stage. So, the node δ_s represents the guess at which of $\mathcal{Q}_0, \ldots \mathcal{Q}_{s-1}$ are infinite which is made at the end of stage s. It easily follows from (6) that the true path is the leftmost path visited infinitely often in the construction.

Claim 1 (True Path Lemma). $f = \liminf_{s \to \infty} \delta_s$, i.e., for any α, $\alpha \sqsubset f$ if and only if $\alpha \sqsubset \delta_s$ for infinitely many s and there are only finitely many s such that $\delta_s <_L \alpha$.

The intervals I_n which might be assigned to the strategies for meeting the nonordering requirements are inductively defined as follows, where the nth interval I_n consists of $n(x_{n,0} + 1)$ subintervals $I_{n,k} = (x_{n,k}, x_{n,k+1}]$.

$$x_{0,0} = \mu x(g(x) \geq x + 2)$$
$$x_{n,k} = \mu x(x > f(x_{n,k-1}) \And g(x) \geq x + k + 2)$$
$$\text{for } n \in \omega \text{ and } 1 \leq k \leq n(x_{n,0} + 1)$$
$$x_{n+1,0} = \mu x(x > x_{n,n(x_{n,0}+1)} + n(x_{n,0} + 1) + 2 \And g(x) \geq x + 2) \text{ for } n \in \omega$$
$$I_{n,k} = (x_{n,k}, x_{n,k+1}] \cdot \text{for } n \in \omega \text{ and } 0 \leq k \leq n(x_{n,0} + 1) - 1$$
$$I_n = \bigcup_{k=0}^{n(x_{n,0}+1)-1} I_{n,k}$$

Note that this definition ensures that $x_{n,k+1} > f(x_{n,k})$, $g(w) \geq w + k + 2$ for $w \in I_{n,k}$ and $g(w) < x_{n+1,0}$ for $w \in I_n$.

For a node α of length e, we call a number $x \in I_n \cup \{x_{n,0}\}$ α-*safe* at stage s if

$$x = x_{n,k} \text{ for some } k \text{ with } 0 \leq k \leq n(x_{n,0} + 1) - 1 \tag{9}$$

$$x \notin B_s, x_{n,k+1} \notin A_{1,s} \text{ and } x_{n,k+1} + k + 2 \notin C_s, \text{ and} \tag{10}$$

$$\forall e'([e' < e \ \& \ \alpha(e') = 0] \Rightarrow I_{n,k} \subseteq W_{e'_0,s}) \tag{11}$$

hold where $e' = \langle e'_0, e'_1, e'_2 \rangle$.

Using the above definitions, the construction of the sets A_0, A_1, B, and C is as follows where stage 0 is vacuous (i.e., $A_{0,0} = A_{1,0} = B_0 = C_0 = \emptyset$).

Stage $s + 1$. A strategy $\mathcal{P}_\alpha$ with $|\alpha| = e$ *requires attention* at stage $s + 1$ if $\alpha \sqsubseteq \delta_s$, $\mathcal{P}_\alpha$ is not satisfied at the end of stage s, and one of the following cases applies.

(i) No interval is assigned to $\mathcal{P}_\alpha$ at the end of stage s.
(ii) Interval $I_n = (x_{n,0}, x_{n,n(x_{n,0}+1)}]$ is assigned to $\mathcal{P}_\alpha$ at the end of stage s,

$$\forall e'([e' < e \ \& \ \alpha(e') = 0] \Rightarrow l(e', s) > x_{n,n(x_{n,0}+1)}) \tag{12}$$

holds, no number $x \in I_n \cup \{x_{n,0}\}$ is α-safe at stage s, and $I_n \nsubseteq A_{0,s}$.
(iii) Interval I_n is assigned to $\mathcal{P}_\alpha$ at the end of stage s, (12) holds, and there is a number $x \in I_n \cup \{x_{n,0}\}$ such that x is α-safe at stage s and $B_s(x) = \Phi_{e,s}^{C_s,f}(x) = 0$.

Fix α minimal such that $\mathcal{P}_\alpha$ requires attention (as $\mathcal{P}_{\delta_s}$ requires attention, there is such an α). Declare that $\mathcal{P}_\alpha$ *receives attention* or *becomes active*, initialize all strategies $\mathcal{P}_\beta$ with $\alpha < \beta$ (i.e., if an interval is assigned to $\mathcal{P}_\beta$ then cancel this assignment and if $\mathcal{P}_\beta$ had been satisfied before, then declare $\mathcal{P}_\beta$ to be unsatisfied), and perform the following action according to the case via which $\mathcal{P}_\alpha$ requires attention.

(i) For the least $n > e, s$ such that the interval I_n has not been assigned to any strategy before, assign I_n to $\mathcal{P}_\alpha$.
(ii) Let y be the greatest number in $I_n \setminus A_{0,s}$. Put y into A_0 and B and, for the unique k such that $y \in I_{n,k}$, put $y + k + 1$ into C.
(iii) Let x be the greatest α-safe number in $I_n \cup \{x_{n,0}\}$ such that $B_s(x) = \Phi_{e,s}^{C_s,f}(x) = 0$. Let k be the unique number such that $x = x_{n,k}$. Put x into B, $x_{n,k+1}$ into A_1, and $x_{n,k+1}+k+2$ into C. Then, declare $\mathcal{P}_\alpha$ to be *satisfied*.

This completes the construction. We will prove a series of claims to show that the construction satisfies all of our requirements. The claims will essentially be the same as in the proof of Theorem 3.2 in [3]. The first of these claims is straightforward and we omit the proof.

Claim 2. Every strategy $\mathcal{P}_\alpha$ on the true path (i.e., $\alpha \sqsubset f$) is initialized only finitely often and requires attention only finitely often. Moreover, for any such strategy, there is an interval I_n which is permanently assigned to it.

Claim 3. The global permitting requirement (5) is satisfied.

Proof. It is crucial to note that numbers from $I_n \cup \{x_{n,0}\} \cup \{g(x) : x \in I_n\}$ can be enumerated into any of the sets under construction at stage $s + 1$ only by the strategy to which I_n is assigned at this stage. So, it follows by a straightforward induction that if a strategy $\mathcal{P}_\alpha$ acts via (ii) at stage $s + 1$ then, for the number

y there, neither y is in B_s nor $y + k + 1$ is in C_s. And, similarly, if a strategy $\mathcal{P}_\alpha$ acts via (iii) at stage $s + 1$ then neither $x_{n,k}$ is in B_s nor $x_{n,k+1}$ is in $A_{1,s}$ nor $x_{n,k+1} + k + 2$ is in C_s where the latter follows from our observations preceding the construction. This easily implies the claim, since a number x is enumerated into A_0 at some stage $s + 1$ only if some strategy $\mathcal{P}_\alpha$ acts at stage $s + 1$ via (ii), hence $x \in I_{n,k}$ for some k and, at stage $s + 1$, x is enumerated into B and $x + k + 1$ is enumerated into C where $x + k + 1 \le g(x)$ by choice of $I_{n,k}$; and since a number x is enumerated into A_1 at some stage $s + 1$ only if some strategy $\mathcal{P}_\alpha$ acts at stage $s + 1$ via (iii), hence $x = x_{n,k+1}$ for some n, k and, at stage $s + 1$, $x_{n,k} < x_{n,k+1}$ is enumerated into B and $x_{n,k+1} + k + 2$ is enumerated into C where by choice of $x_{n,k+1}$, $x_{n,k+1} + k + 2 \le g(x)$.

Claim 4. The join requirements $\mathcal{Q}_e$ are met.

Proof. The argumentation is very similar to the one in the proof of Claim 5 in the proof of Theorem 3.2 in [3]. We fix $e = \langle e_0, e_1, e_2 \rangle$ and assume w.l.o.g. that $\mathcal{Q}_e$ is infinitary, so, $\alpha 0 \sqsubset f$ for $\alpha = f \restriction e$. Hence there are infinitely many $\alpha 0$-stages. By Claims 1 and 2, we can fix an $\alpha 0$-stage $s_0 > e$ such that no strategy $\mathcal{P}_\beta$ with $\beta \le \alpha 0$ becomes active after this stage. Let $S = \{s_n : n \ge 0\}$ be the set of the $\alpha 0$-stages $\ge s_0$. Then, S is computable, $s_0 < s_1 < s_2 < \ldots$, and $l(e, s_0) < l(e, s_1) < l(e, s_2) < \ldots$. So, as explained in the discussion of the strategy for meeting the requirements $\mathcal{Q}_e$, it suffices to show that (7) holds for $s \in S$. But this is immediate by construction since at a stage $s_m + 1$ only a strategy $\mathcal{P}_\beta$ with $\alpha 0 \sqsubseteq \beta$ may act. Namely, if $\mathcal{P}_\beta$ acts via (ii) then the number x enumerated into B is simultaneously enumerated into A_0 and if $\mathcal{P}_\beta$ acts via (iii) then the claim follows from the corresponding action by β-safeness of the number x put into B.

Claim 5. The nonordering requirements $\mathcal{P}_e$ are met.

Proof. For fixed e, assume for a contradiction that $\mathcal{P}_e$ is not met. Exactly as in [3], we can then argue that for $\alpha = f \restriction e$, an interval I_n becomes permanently assigned to $\mathcal{P}_\alpha$ at some stage $s_1 + 1$, that there is no number $x \in I_n \cup \{x_{n,0}\}$ that is α-safe at any stage $s' > s_1$, and that all numbers in I_n are enumerated into A_0 in decreasing order after stage $s_1 + 1$ according to clause (ii) in the definition of requiring and receiving attention. As in [3], for $x \in I_n$, let $t_x > s_1$ be the α-stage such that x is enumerated into A_0 at stage $t_x + 1$. Then (12) holds for $s = t_x$. So, for $x \in I_n$ and for any infinitary higher priority join requirement $\mathcal{Q}_{e'}$, $W_{e'_0, t_x} \restriction x + 1 \ne W_{e'_0, t_x - 1} \restriction x + 1$. So if we let J be the set of the numbers e'_0, such that

$$J = \{e'_0 : \exists e'_1, e'_2 : (\langle e'_0, e'_1, e'_2 \rangle < e \ \& \ \mathcal{Q}_{\langle e'_0, e'_1, e'_2 \rangle} \text{ is infinitary}\},$$

then

$$\forall j \in J \ \forall x \in I_n (W_{j,t_x} \restriction x + 1 \subset W_{j,t_x - 1} \restriction x + 1). \tag{13}$$

Now, for $x \in I_n$ and $j \in J$, let

$$w_j(x) = |W_{j,t_x} \restriction x + 1| \quad \text{and} \quad w_J(x) = \sum_{j \in J} w_j(x),$$

and call x *unsaturated* if $x \notin W_{j,t_x}$ for some $j \in J$. By definition, $|J| \leq e$ and $w_j(x) \leq x + 1$, hence

$$w_J(x_{n,0}) \leq e(x_{n,0} + 1). \tag{14}$$

As in [3], we will now argue that this bound is not compatible with (13) and the fact that there are no α-safe numbers in $I_n \cup \{x_{n,0}\}$. As shown in [3], it follows from (13) that

$$w_J(x_{n,0}) \geq |\{x \in I_n : x \text{ is unsaturated}\}|. \tag{15}$$

Now, it suffices to give a lower bound on the number of unsaturated numbers in I_n that contradicts (14). For a number $x_{n,k} \in I_n \cup \{x_{n,0}\}$ with $0 \leq k \leq n(x_{n,0} + 1) - 1$, (9) and (10) hold for $t_x = s$. So, since there are no α-safe numbers in $I_n \cup \{x_{n,0}\}$ after stage $s_1 + 1$, (11) must fail for $t_x = s$. It follows that at least one number in $I_{n,k}$ must be unsaturated for every k. As there are $n(x_{n,0} + 1)$ many subintervals $I_{n,k}$ in I_n each of which must contain at least one unsaturated number and as $e < n$ by construction, it follows that there are at least $(e + 1)(x_{n,0} + 1)$ unsaturated numbers in I_n, which, together with (15), leads to the desired contradiction.

This completes the proof of Lemma 4. □

4   Meet Preservation

In contrast to Theorem 1, meet preservation holds for the monotone admissible bounded Turing reducibilities in general. This is immediate by the following lemma which generalizes the observation in [4] that ibT-cl and cl-wtt meet preservation hold.

Lemma 5. *Let r and r' be monotone admissible bounded Turing reducibilities induced by $\mathcal{F}$ and $\mathcal{F}'$, respectively, such that r is stronger than r'. Then, r-r' meet preservation holds.*

Proof. The proof is essentially the same as the one for the results in [4]. Let A_0, A_1, and B be c.e. sets such that

$$deg_r(A_0) \wedge deg_r(A_1) = deg_r(B) \tag{16}$$

holds. As r is stronger than r', B is also an upper r'-bound for A_0 and A_1, so, it suffices to show that for a given c.e. set C such that $C \leq_{r'} A_0, A_1$, $C \leq_{r'} B$ holds. Fix functions $f_i \in \mathcal{F}'$ such that $C \leq_{f_i - \mathrm{T}} A_i$ for $i = 0, 1$. Since r' is admissible, as shown in [2], we may assume that $\mathcal{F}'$ is closed under composition, so, $f_0 \circ f_1 = f \in \mathcal{F}'$. As r' is monotone, we may also assume that f_0 and f_1 are strictly increasing, so, $\max(f_0, f_1) \leq f$. It follows that $C \leq_{f - \mathrm{T}} A_0, A_1$. Let $C_f = \{f(x) : x \in C\}$ be the f-shift of C. Then, $C_f \leq_{\mathrm{ibT}} A_0, A_1$. As ibT is stronger than r, $C_f \leq_r A_0, A_1$, so, by (16), $C_f \leq_r B$, hence $C_f \leq_{r'} B$. We know that $C \leq_{f - \mathrm{T}} C_f$, hence by $f \in \mathcal{F}'$, $C \leq_{r'} C_f$, so, by transitivity of r', $C \leq_{r'} B$. □

5   Open Problems

Contrasting previous positive results on join preservation in the bounded Turing degrees (see Lemma 3) we have shown that r-r' join preservation fails for the strongly bounded Turing reducibilities $r = \text{ibT,cl}$ and any monotone admissible uniformy bounded Turing reducibility r' with $cl \prec r'$. This naturally leads to the question of a classification of the monotone admissible bounded Turing reducibilities r and r' for which r-r' join preservation holds. Moreover, one may consider nonmonotone reducibilities, too. For the latter, a classification of the bT-reducibilities for which meet preservation holds is open, too.

References

1. Ambos-Spies, K.: On the strongly bounded Turing degrees of the computably enumerable sets (to appear)
2. Ambos-Spies, K.: Uniformly bounded Turing reducibilities (to appear)
3. Ambos-Spies, K., Bodewig, P., Kräling, T., Yu, L.: Joins and meets in the computably enumerable cl-degrees (to appear)
4. Ambos-Spies, K., Ding, D., Fan, Y., Merkle, W.: Maximal pairs of computably enumerable sets in the computable Lipschitz degrees. Theor. Comput. Syst. **52**(1), 2–27 (2013)
5. Barmpalias, G.: Computably enumerable sets in the solovay and the strong weak truth table degrees. In: Cooper, S.B., Löwe, B., Torenvliet, L. (eds.) CiE 2005. LNCS, vol. 3526, pp. 8–17. Springer, Heidelberg (2005)
6. George, B., Andrew, E.M.L.: The ibT degrees of computably enumerable sets are not dense. Ann. Pure Appl. Log. **141**(1–2), 51–60 (2006)
7. Day, A.: The computable Lipschitz degrees of computably enumerable sets are not dense. Ann. Pure Appl. Log. **161**(12), 1588–1602 (2010)
8. Downey, R.G., Hirschfeldt, D.R.: Algorithmic Randomness And Complexity, vol. XXVIII, 855p. Springer, New York (2010)
9. Downey, R.G., Hirschfeldt, D.R., LaForte, G.: Randomness and reducibility. In: Sgall, J., Pultr, A., Kolman, P. (eds.) MFCS 2001. LNCS, vol. 2136, pp. 316–327. Springer, Heidelberg (2001)
10. Downey, R.G., Hirschfeldt, D.R., LaForte, G.: Randomness and reducibility. (English summary). J. Comput. Syst. Sci. **68**(1), 96–114 (2004)
11. Downey, R.G., Stob, M.: Structural interactions of the recursively enumerable T- and w-degrees. Ann. Pure Appl. Log. **31**(2–3), 205–236 (1986)
12. Fan, Y., Lu, H.: Some properties of sw-reducibility. Nanjing Daxue Xuebao Shuxue Bannian Kan **22**(2), 244–252 (2005)
13. Odifreddi, P.: Strong reducibilities. Bull. Am. Math. Soc. **4**(1), 37–86 (1981)
14. Robert, I.S.: Computability theory and differential geometry. Bull. Symb. Log. **10**(4), 457–486 (2004)

Structured Frequency Algorithms

Kaspars Balodis[1,2]([⊠]), Jānis Iraids[1,2], and Rūsiņš Freivalds[1,2]

[1] Faculty of Computing, University of Latvia,
Raiņa bulvāris 19, Riga 1586, Latvia
[2] Institute of Mathematics and Computer Science,
University of Latvia, Raiņa bulvāris 29, Riga 1459, Latvia
kbalodis@gmail.com

Abstract. B.A. Trakhtenbrot proved that in frequency computability (introduced by G. Rose) it is crucially important whether the frequency exceeds $\frac{1}{2}$. If it does then only recursive sets are frequency-computable. If the frequency does not exceed $\frac{1}{2}$ then a continuum of sets is frequency-computable. Similar results for finite automata were proved by E.B. Kinber and H. Austinat et al. We generalize the notion of frequency computability demanding a specific structure for the correct answers. We show that if this structure is described in terms of finite projective planes then even a frequency $O(\frac{\sqrt{n}}{n})$ ensures recursivity of the computable set. We also show that with overlapping structures this frequency cannot be significantly decreased. We also introduce the notion of graph frequency computation and prove sufficient conditions for a graph G such that a continuum of sets can be G-computed.

1 Introduction

The problem "What is randomness?" has always been interesting not only for philosophers and physicists but also for computer scientists. The term "nondeterministic algorithm" has been deliberately coined to differ from "indeterminism" [13].

Probabilistic (randomized) algorithms form one of central notions in Theory of Computation [12]. However, since long ago computer scientists have attempted to develop notions and technical implementations of these notions that would be similar to but not equal to randomization.

The notion of frequency computation was introduced by G. Rose [14] as an attempt to have an absolutely deterministic mechanism with properties similar to probabilistic algorithms. The definition was as follows. A function $f\colon w \to w$ is (m,n)-computable, where $1 \le m \le n$, iff there exists a recursive function $R\colon w^n \to w^n$ such that, for all n-tuples $(x_1, \cdots, x_n)$ of distinct natural numbers,

K. Balodis—The first author has been supported by the European Social Fund within the project Support for Doctoral Studies at University of Latvia.

R. Freivalds—The research was supported by Co-operation Project "Uzticamas un kontrolētas mobilo ierīču pielietojuma vides izpēte un saistīto ekspertu rīku izveides iespējas" and by Project 271/2012 from the Latvian Council of Science.

© Springer International Publishing Switzerland 2015
R. Jain et al. (Eds.): TAMC 2015, LNCS 9076, pp. 50–61, 2015.
DOI: 10.1007/978-3-319-17142-5_6

$$card\{i : (R(x_1, \cdots , x_n))_i = f(x_i)\} \geq m.$$

McNaughton [11] cites in his survey a problem (posed by Myhill) whether f has to be recursive if m is close to n. This problem was answered by Trakhtenbrot [16] by showing that f is recursive whenever $2m > n$. On the other hand, Trakhtenbrot [16] proved that, if $2m = n$ then nonrecursive functions can be (m, n)-computed. Kinber [8,9] extended the research by considering frequency enumeration of sets. The class of (m, n)-computable sets equals the class of recursive sets if and only if $2m > n$. The notion of frequency computation can be extended to other models of computation. Frequency computation in polynomial time was discussed in full detail by Hinrichs and Wechsung [7].

For resource bounded computations, the behavior of frequency computability is completely different: for example, whenever $n' - m' > n - m$, it is known that under any reasonable resource bound there are sets which are (m', n')-computable, but not (m, n)-computable. However, scaling down to finite automata, the analogue of Trakhtenbrot's [16] result holds again: the class of languages (m, n)-recognizable by deterministic frequency automata equals the class of regular languages if and only if $2m > n$ (cf. Austinat et al. [2]). Conversely, as shown by Austinat et al. [2], for $2m \leq n$, the class of languages (m, n)-recognizable by deterministic frequency automata is uncountable for a two-letter alphabet. A stronger result concerning sets separable by finite automata was claimed by Kinber [9], and this result would imply the results mentioned above as a corollary. However, as shown by Tantau [15], who gave a counter-example, Kinber's [9] Theorem 3 does not hold. When restricted to a one-letter alphabet, then every (m, n)-recognizable language is regular. This was shown by Kinber [8] and also by Austinat et al. [2].

Frequency computations became increasingly popular when relations between frequency computation and computation with a small number of queries was discovered [1–4,6]. Many papers have been written to distinguish properties of frequency algorithms from the properties of probabilistic algorithms [1,4,8,9,16].

2   Definitions

By $\mathbb{N} = \{0, 1, 2, \dots\}$ we denote the set of nonnegative integers and $\mathbb{B} = \{0, 1\}$. $[n] = \{0, 1, 2, \dots, n - 1\}$. We use $|X|$ to denote the cardinality of a set X.

Let $A \subseteq \mathbb{N}$ be a set. By $\chi_A : \mathbb{N} \to \mathbb{B}$ we denote the *characteristic function* of A:

$$\chi_A(x) = \begin{cases} 1, & \text{if } x \in A \\ 0, & \text{if } x \notin A \end{cases}$$

We say that a function f is *recursive* if there is an algorithm (Turing machine) that computes f. If χ_A is a total recursive function then we call the set A *recursive*.

Definition 1. *A set A is (m, n)-computable iff there is a total recursive function f which assigns to all distinct inputs $x_1, x_2, \dots, x_n$ a binary vector $(y_1, y_2, \dots, y_n)$ such that at least m of the equations $\chi_A(x_1) = y_1, \chi_A(x_2) = y_2, \dots, \chi_A(x_n) = y_n$ hold.*

By a *structure* of a finite set K we call a set of K's subsets $S \subseteq 2^K$.

We assume that the elements of K are ordered under some fixed ordering $\phi : K \to [n]$ where $n = |K|$.

Definition 2. *A set A is (S, K)-computable (or computable with a structure S) iff there is a total recursive function f which assigns to all distinct inputs $x_1, x_2, \ldots, x_n$ a binary vector $(y_1, y_2, \ldots, y_n)$ such that $\exists B \in S \, \forall b \in B \, \chi_A(x_{\phi(b)}) = y_{\phi(b)}$*

It can be seen that (m, n)-computability is a special case of (S, K)-computability by taking S to be the set of all subsets of K of size m.

3 Projective Plane Frequency Computation

In finite geometry, the *Fano plane* (named after Gino Fano) is the finite projective plane of order 2, having the smallest possible number of points and lines. This plane has 7 points and 7 lines with 3 points on every line and 3 lines through every point. Every two points are on a unique line and every two lines intersect in a unique point (Fig. 1).

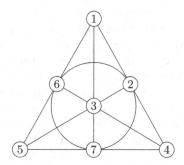

Fig. 1. The Fano Plane

We consider the first example of a structured frequency computability using the Fano plane.

Definition 3. *A set A is* Fano-computable *iff there exists a total recursive function $f : \mathbb{N}^7 \to \mathbb{B}^7$ which assigns to all 7-tuples $(x_1, x_2, \ldots, x_7) \in \mathbb{N}^7$ of distinct inputs a binary vector $\{y_1, y_2, \ldots, y_7\}$ such that*

$$(y_1 = \chi_A(x_1) \wedge y_2 = \chi_A(x_2) \wedge y_4 = \chi_A(x_4)) \vee$$

$$\vee (y_2 = \chi_A(x_2) \wedge y_3 = \chi_A(x_3) \wedge y_5 = \chi_A(x_5)) \vee$$

$$\vee (y_3 = \chi_A(x_3) \wedge y_4 = \chi_A(x_4) \wedge y_6 = \chi_A(x_6)) \vee$$

$$\vee (y_4 = \chi_A(x_4) \wedge y_5 = \chi_A(x_5) \wedge y_7 = \chi_A(x_7)) \vee$$

$$\lor (y_5 = \chi_A(x_5) \land y_6 = \chi_A(x_6) \land y_1 = \chi_A(x_1)) \lor$$
$$\lor (y_6 = \chi_A(x_6) \land y_7 = \chi_A(x_7) \land y_2 = \chi_A(x_2)) \lor$$
$$\lor (y_7 = \chi_A(x_7) \land y_1 = \chi_A(x_1) \land y_3 = \chi_A(x_3))$$

It can be seen that the required fraction of correct answers for *Fano*-comput-ability is $\frac{3}{7} < \frac{1}{2}$. Contrary to the (m, n)-computability however only recursive sets are *Fano*-computable.

Theorem 1. *A set A is* Fano-*computable iff it is recursive.*

The proof of this theorem is a special case of Theorem 3 below.

We want to explore further how much smaller can we get this fraction, i.e., for how small fraction of the inputs can we require the algorithm to give the correct answers so that the computed set can still only be recursive? Recall, that for "unstructured" frequency computations the answer is $\frac{1}{2}$ – if $\frac{m}{n} \leq \frac{1}{2}$ then a continuum of sets can be (m, n)-computed even with a finite automaton, but if $\frac{m}{n} > \frac{1}{2}$ then every (m, n)-computable set is recursive. Surprisingly for the structured frequency computation we can get this fraction close to $\frac{\sqrt{n}}{n}$ as n tends to infinity by extending the *Fano*-computability example.

Of course, it is possible "cheat", for example, by requiring that the algorithm on every input $(x_1, x_2, \ldots, x_n)$ outputs $(y_1, y_2, \ldots, y_n)$ such that $y_1 = \chi_A(x_1)$ and have no requirements for the other y_i's therefore attaining even a fraction of $\frac{1}{n}$ while every such computable set A is recursive (as on the first input the algorithm always has to output the correct answer). However we would like to avoid such cases because if we only look at the first input x_1 the fraction of correct answers there is $\frac{1}{1}$ which is the maximal possible. To avoid such "cheating" we introduce the notion of size consistency.

Definition 4. *By the* size *of a structure $S \subseteq 2^K$ we denote the size of the smallest subset* - $\min_{A \in S} |A|$. *We call the structure* size consistent *iff* $\neg \exists K' \subseteq K$ $\min_{A' \in S} \frac{|A' \cap K'|}{|K'|} > \min_{A \in S} \frac{|A|}{|K|}$.

The size consistency means that there is no smaller subset $K' \subseteq K$ such that the minimal subset $A' \in S$ (respective to K') contains a larger fraction of elements of K' than the minimal subset $A \in S$ does for K. Therefore this excludes the unwanted cases mentioned earlier.

Now we introduce a new type of structures for which we prove that the computed set is guaranteed to be recursive.

Definition 5. *We call a structure $S \subseteq 2^K$* overlapping *iff* $\forall A, B \in S$ $A \cap B \neq \emptyset$.

Theorem 2. *For any set K of size $n = q^2 + q + 1$ where q is a prime power there exists a size consistent overlapping structure of size $q + 1$.*

Proof. The reader might already be familiar with the concept of finite projective geometry. A finite projective plane is a finite set of points P and lines $L \subseteq 2^P$, such that

(A1) For any two distinct points, there is exactly one line containing these points.
(A2) For any two distinct lines, there is exactly one point common to these lines.
(A3) There exist four points, no three of which are on a line.

For all $q \geq 2$ these axioms imply

(B1) Each line contains exactly $q + 1$ points.
(B2) Each point is on exactly $q + 1$ lines.
(B3) There are exactly $q^2 + q + 1$ points in the projective plane.

It is known, that if q is a prime power there exists a finite projective plane denoted by $PG(2, q)$ with $|L| = |P| = q^2 + q + 1$ based on the finite field $\mathbb{F}_q$ with q elements. For a more detailed overview see, for example, [5].

Note, that $S = PG(2, q)$ is an overlapping structure with the points P playing the role of K and the lines L playing the role of S. For $PG(2, q)$: $|K| = |P| = q^2 + q + 1$ and all the lines contain exactly $q + 1$ points. From A2) it follows that the lines of $PG(2, q)$ indeed make an overlapping structure.

To show that the projective plane $S = PG(2, q)$ is a size consistent structure, it is sufficient to count $\sum_{A \in S} |A \cap K'|$.; For any integer $m : 0 < m < n$, if we consider a subset K' of size $n - m$, from B2) follows that $\sum_{A \in S} |A \cap K'| = (q+1)n - (q+1)m$. By the pigeonhole principle, there exists a subset A, such that $|A \cap K'| \leq \frac{(q+1)n - (q+1)m}{n} = \frac{(q+1)(n-m)}{n}$. Therefore $\min_{A' \in S} \frac{|A' \cap K'|}{|K'|} \leq \frac{(q+1)(n-m)}{n(n-m)} = \frac{q+1}{n}$. □

Theorem 3. *If A is computable with an overlapping structure then A is recursive.*

Proof. We will use infinite binary trees whose vertices correspond to binary strings. The root corresponds to the empty string and for every other vertex v the corresponding string $s(v)$ is the string of its parent vertex concatenated with a 0 or 1 depending on which child is v:

$$s(v) = \begin{cases} s(v_p)0, & \text{if } v \text{ is the left child of } v_p \\ s(v_p)1, & \text{if } v \text{ is the right child of } v_p \end{cases}$$

We will use $v(x)$ to denote the x-th (0-based) symbol of $s(v)$. $dom(v) = [|s(v)|]$ Therefore an infinite branch $B = B_0 B_1 B_2 \ldots$ defines a set whose characteristic function $\chi_B(x)$ is given by $B(x) = \lim_{n \to \infty} B_n(x)$. We use the same name for the set as for the branch.

Let $f : \mathbb{N}^n \to \mathbb{B}^n$ be the function that (S, K)-computes A with $|K| = n$ and some overlapping structure $S \subseteq 2^K$.

Consider a tree T which contains all $\sigma \in \{0, 1\}^*$ satisfying the property that for all distinct $x_1, \ldots, x_n \in dom(\sigma)$ there exists $P \in S$ such that $(\sigma(x_1), \ldots, \sigma(x_n))$ coincides with $f(x_1, \ldots, x_n)$ in positions P.

T contains A as an infinite branch because f (S, K)-computes A.

Assume that another infinite branch B in T differs from A in n positions $x_1, x_2, \ldots, x_n$. Then $(B(x_1), B(x_2), \ldots, B(x_n))$ coincides with $f(x_1, x_2, \ldots, x_n)$ in some positions $P_1 \in S$. But $(A(x_1), A(x_2), \ldots, A(x_n))$ also coincides with

$f(x_1, x_2, \ldots, x_n)$ in some positions $P_2 \in S$. As $\forall P_1, P_2 \in S\ P_1 \cap P_2 \neq \emptyset$ there is an x_i such that $A(x_i) = B(x_i)$ contradicting the assumption that A and B differs in $x_1, x_2, \ldots, x_n$. Therefore every infinite branch of T differs from A in at most $n-1$ positions.

Let B be an infinite branch that differs from A in maximum number of positions and let D be the finite set on which A and B differs. B is also an infinite branch in the subtree

$$T' = \{\sigma \in T \mid \forall x \in dom(\sigma) \cap D\ \ \sigma(x) = B(x)\}$$

Assume that C is another infinite branch in T'. Let x be such that $B(x) \neq C(x)$. From the definition of T' follows that $x \notin D$. But then C and A differ on $D \cup \{x\}$ thus contradicting the choice of B as an infinite branch that differs from A in maximum number of positions. Therefore B is the only infinite branch in T'.

For any vertex v the procedure of deciding whether v is in T' is recursive. The following algorithm computes $B(x)$ for any x:

Search for the first $t > x$ such that all $\sigma \in T' \cap \{0,1\}^t$ take only a unique value $y = \sigma(x)$ at x. Output this y as the value of $B(x)$.

The returned value cannot be different from $B(x)$ as T' has at every length $t > x$ a string σ with $\sigma(x) = B(x)$. If the algorithm wouldn't terminate for some x then for every $t > x$ there would be σ with $\sigma(x) \neq B(x)$ and there would be an infinite subtree $T'' = \{\sigma \in T' \mid x \in dom(\sigma) \to \sigma(x) \neq B(x)\}$. By König's Lemma this subtree would contain an infinite branch C different from B contradicting the fact that B is the only infinite branch of T'.

Therefore B is recursive and as A differs from B in a finite set of positions A is also recursive. $\qquad \square$

The following theorem shows that for overlapping structures the fraction obtained by the finite planes is close to the best possible.

Theorem 4. *Every size consistent overlapping structure* $S \subseteq 2^K$ *has size at least* $\sqrt{n}$ *where* $n = |K|$.

Proof. If S is size consistent, $\forall K' \subseteq K\ \min_{A' \in S} \frac{|A' \cap K'|}{|K'|} \leq \min_{A \in S} \frac{|A|}{|K|}$. In particular, take K' equal to a set of minimal size in S. Then

$$\frac{|K'|}{|K|} = \min_{A \in S} \frac{|A|}{|K|} \geq \min_{A' \in S} \frac{|A' \cap K'|}{|K'|} \geq \frac{1}{|K'|},$$

where the second inequality follows from the fact that S is overlapping, hence even K' has at least one element common with any other set from S. The size of the structure S is equal to the size of the smallest set in it – $|K'|$ and $|K'|^2 \geq |K| = n$. Therefore the size of the structure is at least $\sqrt{n}$. $\qquad \square$

4   Graph Frequency Computation

If we consider structures with the sizes of all subsets equal to some $k \geq 1$, the first interesting case is with $k = 2$ (with $k = 1$ either there are some inputs for which the outputs are not taken into account or it is the same as $(1, n)$-computability). A convenient and well-known way to represent such structures is using graphs.

Definition 6. *We call a structure $S \subseteq 2^K$ a graph structure iff $\forall A \in S \ |A| = 2$. For a graph $G = (V, E)$ by saying that a set A is G-computable we mean that A is (E, V)-computable.*

A natural question arises – for which graphs G are the G-computable sets recursive?

For some graphs G it is very easy to show that only recursive sets are G-computable.

Proposition 1. *If the graph G is either a triangle (C_3) or a star graph (S_k) then every G-computable set is recursive.*

Proof. The internal vertex of a star graph S_k is involved in every edge therefore on the input corresponding to this vertex the algorithm must always output the correct answer (on this vertex the algorithm $(1, 1)$-computes the set).

For a triangle graph C_3 if an algorithm C_3-computes a set A then it also $(2, 3)$-computes A.   □

The following theorem shows a sufficient condition for a graph G to allow computability of non-recursive sets.

Theorem 5. *If a graph G contains as a subgraph a cycle of length 4 (C_4) or two vertex-disjoint paths of length 3 $(2P_3)$ then there is a continuum of G-computable sets, namely, every $(1, 2)$-computable set is also G-computable.*

Proof. Assume there is an algorithm $\mathcal{A}_1$ that $(1, 2)$-computes a set A. For a graph G that contains a cycle of length 4 - $\{(1, 2), (2, 3), (3, 4), (4, 1)\}$ for some vertices $1, 2, 3, 4$ (see Fig. 2) consider the following algorithm – on inputs x_1 and x_3 output the values $(y_1, y_3) = \mathcal{A}_1(x_1, x_3)$ and on inputs x_2 and x_4 output the values $(y_2, y_4) = \mathcal{A}_1(x_2, x_4)$. At least one of the outputs y_1 and y_3 is correct and at least one of the outputs y_2 and y_4 is correct, therefore on at least one of the pairs of inputs $\{(1, 2), (2, 3), (3, 4), (4, 1)\}$ the outputs are correct.

Similarly for a graph G containing two vertex-distinct paths of length 3 - $\{(1, 2), (2, 3), (4, 5), (5, 6)\}$ for some vertices $1, 2, 3, 4, 5, 6$. Now the algorithm is to use $\mathcal{A}_1$ on pairs of inputs – (x_1, x_3), (x_4, x_6) and (x_2, x_5). In this case also there exists at least one pair of correct outputs corresponding to an edge of G.   □

As shown by the following corollaries Theorem 5 discards many graphs as the potential candidates for a structure that allows only recursive functions.

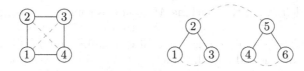

Fig. 2. Continuum implying subgraphs. Dashed lines show on which pairs of vertices (inputs) apply the $(1, 2)$-algorithm.

Corollary 1. *If G contains more than one connected component of size ≥ 3 then every $(1, 2)$-computable set is also G-computable.*

Corollary 2. *If G contains as a subgraph a cycle of length other than 3 or 5 then every $(1, 2)$-computable set is also G-computable.*

The following two theorems show that graph structures two pairs(**: :**) and three pairs(**: : :**) differ very much.

First we will need some lemmas. Let H be the 4-vertex graph with vertices $1, 2, 3, 4$ and edges $(1, 2)$, $(3, 4)$.

Lemma 1. *If M is a Turing machine with 4 inputs and 4 outputs H-computes two distinct total functions $f(x)$ and $g(x)$ such that there exist d_0 and d_1 with properties $d_0 \neq d_1$, $f(d_0) \neq g(d_0)$ and $f(d_1) \neq g(d_1)$ then there exists an algorithmic procedure computing all the values of the functions f and g with at most one exception.*

Proof. With no restriction to generality, we can assume that $d_0 = 0$ and $d_1 = 1$.

We start with considering $(x_1 = 0, x_2 = 2, x_3 = 1, x_4 = 3)$. Since $f(0) \neq g(0)$ and $f(1) \neq g(1)$ but the two functions f and g are computed correctly, the values y_1, y_2 are to be correct values of one of these functions, and y_3, y_4 are to be correct values of the other function. If $y_1 = f(0)$ then $y_2 = f(2), y_3 = g(1), y_4 = g(3)$. If $y_1 = g(0)$ then $y_2 = g(2), y_3 = f(1), y_4 = f(3)$.

Next, we consider $(x_1 = 0, x_2 = 1, x_3 = 2, x_4 = 3)$. Three cases are possible. First, if $y_1 = f(0), y_2 = f(1)$ then $y_3 = g(2), y_4 = g(3)$. Second, if $y_1 = g(0), y_2 = g(1)$ then $y_3 = f(2), y_4 = f(3)$. Third, if neither $y_1 = f(0), y_2 = f(1)$ nor $y_1 = g(0), y_2 = g(1)$ then $y_3 = f(2) = g(2)$ and $y_4 = f(3) = g(3)$.

In any of these cases we have found either both $f(2)$ and $g(2)$, or both $f(3)$ and $g(3)$. Denote by $a \in \{2, 3\}$ the value of x such that we have not yet found both $f(a)$ and $g(a)$. Then we go on considering the 4-tuples $(x_1 = 0, x_2 = a, x_3 = 1, x_4 = 4)$ and $(x_1 = 0, x_2 = 1, x_3 = a, x_4 = 4)$. This way, gradually we get all the values of the functions f and g l values of with at most one exception. □

Lemma 2. *If a Turing machine M with 4 inputs and 4 outputs is not correctly H-computing some total recursive function $f(x)$ then this property of M can be discovered considering only a finite number of 4-tuples $(x_1, x_2, x_3, x_4) \in \mathbb{N}^4$.*

Proof. By the definition, the machine M produces some result on arbitrary 4-tuple $(x_1, x_2, x_3, x_4) \in \mathbb{N}^4$. Since all such 4-tuples can be algorithmically enumerated, either a contradiction is found after a finite number of steps, or no contradiction is ever found and the function f is computed correctly. □

Lemma 3. *If $\alpha \in \{0,1\}^n$ is a finite binary word and if a Turing machine M with 4 inputs and 4 outputs is not correctly H-computing any total function $f(x)$ with values $f(0) = \alpha(0), f(1) = \alpha(1), \ldots, f(n-1) = \alpha(n-1)$, then this property of M can be discovered considering only a finite number of 4-tuples $(x_1, x_2, x_3, x_4) \in \mathbb{N}^4$.*

Proof. We consider an infinite binary tree representing all infinite binary sequences. If $\alpha \in \{0,1\}^n$ is a prefix of a function that is not correctly H-computed by M, then, by Lemma 2, this can be discovered considering only a finite number of 4-tuples $(x_1, x_2, x_3, x_4) \in \mathbb{N}^4$. In our infinite binary tree we make a cut corresponding to this prefix α. By formulation of our Lemma, these cuts leave no infinite binary path in the tree. By König's lemma [10], every tree that contains infinitely many vertices, each having finite degree, has at least one infinite simple path. Hence after all the cuts in our infinite binary tree, there remain only a finite number of vertices. □

Now we consider a tree $\mathbb{T}$ of all the total functions H-computed by M. (Since we consider only functions $\mathbb{N} \to \{0,1\}$, all the vertices of this tree have finite degree.) By Lemma 1, every function $g(x)$ correctly H-computed by the machine M differs from $f(x)$ at most for one value of x.

Lemma 4. *If M is a Turing machine with 4 inputs and 4 outputs H-computes at least one total nonrecursive function $f(x)$, then the tree $\mathbb{T}$ either contains only a finite number of functions or $\mathbb{T}$ has only one accumulation point.*

Proof. Accumulation point of the tree $\mathbb{T}$ is an infinite path P such that for every prefix π of the path P there exists an infinite path Q distinct from P but also having the prefix π. Had there been two distinct accumulation points P and Q in $\mathbb{T}$, there would be two functions $f(x)$ and $g(x)$ and values d_0 and d_1 with properties $d_0 \neq d_1$, $f(d_0) \neq g(d_0)$ and $f(d_1) \neq g(d_1)$. However, then, by Lemma 1, all the functions H-computed by M are recursive. □

Theorem 6. *If a Turing machine M with 4 inputs and 4 outputs correctly H-computes a total function then this function is recursive.*

Proof. Consider a tree $\mathbb{T}$ of all the total functions H-computed by M. If $\mathbb{T}$ contains only a finite number of functions then for each of these functions there is a prefix π which is not not a prefix of any other total function H-computed by M. If $\mathbb{T}$ has only one accumulation point then, by the construction of the tree $\mathbb{T}$ described in the proof of Lemma 3, we gradually construct initial fragments of $\mathbb{T}$. Since $\mathbb{T}$ has exactly one accumulation point, the accumulation point is always the path with the maximum other functions branching off this initial fragment of the path. Hence this path can be algorithmically constructed, and the function is recursive. □

Theorem 7. *If a graph G contains as a subgraph three vertex-disjoint paths of length 2 ($3P_2$) then there is a continuum of G-computable sets.*

Proof (Sketch of). Consider a complete infinite binary tree T whose vertices are labeled with nonnegative integers. The root is labeled with 0. For each vertex labeled x its right child is labeled $2x + 1$ and its left child is labeled $2x + 2$. Therefore T contains all numbers in $\mathbb{N}$. If we fix an infinite branch B in T, it defines the set $L_B = \{x \mid x \in B\}$.

We will show that if G contains as a subgraph $3P_2$ then there is an algorithm which G-computes any language L_B, irrespective of which branch B is chosen. As there is a continuum different ways to choose a branch B, it will follow that there is a continuum of G-computable sets.

As a side note, we should note that this is also the way how to prove that there is a continuum of $(1,2)$-computable sets. The algorithm $(1,2)$-computing L_B is the following:

On inputs (x_1, x_2):

- if there is a branch which goes through both x_1 and x_2, then output $(1,0)$, if $x_1 < x_2$, and $(0,1)$, if $x_1 > x_2$
- otherwise, output $(0,0)$

It can be checked that no matter how the branch B is chosen, at least one of these outputs will be correct.

If, instead of $(1,2)$-computing, we consider computing with a graph with 3 pairs of connected vertices, the idea of the proof is the same, only now we have to deal with a larger number of different possibilities for the input instances. A single input instance can be represented as a 7-vertex rooted tree I in which all vertices except the root are divided into three pairs. The root of I represents a vertex prepended to the root of T and the 3 pairs of vertices represent the 3 pairs of inputs for the algorithm. For any vertices x_1, x_2 if x_1 is a descendant of x_2 in T then x_1 is also a descendant of x_2 in I. See Fig. 3 for an example instance.

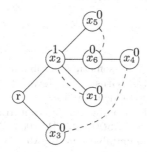

Fig. 3. An example instance. The tree shows relationships between the inputs in the tree T. The dashed lines show how the inputs are distributed into pairs. The attached output labels $y_i \in \{0,1\}$ show one possible assignment such that no matter how a branch B is chosen, there exists a pair for which both outputs are correct.

To correctly solve an instance I means to assign outputs $y_i \in \{0,1\}$ to all non-root vertices so that, no matter which branch B is chosen, on at least one pair both outputs will be correct.

It was checked with a computer program that for each possible instance I there exists an assignment that solves it.

The pseudocode of the program:

for each 7-vertex tree T **do**
    for each possible division of T into root and 3 pairs of vertices **do**
        if there exists no assingment such that no matter which branch B is
        chosen, there exists a pair with both correct outputs **then**
            return fail
        end if
    end for
end for
return success   □

5 Conclusions and Open Problems

We have introduced a new model of computability by extending the previously known frequency (m,n)-computability. We have shown some structures which lead to computability of only recursive sets and some structures which allow a continuum of computable sets. However, we are still far from a complete characterization of all structures.

Some open problems are:

- Are there any size consistent non-overlapping structures of size less than $\sqrt{n}$ that allow only computability of recursive sets? If so then what is the smallest possible fraction of correct answers attainable?
- For graph frequency computation obtain a complete classification of all graphs G and classes of G-computable sets.
- What other types of structures are interesting and worth considering and what classes of sets are computable with them?

References

1. Ablaev, F., Freivalds, R.: Why sometimes probabilistic algorithms can be more effective. In: Wiedermann, J., Gruska, J., Rovan, B. (eds.) MFCS 1986. LNCS, vol. 233, pp. 1–14. Springer, Heidelberg (1986)
2. Austinat, H., Diekert, V., Hertrampf, U., Petersen, H.: Regular frequency computations. Theoret. Comput. Sci. **330**(1), 15–21 (2005). Insightful Theory
3. Degtev, A.N.: On (m,n)-computable sets. In: Moldavanskij, D.I. (ed.), Algebraic Systems, pp. 88–99. Ivanovo Gos. Universitet, (1981) (In Russian)
4. Freivalds, R.: Inductive inference of recursive functions: qualitative theory. In: Barzdins, J., Bjorner, D. (eds.) Baltic Computer Science. LNCS, vol. 502, pp. 77–110. Springer, Heidelberg (1991)
5. Hall Jr., M.: Combinatorial Theory, 2nd edn. Wiley, New York (1986)

6. Harizanov, V., Kummer, M., Owings, J.: Frequency computations and the cardinality theorem. J. Symb. Log. **57**, 682–687 (1992)
7. Hinrichs, M., Wechsung, G.: Time bounded frequency computations. In: Proceedings of Twelfth Annual IEEE Conference on Computational Complexity, 1997 (Formerly: Structure in Complexity Theory Conference), pp. 185–192. IEEE (1997)
8. Kinber, E.B.: Frequency calculations of general recursive predicates and frequency enumerations of sets. Sov. Math. **13**, 873–876 (1972)
9. Kinber, E.B.: Frequency computations in finite automata. Cybern. Sys. Anal. **12**(2), 179–187 (1976)
10. König, D.: Sur les correspondances multivoques des ensembles. Fundamenta Math. **8**(1), 114–134 (1926)
11. McNaughton, R.: The theory of automata, a survey. Adv. Comput. **2**, 379–421 (1961)
12. Rabin, M.O.: Probabilistic automata. Inf. Control **6**(3), 230–245 (1963)
13. Rabin, M.O., Scott, D.: Finite automata and their decision problems. IBM J. Res. Dev. **3**(2), 114–125 (1959)
14. Rose, G.F.: An extended notion of computability. In: International Congress for Logic, Methodology and Philosophy of Science, Stanford, California (1960)
15. Tantau, T.: Towards a cardinality theorem for finite automata. In: Diks, K., Rytter, W. (eds.) MFCS 2002. LNCS, vol. 2420, pp. 625–636. Springer, Heidelberg (2002)
16. Trakhtenbrot, B.A.: On the frequency computation of functions. Algebra i Logika **2**(1), 25–32 (1964). In Russian

Asymptotic Properties of Combinatory Logic

Maciej Bendkowski, Katarzyna Grygiel[⊠], and Marek Zaionc

Theoretical Computer Science Department, Faculty of Mathematics and Computer
Science, Jagiellonian University, ul. Łojasiewicza 6, 30-348 Kraków, Poland
{bendkowski,grygiel,zaionc}@tcs.uj.edu.pl

Abstract. We present a quantitative analysis of random combinatory
logic terms. Our main goal is to investigate likelihood of semantic properties of random combinators. We show that asymptotically almost all
weakly normalizing terms are not strongly normalizing. Moreover, we
present a proof that asymptotically almost all strongly normalizing terms
are not in normal form. We also prove that asymptotically almost all normal forms in combinatory logic are not typeable.

Keywords: Models of computation · Combinatory logic · Asymptotic
probability in logic

1 Introduction

Over the last decade quantitative aspects of logic have attracted increasing
attention from researchers working on the border of combinatorics, logic, and
computer science. Probabilistic methods used in the paper appear to be very
powerful in computer science investigations. From a point of view of these methods we study typical objects chosen from a given set. In recent years we have
investigated sets of syntactic objects of logical flavor in order to estimate likelihood of the fact that a randomly chosen syntactic object belongs to a given
set. There is a long history of using this kind of asymptotic approach applied to
logic and computability. Probability of truth of logical formulas has been investigated in several papers. For the purely implicational logic of one variable (and
at the same time simply typed system), the likelihood of finding true formulas
was computed by Moczurad, Tyszkiewicz, and Zaionc in [14]. The classical logic
of one variable and two connectives of implication and negation was studied in
Zaionc [20]; over the same language, the exact proportion between intuitionistic
and classical logics was determined by Kostrzycka and Zaionc in [11].

Asymptotic id entity between classical and intuitionistic logic of implication
has been proved in Fournier, Gardy, Genitrini, and Zaionc in [6]. Some variants
involving expressions with other logical connectives have also been considered.

This work was supported within the grant 2013/11/B/ST6/0095 funded by the Pol-
ish National Science Center.

K. Grygiel—This author was supported by funding from the Jagiellonian University
within the SET project. The project is co-financed by the European Union.

© Springer International Publishing Switzerland 2015
R. Jain et al. (Eds.): TAMC 2015, LNCS 9076, pp. 62–72, 2015.
DOI: 10.1007/978-3-319-17142-5_7

Genitrini and Kozik in [9] have studied asymptotic behavior of full propositional system. For two connectives again, the *and/or* case has already received much attention—see Lefmann and Savický [13], Chauvin, Flajolet, Gardy, and Gittenberger [2], Gardy and Woods [8], Woods [19] and Kozik [12]. Let us also mention the survey [7] of Gardy on probability distributions on Boolean functions induced by random Boolean expressions.

In [4] investigations of computational objects from lambda calculus and combinatory logic were started. It was shown that a randomly chosen λ-term is strongly normalizing. In the case of combinatory logic (the equivalent translation of the λ-calculus), the situation is exactly opposite—a random combinator does not strongly normalize. Since every strongly normalizing term (in both models) is weakly normalizing, the obtained results imply that a random lambda term satisfies the weak normalization property, however, they do not allow us to claim anything about weakly normalizing combinators. The counting problem for lambda terms is still a very hot open research subject. Some variants of lambda calculus have also been considered. Bodini, Gardy, Gittenberger and Jacquot in [15] studied enumeration of BCI lambda terms. John Tromp in [17], as well as Grygiel and Lescanne in [10], considered the enumeration problem in the so called binary lambda calculus.

The syntax of combinators is very simple, as the terms in question can be uniquely represented by finite binary planar trees whose leaves are labeled by constants. In contrast to lambda calculus terms, whose unusual tree representation makes the combinatorial analysis very difficult (see, e.g. [4]), the analysis of combinators satisfying a given syntactic property is usually simple. However, in the case of properties that are undecidable, the enumeration problems become hard or even impossible, as for any nonrecursive set it is impossible to find a finite pattern collection defining the whole set. For example, any nonrecursive set of combinators cannot be defined by a context-free grammar. Therefore the only possible approach to find asymptotic behavior of nonrecursive sets of combinators is to construct proper recursive subsets and proper recursive supersets.

In this paper we give a simple argument that the density of weakly normalizing combinators is neither zero nor one. Moreover, we present lower and upper bounds for the density in question. This result allows us to compare two basic nonrecursive sets of combinators, one being the subset of another: the set of all weakly normalizing combinators and its proper subset—the set of all strongly normalizing combinators. It turns out, that the set of strongly normalizing terms can be seen as a tiny fragment of the set of weakly normalizing combinators. In other words, we prove that the asymptotic probability of finding strongly normalizing terms chosen from the set of weakly normalizing ones is zero.

Another part of the paper is oriented toward terms in so called normal forms. At the same time we are interested in typeable terms in combinatory logic which form an important subclass motivated by programming languages. Both typeable terms (in the simple type system) and terms in normal forms form recursive sets of combinators. In the paper we present a result concerning typeable normal forms in the setting of all normal forms.

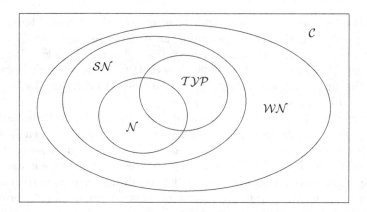

Fig. 1. Partition of combinatory logic

Figure 1 illustrates the topography of all major classes of combinatory logic terms used in this paper. $\mathcal{C}$ denotes the set of all combinators, $\mathcal{WN}$ stands for weakly normalizing terms, $\mathcal{SN}$ for strongly normalizing ones, $\mathcal{N}$ for normal forms and, finally, $\mathcal{TYP}$ stands for typeable terms.

2   Combinators

The set $\mathcal{C}$ of *combinators* is defined inductively as follows. Constants **K** and **S** are combinators. If M and N are combinators, then $(M\ N)$ is a combinator. Terms built as in the third case are called *applications*. Following standard notational conventions, we omit outermost parentheses and drop parentheses from left-associated terms, e.g., instead of $((MN)(PQ))$ we write $MN(PQ)$. We define a one-step reduction relation $\rightarrow$ on the set of combinators in the following way. Let P, Q, R be arbitrary combinators. Then

- $\mathbf{K}PQ \rightarrow P$,
- $\mathbf{S}PQR \rightarrow PR(QR)$,
- if $P \rightarrow Q$ then $PR \rightarrow QR$ and $RP \rightarrow RQ$.

Let P be a combinator. If there exists no combinator Q such that $P \rightarrow Q$, then P is said to be in *normal form*. If there exists a finite sequence of combinators $P_0, P_1, \ldots, P_k$ such that $P = P_0 \rightarrow P_1 \rightarrow \ldots \rightarrow P_k$ and P_k is in normal form, then P is *weakly normalizing*. If there does not exist an infinite sequence of combinators $P_0, P_1, \ldots$ such that $P = P_0 \rightarrow P_1 \rightarrow \ldots$, then we say that P is *strongly normalizing*. Of course, strong normalization implies weak normalization.

3   Densities of Sets of Combinators

With the set of all combinators $\mathcal{C}$ we associate the size function defined as the number of all applications occurring in a given combinator, i.e.,

$$|\mathbf{S}| = |\mathbf{K}| = 0 \quad \text{and} \quad |PQ| = 1 + |P| + |Q|.$$

Given a subset $\mathcal{X} \subseteq \mathcal{C}$ of combinators we define the *asymptotic density* $\mu(\mathcal{X})$ as

$$\mu(\mathcal{X}) = \lim_{n \to \infty} \frac{\#\{t \in \mathcal{X} : |t| = n\}}{\#\{t \in \mathcal{C} : |t| = n\}}$$

if the limit exists. The number $\mu(\mathcal{X})$ is an asymptotic probability of finding a combinator from the class $\mathcal{X}$ among all combinators. It can be immediately seen that the density μ is finitely additive, but not countably additive. Finally, we define:

$$\mu^-(\mathcal{X}) = \liminf_{n \to \infty} \frac{\#\{t \in \mathcal{X} : |t| = n\}}{\#\{t \in \mathcal{C} : |t| = n\}}$$

$$\mu^+(\mathcal{X}) = \limsup_{n \to \infty} \frac{\#\{t \in \mathcal{X} : |t| = n\}}{\#\{t \in \mathcal{C} : |t| = n\}}$$

These two numbers are well defined for any set $\mathcal{X}$ of combinators, even when the limiting ratio $\mu(\mathcal{X})$ is not known to exist. Given two classes of combinators $\mathcal{X}$ and $\mathcal{Y}$, assuming that $\mathcal{X}$ is a subset of $\mathcal{Y}$, we define *relative density* $\mu\left(\frac{\mathcal{X}}{\mathcal{Y}}\right)$ in the usual way by allowing:

$$\mu\left(\frac{\mathcal{X}}{\mathcal{Y}}\right) = \lim_{n \to \infty} \frac{\#\{t \in \mathcal{X} : |t| = n\}}{\#\{t \in \mathcal{Y} : |t| = n\}}$$

The relative $\mu^-\left(\frac{\mathcal{X}}{\mathcal{Y}}\right)$ and $\mu^+\left(\frac{\mathcal{X}}{\mathcal{Y}}\right)$ functions are defined in the very same way as in general case, i.e.,

$$\mu^-\left(\frac{\mathcal{X}}{\mathcal{Y}}\right) = \liminf_{n \to \infty} \frac{\#\{t \in \mathcal{X} : |t| = n\}}{\#\{t \in \mathcal{Y} : |t| = n\}}$$

$$\mu^+\left(\frac{\mathcal{X}}{\mathcal{Y}}\right) = \limsup_{n \to \infty} \frac{\#\{t \in \mathcal{X} : |t| = n\}}{\#\{t \in \mathcal{Y} : |t| = n\}}$$

For technical reasons, we assume $\frac{0}{0} := 1$. Given any subclass $\mathcal{X} \subseteq \mathcal{C}$ and $n \in \mathbb{N}$, we denote by $\mathcal{X}_n$ the set of all combinators from $\mathcal{X}$ that are of size n. Obviously $\mathcal{X}_n$ is always finite.

4   Generating Functions

Many questions concerning the asymptotic behavior of sequences of real non-negative numbers can be efficiently resolved by analyzing the behavior of their generating functions (see [18] for introductory reference). This is the approach we take to determine the asymptotic fraction of certain combinatory logic terms. Let $(a_n)_{n \in \mathbb{N}}$ be a sequence of non-negative numbers. The power series $A(z) = \sum_{n \in \mathbb{N}} a_n z^n$ is called the *generating function* enumerating the sequence $(a_n)_{n \in \mathbb{N}}$. We denote by $[z^n]\{A(z)\}$ the coefficient of z^n in the expansion of $A(z)$. We say that two sequences $(A_n)_{n \in \mathbb{N}}$ and $(B_n)_{n \in \mathbb{N}}$ are *asymptotically equivalent* if $\lim_{n \to \infty} \frac{A_n}{B_n} = 1$. In such a case we write $A_n \sim B_n$. The following theorem is a well-known result in the theory of generating functions. Its derivation from Szegö Lemma (see [16]) can be found, e.g., in [21, Theorem 22].

Theorem 1 (Generating Function Method via Szegö Lemma). *Let A, B be functions satisfying the following conditions:*

1. *A, B are analytic in $|z| < 1$ with $z = 1$ being the only singularity on the circle $|z| = 1$,*
2. *A, B have the following expansions in the vicinity of $z = 1$:*

$$A(z) = \sum_{p \geq 0} a_p (1 - z)^{\frac{p}{2}}, \qquad B(z) = \sum_{p \geq 0} b_p (1 - z)^{\frac{p}{2}},$$

where $b_1 \neq 0$.

Let $\tilde{A}$ and $\tilde{B}$ be functions satisfying $\tilde{A}(\sqrt{1 - z}) = A(z)$ and $\tilde{B}(\sqrt{1 - z}) = B(z)$. Then

$$\lim_{n \to \infty} \frac{[z^n]\{A(z)\}}{[z^n]\{B(z)\}} = \frac{a_1}{b_1} = \frac{\tilde{A}'(0)}{\tilde{B}'(0)}.$$

Theorem 2 (Pringsheim, see [5, Theorem IV.6]). *If $A(z)$ is representable at the origin by a series expansion that has non-negative coefficients and radius of convergence R, then the point $z = R$ is a singularity of $A(z)$.*

Theorem 3 (Exponential Growth Formula, see [5, Theorem IV.7]). *If $A(z)$ is analytic at 0 and R is the modulus of a singularity nearest to the origin in the sense that*

$$R = \sup\{r \geq 0 \ : \ A \text{ is analytic in } |z| < r\},$$

then the coefficient $a_n = [z^n]\{A(z)\}$ satisfies

$$a_n = R^{-n} \theta(n) \quad \text{with} \quad \limsup |\theta(n)|^{\frac{1}{n}} = 1.$$

By T_n we denote n-th Catalan number, i.e., the number of expressions (or equivalently trees) containing n pairs of parentheses which are correctly matched. It is well-known that $T_n = \frac{1}{n+1}\binom{2n}{n}$ and that

$$\lim_{n \to \infty} \frac{T_{n+1}}{T_n} = 4. \tag{1}$$

Since the set of all combinators is defined by a very simple grammar, we can easily count all combinators of a given size.

Fact 1 (see, e.g. [4]). *Let C be the generating function enumerating combinators of a given size. Then*

$$C(z) = \frac{1 - \sqrt{1 - 8z}}{2z} \quad \text{and} \quad |\mathcal{C}_n| = [z^n]\{C(z)\} = 2^{n+1} \cdot T_n$$

5 Weakly Normalizing Combinators

Let us start with the well-known classical fact observed already in [3].

Theorem 4 (Standarization Theorem). *If a combinator is weakly normalizing, then the leftmost outermost reduction always leads to a normal form.*

Another classical observation is that the set of weakly normalizing combinators $\mathcal{WN}$ is undecidable. It follows that there is no purely syntactic formula enumerating $\mathcal{WN}_n$ and thus we cannot find the cardinality of $\mathcal{WN}_n$ explicitly. For that reason we take the following approach. We find feasible subclasses of $\mathcal{WN}$ and $\mathcal{C} \setminus \mathcal{WN}$ and use them to bound the density of $\mathcal{WN}$ in $\mathcal{C}$.

Lemma 1. *Asymptotically at least $\frac{1}{32}$ of combinators are weakly normalizing i.e., $\mu^- \left(\frac{\mathcal{WN}}{\mathcal{C}} \right) \geq \frac{1}{32}$.*

Proof. Let $\mathcal{L}$ be the class of combinators which are either of the form $\mathbf{KK}M$ or $\mathbf{KS}M$, where $M \in \mathcal{C}$ is an arbitrary combinator. Let us notice that in just one reduction step every combinator from this class is reducible either to $\mathbf{K}$ or to $\mathbf{S}$. Therefore $\mathcal{L}$ is a subset of all weakly normalizing combinators. Moreover, we have $|\mathcal{L}_n| = 2|\mathcal{C}_{n-2}|$ for $n \geq 2$ and so

$$\mu \left(\frac{\mathcal{L}}{\mathcal{C}} \right) = \lim_{n \to \infty} \frac{|\mathcal{L}_n|}{|\mathcal{C}_n|} = \lim_{n \to \infty} \frac{2|\mathcal{C}_{n-2}|}{|\mathcal{C}_n|} = \lim_{n \to \infty} \frac{2^n \cdot T_{n-2}}{2^{n+1} \cdot T_n} \overset{(1)}{=} \frac{1}{2} \cdot \frac{1}{4^2} = \frac{1}{32}. \qquad \square$$

Lemma 2. *Asymptotically at most $1 - \frac{1}{2^{18}}$ of combinators are weakly normalizing, i.e., $\mu^+ \left(\frac{\mathcal{WN}}{\mathcal{C}} \right) \leq 1 - \frac{1}{2^{18}}$.*

Proof. Let $\omega_1 = \mathbf{S}(\mathbf{SS})\mathbf{SSSS}$ and $\omega_2 = \mathbf{SSS}(\mathbf{SS})\mathbf{SS}$. Consider the class $\mathcal{U}$ of combinators that are in form of $\omega_1 M_1 \ldots M_k$ or $\omega_2 M_1 \ldots M_k$ for arbitrary $k \geq 0$ and $M_1, \ldots, M_k \in \mathcal{C}$. In [1] it was shown that ω_1 is not normalizable, which implies that so is ω_2 since ω_1 reduces to ω_2. By the standarization theorem for combinatory logic, we obtain that $\mathcal{U} \subseteq (\mathcal{C} \setminus \mathcal{WN})$. Since both ω_1 and ω_2 are of size 6, we get $|\mathcal{U}_n| = 2^{n-6} \cdot T_{n-6}$ for $n \geq 7$. Finally,

$$\mu \left(\frac{\mathcal{U}}{\mathcal{C}} \right) = \lim_{n \to \infty} \frac{|\mathcal{U}_n|}{|\mathcal{C}_n|} = \lim_{n \to \infty} \frac{|\mathcal{C}_{n-6}|}{|\mathcal{C}_n|} = \lim_{n \to \infty} \frac{T_{n-6}}{2^6 \cdot T_n} \overset{(1)}{=} \frac{1}{2^6} \cdot \frac{1}{4^6} = \frac{1}{2^{18}}. \qquad \square$$

The two above lemmas show that the density of weakly normalizing combinators, provided it exists, is neither zero nor one. In [4] it was shown that a random combinator is not strongly normalizing. This immediately implies the following result.

Theorem 5. *Asymptotically almost all weakly normalizing terms are not strongly normalizing i.e., $\mu \left(\frac{\mathcal{SN}}{\mathcal{WN}} \right) = 0$.*

6   Combinatorial Results

Lemma 3. *Let $\mathcal{N}$ be the set of combinators in normal form. The generating function $F_{\mathcal{N}}$ enumerating cardinality of $\mathcal{N}$ is given by*

$$F_{\mathcal{N}}(z) = \frac{1 - 2z - \sqrt{1 - 4z - 4z^2}}{2z^2}.$$

This implies

$$[z^n]\{F_{\mathcal{N}}(z)\} \sim (2 + 2\sqrt{2})^n \theta(n) = (4.82843\ldots)^n \theta(n)$$

with $\limsup_{n \to \infty} |\theta(n)|^{\frac{1}{n}} = 1$.

Proof. The grammar for $\mathcal{N}$ is given by

$$\mathcal{N} := \mathbf{S} \mid \mathbf{K} \mid \mathbf{K}\,\mathcal{N} \mid \mathbf{S}\,\mathcal{N} \mid \mathbf{S}\,\mathcal{N}\,\mathcal{N}.$$

It follows that the generating function $F_{\mathcal{N}}$ satisfies

$$F_{\mathcal{N}}(z) = 2 + 2zF_{\mathcal{N}}(z) + z^2(F_{\mathcal{N}}(z))^2.$$

Solving the equation for $F_{\mathcal{N}}(z)$ we obtain two solutions $\frac{1 - 2z \pm \sqrt{1 - 4z - 4z^2}}{2z^2}$. Because $\lim_{n \to \infty} F_{\mathcal{N}}(0) = 2$ we conclude that $F_{\mathcal{N}}(z) = \frac{1 - 2z - \sqrt{1 - 4z - 4z^2}}{2z^2}$. In order to compute the asymptotic growth of $[z^n]\{F_{\mathcal{N}}(z)\}$, we start with the observation that $F_{\mathcal{N}}(z)$ has an analytic continuation in 0 and its radius of convergence R is equal to $\frac{1}{2}(\sqrt{2} - 1)$. By Pringsheim's theorem be obtain that R is also the modulus of the dominating singularity of $F_{\mathcal{N}}(z)$ and thus applying the Exponential Growth Formula we obtain that

$$[z^n]\{F_{\mathcal{N}}(z)\} = R^{-n}\theta(n) \sim (4.82843\ldots)^n \theta(n) \text{ with } \limsup_{n \to \infty} |\theta(n)|^{\frac{1}{n}} = 1. \qquad \square$$

In order to determine the density of normal forms in the set of all strongly normalizing combinators, we define a class $\mathcal{G}$ as the set of combinators defined by the following grammar:

$$\mathcal{G} := \mathbf{S} \mid \mathbf{K} \mid \mathbf{KKK} \mid \mathbf{K}\,\mathcal{G} \mid \mathbf{S}\,\mathcal{G} \mid \mathbf{S}\,\mathcal{G}\,\mathcal{G}.$$

Since $\mathcal{G}$ contains all productions of $\mathcal{N}$, we have $\mathcal{N} \subseteq \mathcal{G}$. Moreover, the only redexes in $\mathcal{G}$ are of the form $\mathbf{KKK}$, which implies that $\mathcal{G} \subseteq \mathcal{SN}$.

Lemma 4. *The generating function $F_{\mathcal{G}}$ enumerating cardinality of $\mathcal{G}$ is given by*

$$F_{\mathcal{G}}(z) = \frac{1 - 2z - \sqrt{1 - 4z - 4z^2 - 4z^4}}{2z^2},$$

which yields

$$[z^n]\{F_{\mathcal{G}}(z)\} \sim (4.85823\ldots)^n \theta(n)\, with \limsup |\theta(n)|^{\frac{1}{n}} = 1.$$

Proof. Given the grammar for $\mathcal{G}$ we obtain that $F_{\mathcal{G}}$ satisfies the following equation $F_{\mathcal{G}}(z) = 2 + z^2 + 2zF_{\mathcal{G}}(z) + z^2(F_{\mathcal{G}}(z))^2$. Solving for $F_{\mathcal{G}}(z)$ we find two possible solutions $\frac{1-2z\pm\sqrt{1-4z-4z^2-4z^4}}{2z^2}$. Since $\lim_{n\to\infty} F_{\mathcal{G}}(0) = 2$ we conclude that $F_{\mathcal{G}}(z) = \frac{1-2z-\sqrt{1-4z-4z^2-4z^4}}{2z^2}$. In order to find the dominating singularity of $F_{\mathcal{G}}(z)$, we examine the real roots of $\sqrt{1 - 4z - 4z^2 - 4z^4}$. This expression yields two real roots $z_1 \approx -0.800151$ and $z_2 \approx 0.205836$. Because z_2 lies closer to the origin, it follows that $z \approx 0.205836$ dictates the asymptotic growth of $[z^n]\{F_{\mathcal{G}}(z)\}$. Applying the Exponential Growth Formula we obtain that

$$[z^n]\{F_{\mathcal{G}}(z)\} \sim (4.85823\ldots)^n\theta(n) \text{ with } \limsup |\theta(n)|^{\frac{1}{n}} = 1. \qquad \square$$

Theorem 6. *Asymptotically almost all strongly normalizing terms are not in normal form, i.e.,* $\mu\left(\frac{\mathcal{N}}{\mathcal{SN}}\right) = 0$.

Proof. Similarly to $\mathcal{WN}$, the set of strongly normalizing combinators $\mathcal{SN}$ is undecidable and therefore we cannot enumerate $\mathcal{SN}_n$ explicitly. Fortunately, it suffices to prove that $\mu\left(\frac{\mathcal{N}}{\mathcal{H}}\right) = 0$ for some sufficiently large subclass $\mathcal{H} \subseteq \mathcal{SN}$. Let $\mathcal{G}$ be the set of combinators as defined in Lemma 4. We claim that $\mu\left(\frac{\mathcal{N}}{\mathcal{G}}\right) = 0$. Indeed, using Lemmas 3 and 4 we obtain

$$\mu\left(\frac{\mathcal{N}}{\mathcal{G}}\right) = \lim_{n\to\infty} \frac{[z^n]\{F_{\mathcal{N}}(z)\}}{[z^n]\{F_{\mathcal{G}}(z)\}} = \lim_{n\to\infty} \frac{(4.82843\ldots)^n}{(4.85823\ldots)^n} = 0. \qquad \square$$

Lemma 5. *Let $t_0 \in \mathcal{N}$ be a combinator of size $|t_0| \geq 1$. Let $\mathcal{N}_{t_0}$ be the set of combinators in normal form which contain t_0 as a subterm. The generating function $F_{\mathcal{N}_{t_0}}$ enumerating cardinality of $\mathcal{N}_{t_0}$ is given by*

$$F_{\mathcal{N}_{t_0}}(z) = \frac{-\sqrt{1 - 4z - 4z^2} + \sqrt{1 - 4z - 4z^2 + 4z^{|t_0|+2}}}{2z^2}.$$

Proof. Note that if $Q \in \mathcal{N}_{t_0}$ then either:

1. $Q = t_0$, or
2. $Q = \mathbf{K}M$ and t_0 is a subterm of M, or
3. $Q = \mathbf{S}M$ and t_0 is a subterm of M, or
4. $Q = \mathbf{S}MP$ and t_0 is a subterm of M but not P, or
5. $Q = \mathbf{S}MP$ and t_0 is a subterm of P but not M, or
6. $Q = \mathbf{S}MP$ and t_0 is a subterm of both M and P.

It follows that $F_{\mathcal{N}_{t_0}}$ satisfies the following equation

$$F_{\mathcal{N}_{t_0}}(z) = z^{|t_0|} + 2zF_{\mathcal{N}_{t_0}}(z) + 2z^2(F_{\mathcal{N}}(z) - F_{\mathcal{N}_{t_0}}(z))F_{\mathcal{N}_{t_0}}(z) + z^2(F_{\mathcal{N}_{t_0}}(z))^2.$$

Using the generating function for $\mathcal{N}$ we solve this equation for $F_{\mathcal{N}_{t_0}}(z)$ and obtain two solutions

$$\frac{-\sqrt{1 - 4z - 4z^2} \pm \sqrt{1 - 4z - 4z^2 + 4z^{|t_0|+2}}}{2z^2}.$$

Since there is no $Q \in \mathcal{N}_{t_0}$ of size 0, we get that $\lim_{z \to 0} F_{\mathcal{N}_{t_0}}(z) = 0$ and finally

$$F_{\mathcal{N}_{t_0}}(z) = \frac{-\sqrt{1 - 4z - 4z^2} + \sqrt{1 - 4z - 4z^2 + 4z^{|t_0|+2}}}{2z^2}.$$

$\square$

Theorem 7. *Let $t_0 \in \mathcal{N}$. The density of combinators in normal form which contain t_0 as a subterm is 1.*

Proof. We prove this result applying Theorem 1. We start with normalizing $F_{\mathcal{N}_{t_0}}$ and $F_{\mathcal{N}}$ in such a way that both generating functions are analytic in the disc $|z| < 1$ with $z = 1$ being their only singularity on the circle $|z| = 1$. For convenience, let us shift both generating functions by two positions obtaining $\hat{F}_{\mathcal{N}}(z) := z^2 F_{\mathcal{N}}(z)$ and $\hat{F}_{\mathcal{N}_{t_0}}(z) := z^2 F_{\mathcal{N}_{t_0}}(z)$. Since $R = \frac{1}{2}(\sqrt{2} - 1)$ is the dominating singularity of both $F_{\mathcal{N}}$ and $F_{\mathcal{N}_{t_0}}$, we define $\overline{F}_{\mathcal{N}}(z) := \hat{F}_{\mathcal{N}}(Rz)$ and $\overline{F}_{\mathcal{N}_{t_0}}(z) := \hat{F}_{\mathcal{N}_{t_0}}(Rz)$.

First, we examine $\overline{F}_{\mathcal{N}}(z)$. Simplifying, we get

$$\overline{F}_{\mathcal{N}}(z) = \frac{1}{2}\left(-\sqrt{2}z + z - \sqrt{(1 - z)\left(1 - (2\sqrt{2} - 3)z\right)} + 1\right).$$

Note that the inner expression $\sqrt{(1 - z)\left(1 - (2\sqrt{2} - 3)z\right)}$, carrying the singularities of $\overline{F}_{\mathcal{N}}(z)$, has exactly two roots, i.e., $z_1 = 1$ and $z_2 = \frac{1}{2\sqrt{2}-3} \approx -5.82843$. It follows that $\overline{F}_{\mathcal{N}}(z)$ must be analytic in the disc $|z| < 1$ with $z = 1$ being the only singularity on the circle $|z| = 1$. Moreover, $\overline{F}_{\mathcal{N}}(z)$ yields an expansion in the vicinity of $z = 1$ in form of $\sum_{p \geq 0} w_p (1 - z)^{p/2}$ with $w_1 = -\frac{1}{2}\sqrt{4 - 2\sqrt{2}} \neq 0$.

Simplifying $\overline{F}_{\mathcal{N}_{t_0}}$, we obtain

$$\overline{F}_{\mathcal{N}_{t_0}} = -\frac{1}{2}\sqrt{(1 - z)\left(1 - (2\sqrt{2} - 3)z\right)}$$
$$+ \frac{1}{2}\sqrt{1 - 2(\sqrt{2} - 1)z - (3 - 2\sqrt{2})z^2 + 2^{-|t_0|}\left(\sqrt{2} - 1\right)^{|t_0|+2}}.$$

Since $1 - 2(\sqrt{2} - 1)z - (3 - 2\sqrt{2})z^2 + 2^{-|t_0|}\left(\sqrt{2} - 1\right)^{|t_0|+2}$ is decreasing in $[0, 1]$ attaining values 1 and 0 for $z = 0$ and $z = 1$ respectively, we obtain that $\overline{F}_{\mathcal{N}_{t_0}}$ is analytic in the disc $|z| < 1$ with $z = 1$ being the only singularity on the circle $|z| = 1$. Moreover, $\overline{F}_{\mathcal{N}_{t_0}}$ has an expansion in the vicinity of $z = 1$ in form of $\sum_{p \geq 0} v_p (1 - z)^{p/2}$ with $v_1 = -\frac{1}{2}\sqrt{4 - 2\sqrt{2}}$.

Next, let us consider functions $\widetilde{F}_{\mathcal{N}}$ and $\widetilde{F}_{\mathcal{N}_{t_0}}$ such that

$$\widetilde{F}_{\mathcal{N}}(\sqrt{1 - z}) = \overline{F}_{\mathcal{N}}(z) \quad \text{and} \quad \widetilde{F}_{\mathcal{N}_{t_0}}(\sqrt{1 - z}) = \overline{F}_{\mathcal{N}_{t_0}}(z).$$

By the analyticity of $\overline{F}_{\mathcal{N}}$ and $\overline{F}_{\mathcal{N}_{t_0}}$ in the disc $|z| < 1$, we obtain that both $\widetilde{F}_{\mathcal{N}}(z)$ and $\widetilde{F}_{\mathcal{N}_{t_0}}(z)$ yield derivatives in this disc and thus $\widetilde{F}'_{\mathcal{N}}(0)$ and $\widetilde{F}'_{\mathcal{N}_{t_0}}(0)$ exist.

Finally, computing those derivatives we get $\widetilde{F}'_{\mathcal{N}}(0) = \widetilde{F}'_{\mathcal{N}_{t_0}}(0) = -\frac{1}{2}\sqrt{4 - 2\sqrt{2}}$ and so

$$\lim_{n\to\infty} \frac{[z^n]\{F_{\mathcal{N}_{t_0}}\}}{[z^n]\{F_{\mathcal{N}}\}} = \lim_{n\to\infty} \frac{R^{-n-2}[z^{n+2}]\{\overline{F}_{\mathcal{N}_{t_0}}\}}{R^{-n-2}[z^{n+2}]\{\overline{F}_{\mathcal{N}}\}} = \lim_{n\to\infty} \frac{\widetilde{F}'_{\mathcal{N}_{t_0}}(0)}{\widetilde{F}'_{\mathcal{N}}(0)} = 1. \qquad \square$$

Theorem 8. *Asymptotically almost all normal forms are not typeable, i.e.,* $\mu\left(\frac{\mathcal{TYP} \cap \mathcal{N}}{\mathcal{N}}\right) = 0.$

Proof. Note that $\Omega = \mathbf{S}(\mathbf{SKK})(\mathbf{SKK})$ is in normal form and is not typeable. Directly from Theorem 7 we obtain that asymptotically almost every combinator in normal form contains Ω as a subterm and is thus not typeable. $\qquad \square$

References

1. Barendregt, H.P., Bergstra, J., Klop, J.W., Volken, H.: Some notes on lambda reduction, in: Degrees, reductions and representability in the lambda calculus. Preprint no. 22, University of Utrecht, Department of mathematics, pp. 13–53 (1976)
2. Chauvin, B., Flajolet, P., Gardy, D., Gittenberger, B.: And/Or trees revisited. Comb. Probab. Comput. **13**(4–5), 475–497 (2004)
3. Curry, H., Feys, R.: Combinatory Logic, vol. I. North Holland, Amsterdam (1958)
4. David, R., Grygiel, K., Kozik, J., Raffalli, C., Theyssier, G., Zaionc, M.: Asymptotically almost all λ-terms are strongly normalizing. Logical Methods Comput. Sci. **9**, 1–30 (2013)
5. Flajolet, P., Sedgewick, R.: Analytic Combinatorics. Cambridge University Press, Cambridge (2009)
6. Fournier, H., Gardy, D., Genitrini, A., Zaionc, M.: Classical and intuitionistic logic are asymptotically identical. In: Duparc, J., Henzinger, T.A. (eds.) CSL 2007. LNCS, vol. 4646, pp. 177–193. Springer, Heidelberg (2007)
7. Gardy, D. Random boolean expressions. In: Discrete Mathematics and Theoretical Computer Science Proceedings AF, pp.1–36 (2005)
8. Gardy, D., Woods, A.: And/Or tree probabilities of boolean functions. Discrete Math. Theor. Comput. Sci. **6**, 139–146 (2005)
9. Genitrini, A., Kozik, J.: In the full propositional logic, 5/8 of classical tautologies are intuitionistically valid. Ann. Pure Appl. Logic **163**(7), 875–887 (2012)
10. Grygiel, K., and Lescanne, P. Counting terms in the binary lambda calculus. In: DMTCS 25th International Conference on Probabilistic, Combinatorial and Asymptotic Methods for the Analysis of Algorithms (2014)
11. Kostrzycka, Z., Zaionc, M.: Statistics of intuitionistic versus classical logic. Stud. Logica **76**(3), 307–328 (2004)
12. Kozik, J.: Subcritical pattern languages for And/Or trees. In: DMTCS Proceedings from Fifth Colloquium on Mathematics and Computer Science Algorithms Trees, Combinatorics and Probabilities, pp. 437–448 (2008)
13. Lefmann, H., Savický, P.: Some typical properties of large And/Or Boolean formulas. Random Struct. Algorithms **10**, 337–351 (1997)
14. Moczurad, M., Tyszkiewicz, J., Zaionc, M.: Statistical properties of simple types. Math. Struct. Comput. Sci. **10**(5), 575–594 (2000)

15. Olivier, B., Danièle, G., Bernhard, G., Alice, J.: Enumeration of generalized BCI lambda-terms. Electr. J. Comb. **20**, 4 (2013)
16. Szegö, G.: Orthogonal polynomials. Am. Math. Soc. Colloquium Ser. Publ. **23**, 413–421 (1967)
17. Tromp, J. Binary lambda calculus and combinatory logic. Unpublished manuscript (2014). http://tromp.github.io/cl/LC.pdf.
18. Wilf, H.: Generating Functionology. Academic Press, Boston (1994)
19. Woods, A.: On the probability of absolute truth for And/Or formulas. Bull. Symbolic Logic **12**, 3 (2006)
20. Zaionc, M.: On the asymptotic density of tautologies in logic of implication and negation. Rep. Math. Logic **39**, 67–87 (2005)
21. Zaionc, M.: Probability distribution for simple tautologies. Theor. Comput. Sci. **355**(2), 243–260 (2006)

Computational Complexity
and Boolean Functions

Some New Consequences of the Hypothesis That P Has Fixed Polynomial-Size Circuits

Ning Ding[1,2](✉)

[1] Department of Computer Science and Engineering,
Shanghai Jiao Tong University, Shanghai, China
dingning@sjtu.edu.cn
[2] NTT Secure Platform Laboratories, Tokyo, Japan

Abstract. We present some new consequences of the hypothesis that **P** can be computed by fixed polynomial-size circuits since [Lipton SCTC 94]. For instance, we show that the hypothesis implies that some small circuit family and BPP machines cannot be fooled by any complexity-theoretic pseudorandom generator $G : \{0,1\}^{\Theta(\log n)}$ to $\{0,1\}^n$, which means the known derandomization argument of **BPP** = **P** no longer works. It also implies the existence of 2-round public-coin zero-knowledge proofs for **NP**.

1 Introduction

Proving non-uniform general circuit lower bounds for complexity classes is one of the most fundamental and challenging tasks in complexity theory. Let $\mathbf{SIZE}(n^c)$ denote the class of languages that can be determined by $O(n^c)$-size circuit families. Let $\mathbf{P}/poly = \cup_c \mathbf{SIZE}(n^c)$. With the notions of $\mathbf{SIZE}(n^c)$ and $\mathbf{P}/poly$, a typical lower bound result is of the form that some uniform class $\mathcal{C}$ cannot be compute by $\mathbf{SIZE}(n^c)$) or $\mathbf{P}/poly$.

For $\mathbf{P}/poly$ lower bounds, the best separation result we do know so far is the exponential-time version of Merlin-Arthur games is not in $\mathbf{P}/poly$ due to Buhrman *et al.* [3]. Karp and Lipton [13] showed that if $\mathbf{NP} \subset \mathbf{P}/poly$, the polynomial hierarchy collapses. However, currently we do not have any techniques for proving $\mathbf{NEXP} \not\subseteq \mathbf{P}/poly$. Williams [21] showed any algorithm for Circuit-SAT or for Circuit Acceptance Probability Problem slightly faster than exhaustive search implies $\mathbf{NEXP} \not\subseteq \mathbf{P}/poly$.

As for $\mathbf{SIZE}(n^c)$ lower bounds, Kannan [12] showed that $\Sigma_2 \cap \Pi_2 \not\subseteq \mathbf{SIZE}(n^c)$ for any constant c, Vinodchandran [20] showed $\mathbf{PP} \not\subseteq \mathbf{SIZE}(n^c)$ and Santhanam [17] showed $\mathbf{promiseMA} \not\subseteq \mathbf{SIZE}(n^c)$ for any $c \in \mathbb{N}$. When considering lower bounds for $\mathbf{P}$ and $\mathbf{NP}$, however, currently the best known lower bound is $5n - o(n)$ due to Iwama and Morizumi [11].

After long-time failure to present non-linear lower bounds for $\mathbf{P}$, some researchers thought possibly $\mathbf{P} \subseteq \mathbf{SIZE}(n^c)$. As mentioned in [16] Levin pointed out that Kolmogorov even believed $\mathbf{P} \subseteq \mathbf{SIZE}(n)$, and Lipton then investigated what can be implied if $\mathbf{P} \subseteq \mathbf{SIZE}(n^c)$ and provided some interesting results e.g. $\mathbf{P} \subseteq \mathbf{SIZE}(n^c)$ implies $\mathbf{NP} \neq \mathbf{P}$. Two decades passed since then and we still cannot prove or disprove the hypothesis.

© Springer International Publishing Switzerland 2015
R. Jain et al. (Eds.): TAMC 2015, LNCS 9076, pp. 75–86, 2015.
DOI: 10.1007/978-3-319-17142-5_8

Our Results. We continue the research of [16] by presenting some new consequences of the hypothesis $\mathbf{P} \subseteq \mathbf{SIZE}(n^c)$, some of which are about the topics emerging posterior to [16]. More concretely, our results are as follows.

Basic Consequences. If $\mathbf{P} \subseteq \mathbf{SIZE}(n^c)$, we have the following two conclusions (which are elementary but did not appear in literature to our knowledge).

1. $\mathbf{E} \subseteq \mathbf{SIZE}(2^{o(n)})$. It follows from this result that the assumption that $\mathbf{E}$ has a language that requires $2^{\Omega(n)}$ circuit lower bound is false. Recall that the known derandomization argument of $\mathbf{BPP} = \mathbf{P}$ in many works e.g. [10,18,19] requires this assumption. So our result means the known derandomization of $\mathbf{BPP} = \mathbf{P}$ no longer works under the hypothesis.
2. $\mathbf{BPP} \subseteq \mathbf{SIZE}(n^{c+\epsilon})$ for any constant $0 < \epsilon < 1$ if one-way functions exist.

$\mathbf{P} \subseteq \mathbf{SIZE}(n^c)$ vs Pseudorandom Generators. We show $\mathbf{P} \subseteq \mathbf{SIZE}(n^c)$ implies the following negative results on complexity-theoretic pseudorandom generators $G : \{0,1\}^{l(n)=d\log n}$ to $\{0,1\}^n$ in polynomial-time for any $d \in \mathbb{N}$. This kind of generators is used to derandomize $\mathbf{BPP}$ in literature.

1. General such pseudorandom generators fooling small circuits do not exist. That is, there is no such G such that for all circuits D of size n, $|\Pr[D(G(U_{l(n)})) = 1] - \Pr[D(U_n) = 1]| \leq \frac{1}{n}$. Note that such generators are required in many works e.g. [10,18,19].
2. Some small circuit family $\{D_n\}_{n \in \mathbb{N}}$ is unfoolable against all G. That is, for each such G it holds for any constant $0 < \epsilon < 1$, $|\Pr[D_n(G(U_{l(n)})) = 1] - \Pr[D_n(U_n) = 1]| \geq \epsilon$ for infinitely many n. Note that this result is stronger than the first one.
3. Some BPP machines are unfoolable against all G if one-way functions exist. That is, for each $L \in \mathbf{BPP}$, there is a BPP machine M for L such that for each G there are instances x satisfying $M(x, G(U_{l(n)}))$ outputs wrong decisions with high probability (but in contrast $M(x, U_n)$ outputs wrong decisions with small probability).

The first result eliminates the existence of such general G which can fool all small circuits, but it does not eliminate the possibility that for any specific small circuit, there may exist a specific G which can fool the circuit (and may not fool other small circuits). However, the second result eliminates such possibility. Despite these two results, there is still a possibility that for each $L \in \mathbf{BPP}$ and some BPP machine M for L, there is such G such that we can derandomize M with G. The third result says for any $L \in \mathbf{BPP}$, some BPP machine for it cannot be derandomized by any G.

2-round Public-coin Zero-knowledge Proofs for NP. Zero-knowledge proofs [8] are of extreme importance in cryptography. Currently we have a 5-round construction in [6] and some impossibilities on fewer round numbers in [6,7,14]. There is no constant-round *public-coin* zero-knowledge proof for **NP** ever known. We show under the hypothesis there is a 2-round public-coin zero-knowledge proof for **NP**. The simulator of the protocol is non-uniform. The non-triviality of such a simulator is despite being non-uniform, it is able to simulate the interaction for all public inputs.

Then we present a witness-extractor for the protocol from program obfuscation, i.e. indistinguishability obfuscators recently proposed by e.g. [5,15], which can work for all bounded-size provers.

Our Techniques. Basically, the core technique in each consequence is to first define a problem/function and then show it is in $\mathbf{P}$ and thus gain an $O(n^c)$-size circuit family solving the problem which can then be used to establish the consequence. Here we sketch it in more detail with respect to unfoolable circuits against all pseudorandom generators.

Recall that our goal is to present some circuit family that can tell U_n from $G(U_{l(n)})$ for any G. So we first define a problem L_i: given an n-bit string r, decide if there is a string s with length $|s| \leq i \log n$ such that there is a G among the first n^i machines (in lexicographical order, say) within $n/2$-bit size such that $G(s)$ halts in n^i-time and $r = G(s)$. It can be seen $L_i \in \mathbf{P}$ for any i. So there is an $O(n^c)$-size circuit family $\{C_n^i\}_{n \in \mathbb{N}}$ determining $L_i, i \in \mathbb{N}$.

Let us investigate the output of C_n^i on input U_n or $G(U_{l(n)})$. First we can show C_n^i's output is almost always 0 when the input is U_n. On the other hand, for $G(U_{l(n)})$, for large enough i C_n^i can indeed output 1, indicating it can tell $G(U_{l(n)})$ from U_n. Lastly, we carefully choose such circuits over infinitely many n such that the circuit family can tell U_n from $G(U_{l(n)})$ for all G.

Organizations. Section 2 presents very short preliminaries. In Sects. 3 to 5 we present the consequences of the three parts respectively.

2   Preliminaries

Let $T : \mathbb{N} \to \mathbb{N}$ be some function. A language L is in $\mathbf{DTIME}(T(n))$ iff there is a Turing machine that runs in time $O(T(n))$ and determines L. Let $\mathbf{P} = \cup_{c \geq 1} \mathbf{DTIME}(n^c)$ and $\mathbf{E} = \cup_{c \geq 1} \mathbf{DTIME}(2^{cn})$.

Let $\mathbf{SIZE}(T(n))$ denote the class of languages satisfying for each L in it there is a circuit family $\{C_n\}_{n \in \mathbb{N}}$ such that $|C_n| = O(T(n))$ and for every $x \in \{0,1\}^n$, $x \in L \Leftrightarrow C_n(x) = 1$.

Let $L(x)$ denote the indicator function that outputs 1 if $x \in L$ and outputs 0 otherwise. Let $\mathbf{BPP}$ denote the class in which each language L admits a PPT machine M such that for each x, $\Pr[M(x) = L(x)] > \frac{1}{10}$ where the probability is taken over all choices of the coins of M. We call M a BPP machine for L.

3   Some Basic Consequences of $\mathbf{P} \subseteq \mathbf{SIZE}(n^c)$

3.1   E and $\mathbf{SIZE}(2^{o(n)})$

Theorem 1. *If* $\mathbf{P} \subseteq \mathbf{SIZE}(n^c)$ *for some* $c \in \mathbb{N}$, *then* $\mathbf{E} \subseteq \mathbf{SIZE}(2^{o(n)})$.

Proof. We use the padding argument to show this. Suppose $\mathbf{E} - \mathbf{SIZE}(2^{o(n)}) \neq \phi$ and L is a language in it. This means $L \in \mathbf{DTIME}(2^{c_1 n})$ for some $c_1 \in \mathbb{N}$, but there exists $0 < \epsilon < 1$ such that L requires circuit lower bound $\Omega(2^{\epsilon n})$ for infinitely many n. Choose a sufficiently small constant δ satisfying $\epsilon/\delta > c$.

Consider the language L' that consists of all instances of form $x \circ 0^{2^{\delta n} - n}$ for $x \in L$ where $n \leftarrow |x|$. Then L' can be determined in $O(2^{\delta n \cdot c_1/\delta})$-time when inputs are of $2^{\delta n}$ bits. Thus by translation $L' \in \mathbf{DTIME}(n^{c_1/\delta}) \subseteq \mathbf{P}$. On the other hand, L' requires circuit lower bound $\Omega(2^{\epsilon n})$ when inputs are of $2^{\delta n}$ bits for infinitely many n. Thus by translation L' requires circuit lower bound $\Omega(n^{\epsilon/\delta})$ for infinitely many n and so it is not in $\mathbf{SIZE}(n^c)$. This is a contradiction. $\square$

The theorem immediately asserts the following assumption is conditionally false which is used to establish the derandomization result $\mathbf{BPP} = \mathbf{P}$.

Assumption 2. E *has a language of deterministic circuit complexity* $2^{\Omega(n)}$.

Corollary 1. *If* $\mathbf{P} \subseteq \mathbf{SIZE}(n^c)$ *for some* $c \in \mathbb{N}$, *then Assumption 2 is false.*

3.2   BPP and SIZE($n^{c+\epsilon}$)

Theorem 3. *If* $\mathbf{P} \subseteq \mathbf{SIZE}(n^c)$ *for some* $c \in \mathbb{N}$ *and one-way functions exist, then* $\mathbf{BPP} \subseteq \mathbf{SIZE}(n^{c+\epsilon})$ *for any constant* $0 < \epsilon < 1$.

Proof. First if one-way functions exist, for any constant $0 < \delta < 1$ there exists a pseudorandom generator $G : \{0,1\}^{n^\delta} \rightarrow \{0,1\}^{\mathrm{poly}(n)}$ such that G is computable in time $\mathrm{poly}(n)$ and for all polynomial-size circuits D, $|\Pr[D(U_{\mathrm{poly}(n)}) = 1] - \Pr[D(G(U_{n^\delta})) = 1]| \leq 1/\mathrm{poly}(n)$ [9]. Thus for any $L \in \mathbf{BPP}$ and a BPP machine for L, there is another BPP machine for L which uses only n^δ coins: The machine first runs G with n^δ coins to get polynomial pseudorandom coins and then feeds the original BPP machine the pseudorandom coins to make decisions. Let M_1 denote such a BPP machine using n^δ coins with error $\frac{1}{n}$.

Let M denote a machine that runs M_1 $8n$ times independently and outputs the majority. Then there exists a specific value for all the coins used by M, denoted r_n, such that $M_{r_n}(x)$ outputs the correct decision for all $x \in \{0,1\}^n$ (as the proof of $\mathbf{BPP} \subset \mathbf{P}/poly$ shows). Note that $|r_n| = 8n^{1+\delta}$.

Now we define a language L_1 which consists of all instances (x, y) satisfying $M_y(x) = 1$. Thus $L_1 \in \mathbf{P}$. Then there is an $O(m^c)$-size circuit family $\{C_n\}_{n \in \mathbb{N}}$ deciding $m = |(x, y)| = O(n^{1+\delta})$-bit instances of L_1. Then we construct an $O(n^{(1+\delta)c})$-size circuit family $\{C'_n\}_{n \in \mathbb{N}}$ determining L. Actually, C'_n has r_n hardwired and on input x outputs $C_n(x, r_n)$. Since $C_n(x, r_n) = M_{r_n}(x)$ that equals the correct decision and $|C'_n| = O(n^{c+\epsilon})$ for $\epsilon = c\delta$, $L \in \mathbf{SIZE}(n^{c+\epsilon})$. $\square$

4   P ⊆ SIZE(n^c) vs Pseudorandom Generators

In this section we investigate the relations between the hypothesis and complexity-theoretic pseudorandom generators. We focus on the polynomial-time generators $G : \{0,1\}^{\Theta(\log n)}$ to $\{0,1\}^n$, which are used to derandomize $\mathbf{BPP}$ and can result in $\mathbf{BPP} = \mathbf{P}$.

4.1   On General Pseudorandom Generators Fooling Small Circuits

Recall the derandomization argument of **BPP** in [10, 18, 19] that basically proceeds in two steps: first assume Assumption 2 to deduce Assumption 4 in the following is true; second use the pseudorandom generator G to derandomize any BPP machine for a language in **BPP**. Conversely, we also know Assumption 4 implies Assumption 2.

Assumption 4. *There exists a pseudorandom generator $G : \{0,1\}^{l(n)} \to \{0,1\}^n$ such that G maps inputs of length $l(n) = \Theta(\log n)$ to length n in time $poly(n)$, and for all circuits D of size n, $|\Pr[D(G(U_{l(n)})) = 1] - \Pr[D(U_n) = 1]| \leq \frac{1}{n}$.*

However, due to Corollary 1, we immediately have the following result.

Proposition 1. *If $\mathbf{P} \subseteq \mathbf{SIZE}(n^c)$ for some $c \in \mathbb{N}$, Assumption 4 is false.*

Proposition 1 eliminates the existence of such general G which can fool all small circuits. However, it does not eliminate the possibility that for any specific small circuit, there may exist a specific G which can fool the circuit (and may not fool other small circuits). So a further question is whether for each small circuit there is such a specific generator G that can fool it. In the next subsection, unexpectedly, we will answer this question negatively.

4.2   Unfoolable Circuit Families Against All Pseudorandom Generators

We now present a circuit family that cannot be fooled by any pseudorandom generator that stretches $\Theta(\log n)$-bit coins to n-bit pseudorandom coins.

Theorem 5. *If $\mathbf{P} \subseteq \mathbf{SIZE}(n^c)$ for some $c \in \mathbb{N}$, there is an n^{c+1}-size circuit family $\{D_n\}_{n \in \mathbb{N}}$ such that for any pseudorandom generator G that maps inputs of length $l(n) = d \log n$ for arbitrary $d \in \mathbb{N}$ to length n in time $poly(n)$, it holds for any constant $0 < \epsilon < 1$, $|\Pr[D_n(G(U_{l(n)})) = 1] - \Pr[D_n(U_n) = 1]| \geq \epsilon$ for infinitely many n.*

Proof. To present the circuit family $\{D_n\}_{n \in \mathbb{N}}$ such that for any generator G the result holds, we first define the following problems.

Problems L_i. For each $i \in \mathbb{N}$, we define problem L_i as follows. Given $r \in \{0,1\}^n$, decide if there is s of length no more than $i \log n$ such that for at least one machine G among the first n^i machines (in lexicographical order, say), G is at most $n/2$-bit long and $G(s)$ halts in n^i-time and r is equal to $G(s)$.

It can be seen that an exhaustive search algorithm can run each one of the first n^i machines at most n^i steps on input a string s of length no more than $i \log n$ and check if there are G and s satisfying the requirement. Since we only need to check the first n^i machines and emulate $G(s)$ n^i steps and the number of all s is $O(n^i)$, the algorithm can output a correct decision in polynomial-time. So L_i is in **P** for all $i \in \mathbb{N}$. Thus if $\mathbf{P} \subseteq \mathbf{SIZE}(n^c)$, there is an $O(n^c)$-size circuit family $\{C_n^i\}_{n \in \mathbb{N}}$ determining $L_i, i \in \mathbb{N}$.

Note that C_n^i is of $O(n^c)$-size. Set $k = c + 1$, which means the size of $C_n^i, i \in \mathbb{N}$, is bounded by n^k for sufficiently large n. Let n_1 denote the least integer satisfying $|C_n^1| < n^k$ for each $n \geq n_1$ and for $i = 2, 3, \cdots$, let n_i denote the least integer satisfying $|C_n^i| < n^k$ for each $n \geq n_i$ and $n_i > n_{i-1}$. Define the following distinguisher $\{D_n\}_{n \in \mathbb{N}}$: for each $n \in \{n_1, \cdots, n_i, \cdots\}$, let D_n be $C_{n_i}^i$; for all other n, let it be any n^k-size circuit. Thus $|D_n| \leq n^k$.

For $\{D_n\}_{n \in \mathbb{N}}$, let us consider an arbitrary pseudorandom generator G which stretches $d \log n$ bits to n bits in polynomial-time for some d. It can be first seen that for all machines of length $n/2$ and all inputs s of length no more than $i \log n$, there are at most $\text{poly}(n) \cdot 2^{n/2}$ different outputs of all these machines with input s. So for truly random U_n, $U_n \notin L_i$ except for probability $\text{poly}(n)/2^{n/2}$ for any i. Thus $\Pr[D_n(U_n) = 1] = \text{poly}(n)/2^{n/2}$.

On the other hand, for this G, the order number of G in the enumeration of all machines is a constant and G's running-time is a fixed polynomial. When the input is $r = G(U_{d \log n})$, we have for each large enough i and for each $n \in \{n_i, n_{i+1}, \cdots\}$ the order number G is less than n^i and $G(s)$ outputs r for some s of length $d \log n$ (no more than $i \log n$) in n^i-time, which shows $C_{n_i}^i(r)$ outputs 1 always. Thus $|\Pr[D_n(G(U_{l(n)})) = 1] - \Pr[D_n(U_n) = 1]| \geq \epsilon$ for infinitely many n for any constant $0 < \epsilon < 1$. The theorem holds.   □

4.3   Unfoolable BPP Machines Against All Pseudorandom Generators

The previous subsections show under the hypothesis, not only the general pseudorandom generator for all small circuits, but also specific generators for all specific circuits do not exist. But for the purpose of derandomizing **BPP**, both the two-type generators are not necessary. Actually, a specific pseudorandom generator that can fool a specific BPP machine for any language in **BPP** suffices. More precisely, let $L \in$ **BPP** and M be a BPP machine for L. A pseudorandom generator G satisfying for any instance x, $G(U_{l(n)})$ can fool M with x suffices to induce a deterministic polynomial-time machine for L. Since intuitively $M(x)$ could not be the $\{D_n\}_{n \in \mathbb{N}}$ in Theorem 5, there is a possibility that for each $L \in$ **BPP** and some M for L, there is such G such that we can derandomize M with G.

However, to do this we need to select a derandomizable one instead of any BBP machine for L, since the following theorem says that some M for L cannot be derandomized by any generator G, in the sense that on one hand M with truly random coins can decide all instances correctly with high probability and on the other hand M with pseudorandom coins from any G will output wrong decisions for some instances with high probability. When errors occur, we cannot be aware of this.

In the following for any BPP machine M, we use notation $M(x, U_{\text{poly}(n)})$ to denote the computation of M with input instance x and coins $U_{\text{poly}(n)}$.

Theorem 6. *If* $\mathbf{P} \subseteq \mathbf{SIZE}(n^c)$ *for some* $c \in \mathbb{N}$ *and one-way functions exist, then for all* $L \in$ **BPP** *there is a BPP machine* M *for* L *which needs no more than* n^k *coins for* $k \in \mathbb{N}$ *such that for any pseudorandom generator* G *that maps* inputs of length $l(n) = d \log n$ for arbitrary $d \in \mathbb{N}$ to length n^k in time $poly(n)$, there is an instance serial $\{x_n\}_{n \in \mathbb{N}}$ satisfying $\Pr[M(x_n, G(U_{l(n)})) \neq L(x_n)] \geq 1 - \frac{poly(n)}{2^n} - \frac{1}{n}$ for all sufficiently large n.

Proof. Let M_L be a BPP machine for L with error $\epsilon = \frac{1}{n}$ which uses no more than n coins (using a cryptographically pseudorandom generator constructed from one-way functions to generate $poly(n)$ coins). We construct a BPP machine M for L that uses n^k coins and cannot be derandomized. On input any instance $x \in \{0,1\}^{n^{c+1}}$ and coins U_{n^k}, M does the following.

1. If x cannot be parsed to the form (C, r, r') where C denotes a boolean circuit of n-bit input and $|C| = n^{c+1/2}$ and $|r| = n$ and $|r'| = n$, output $M_L(x, r' \oplus r_1)$ where r_1 denotes n coins in U_{n^k}. Otherwise, move to the next step.
2. Set $t = n^3$ and let $r_1, \cdots, r_t, r_{t+1}, r_{t+2}$ be the first $t + 2$ n-bit blocks in U_{n^k}. Compute $C(r_1), \cdots, C(r_t)$ and count the fraction of 1 among all outputs. If the fraction is less than $1 - \epsilon$, output $M_L(x, r' \oplus r_{t+2})$. Otherwise, output $M_L(x, r' \oplus r_{t+2})$ if $C(r \oplus r_{t+1}) = 1$ and output $1 - M_L(x, r' \oplus r_{t+2})$ otherwise.

We now show M is indeed a BPP machine for L. First consider x that is not of form (C, r, r'). Then M outputs $M_L(x, U_n)$. Thus it has error ϵ. Second consider $x = (C, r, r')$ with $\Pr[C(U_n) = 1] < 1 - 2\epsilon$. Due to the Chernoff bound, $\frac{1}{t} \sum_{i=1}^{t} C(r_i) < 1 - 2\epsilon + \delta < 1 - \epsilon$ except for probability $e^{-2\delta^2 t} = e^{-2n}$ for $\delta = \frac{1}{n}$. This shows M's error is at most $\epsilon + e^{-2n}\epsilon < 2\epsilon$. Third consider $x = (C, r, r')$ with $\Pr[C(U_n) = 1] \geq 1 - 2\epsilon$. Then M's error is at most $\epsilon + \Pr[C(U_n) = 0] + \epsilon < 4\epsilon$. So for any instance M's error is at most 4ϵ. That shows M is a BPP machine for L.

Consider an arbitrary G that maps inputs of length $d \log n$ to length n^k in time $poly(n)$. We now define the following function.

Function f. Given $r \in \{0,1\}^n$, output 1 if there is $s \in \{0,1\}^{d \log n}$ such that r equals any one of the first n^3 n-bit blocks in the output of $G(s)$, and output 0 otherwise.

Similarly, viewed as a language, $f^{-1}(1)$ is in **P**. Thus there is an $O(n^c)$-size circuit family $\{C_n\}_{n \in \mathbb{N}}$ computing f. Note that $|C_n| < n^{c+1/2}$ and can be padded to $n^{c+1/2}$-size for large enough n. Similarly, $G(s)$ has at most $poly(n)$ different outputs, one of which happens to contain U_n as a block with probability $\frac{poly(n)}{2^n}$. Thus $\Pr[C_n(U_n) = 1] = \frac{poly(n)}{2^n}$.

First consider the instance $x_n = (C_n, r, r')$ for uniformly random r, r'. When the coins for M is $G(U_{d \log n})$, letting $r_1, \cdots, r_{t+1}$ denote the first $n^3 + 1$ n-bit blocks in the output of $G(U_{d \log n})$, we have $C_n(r_1) = \cdots = C_n(r_t) = 1$ and $\Pr[C_n(r \oplus r_{t+1}) = 0] = 1 - \frac{poly(n)}{2^n}$ since r is uniformly random. Due to M's strategy, $M(x_n, G(U_{d \log n}))$ outputs $1 - M_L(x_n, r' \oplus r_{t+2})$ almost all the time. Thus $\Pr[M(x_n, G(U_{d \log n})) \neq L(x_n)] \geq 1 - \frac{poly(n)}{2^n} - \epsilon$. Thus there exist specific r, r' such that fixing $x_n = (C_n, r, r')$, the probability formula still holds. The theorem holds. $\qquad \square$

We remark that even in the case $\mathbf{BPP} = \mathbf{P}$, L which admits a deterministic polynomial-time machine still admits a BPP machine that cannot be derandomized. Actually with a similar argument to that of Theorem 6 (where M_L changes to be a deterministic polynomial-time machine for L and consider x of form (C, r)) we have the following proposition which does not need one-way functions and achieves stronger probability result.

Proposition 2. *If $\mathbf{P} \subseteq \mathbf{SIZE}(n^c)$ for some $c \in \mathbb{N}$, then for all $L \in \mathbf{P}$ there is a BPP machine M for L which needs no more than n^k coins for $k \in \mathbb{N}$ such that for any pseudorapndom generator G described in Theorem 6, there is an instance serial $\{x_n\}_{n \in \mathbb{N}}$ satisfying $\Pr[M(x_n, G(U_{l(n)})) \neq L(x_n)] \geq 1 - \frac{poly(n)}{2^n}$ for all sufficiently large n.*

5   Two-Round Public-Coin Zero-Knowledge Proofs

In this section we investigate the question of constructing constant-round public-coin zero-knowledge proofs for $\mathbf{NP}$ if $\mathbf{P} \subseteq \mathbf{SIZE}(n^c)$. An interactive proof is zero-knowledge if for any polynomial-time verifier there is a polynomial-time simulator such that what the verifier sees, i.e. random coins, the public input and prover's messages, can be computationally indistinguishably reconstructed by the simulator [8].

Currently we have a 5-round private-coin construction due to [6] and some impossibilities on fewer round numbers in e.g. [6,7,14]. Reference [2] presents a negative result on 2-round public-coin zero-knowledge proofs, but it assumes that $\mathbf{E}$ has a language of non-deterministic circuit complexity $2^{\Omega(n)}$, which is even stronger than Assumption 2. So due to the hypothesis $\mathbf{P} \subseteq \mathbf{SIZE}(n^c)$, this assumption is false and the negative result in [2] no longer works.

So there is no constant-round public-coin zero-knowledge proofs for $\mathbf{NP}$ ever known. However, we show that based on the hypothesis there exists a 2-round public-coin zero-knowledge proof for $\mathbf{NP}$ with respect to a relaxed requirement that the simulator can be non-uniform. Despite being non-uniform the simulator is able to simulate the interaction for all public inputs.

5.1   The Protocol

We first present some preparations as follows.

Definition 1. *For each polynomial-time machine M, we define a function f_M as follows. Given $x \in \{0,1\}^n, u \in \{0,1\}^n, i \in [1, n^{c+2}]$, output r_i that is the ith bit of $r \leftarrow M(x, u)$.*

Note that in the definition i can be represented by a $\lceil (c+2) \log n \rceil$-bit string. The function f_M induces a problem L_M that consists all instances (x, u, i) satisfying $f_M(x, u, i) = 1$. Since M is polynomial-time, $L_M \in \mathbf{P}$. Due to the hypothesis, L_M can be determined by an $O(n^c)$-size circuit family $\{C_n\}_{n \in \mathbb{N}}$. Namely, f_M can be computed by $\{C_n\}_{n \in \mathbb{N}}$.

Let L be any language in $\mathbf{NP}$. Then we define the following language Λ.

Public input: x;
Prover's auxiliary input: w, (a witness for $x \in L$).

---

1. $V \to P$: Send $r \in_R \{0,1\}^{n^{c+2}}$, ZAP_1.
2. $P \to V$: Send ZAP_2 generated using witness w for the statement that $(x,r) \in \Lambda$.

Protocol 1 *The 2-round public-coin zero-knowledge proof for L.*

Definition 2. *We define the following language Λ: $(x,r) \in \Lambda$ where $|x| = n$, $|r| = n^{c+2}$ iff either there is a witness w for $x \in L$ or there are a boolean circuit C of size at most n^{c+1} and $u \in \{0,1\}^n$ such that $C(x,u,i) = r_i$ for all $1 \leq i \leq n^{c+2}$.*

Then $\Lambda \in \mathbf{NP}$ and a witness for $(x,r) \in \Lambda$ is either w for $x \in L$ or a circuit C and u satisfying the second condition.

Let ZAP denote the 2-round public-coin witness-indistinguishable (WI) proof for $\mathbf{NP}$ in [4], $(\mathsf{ZAP}_1, \mathsf{ZAP}_2)$ denote the two messages of ZAP. Let PRG denote a cryptographically pseudorandom generator in [9]. Our protocol for L is shown in Protocol 1.

Theorem 7. *Assuming $\mathbf{P} \subseteq \mathbf{SIZE}(n^c)$ for some $c \in \mathbb{N}$ and the existence of ZAP, PRG, Protocol 1 is a 2-round public-coin zero-knowledge proof for L.*

Proof. We show the completeness, soundness and zero-knowledge properties are satisfied.

Completeness. For $x \in L$ P can always convince V using w.

Soundness. For each $x \notin L$ and all possible n^{c+1}-size boolean circuits C and $u \in \{0,1\}^n$, the string of $C(x,u,1) \circ \cdots \circ C(x,u,n^{c+2})$ in which "$\circ$" means concatenation has at most $2^{n^{c+1}+n}$ different values. Now r is randomly chosen from $\{0,1\}^{n^{c+2}}$. So one of these values equals r with probability $2^{-\Omega(n^{c+2})}$, which shows $(x,r) \notin \Lambda$ with probability $1 - 2^{-\Omega(n^{c+2})}$, i.e. the statement that ZAP proves is false. Thus the soundness follows from the soundness of ZAP.

Zero-Knowledge. For each PPT verifier V^*, we present a polynomial-size simulator S which is constructed as follows.

1. Consider the following machine M. On input (x,u), $M(x,u)$ runs $V^*(x)$ and when V^* needs random coins, run $\mathsf{PRG}(u)$ and provide the output to it. For this machine M, let f_M and L_M be defined previously. Then there is an $O(n^c)$-circuit family $\{C_n\}_{n \in \mathbb{N}}$ computing f_M.
2. Sample coins $u \in \{0,1\}^n$. Let S have V^*, C_n, u hardwired. $S(x)$ runs as follows. It runs $V^*(x)$ to output $r \in \{0,1\}^{n^{c+2}}$ and ZAP_1 in which when V^* needs coins, run $\mathsf{PRG}(u)$ and provide the pseudorandom coins to it. Then S computes ZAP_2 using witness (C_n, u) and sends it to V^*.

We first show that (C_n, u) is a witness for $(x, r) \in \Lambda$. It can be seen that r is the output of $V^*(x)$ with coins from $\mathsf{PRG}(u)$. This means $r = M(x, u)$. Due to the definition of C_n, we have $C_n(x, u, i) = r_i$ for all $1 \leq i \leq n^{c+2}$. And $|C_n| < n^{c+1}$ for large enough n. This shows (C_n, u) is a witness for $(x, r) \in \Lambda$. So S can finish the interaction.

Then we show S can reconstruct indistinguishably V^*'s view (random tape, prover's messages). Since V^*'s coins are now $\mathsf{PRG}(u)$ and S differs from $P(w)$ only in the witnesses they use, the indistinguishability is ensured by the pseudo-randomness of PRG and WI of ZAP. The zero-knowledge property holds.   □

5.2   Obtaining Witness Extraction from Program Obfuscation

In this subsection we consider an enhanced property of witness extraction, which claims an extractor E such that for any polynomial-time prover P' that can convince V some $x \in L$, then $E(P', x)$ can output a witness for $x \in L$ in polynomial-time. A proof system admitting an extractor is called a proof of knowledge in cryptography. Our result is that we present a witness extractor from program obfuscation for Protocol 1 which works for bounded-size provers. For lack of space, we only sketch the construction.

Informally a program obfuscator is a PPT algorithm that given a program can output a new program such that the output program is of same functionality as the input program but hides some secrets. In particular, an indistinguishability obfuscator, denoted $i\mathcal{O}$, which was first introduced by [1] and which candidate constructions were recently proposed by [5,15] etc. is such that for any two machines (M_1, M_2) of same functionality (and same size and same running-time), $i\mathcal{O}(M_1)$ and $i\mathcal{O}(M_2)$ are computationally indistinguishable. We will employ $i\mathcal{O}$ to achieve our result.

We modify Protocol 1 with $i\mathcal{O}$. That is, we let P send a random $r_1 \in \{0, 1\}^{\mathrm{poly}(n)}$ for a sufficiently large $\mathrm{poly}(n)$ (e.g. $\geq n^{c+3}$) and $\widetilde{Q}_1 \leftarrow i\mathcal{O}(Q_1)$ in Step 2, where Q_1 denotes the program that on input a program Π with $|\Pi| < |r_1|/2$ outputs w for $x \in L$ if Π outputs r_1 within $n^{\log \log n}$ steps and outputs 0^n otherwise. And accordingly, the first condition in Definition 2 changes to that $\widetilde{Q}_1$ is honestly generated. The modified protocol is shown in Protocol 2.

It can be seen that Protocol 2 is complete and sound. Moreover, the simulator S needs slight modification. That is, it samples r_1 and computes $\widetilde{Q}_2 \leftarrow i\mathcal{O}(Q_2)$ in Step 2, where Q_2 is equal to Q_1 except that it always outputs 0^n, and computes ZAP_2 as before. Note that Q_1, Q_2 are of same functionality except on input a program Π satisfying Π outputs r_1. However, for random r_1, since $|\Pi| < |r_1|/2$, the Π does not exist except for exponentially small probability. Thus the two programs are of same functionality and thus $\widetilde{Q}_1, \widetilde{Q}_2$ are indistinguishable. So the zero-knowledge property still holds.

Finally, let us sketch the construction of the extractor. Actually, as shown in the soundness, if some prover P' can convince V $x \in L$, then due to the soundness of ZAP, $\widetilde{Q}_1$ is honestly generated. If P''s size is bounded by $|r_1|/2$, basically its code is a valid input Π such that $\widetilde{Q}_1(\Pi)$ outputs w. So an extractor E can adopt V's strategy to send the message of Step 1 and emulates P''s computation where

---

Public input: x;
Prover's auxiliary input: w, (a witness for $x \in L$).

---

1. $V \to P$: Send $r \in_R \{0,1\}^{n^{c+2}}$, ZAP_1.
2. $P \to V$: Send $r_1 \in_R \{0,1\}^{\text{poly}(n)}$, $\widetilde{Q}_1$, ZAP_2.

Protocol 2 *The 2-round public-coin zero-knowledge proof of knowledge for L.*

providing P' pseudorandom coins from $\mathsf{PRG}(u')$ for random $u' \in \{0,1\}^n$. Thus P''s code, u' and PRG constitute a valid Π which size is bounded. On receiving P''s message, E runs $\widetilde{Q}_1(\Pi)$ to gain w. Thus we have the following result.

Theorem 8. *Assuming* $\mathbf{P} \subseteq \mathbf{SIZE}(n^c)$ *for some* $c \in \mathbb{N}$ *and the existence of* $\mathsf{ZAP}, \mathsf{PRG}, i\mathcal{O}$, *Protocol 2 is a 2-round public-coin zero-knowledge proof for L which admits an extractor for all bounded-size provers ($< \frac{|r_1|}{2}$).*

Acknowledgments. The author is grateful to the reviewers of TAMC 2015 for their detailed and useful comments. This work is supported by the National Natural Science Foundation of China (Grant No. 61100209) and Doctoral Fund of Ministry of Education of China (Grant No. 20120073110094).

References

1. Barak, B., Goldreich, O., Impagliazzo, R., Rudich, S., Sahai, A., Vadhan, S.P., Yang, K.: On the (im)possibility of obfuscating programs. J. ACM **59**(2), 6 (2012)
2. Barak, B., Lindell, Y., Vadhan, S.P.: Lower bounds for non-black-box zero knowledge. J. Comput. Syst. Sci. **72**(2), 321–391 (2006)
3. Buhrman, H., Fortnow, L., Thierauf, T.: Nonrelativizing separations. In: IEEE Conference on Computational Complexity, pp. 8–12. IEEE Computer Society (1998)
4. Dwork, C., Naor, M.: Zaps and their applications. In: FOCS, pp. 283–293. IEEE Computer Society (2000)
5. Garg, S., Gentry, C., Halevi, S., Raykova, M., Sahai, A., Waters, B.: Candidate indistinguishability obfuscation and functional encryption for all circuits. In: FOCS, pp. 40–49. IEEE Computer Society (2013)
6. Goldreich, O., Kahan, A.: How to construct constant-round zero-knowledge proof systems for np. J. Cryptol. **9**(3), 167–190 (1996)
7. Goldreich, O., Oren, Y.: Definitions and properties of zero-knowledge proof systems. J. Cryptol. **7**(1), 1–32 (1994)
8. Goldwasser, S., Micali, S., Rackoff, C.: The knowledge complexity of interactive proof systems. SIAM J. Comput. **18**(1), 186–208 (1989)
9. Håstad, J., Impagliazzo, R., Levin, L.A., Luby, M.: A pseudorandom generator from any one-way function. SIAM J. Comput. **28**(4), 1364–1396 (1999)

10. Impagliazzo, R., Wigderson, A.: P = BPP if e requires exponential circuits: derandomizing the xor lemma. In: Leighton, F.T., Shor, P.W. (eds.) STOC, pp. 220–229. ACM (1997)
11. Iwama, K., Morizumi, H.: An explicit lower bound of $5n\text{-}o(n)$ for boolean circuits. In: Diks, K., Rytter, W. (eds.) MFCS 2002. LNCS, vol. 2420, pp. 353–364. Springer, Heidelberg (2002)
12. Kannan, R.: Circuit-size lower bounds and non-reducibility to sparse sets. Inf. Control 55(1–3), 40–56 (1982)
13. Karp, R.M., Lipton, R.J.: Some connections between nonuniform and uniform complexity classes. In: Miller, R.E., Ginsburg, S., Burkhard, W.A., Lipton, R.J. (eds.) STOC, pp. 302–309. ACM (1980)
14. Katz, J.: Which languages have 4-round zero-knowledge proofs? In: Canetti, R. (ed.) TCC 2008. LNCS, vol. 4948, pp. 73–88. Springer, Heidelberg (2008)
15. Koppula, V., Lewko, A.B., Waters, B.: Indistinguishability obfuscation for turing machines with unbounded memory. Cryptology ePrint Archive, Report 2014/925 (2014). http://eprint.iacr.org/
16. Lipton, R.J.: Some consequences of our failure to prove non-linear lower bounds on explicit functions. In: Structure in Complexity Theory Conference, pp. 79–87. IEEE Computer Society (1994)
17. Santhanam, R.: Circuit lower bounds for merlin-arthur classes. In: Johnson, D.S., Feige, U. (eds.) STOC, pp. 275–283. ACM (2007)
18. Shaltiel, R., Umans, C.: Simple extractors for all min-entropies and a new pseudo-random generator. In: FOCS, pp. 648–657. IEEE Computer Society (2001)
19. Umans, C.: Pseudo-random generators for all hardnesses. J. Comput. Syst. Sci. 67(2), 419–440 (2003)
20. Vinodchandran, N.V.: A note on the circuit complexity.In: Electronic Colloquium on Computational Complexity (ECCC) (056) (2004)
21. Williams, R.: Improving exhaustive search implies superpolynomial lower bounds. In: Schulman, L.J. (ed.) STOC, pp. 231–240. ACM (2010)

Computational Complexity Studies of Synchronous Boolean Finite Dynamical Systems

Mitsunori Ogihara[1] and Kei Uchizawa[2]([⊠])

[1] Department of Computer Science, University of Miami, 1365 Memorial Drive,
Coral Gables, FL 33146, USA
ogihara@cs.miami.edu
[2] Faculty of Engineering, Yamagata University, Jonan 4-3-16, Yonezawa,
Yamagata 992-8510, Japan
uchizawa@yz.yamagata-u.ac.jp

Abstract. The finite dynamical system is a system consisting of some finite number of objects that take upon a value from some domain as a state, in which after initialization the states of the objects are updated based upon the states of the other objects and themselves according to a certain update schedule. This paper studies the subclass of finite dynamical systems the *synchronous boolean finite dynamical system* (*synchronous BFDS*, for short), where the states are boolean and the state update takes place in discrete time and at the same on all objects. The present paper is concerned with some problems regarding the behavior of synchronous BFDS in which the state update functions (or the local state transition functions) are chosen from a predetermined finite basis of boolean functions $\mathcal{B}$. Specifically the following three behaviors are studied:

- *Convergence.* Does a system at hand converge on a given initial state configuration?
- *Path Intersection.* Will a system starting in given two state configurations produce a common configuration?
- *Cycle Length.* Since the state space is finite, every BFDS on a given initial state configuration either converges or enters a cycle having length greater than 1. If the latter is the case, what is the length of the loop? Or put more simply, for an integer t, is the length of loop greater than t?

The paper studies these questions in terms of computational complexity (in the case of Cycle Length using the decision version of the problem) and shows the following:

1. The three problems are each PSPACE-complete if the boolean function basis contains NAND, NOR or both AND and OR.
2. The Convergence Problem is solvable in polynomial time if the set B is one of {AND}, {OR} and {XOR, NXOR}.
3. If the set B is chosen from the three sets as in the case of the Convergence Problem, the Path Intersection Problem is in UP, and the Cycle Length Problem is in UP $\cap$ coUP; thus, these are unlikely to be NP-hard.

© Springer International Publishing Switzerland 2015
R. Jain et al. (Eds.): TAMC 2015, LNCS 9076, pp. 87–98, 2015.
DOI: 10.1007/978-3-319-17142-5_9

1   Introduction

The finite dynamical system is a system consisting of some finite number of objects that each take upon a value from some domain D. After receiving an initial state assignment the system evolves over time by means of state updates, where the updates occur in discrete time and are governed by a global state-update schedule and a local (meaning assigned to each node individually) state-update functions (or local state-transition functions) that take as input the states of the objects in the system.

Because of its flexibility the finite dynamical system has been used as a mathematical model for time-dependent systems and can contain in itself other multi-object computational models, such as cellular and graph automata and Hopfield networks.

Classes of finite dynamical systems can be defined by giving certain requirements to their operation. First, classes can be defined by specifying the domain, that is, the set of permissible states: *infinite*, *finite*, and *boolean*. Next, classes can be defined based upon the types of the state update functions.

It is usually assumed that at each time step, all the objects conduct their state updates exactly once, and so, classes can be defined depending on the order in which the state updates occur in the objects. Specifically we have the *asynchronous* (any update order), the *sequential* (a fixed predetermined order), and the *synchronous* (all at the same time) finite dynamical systems.

For each $n \geq 1$, the underlying structure of an n-object dynamical system over domain D can be represented as a node- and edge-labeled directed graph G of n nodes. The nodes of G represent the objects, the edges of G represent the direct dependencies among the objects in updating their states in a natural way: an edge from a node u to a node v indicates that the state updating function of v takes the state of u as input. Also, for each node v, v is labeled by the state update function of v and the incoming edges of v are labeled by the input positions of the source node in the state update function. Because of this representation, classes of the finite dynamical systems can be defined in terms of the properties of the underlying graph, e.g., whether the graph is planar, whether the graph is regular, and whether the edges are undirected in the sense, that if there is an edge from node u to node v, there is an edge from v to u.

The subject of this paper is the *synchronous boolean finite dynamical systems* (*synchronous BFDS*, for short). A synchronous BFDS is the subclass of BFDS in which the domain is boolean and the update is synchronous.

Given a finite dynamical system we are naturally interested in its behavior. For example, we may ask questions about fixed points, such as whether the system has a fixed point (that is, whether there is a state configuration in which the state update of the system produces no change). In the case where the state domain is finite, there are a finite number of state configurations, and so we can ask such questions how many fixed points the system has and how many initial state configurations lead to fixed points. Furthermore, we can ask about the behavior of the system on a particular initial state configuration, such as, whether a given initial state configuration leads the system to a fixed point, and if not, since the system eventually enters a cycle of state configurations, how many steps it will take for the system to enter a cycle and how long the cycle is.

That the underlying structure of finite dynamical systems can be represented as a graph suggests that the classes of finite dynamical systems can be studied using the number of objects as the size parameter and so the behavioral properties of a class of finite dynamical systems can be studied in terms of its computational complexity. In other words, for a class of finite dynamical systems C and for a question Q, we ask how computationally hard it is to answer Q for class C: *Is it polynomial time solvable? If not, is the problem hard for a known complexity class, such as NP and PSPACE?*

Much work has been done to explore the computational complexity of behavioral properties of finite dynamical systems. Barrett *et al.* [3] study the computational complexity of the sequential finite dynamical systems, the model first introduced by Barrett, Mortveit, and Reidys in [1]. Barrett *et al.* [3] study particularly the sequential boolean finite dynamical systems regarding the existence of fixed points. For a variety of permissible state update functions, they ask which combinations of the functions make the problem easy or difficult. They show that the problem is NP-complete if the set of permissible local transition functions is either {NAND, XNOR}, {NAND, XOR}, {NOR, XNOR} or {NOR, XOR}. They also show that the problem is solvable in polynomial time if the functions are chosen from {AND, OR, NAND, NOR}.

The above results have been strengthened by Kosub [7], who shows a dichotomy result in the sense of Schaefer [11]; i.e., the problem in question is either NP-hard or polynomial time solvable. Kosub obtains a complete complexity-theoretic characterization of the fixed-point problem about boolean finite dynamical systems with respect to the state update function classes, which Kosub calls *Post Classes*, as well as with respect to the structure of the underlying graph. He shows exactly in which case the problem is NP-hard and for all the remaining cases the problem is polynomial-time solvable. Kosub and Homan [9] prove a dichotomy result on the counting version of the fixed point problem, in the sense that the problem is either #P-complete or polynomial-time solvable.

Another set of natural problems that arise in finite dynamical systems is the reachability; that is, given a system and two state configurations a and b, can b be reached from a? A variant of this problem is whether any fixed point can be reached from a given configuration a. Barrett *et al.* [2] study these problems for the sequential and synchronous dynamical systems in which the underlying graph is an undirected graph. They show that the problems are PSPACE-complete in general but polynomial time solvable if the state update functions are symmetric and monotone boolean functions.

In this paper, as a follow-up of the aforementioned prior work [2,3,7,9], we study the computational complexity of the synchronous boolean finite dynamical systems in which the basis B of the state update functions is finite. We are particularly interested in three questions:

1. CONVERGENCE(B): Given a system F and an initial state configuration a, decide whether the system converges to any fixed point.

2. PATHINTERSECTION($\mathcal{B}$): Given an n-object system $\mathcal{F}$ and two state configurations $\boldsymbol{a}$ and $\boldsymbol{b}$, do there exist time steps s and t, such that the state configuration of $\mathcal{F}$ on $\boldsymbol{a}$ at step s is equal to the state configuration of $\mathcal{F}$ on $\boldsymbol{b}$ at step t?
3. CYCLELENGTH($\mathcal{B}$): Given a system $\mathcal{F}$, an initial state configuration $\boldsymbol{a}$, and an integer t, decide whether the state configuration sequence generated by the system starting from $\boldsymbol{a}$ contains a cycle having length greater than or equal to t. Note that the complement of this problem with $t = 2$ is CONVERGENCE($\mathcal{B}$).

Although our work may seem reminiscent of the previous work, our focus is on the dynamical systems whose underlying graph is directed, not undirected. It is known that the dynamical behavior of the Hopfield networks is different depending on whether they are symmetric or not [10]. There is thus no *a priori* reason to believe that the results regarding directed graph structures are derived from the results regarding undirected graph structures.

We first show that the above three problems are all PSPACE-complete if $\mathcal{B}$ contains NAND or NOR. While we provide a proof for the result, these follow from an earlier paper by Floréen and Orponen [5]. We note that Barrett *et al.* [2] study this problem too, but in their setting the underlying graph is undirected.

We then prove that if $\mathcal{B}$ is one of {AND}, {OR}, and {XOR, NXOR}, CONVERGENCE is solvable in polynomial time and that the same assumption implies that PATHINTERSECTION belongs to UP and CYCLELENGTH belongs to UP ∩ coUP. We suspect that the latter two problems are polynomial time solvable, but we do not have at hand yet proofs that the problems are in P.

The rest of the paper is organized as follows. In Sect. 2, we formally define the dynamical systems and the problems we will study. In Sect. 3, we give algorithms for the convergence problem and the path intersection problem. In Sect. 4, we show that the cycle length problem is in UP ∩ coUP.

2 Preliminaries

2.1 Definitions

Below, following the definition of the sequential dynamical systems by Laubenbacher and Pareigis [8] we define synchronous boolean finite dynamical systems.

Let $n \geq 1$ be an integer. A *synchronous boolean finite dynamical system* (synchronous BFDS, for short) of n variables is an n-tuple $\mathcal{F} = (f_1, f_2, \ldots, f_n)$ such that $f_1, \ldots, f_n$ are boolean functions of n variables.

Let $\mathcal{F} = (f_1, f_2, \ldots, f_n)$ be an n-variable synchronous BFDS. A *state configuration* (or simply a *configuration*) of $\mathcal{F}$ is an n-dimensional boolean vector. We use the vector notation $\boldsymbol{x} = (x_1, x_2, \ldots, x_n)$ to denote a state configuration, where $x_1, \ldots, x_n$ are boolean variables.

The action of $\mathcal{F}$ on an state configuration $\boldsymbol{x}$ is defined as:

$$\mathcal{F}(\boldsymbol{x}) = (f_1(\boldsymbol{x}), f_2(\boldsymbol{x}), \ldots, f_n(\boldsymbol{x}))$$

In other words, the elements of $\mathcal{F}(\boldsymbol{x})$ are obtained by applying the n boolean functions $f_1, \ldots, f_n$ concurrently on the variables $x_1, \ldots, x_n$. Given an initial state configuration $\boldsymbol{x}^0 = (x_1^0, x_2^0, \ldots, x_n^0)$, the synchronous BFDS defines n sequences of boolean values $\{x_i^t\}$, $1 \leq i \leq n$ and $t \geq 0$ by iterative applications of $\mathcal{F}$ on the initial state configuration vector:

$$\text{for all } t \geq 0, \boldsymbol{x}^{t+1} = \mathcal{F}(\boldsymbol{x}^t),$$

where for all $t \geq 0$, $\boldsymbol{x}^t = (x_1^t, x_2^t, \ldots, x_n^t)$. In other words, for all $t \geq 0$,

$$\boldsymbol{x}^t = \mathcal{F}^t(\boldsymbol{x}^0).$$

For an n-state boolean finite dynamical system, there are exactly 2^n possible state configurations. This implies that in an n-state synchronous BFDS, regardless of which initial state configuration $\boldsymbol{x}^0$ it starts, the state configuration sequence generated from $\boldsymbol{x}^0$ *enters a cycle*; that is, in the sequence there exist indices s and t, $0 \leq s < t$, such that $\boldsymbol{x}^s = \boldsymbol{x}^t$. Clearly, for all such pairs (s, t), it holds:

$$\text{for all } i \geq 0, \boldsymbol{x}^{s+i} = \boldsymbol{x}^{t+i}.$$

This implies that there is the smallest value of s for which there exists some $t > s$ such that $\boldsymbol{x}^s = \boldsymbol{x}^t$ and that, for that smallest value of s, there exists the smallest value of $t > s$ such that $\boldsymbol{x}^s = \boldsymbol{x}^t$. Let s_0 and t_0 respectively be the values of s and t thus defined. Then we have:

- $t_0 \leq 2^n$ and
- for all i and j, $0 \leq i < j \leq t_0 - 1$, $\boldsymbol{x}^i \neq \boldsymbol{x}^j$.

We say that $\mathcal{F}$ on $\boldsymbol{x}$ enters a cycle (or *enters a loop*) at step s_0 and its cycle has *length* $t_0 - s_0$. We call s_0 the *tail length* of $\mathcal{F}$ on $\boldsymbol{x}$. We define $L_{\mathcal{F}}(\boldsymbol{x}^0)$ to be the length of the cycle $t_0 - s_0$.

In the case where $t_0 = s_0 + 1$, the cycle length is 1, and so, for all $s \geq s_0$ it holds that $\boldsymbol{x}^{s_0} = \boldsymbol{x}^s$. In such a case we say that the vector $\boldsymbol{x}^{s_0}$ is a *fixed point* of $\mathcal{F}$; we also say that $\mathcal{F}$ *converges* on the initial state configuration $\boldsymbol{x}^0$.

A *function family* is a collection of boolean functions $\mathcal{H} = \{h_i\}_{i \geq 1}$ such that for each $i \geq 1$, h_i takes i inputs. For example, the disjunction of any input size, which can be described as

$$\{h_i\}_{i \geq 1}, h_i(x_1, \ldots, x_i) = x_1 \vee \cdots \vee x_i,$$

is a function family. For a function family $\mathcal{H}$, we write $\mathcal{H}_k$ to mean the element of $\mathcal{H}$ for input size k. For example, OR is the family of the disjunction functions while OR_2 is the binary disjunction function.

A *basis boolean function* is either a single boolean function or a function family. Let f be a boolean function of n variables and let g be a boolean function of m variables for some $m < n$. We say that g is *equivalent* to f if there exist indices $x_{i_1}, \ldots, x_{i_m}$ such that for all $x_1, \cdots, x_n \in \{0, 1\}$, it holds that

$$f(x_1, \ldots, x_n) = g(x_{i_1}, \ldots, x_{i_m}).$$

In other words, f is a function that depends only on the variables $x_{i_1}, \ldots, x_{i_m}$ and g characterizes the behavior of f on those m inputs.

Let $\mathcal{B}$ be a finite set of basis functions. We say that a synchronous BFDS $\mathcal{F} = (f_1, \ldots, f_n)$ has basis $\mathcal{B}$ if each function of $\mathcal{F}$ is either a function family in $\mathcal{B}$ or equivalent to a boolean function in $\mathcal{B}$. In this paper, we consider specifically the bases that are chosen from function families AND, NAND, OR, NOR, XOR, and NXOR.

We say that a function family $H = \{h_i\}_{i \geq 1}$ is *polynomial-time (respectively, polynomial-space) computable* if there exists an algorithm for computing, given an integer $i \geq 1$ and $a_1, \ldots, a_i \in \{0, 1\}$, the value of $h_i(a_1, \ldots, a_i)$ in time (respectively, space) polynomial in i. We say that a function base $\mathcal{B}$ is *polynomial-time (respectively, polynomial-space) computable* if each function family in $\mathcal{B}$ is polynomial-time (respectively, polynomial-space) computable.

Given the above formulation it is now possible to discuss how to encode the synchronous BFDS $\mathcal{F}$ over a basis $\mathcal{B}$. An n-object synchronous BFDS f over a basis $\mathcal{B}$ is encoded as a labeled directed graph $G = (V, E)$ in which V is the object set and E represents the dependency of the objects in terms of their state update. The nodes are labeled with their basis function. The number of incoming edges to each node is no more than the number of inputs to the basis function it is associated with, and those edges are labeled to indicate the positions of the variables in the input of the basis functions. Thus, any basis $\mathcal{B}$, the synchronous BFDS $\mathcal{F}$ over $\mathcal{B}$ has an encoding whose length is bounded by a fixed polynomial in the number of objects. Note that such an encoding may not exist if $\mathcal{B}$ contains a function family that does not have a polynomial-size encoding.

We now formally define the three decision problems we consider in the paper. Let $\mathcal{B}$ be a boolean function basis.

1. CONVERGENCE($\mathcal{B}$) is the problem of deciding, given a synchronous BFDS $\mathcal{F}$ having basis $\mathcal{B}$ and an initial state configuration a of $\mathcal{F}$, whether $\mathcal{F}$ converges on a.
2. PATHINTERSECTION($\mathcal{B}$) is the problem of deciding, given a synchronous BFDS $\mathcal{F}$ having basis $\mathcal{B}$ and two initial state configurations a and b of $\mathcal{F}$, whether there exist some s and t, $0 \leq s, t \leq 2^n - 1$, such that $\mathcal{F}^s(a) = \mathcal{F}^t(b)$.
3. CYCLELENGTH($\mathcal{B}$) is the problem of deciding, given a synchronous BFDS $\mathcal{F}$ having basis $\mathcal{B}$, an initial state configuration a of $\mathcal{F}$, and an integer t, whether the cycle length of $\mathcal{F}$ on a, i.e., $L_{\mathcal{F}}(a)$, is greater than t.

We assume that the reader is familiar with introductory-level complexity classes (see, e.g., Hemaspaandra and Ogihara [6], for reference). The class PSPACE consists of all decision problems that can be decided by a polynomial space-bounded Turing machines. The class UP consists of all decision problems that can be decided by a polynomial time-bounded nondeterministic Turing machines with a special property that given as input each positive (respectively, negative) instance, the number of accepting computation paths of the machine is 1 (respectively, 0). The class coUP is the class of all decision problems that are the complement of some decision problem in UP.

2.2 PSPACE-Completeness

Here we prove that the aforementioned three problems are PSPACE-complete if the basis contains NAND, NOR or both AND and OR.

Proposition 1. *For all polynomial-space computable bases* $\mathcal{B}$, CONVERGENCE $(\mathcal{B})$, PATHINTERSECTION$(\mathcal{B})$ *and* CYCLELENGTH$(\mathcal{B})$ *are in* PSPACE.

Proof. Let M be a Turing machine that, given as input a synchronous BFDS S of some n objects, an initial state configuration $\boldsymbol{a}$, and an integer $t, 0 \leq t \leq 2^n$, outputs $S^t(\boldsymbol{a})$. Since the basis $\mathcal{B}$ is polynomial-space computable, M can be made to run in polynomial space. Using this machine M as a subroutine, the three problems can be solved as follows:

- CONVERGENCE$(\mathcal{B})$: Test whether there exists a $t, 0 \leq t \leq 2^n$, such that $M(S, \boldsymbol{a}, t) = M(S, \boldsymbol{a}, t+1)$.
- PATHINTERSECTION$(\mathcal{B})$: Test whether there exist s and t, $0 \leq s, t \leq 2^n - 1$, such that $M(S, \boldsymbol{a}, s) = M(S, \boldsymbol{b}, t)$.
- CYCLELENGTH$(\mathcal{B})$: Test whether there are no k and l, $0 \leq k, l \leq 2^n$ and $l - k \leq t$, such that $M(S, \boldsymbol{a}, k) = M(S, \boldsymbol{a}, l)$.

Clearly, each of the above search can be run using $O(n)$ space. Thus, all three problems are in PSPACE. □

The following theorem follows from [5, Corollary 3.2]; we omit the proof due to the page limitation.

Theorem 1. *If the basis* $\mathcal{B}$ *contains either* NAND, NOR *or* {AND, OR}, *the problems* CONVERGENCE$(\mathcal{B})$, PATHINTERSECTION$(\mathcal{B})$ *and* CYCLELENGTH$(\mathcal{B})$ *are* PSPACE-*hard.*

The theorem immediately implies the following corollaries.

Corollary 1. *If the basis* $\mathcal{B}$ *contains either* NAND *or* NOR, *the problems* CONVERGENCE$(\mathcal{B})$, PATHINTERSECTION$(\mathcal{B})$ *and* CYCLELENGTH$(\mathcal{B})$ *are* PSPACE-*complete.*

Corollary 2. *If the basis* $\mathcal{B}$ *contains both* AND *and* OR, *the problems* CONVERGENCE$(\mathcal{B})$, PATHINTERSECTION$(\mathcal{B})$ *and* CYCLELENGTH$(\mathcal{B})$ *are* PSPACE-*complete.*

3 Algorithms for Convergence and PathIntersection

In this section, we prove the following theorem.

Theorem 2. *If* $\mathcal{B}$ *is one of* {AND}, {OR}, *and* {XOR, NXOR}, CONVERGENCE $(\mathcal{B})$ *is polynomial-time computable and* PATHINTERSECTION$(\mathcal{B})$ *belongs to* UP.

The theorem is built upon the following lemma, which states that the state configuration at any time step $t, 0 \leq t \leq 2^n$, of a synchronous BFDS of n objects can be computed in time polynomial in n for a basis chosen from $\{\text{AND}\}$, $\{\text{OR}\}$, or $\{\text{XOR}, \text{NXOR}\}$.

Lemma 1. *Let $\mathcal{B}$ be one of $\{\text{AND}\}$, $\{\text{OR}\}$, and $\{\text{XOR}, \text{NXOR}\}$. Given an n-object synchronous BFDS $\mathcal{F}$ over basis $\mathcal{B}$, a state configuration $\boldsymbol{a} \in \{0,1\}^n$, and an integer $k \geq 0$, we can compute $\mathcal{F}^k(\boldsymbol{a})$ in time polynomial in $n + \log k$.*

Proof. In this proof we will think of the state configurations to be column vectors. We first consider the case where $\mathcal{B} = \{\text{OR}\}$. Let $\mathcal{F} = (f_1, f_2, \ldots, f_n)$ and $\boldsymbol{a} \in \{0,1\}^n$ be respectively an n-object BFDS over $\mathcal{B}$ and its state configuration. Let A be the adjacency matrix of the system $\mathcal{F}$ in terms of its graph-based encoding; that is, for all i and j, $1 \leq i, j \leq n$, the entry (i, j) of A is 1 if there is an edge from node j to node i and 0 otherwise. We then have

$$\mathcal{F}(\boldsymbol{a}) = A\boldsymbol{a},$$

where the multiplication is interpreted as AND and the addition as OR. It follows from this that for all $k \geq 0$

$$\mathcal{F}^k(\boldsymbol{a}) = A^k \boldsymbol{a}$$

and that $\mathcal{F}^k$ therefore can be computed by way of the standard iterated multiplication. Thus, for all k, $\mathcal{F}^k(\boldsymbol{a})$ can be computed in time polynomial in $n + \log k$.

Next we consider the case where $\mathcal{B} = \{\text{AND}\}$. For each vector $\boldsymbol{a}$, $\boldsymbol{a}^c$ be the component-wise complement of $\boldsymbol{a}$, that is, the vector constructed from $\boldsymbol{a}$ by flipping each element. Then we have

$$\mathcal{F}(\boldsymbol{a})^c = A\boldsymbol{a}^c.$$

This implies that for all $k \geq 0$,

$$(\mathcal{F}(\boldsymbol{a})^k)^c = A^k \boldsymbol{a}^c$$

and so

$$\mathcal{F}(\boldsymbol{a})^k = (A^k \boldsymbol{a}^c)^c.$$

Thus, from the previous discussion, the lemma holds in the case where the basis is $\{\text{AND}\}$.

Finally we consider the case where $\mathcal{B} = \{\text{XOR}, \text{NXOR}\}$. We will consider each state to be an element of Z_2 and perform the arithmetic over Z_2. For each $i, 1 \leq i \leq n$, f_i can be represented by a linear function over Z_2:

$$f_i(\boldsymbol{x}) = \left(\bigoplus_{j \in X_i} x_j \right) \oplus b_i$$

where $\oplus$ is the addition over Z_2, X_i is a set of all indices of variables involved in f_i, and $b_i = 1$ if f_i is equivalent to NXOR and 0 otherwise. By using the same adjacency matrix as before and using the column vector $b = (b_1, b_2, \ldots, b_n)^T$, we have:

$$\mathcal{F}(a) = Aa \oplus b,$$

and so for all $k \geq 0$,

$$\mathcal{F}^k(a) = A^k a \oplus (A^{k-1} \oplus A^{k-2} \oplus \cdots \oplus I)b, \tag{1}$$

where I is the $n \times n$ identity matrix. By the standard iterated multiplication, for each $k, 0 \leq k \leq 2^n$, we can compute A^k in polynomial time. Thus, it suffices to show that the second term of Eq. (1) is computable in time polynomial in $n + \log k$.

Let $Q(k)$ denote the summation in question. Suppose k is a power of 2. Let $p = \log k$. We have $k = 2^p$ and

$$Q(k) = (A^{2^{p-1}} \oplus I)(A^{2^{p-2}} \oplus I) \cdots (A \oplus I).$$

Since $p = \log k$, by the iterative multiplication, we can compute all the components on the right-hand side in time polynomial in $n + \log k$, and so the left-hand side can be obtained in time polynomial in $n + \log k$.

Now suppose k is not a power of 2. There exist p and k' such that $2^p < k < 2^{p+1}$ and $k' = k - 2^p \leq k/2$. We have

$$Q(k) = Q(2^p) \oplus A^{2^p} Q(k').$$

Since $1 \leq k' < k/2$, this allows us to establish a recursive method for computing $Q(k)$. The depth of recursion is at most $\log k$, and each term of the form either $Q(2^m)$ or A^{2^m} during the recursion can be computed in time polynomial in $n + \log k$. Thus, $Q(k)$ can be computed in time polynomial in $n + \log k$. Hence, the claim holds, and the proof is complete.   □

Theorem 2 can be proven using Lemma 1 as follows.

Proof of Theorem 2. To show that CONVERGENCE is polynomial-time computable, let $\mathcal{F}$ be an n-object synchronous BFDS over one of the three bases and let a be an initial state configuration. By the definition of convergence, we have that $\mathcal{F}$ converges on a if and only if $\mathcal{F}^{2^n-1}(a) = \mathcal{F}^{2^n}(a)$ holds. By Lemma 1, we can compute $\mathcal{F}^{2^n-1}(a)$ and $\mathcal{F}^{2^n}(a)$ in polynomial time, and thus we complete the proof.

The following algorithm shows that PATHINTERSECTION($\mathcal{B}$) is in UP: Given $\mathcal{F}$, a, and b,

Step 1. Nondeterministically choose s, $0 \leq s \leq 2^n - 1$.
Step 2. Nondeterministically choose t, $0 \leq t \leq 2^n - 1$.
Step 3. Test whether $\mathcal{F}^s(a) = \mathcal{F}^t(b)$. If the test fails, reject.
Step 4. If either $s = 0$ or $t = 0$, then accept. Otherwise, test whether $\mathcal{F}^{s-1}(a) \neq \mathcal{F}^{t-1}(b)$. If the inequality holds, accept; otherwise, reject.

Clearly the algorithm runs in time polynomial in n. If the two state configuration paths intersect, then there is a unique combination of s and t for which the tests pass. Thus, the algorithm runs in UP.   □

4   Algorithm for CycleLength

In this section we prove the following theorem.

Theorem 3. *If $\mathcal{B}$ is one of $\{$AND$\}$, $\{$OR$\}$, and $\{$XOR, NXOR$\}$, then* CYCLE-LENGTH$(\mathcal{B})$ *belongs to* UP $\cap$ coUP.

This result together with the latter statement of Theorem 2 can be used as evidence that for the bases mentioned in the theorems PATHINTERSECTION$(\mathcal{B})$ and CYCLELENGTH$(\mathcal{B})$ are unlikely to be NP-hard.

We first prove the following proposition.

Proposition 2. *Let $\mathcal{F}$ be an n-object BFDS and let $\boldsymbol{a}$ be an initial state configuration. For all integers $p \geq 0$ and $q \geq 1$, $\mathcal{F}$ on $\boldsymbol{a}$ has tail length p and cycle length q if and only if the following properties hold:*

1. $\mathcal{F}^p(\boldsymbol{a}) = F^{p+q}(\boldsymbol{a})$.
2. *If $p > 0$, then $\mathcal{F}^{p-1}(\boldsymbol{a}) \neq F^{p+q-1}(\boldsymbol{a})$.*
3. *For all prime numbers d dividing q, $\mathcal{F}^p(\boldsymbol{a}) \neq F^{p+q/d}(\boldsymbol{a})$.*

Proof. Let $\mathcal{F}$ and $\boldsymbol{a}$ be as in the statement of the proposition. Suppose $\mathcal{F}$ on $\boldsymbol{a}$ enters a cycle at step p and the cycle length is q. Then, we have $\mathcal{F}^p(\boldsymbol{a}) = F^{p+q}(\boldsymbol{a})$. This is identical to Property 1 in the above. Also, by the minimality of p, we have: for all i, $0 \leq i \leq p-1$ and for all $j \geq i$, $\mathcal{F}^i(\boldsymbol{a}) \neq \mathcal{F}^j(\boldsymbol{a})$. By setting $i = p-1$ and $j = p+q-1$, we get Property 2. Finally, by the minimality of q, we have for all $i \geq 0$ and $s, 1 \leq s \leq q-1$, $\mathcal{F}^i(\boldsymbol{a}) \neq \mathcal{F}^{i+s}(\boldsymbol{a})$. In particular, if d is a prime number dividing q, then $q/d < q$, and so by setting $i = p$ and $s = q/d$, we have Property 3.

Conversely, suppose that one of the three properties in the statement of the proposition fails to hold for p and q. If Property 1 fails to hold, clearly q is not the cycle length. If Property 2 fails to hold, $\mathcal{F}^{p-1}(\boldsymbol{a}) = F^{p+q-1}(\boldsymbol{a})$, and so $\mathcal{F}$ on $\boldsymbol{a}$ enters a cycle earlier than step p. If Property 3 fails to hold, there is a divisor $e = p/d$ for some prime number d such that $\mathcal{F}^p(\boldsymbol{a}) = F^{p+e}(\boldsymbol{a})$. This implies that the cycle length is smaller than q.

This proves the proposition. $\square$

For a total function g, we say that g *is* UP-*computable* if there exists a polynomial-time nondeterministic Turing machine M such that for all inputs x, M on x accepts along exactly one computational path and in that unique computation path M on x outputs $g(x)$.

In the following lemma, we show that the cycle length is UP-computable, which immediately implies Theorem 3.

Lemma 2. *Suppose $\mathcal{B}$ is one of $\{$AND$\}$, $\{$OR$\}$, and $\{$XOR, NXOR$\}$. Then for all synchronous BFDS $\mathcal{F}$ and initial configurations $\boldsymbol{a}$, the tail length and the cycle length of $\mathcal{F}$ on $\boldsymbol{a}$ are UP-computable.*

Proof. Let $\mathcal{B}$ be one of $\{\text{AND}\}$, $\{\text{OR}\}$, and $\{\text{XOR}, \text{NXOR}\}$. Since the tail length p and the cycle length q are uniquely determined for each combination of $\mathcal{F}$ and $\boldsymbol{a}$ and since the prime factorization is in UP $\cap$ coUP [4], we can design a UP-algorithm for calculating p and q given $\mathcal{F}$ and $\boldsymbol{a}$ as follows:

Step 1. Our algorithm nondeterministically guesses p and q such that $0 \leq p < 2^n$ and $1 \leq q \leq 2^n - p$.

Step 2. Using the algorithm presented in [4], we compute the prime factorization of q in UP. If the factorization is successful, the algorithm proceeds to the next step.

Step 3. Our algorithm tests the three properties in Proposition 2.

Step 4. Our nondeterministic algorithm accepts and outputs p and q if and only if all the tests pass.

The prime factorization part is carried out nondeterministically and since it is in UP, there is exactly one computation path along which the factorization is successfully obtained. Since $q \leq 2^n$, the number of distinct prime factors of q is at most n. This implies that there will be at most $n + 2$ equalities to be tested in Step 3. Since both p and q are bounded from above by 2^n, we have from Lemma 1 that each equality can be tested in time polynomial in n. Thus, the above algorithm runs in time polynomial in n. The algorithm has exactly one accepting computation path for all $\mathcal{F}$ and $\boldsymbol{a}$, and on that unique accepting computation path computes p and q. Thus, the algorithm is an UP-algorithm.

This proves the lemma. □

5   Conclusion

In this paper, we consider the convergence, path intersection, and cycle length problems for the synchronous BFDS on various fixed function bases $\mathcal{B}$ and show that while the three problems are PSPACE-complete for $\mathcal{B} \in \{\text{NAND}\}, \{\text{NOR}\}$, they are solvable in polynomial time or belongs to UP (or UP $\cap$ coUP) if $\mathcal{B}$ is $\{\text{AND}\}$, $\{\text{OR}\}$, or $\{\text{XOR}, \text{NXOR}\}$. An interesting question is whether the complexity upper bound of UP can be reduced to P.

References

1. Barrett, C.L., Mortveit, H.S., Reidys, C.M.: Elements of a theory of simulation II: sequential dynamical systems. Appl. Math. Comput. **107**(2–3), 121–136 (2000)
2. Barrett, C.L., Hunt III, H.B., Marathe, M.V., Ravi, S.S., Rosenkrantz, D.J., Stearns, R.E.: Complexity of reachability problems for finite discrete dynamical systems. J. Comput. Syst. Sci. **72**(8), 1317–1345 (2006)
3. Barrett, C.L., Hunt III, H.B., Marathe, M.V., Ravi, S.S., Rosenkrantz, D.J., Stearns, R.E., Tošić, P.T.: Gardens of eden and fixed points in sequential dynamical systems. In: Proceedings of Discrete Models: Combinatorics, Computation, and Geometry, pp. 95–110 (2001)

4. Fellows, M.R., Koblitz, N.: Self-witnessing polynomial-time complexity and prime factorization. In: Proceedings of the Seventh Annual Conference on Structure in Complexity Theory, pp.107–110 (1992)
5. Floréen, P., Orponen, P.: Complexity issues in discrete Hopfield networks. Neuro-COLT Technical report Series, NC-TR-94-009 (1994)
6. Hemaspaandra, L.A., Ogihara, M.: A Complexity Theory Companion. Springer, Berlin (2001)
7. Kosub, S.: Dichotomy results for fixed-point existence problems for boolean dynamical systems. Math. Comput. Sci. 1(3), 487–505 (2008)
8. Laubenbacher, R., Pareigis, B.: Equivalence relations on finite dynamical systems. Adv. Appl. Math. 26(3), 237–251 (2001)
9. Kosub, S., Homan, C.M.: Dichotomy results for fixed point counting in boolean dynamical systems. In: Proceedings of the Tenth Italian Conference on Theoretical Computer Science (ICTCS 2007), pp. 163–174 (2007)
10. Parberry, I.: Circuit Complexity and Neural Networks. MIT Press, Cambridge (1994)
11. Schaefer, T.J.: The complexity of satisfiability problems. In: Proceedings of the Tenth ACM Symposium on Theory of Computing, pp. 216–226 (1978)

On the Power of Parity Queries in Boolean Decision Trees

Raghav Kulkarni[1], Youming Qiao[1](✉), and Xiaoming Sun[2]

[1] Centre for Quantum Technologies,
The National University of Singapore, Singapore, Singapore
{kulraghav86,jimmyqiao86}@gmail.com
[2] Institute of Computing Technology, Chinese Academy of Sciences, Beijing, China
sunxiaoming@ict.ac.cn

Abstract. In an influential paper, Kushilevitz and Mansour (1993) introduced a natural extension of Boolean decision trees called *parity decision tree* (PDT) where one may query the sum modulo 2, i.e., the *parity*, of an arbitrary subset of variables. Although originally introduced in the context of learning, parity decision trees have recently regained interest in the context of communication complexity (cf. Shi and Zhang 2010) and property testing (cf. Bhrushundi, Chakraborty, and Kulkarni 2013). In this paper, we investigate the power of parity queries. In particular, we show that the parity queries can be replaced by ordinary ones at the cost of the *total influence* aka *average sensitivity* per query. Our simulation is tight as demonstrated by the parity function.

At the heart of our result lies a qualitative extension of the result of O'Donnell, Saks, Schramme, and Servedio (2005) titled: *Every decision tree has an influential variable*. Recently Jain and Zhang (2011) obtained an alternate proof of the same. Our main contribution in this paper is a simple but surprising observation that the query elimination method of Jain and Zhang can indeed be adapted to eliminate, seemingly much more powerful, parity queries. Moreover, we extend our result to *linear* queries for Boolean valued functions over arbitrary finite fields.

1 Introduction

The decision tree model [8], perhaps due to its simplicity and fundamental nature has been extensively studied over decades, yet remains a fascinating source of some of the outstanding open questions. In the first part of this paper we focus on decision trees for Boolean functions, i.e., functions of the

Raghav Kulkarni—Research at the Centre for Quantum Technologies is funded by the Singapore Ministry of Education and the National Research Foundation.

Xiaoming Sun—Part of this work was done while the author was visiting the Centre for Quantum Techologies, National University of Singapore. He is supported in part by the National Natural Science Foundation of China Grant 61170062, 61222202, 61433014 and the China National Program for support of Top-notch Young Professionals.

R. Jain et al. (Eds.): TAMC 2015, LNCS 9076, pp. 99–109, 2015.
DOI: 10.1007/978-3-319-17142-5_10

form $f : \{0,1\}^n \rightarrow \{0,1\}$. In later section, we extend our results for decision trees over any finite field, i.e., for functions of the form $\mathbb{F}_q^n \rightarrow \{0,1\}$. A deterministic decision tree D_f for f takes $x = (x_1, \ldots, x_n)$ as an input and determines the value of $f(x_1, \ldots, x_n)$ using queries of the form "is $x_i = 1$?". Let $C(D_f, x)$ denote the cost of the computation, i.e., the number of queries made by D_f on input x. The *deterministic decision tree complexity* of f is defined as $D(f) = \min_{D_f} \max_x C(D_f, x)$.

Variants of decision tree model are fundamental for several reasons including their connection to other models such as communication complexity, their usability in analyzing more complicated models such as circuits, their mathematical elegance and richness, and finally the notoriety of some simple yet fascinating open questions about them such as the Evasiveness Conjecture [3,14,15,19,22] that have caught the imagination of generations of researchers over decades. In this paper we study a variant of decision trees called *parity decision tree* (PDT) and its extension over finite fields, which we call *linear decision tree* (LDT).

Motivation for Studying PDTs and LDTs

A parity decision tree may query "is $\sum_{i \in S} x_i \equiv 1 \pmod 2$?" for an arbitrary subset $S \subseteq [n] = \{1, 2, \ldots, n\}$. We call such queries *parity queries*. For a PDT P_f for f, let $C(P_f, x)$ denote the number of parity queries made by P_f on input x. The *parity decision tree complexity* of f is $D^{\oplus}(f) = \min_{P_f} \max_x C(P_f, x)$. Note that $D^{\oplus}(f) \leq D(f)$ as "is $x_i = 1$?" can be treated as a parity query.

The PDTs were introduced by Kushilevitz and Mansour [17] in the context of learning Boolean functions by estimating their Fourier coefficients. Several other models such as circuits and branching programs have been also been analysed in the past after augmenting their power by allowing counting operations.

In spite of being combinatorially rich and beautiful model, the PDT somehow remained dormant until recently where it was brought back into light in an entirely different context, namely the *communication complexity* of XOR functions [23,31]. Shi and Zhang [31] and Montanaro and Osborne [23] have observed that the deterministic communication complexity $CC(f^{\oplus})$ of computing $f(x \oplus y)$, when x and y are distributed between the two parties, is upper bounded by $D^{\oplus}(f)$. The importance for communication complexity comes from the conjecture [23,31] that for some positive constant c, every Boolean function f satisfies $D^{\oplus}(f) = O((\log \|\widehat{f}\|_0)^c)$; where $\|\widehat{f}\|_0$ is the *sparsity* (number of non-zero Fourier coefficients) of f. Settling this conjecture in affirmative would confirm the famous Log-rank Conjecture [24] in the important special case of XOR functions. Recently Tsang et al. [36] confirm it for functions with constant degree over $\mathbb{F}_2$ and Kulkarni and Santha [18] confirm it for AC^0 functions.

Very recently, Bhrushundi, Chakraborty, and Kulkarni [4] connected parity decision trees to property testing of linear and quadratic functions. Their approach for instance can potentially be used to solve a long-standing open question of closing the gap for k-linearity by analysing the randomized PDT complexity of the function E_k that evaluates to 1 iff the number of 1s in the input is exactly k. Recently PDTs were analysed further in several papers including [18,32,34,36] and many more to come.

Similar to PDTs, the LDTs are closely related to the Fourier spectrum of functions over $\mathbb{Z}_p$. In recent paper by Shpilka, Tal, and Volk [32] the authors derive various structural results of the Fourier spectrum by analysing LDTs. Given the evidence of abundance of connections to other models and mathematics, and given the rich combintaorial structure of PDTs and LDTs, we believe that they deserve a systematic and independent study at this point. Our paper is a step in this direction.

Motivation for Studying Influence Lower Bounds

Proving lower bounds on the influence of Boolean functions has had a long history in Theoretical Computer Science. It is nicely summerized in the paper [29], we restate a part from that for illustration. Influence lower bounds have been crucial part of several fundamental results such as threshold phenomenon, lower bound on randomized query complexity of graph properties, quantum and classical equivalence etc. Ben-Or and Linial [6], in their 1985 paper on collective coin flipping, observe that the maximum influence $\mathrm{Inf}_{max}(f) \geq 1/n$ for any balanced function and conjectured $\Theta(\log n/n)$ bound. The seminal paper by Kahn, Kalai, Linial [16] confirmed the conjecture via an application of the *Hypercontractive Inequality*. This result was subsequently generalized by Talagrand [35] in order to show *sharp threshold behaviour* for monotone functions.

In their celebrated paper *Every decision tree has an influential variable*, O'Donnell, Saks, Schramme, and Servedio [29] showed a crucial inequality lower bounding the maximum influence: $\mathrm{Inf}_{max}(f) \geq \mathrm{Var}(f)/\Delta(f)$, where $\Delta(f)$ denotes the minimum possible average depth of a decision tree for f. This inequality found application in the lower bounds on randomized query complexity of monotone graph properties. Homin Lee [20] found a simple inductive proof of the OSSS result. Recently Jain and Zhang [13] found another simple and conceptually different proof via the method of query elimination, which we use here.

Aaronson and Ambainis [1] study a conjecture lower bounding the maximum influence of real valued polynomials in terms of their degree. This conjecture, if true, would imply polynomial equivalence between bounded-error quantum and classical query complexity. These previous results seems to indicate the importance of lower bounds on influence in terms of several complexity measures. In this paper, we present such new lower bounds in terms of PDT and LDT complexity.

Our Results

Let $D_\epsilon(f)$ and $D_\epsilon^\oplus(f)$ denote the minimum depth of a DT and a PDT (resp.) computing f correctly on at least $1 - \epsilon$ fraction of the inputs.

Theorem 1. *For any Boolean function f and any $\epsilon \geq 0$:*

$$\mathrm{Inf}_{max}(f) \geq \frac{\mathrm{Var}(f) - \epsilon}{D_\epsilon^\oplus(f)}.$$

Corollary 1. *For any Boolean function f and any $\epsilon > 0$:*

$$D_\epsilon(f) \leq \frac{1}{\epsilon^2} \cdot D^\oplus(f) \cdot \text{Inf}(f).$$

Corollary 2. *If f is computable by a polynomial size constant depth circuit, i.e., $f \in AC^0$, then:[1]*

$$D_\epsilon(f) = \tilde{O}_\epsilon(D^\oplus(f)).$$

To prove Theorem 1 we use an adaptation of the query elimination method of Jain and Zhang. Our main observation is that assuming the uniform distribution on the inputs, one can eliminate seemingly powerful parity queries at the expense of $\text{Inf}_{max}(f)$ error per elimination. Corollary 1 is obtained by analysing the *'query the most influential variable'* strategy using our new bound. We extend Theorem 1 for LDTs over arbitrary fields (see Sect. 4). The Corollary 1 can also be extended with similar techniques; we omit its simple proof.

Theorem 2. *Let q be a prime power. For any $f : \mathbb{F}_q^n \rightarrow \{0, 1\}$ and any $\epsilon \geq 0$:*

$$\text{Inf}_{max}(f) \geq \frac{1}{q-1} \cdot \frac{\text{Var}(f) - \epsilon}{D_\epsilon^{\oplus_q}(f)}.$$

Further we explore the power of PDTs for monotone functions and show:

Theorem 3. *For any monotone Boolean function f and any $\epsilon > 0$:*

$$D_\epsilon(f) \leq \frac{3}{\epsilon^2} \cdot D^\oplus(f)^{3/2}.$$

To prove Theorem 3 we show an upper bound on L_1 norm of Fourier spectrum in terms of PDT depth, which in turn gives an upper bound on sum of linear Fourier coefficients restricted to monotone functions. We adapt the proof of the same for ordinary decision trees by O'Donnell and Servedio. Our main observation is that under the uniform distribution on inputs their proof can be extended for PDTs as well. Our result naturally raises the following question:

Question 1. Is it true that for every monotone Boolean function f and for every $\epsilon > 0$ we have:

$$D_\epsilon(f) = \tilde{O}_\epsilon(D^\oplus(f))?$$

It is also interesting to see if our results can be strengthened to $D_\epsilon^\oplus$ rather than just $D^\oplus$ as zero-error and bounded error complexities may behave differently.

We believe that our observations, although might appear simple, are indeed surprising. They seem to make a crucial qualitative point, that under the uniform distribution, the method of lower bounding the ordinary (randomized) decision tree complexity by $\text{Var}(f) / \text{Inf}_{max}(f)$ works equally well for seemingly much more powerful PDTs and LDTs as well. For non-balanced functions the uniform distribution does not seem to be an optimal choice for maximizing $\text{Var}(f) / \text{Inf}_{max}(f)$ but for balanced functions it does. As an application, finally we exhibit a gap between randomized PDT complexity and approximate L_1, both of which are relevant for communication complexity of XOR functions.

---

[1] The O_ϵ notation hides a multiplicative constant depending on ϵ and the $\tilde{O}_\epsilon$ notation hides a further poly-logarithmic multiplicative factor.

Organization. Section 2 contains preliminaries. Section 3 contains the proof of Theorem 1. Section 4 contains the proof of Theorem 2. Unfortunately, we had to move the other proofs to appendix and hence omit it from this version due to space constraint.

2   Preliminaries

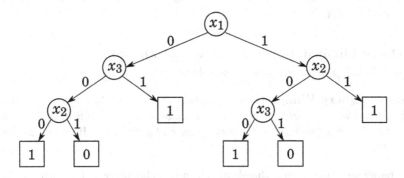

Fig. 1. A boolean decision tree

Randomized Decision Trees

A bounded error randomized decision tree R_f is a probability distribution over all deterministic decision trees such that for every input, the expected error of the algorithm is bounded by some fixed constant less than $1/2$ (say $1/3$). The cost $C(R_f, x)$ is the highest possible number of queries made by R_f on x, and the *bounded error randomized decision tree complexity* of f is $R(f) = \min_{R_f} \max_x C(R_f, x)$. Similarly one can define bounded error randomized PDT complexity of f, denoted by $R^\oplus(f)$. Using Yao's min-max principle one may obtain: $D_{1/3}(f) \leq R(f)$ and $D_{1/3}^\oplus(f) \leq R^\oplus(f)$. (Fig. 1)

Variance and Influence

Let μ_p denote the p-biased distribution on the Boolean cube, i.e., each coordinate is independently chosen to be 1 with probability p. The *variance* of a Boolean function is $\mathrm{Var}(f, p) := 4 \cdot \Pr_{x \leftarrow \mu_p}(f(x) = 0)\Pr_{x \leftarrow \mu_p}(f(x) = 1)$. The *influence* of the ith variable under μ_p is $\mathrm{Inf}_i(f, p) := \Pr_{x \leftarrow \mu_p}(f(x) \neq f(x \oplus e_i))$. Let $\mathrm{Inf}_{max}(f) := \max_i \mathrm{Inf}_i(f)$. The *total influence* aka *average sensitivity* of f is $\mathrm{Inf}(f, p) := \sum_i \mathrm{Inf}_i(f, p)$. In this paper we focus on $p = 1/2$ case.

Fourier Spectrum, Polynomial Degree, and Sparsity

Let $f_\pm : \{-1, 1\}^n \to \{-1, 1\}$ be represented by the following polynomial with real coefficients: $f_\pm(z_1, \ldots, z_n) = \sum_{S \subseteq [n]} \widehat{f}(S) \prod_{i \in S} z_i$. The above polynomial is unique and it is called the Fourier expansion of f. The $\widehat{f}(S)$ are called the Fourier coefficients of f. The *polynomial degree* of f is $\deg(f) := max\{|S| \mid \widehat{f}(S) \neq 0\}$. The *sparsity* of a Boolean function f is $||\widehat{f}||_0 := |\{S \mid \widehat{f}(S) \neq 0\}|$. We know that $\deg(f) \leq D(f)$, $\log ||\widehat{f}||_0 \leq D_\oplus(f)$ and $\log ||\widehat{f}||_0 \leq \deg(f)$.

Representing Decision Trees

We represent a decision tree T as $T = (x_i, T_0, T_1)$ where x_i denotes the first variable queried by T, i.e., x_i is the variable at the root of T : if $x_i = 0$ then T_0 is consulted; if $x_i = 1$ then T_1 is consulted. A leaf labeled 1 is represented as $(1, \emptyset, \emptyset)$ and the one labeled 0 is represented as $(0, \emptyset, \emptyset)$. We represent a parity decision tree as $T = (x_S, T_0, T_1)$; if $\sum_{i \in S} x_i = 0 \pmod 2$ then consult T_0, else consult T_1. A leaf labeled 1 is represented as $(1, \emptyset, \emptyset)$ and the one labeled 0 is represented as $(0, \emptyset, \emptyset)$.

The Query Elimination Lemma (Jain and Zhang)

Jain and Zhang prove the following simple yet powerful lemma:

Lemma 1 (Query Elimination Lemma). *If $T = (x_i, T_0, T_1)$ is an ordinary decision tree that computes f correctly on at least $1 - \delta$ fraction of the inputs then either T_0 or T_1 computes f correctly on at least $1 - \delta - \mathrm{Inf}_i(f)$ fraction of the inputs.*

In this paper we observe that the above lemma can be adapted for parity decision trees. This observation is a crucial part of our results.

Overview of the Query Elimination Method

The *query elimination method* of Jain and Zhang works as follows: Suppose we have a decision tree of depth $D_\epsilon(f)$ that computes f correctly on at least $1 - \epsilon$ fraction of the inputs. We repeatedly apply the Query Elimination Lemma to obtain a decision tree that computes f correctly on at least $1 - \epsilon - D_\epsilon(f) \cdot \mathrm{Inf}_{max}(f)$ fraction of the inputs without making any single query. Of course, such (zero-query) decision tree must make error on at least $\mathrm{Var}(f)$ fraction of the inputs. Hence: the error of the zero-query decision tree that we obtained $(\epsilon + D_\epsilon(f) \cdot \mathrm{Inf}_{max}(f))$ can be lower bounded by $\mathrm{Var}(f)$. In other words:

$$D_\epsilon(f) \geq \frac{\mathrm{Var}(f) - \epsilon}{\mathrm{Inf}_{max}(f)}.$$

3 Every PDT Has an Influential Variable

In this section we present the proof of Theorem 1. We start with eliminating queries in PDTs.

Eliminating Ordinary Queries in PDTs

First we note that Jain and Zhang's proof of the Query Elimination Lemma generalizes when T_i are parity decision trees instead of ordinary ones. In other words, if the first query in a parity decision tree is an ordinary query then one can remove it at the expense of $\mathrm{Inf}_i(f)$ increase in the error. We formulate this below.

Lemma 2. *If $T = (x_{\{i\}}, T_0, T_1)$ is a parity decision tree that computes f correctly on at least $1 - \delta$ fraction of the inputs then either T_0 with every occurrence of x_i hard-wired to 0 or T_1 with every occurrence of x_i hard-wired to 1 computes f correctly on at least $1 - \delta - \mathrm{Inf}_i(f)$ fraction of the inputs.*

Eliminating Parity Queries in PDTs

Let T be a parity decision tree that computes f correctly on at least $1 - \delta$ fraction of the inputs. Our idea is to convert the parity queries to an ordinary one and then eliminate the queries at the root of the tree. Let

$$Lf(x) := f(Lx).$$

We apply the linear transformation L on the input space $\mathbb{F}_2^n$ and work with Lf instead of f.

Observation 4. $\mathrm{Var}(f) = \mathrm{Var}(Lf)$ *and* $D_\oplus(f) = D_\oplus(Lf)$.

Rotatating the PDT T: Without loss of generality, let us assume that the first parity query in T is the parity of the first k bits, i.e., $x_1 \oplus \ldots \oplus x_k$ (for some k). Let $g(x_1, \ldots, x_n) := f(x_1 \oplus \ldots \oplus x_k, x_2, \ldots, x_n)$. Note that $g = Lf$ where L is the following *invertible* linear transformation on the vector space $\mathbb{F}_2^n$: $L(x_1, \ldots, x_n) := (x_1 \oplus \ldots \oplus x_k, x_2, \ldots, x_n)$. Also note that: $f(x_1, \ldots, x_n) = g(x_1 \oplus \ldots \oplus x_k, x_2, \ldots, x_n)$. Thus by querying $x_1 \oplus \ldots \oplus x_k$, we know the value of the 'first input bit' of g. Moreover the influence of the first variable remains unchanged.

Observation 5. $\mathrm{Inf}_1(g) = \mathrm{Inf}_1(f)$.

Note however that the influences of the variables $x_2, \ldots, x_k$ might have changed!
A PDT $T = (x_{[k]}, T_0, T_1)$ for f can be easily modified to a PDT LT for $Lf = g$. We call the transformation from T to LT as the *rotation* of T and it is defined as follows:

$$L(x_S, T_0, T_1) := (L(x_S), L(T_0), L(T_1)),$$

$$\text{(base case)} \quad L(0, \emptyset, \emptyset) = (0, \emptyset, \emptyset),$$

$$\text{(base case)} \quad L(1, \emptyset, \emptyset) = (1, \emptyset, \emptyset).$$

Next we observe that the error is preserved by a rotation.

Observation 6. *If T computes f correctly on $1 - \delta$ fraction of the inputs then LT computes $g = Lf$ correctly on $1 - \delta$ fraction of the inputs.*

Moreover: the tree LT has a nice property that the query at the root is not an arbitrary parity query but in fact an ordinary query, i.e., a variable x_1. Hence we can use Lemma 2 to remove the first query at the expense of $\mathrm{Inf}_1(g) = \mathrm{Inf}_1(f)$ increase in the error. Thus we conclude that:

Proposition 1. *If T computes f with error δ then either LT_0 or LT_1 computes LF correctly on at least $1 - \delta - \mathrm{Inf}_{max}(f)$ fraction of inputs.*

Rotating the PDT LT_i back to T_i:

Observation 7. *For the particular L above, $L^{-1} = L$.*

Suppose that LT_i computes Lf correctly on at least $1 - \delta - \text{Inf}_{max}(f)$ fraction of the inputs.

Thus we can rewrite Observation 6 as follows:

Observation 8. *If LT computes Lf correctly on $1 - \delta$ fraction of the inputs then $L(LT)$ computes $f = L(Lf)$ correctly on $1 - \delta$ fraction of the inputs.*

Proof of Theorem 1. Since $L(LT_i) = T_i$ and since LT_i computes Lf correctly on at least $1 - \delta - \text{Inf}_{max}(f)$ fraction of the inputs, T_i computes f with the same error. Notice that T_i makes one less parity query than T. So we have eliminated one parity query with an increase in error at most $\text{Inf}_{max}(f)$. Now we can repeat this process starting from a parity tree T of depth $D_\epsilon^\oplus(f)$ that makes error on at most ϵ fraction of the inputs to obtain a zero-query parity decision tree that makes at most $\epsilon + D_\epsilon^\oplus(f) \cdot \text{Inf}_{max}(f)$ error. The error of any zero-query parity decision tree must be at least $\text{Var}(f)$. This completes the proof of Theorem 1.   □

Remark 1. OR and AND functions on n variables can be computed with error probability at most $1/n$ on every input, using $O(\log n)$ parity queries chosen uniformly at random. Thus our Theorem 1 can be extended (up to a multiplicative poly-logarithmic factor) to the decision trees that use AND, OR, and PARITY queries. More generally, one can extend it to so called 1+ queries (see [10]) involving parities of (say polynomially many) arbitrary subsets.

4 Every Linear Decision Tree Has an Influential Variable

Let q be a prime power and $\mathbb{F}_q$ be the finite field with q elements. In this section we consider computing functions from $\mathbb{F}_q^n$ to $\{0, 1\}$ with the model called linear decision trees, denoted by $\oplus_q$-DT. It is a computation tree, with each internal nodel v labeled by a linear form $\ell : \mathbb{F}_q^n \to \mathbb{F}_q$. v has q children, whose edges connecting to v are labeled by elements from $\mathbb{F}_q$. The branching at node v is based on the evaluation of ℓ on the input vector. It is clear that when $q = 2$, this model becomes the parity decision tree model for computing boolean functions. We use $D_\epsilon^{\oplus_q}(f)$ to denote the smallest $\oplus_q$-DT for computing $f : \mathbb{F}_q^n \to \{0, 1\}$ with error ϵ.

We will focus on the setting of uniform distribution over $\mathbb{F}_q^n$. For $f : \mathbb{F}_q^n \to \{0, 1\}$, its variance is defined the same as $\text{Var}(f) = 4 \cdot \Pr(f(x) = 0) \Pr(f(x) = 1)$. If x and y in $\mathbb{F}_q^n$ differ only at the kth position, $k \in [n]$, we denote this by $x \sim_k y$. The influence of the kth variable is $\text{Inf}_k(f) := \Pr_{x \sim_k y}(f(x) \neq f(y))$. Our main result is the following analogue of Theorem 1.

Theorem 2, restated. For any function $f : \mathbb{F}_q^n \to \{0, 1\}$ and any $\epsilon \geq 0$:

$$\text{Inf}_{max}(f) \geq \frac{1}{q - 1} \cdot \frac{\text{Var}(f) - \epsilon}{D_\epsilon^{\oplus_q}(f)}.$$

We now prove Theorem 2. We shall adapt the proof of the query elimination lemma to $\oplus_q$-DT as follows.

Suppose T is a $\oplus_q$-DT for $f : \mathbb{F}_q^n \to \{0,1\}$. Let $\ell : \mathbb{F}_q^n \to \{0,1\}$ be the first query made by T, and $\ell(x_1, \ldots, x_n) = \alpha_1 x_1 + \alpha_2 x_2 + \cdots + \alpha_n x_n$. As ℓ is not trivial, there exists some $k \in [n]$ s.t. $\alpha_k \neq 0$. Fix such a $k \in [n]$. For $i \in \mathbb{F}_q$, let T_i be the $\oplus_q$-DT to be executed when $\ell(x) = i$.

For every T_i, $i \in \mathbb{F}_q$, construct a new $\oplus_q$-DT T_i', by replacing every occurrence of x_k in T_i with

$$\frac{1}{\alpha_k}(i - (\alpha_1 x_1 + \cdots + \alpha_{k-1} x_{k-1} + \alpha_{k+1} x_{k+1} + \cdots + \alpha_n x_n)).$$

It is clear that T_i' and T_i are related as follows. Let $a = (a_1, \ldots, a_n) \in \mathbb{F}_q^n$. Then $T_i'(a_1, \ldots, a_n) = T_i(a_1, \ldots, a_{k-1}, b_k, a_{k+1}, \ldots, a_n)$, where $b_k \in \mathbb{F}_q$ s.t.

$$\ell(a_1, \ldots, a_{k-1}, b_k, a_{k+1}, \ldots, a_n) = i.$$

For $a = (a_1, \ldots, a_n) \in \mathbb{F}_q^n$, we use $a|_k^{\ell,i}$ to denote $(a_1, \ldots, a_{k-1}, b_k, a_{k+1}, \ldots, a_n) \in \mathbb{F}_q^n$ satisfying the above. Then we have $T_i'(a) = T_i(a|_k^{\ell,i})$.

As T computes f with error ϵ, there exists some $j \in \mathbb{F}_q$, s.t. when restricting to $\{a \in \mathbb{F}_q^n \mid \ell(a) = j\}$, T_j computes f with error $\leq \epsilon$. Fix such T_j, and consider T_j'. We claim that T_j' computes f with error no more that $\epsilon + (q-1) \operatorname{Inf}_k(f)$.

To see this, for $i \in \mathbb{F}_q$, $i \neq j$, define

$$A|_k^{\ell,j}(f, i) = \Pr_{a \in \mathbb{F}_q^n, \ell(a)=i}(f(a) \neq f(a|_k^{\ell,j})).$$

It is obvious that T_j' computes f with error $\leq \epsilon + 1/q \cdot (\sum_{i \in \mathbb{F}_q, i \neq j} A|_k^{\ell,j}(f, i))$. Now we verify that $1/q \cdot (\sum_{i \in \mathbb{F}_q, i \neq j} A|_k^{\ell,j}(f, i)) \leq (q-1) \operatorname{Inf}_k(f)$. Fix $a = (a_1, \ldots, a_n)$ from $\{a \in \mathbb{F}_q^n \mid \ell(a) = j\}$. Then the contribution of $(a_1, \ldots, a_{k-1}, a_{k+1}, \ldots, a_n)$ in $1/q \cdot (\sum_{i \in \mathbb{F}_q, i \neq j} A|_k^{\ell,j}(f, i))$ is $\frac{1}{q} \cdot \frac{1}{q^{n-1}} \cdot s$, where $s \in \{0, \ldots, q-1\}$ is the number of field elements b s.t. $f(a_1, \ldots, a_{k-1}, b, a_{k+1}, \ldots, a_n) \neq f(a_1, \ldots, a_n)$. On the other hand, its contribution in $(q-1) \cdot \operatorname{Inf}_k(f)$ is $(q-1) \cdot \frac{1}{q^{n-1}} \cdot \frac{s(q-s)}{\binom{q}{2}}$. Finally note that $\frac{s}{(q-1)q} \leq \frac{s(q-s)}{\binom{q}{2}}$ for $q \geq 2$ and $s \in \{0, \ldots, q-1\}$.

As eliminating the first query introduces an extra error of at most $(q-1) \operatorname{Inf}_{\max}(f)$, similar to the argument in proving Theorem 1, we have $\epsilon + (q-1) D^{\oplus_q}(f) \cdot \operatorname{Inf}_{\max}(f) \geq \operatorname{Var}(f)$, therefore proving that

$$\operatorname{Inf}_{\max}(f) \geq \frac{1}{q-1} \cdot \frac{\operatorname{Var}(f) - \epsilon}{D^{\oplus_q}(f)}.$$

Acknowledgements. We thank Rahul Jain, Supartha Poddar, Miklos Santha, and Avishay Tal for several helpful discussions. We also thank Ben vee Volk for pointing out that the super-linear separation in [27] works for PDTs as well.

References

1. Aaronson, S., Ambainis, A.: The need for structure in quantum speedups. In: ICS 2011, pp. 338–352 (2011)
2. Ada, A., Fawzi, O., Hatami, H.: Spectral norm of symmetric functions. In: Gupta, A., Jansen, K., Rolim, J., Servedio, R. (eds.) APPROX 2012 and RANDOM 2012. LNCS, vol. 7408, pp. 338–349. Springer, Heidelberg (2012)
3. Babai, L., Banerjee, A., Kulkarni, R., Naik, V.: Evasiveness and the distribution of prime numbers. In: STACS 2010, pp. 71-82 (2010)
4. Bhrushundi, A., Chakraborty, S., Kulkarni, R.: Property testing bounds for linear and quadratic functions via parity decision trees. In: Hirsch, E.A., Kuznetsov, S.O., Pin, J.É., Vereshchagin, N.K. (eds.) CSR 2014. LNCS, vol. 8476, pp. 97–110. Springer, Heidelberg (2014). Electronic colloquium on Computational Complexity (ECCC)
5. Benjamini, I., Kalai, G., Schramm, O.: Noise sensitivity of boolean functions and its application to percolation. Inst. Hautes Etudes Sci. Publ. Math. **90**, 5–43 (1999)
6. Ben- Or, M., Linial, N.: Collective coin flipping. In: Proceedings of the 26th FOCS, pp. 408–416 (1985)
7. Bollobas, B.: Combinatorics: Set Systems, Hypergraphs, Families Of Vectors And Combinatorial Probability. Cambridge University Press, New York (1986)
8. Buhrman, H., de Wolf, R.: Complexity measures and decision tree complexity: a survey. Theor. Comput. Sci. **288**(1), 21–43 (2002)
9. Efron, B., Stein, C.: The jackknife estimate of variance. Ann. Stat. **9**, 586–596 (1981)
10. Gopalan, P., O'Donnell, R., Servedio, R.A., Shpilka, A., Wimmer, K.: Testing fourier dimensionality and sparsity. In: Albers, S., Marchetti-Spaccamela, A., Matias, Y., Nikoletseas, S., Thomas, W. (eds.) ICALP 2009, Part I. LNCS, vol. 5555, pp. 500–512. Springer, Heidelberg (2009)
11. Hayes, T.P., Kutin, S., van Melkebeek, D.: The quantum black-box complexity of majority. algorithmica **34**(4), 480–501 (2002)
12. Hatami, P., Kulkarni, R., Pankratov, D.: Variations on the sensitivity conjecture. Theor. Comput. Grad. Surv. **2**, 1–27 (2011)
13. Jain, R., Zhang, S.: The influence lower bound via query elimination. Theor. Comput. **7**(1), 147–153 (2011)
14. Kulkarni, R.: Evasiveness through a circuit lens. In: ITCS 2013 pp. 139–144 (2013)
15. Kulkarni, R.: Gems in decision tree complexity revisited. SIGACT News **44**(3), 42–55 (2013)
16. Kahn, J., Kalai, G., Linial, N.: The influence of variables on boolean functions (extended abstract). In: FOCS 1988, pp. 68–80 (1988)
17. Kushilevitz, E., Mansour, Y.: Learning decision trees using the fourier spectrum. SIAM J. Comput. **22**(6), 1331–1348 (1993)
18. Kulkarni, R., Santha, M.: Query complexity of matroids. In: Spirakis, P.G., Serna, M. (eds.) CIAC 2013. LNCS, vol. 7878, pp. 300–311. Springer, Heidelberg (2013)
19. Kahn, J., Saks, M.E., Sturtevant, D.: A topological approach to evasiveness. Combinatorica **4**(4), 297–306 (1984)
20. Lee, H.K.: Decision trees and influence: an inductive proof of the OSSS inequality. Theor. Comput. **6**(1), 81–84 (2010)
21. Linial, N., Mansour, Y., Nisan, N.: Constant depth circuits, fourier transform, and learnability. J. ACM **40**(3), 607–620 (1993)

22. Lovasz, L., Young, N. E.: Lecture Notes on Evasiveness of Graph Properties arXiv:cs/020503 (2002)
23. Montanaro, A., Osborne, T.: On the communication complexity of XOR functions. CoRR abs/0909.3392 (2009)
24. Mehlhorn, K., Schmidt, E.: Las Vegas is better than determinism in VLSI and distributed computing. In: Proceedings of the 14th STOC, pp. 330–337. ACM Press, New York (1982)
25. Nisan, N.: CREW PRAMs and decision trees. In: Proceedings of the 21st STOC, pp. 327–335. ACM Press, New York (1989)
26. Nisan, N., Szegedy, M.: On the degree of boolean functions as real polynomials. Comput. Complex. 4, 301–313 (1994)
27. Nisan, N., Wigderson, A.: On rank vs. communication complexity. Combinatorica 15(4), 557–565 (1995)
28. O'Donnell, R., Servedio, R.A.: Learning monotone decision trees in polynomial time. SIAM J. Comput. 37(3), 827–844 (2007)
29. O'Donnell, R., Saks, M.E., Schramm, O., Servedio, R.A.: Every decision tree has an influential variable. In: FOCS, pp. 31-39 (2005)
30. Sherstov, A.A.: Making polynomials robust to noise. In: STOC 2012, pp. 747–758 (2012)
31. Shi, Y., Zhang, Z.: Communication Complexities of XOR functions CoRR abs/0808.1762 (2008)
32. Shpilka, A., Tal, A., Volk, B.L.: On the Structure of Boolean Functions with Small Spectral Norm: arXiv:1304.0371
33. Saks, M.E., Wigderson, A.: Probabilistic boolean decision trees and the complexity of evaluating game trees. In: FOCS, pp. 29–38 (1986)
34. Zhang, Z., Shi, Y.: On the parity complexity measures of boolean functions. Theor. Comput. Sci. 411(26–28), 2612–2618 (2010)
35. Talagrand, M.: On russo's approximate 0-1 law. Ann. Probab. 22(3), 1576–1587 (1994)
36. Tsang, H.Y., Wong, C.H., Xie, N., Zhang, S.: Fourier sparsity, spectral norm, and the Log-rank conjecture. CoRR abs/1304.1245 (2013) FOCS (2014)

Card-Based Protocols for Any Boolean Function

Takuya Nishida[1], Yu-ichi Hayashi[1], Takaaki Mizuki[2(✉)],
and Hideaki Sone[2]

[1] Graduate School of Information Sciences, Tohoku University,
6–3–09 Aramaki-Aza-Aoba, Aoba-ku, Sendai, Miyagi 980–8578, Japan
[2] Cyberscience Center, Tohoku University, 6–3 Aramaki-Aza-Aoba,
Aoba-ku, Sendai, Miyagi 980–8578, Japan
tm-paper+cardany@g-mail.tohoku-university.jp

Abstract. Card-based protocols that are based on a deck of physical cards achieve secure multi-party computation with information-theoretic secrecy. Using existing AND, XOR, NOT, and copy protocols, one can naively construct a secure computation protocol for any given (multivariable) Boolean function as long as there are plenty of additional cards. However, an explicit sufficient number of cards for computing any function has not been revealed thus far. In this paper, we propose a general approach to constructing an efficient protocol so that six additional cards are sufficient for any function to be securely computed. Further, we prove that two additional cards are sufficient for any symmetric function.

1 Introduction

It is known that secure multi-party computation (MPC) can be achieved using a number of physical cards such as black ♣ and red ♡ cards (with identical backs ?). Several card-based cryptographic protocols have been reported in the literature: in addition to the elementary computations, namely, the AND [1,3,7,10,12,15] and XOR [3,10,11] protocols, *efficient* protocols (i.e., protocols that require fewer cards) have been designed for specific functions such as the adder [6] and the 3-variable functions [13]. Whereas previous studies have dealt with specific functions, this paper proposes a general approach to constructing an efficient protocol for any given (multivariable) Boolean function.

We start with introducing some preliminary notations for card-based protocols.

1.1 Preliminary Notations

To deal with Boolean values, we use the following encoding rule based on the order of a pair of cards:

$$♣\,♡ = 0, \quad ♡\,♣ = 1. \tag{1}$$

© Springer International Publishing Switzerland 2015
R. Jain et al. (Eds.): TAMC 2015, LNCS 9076, pp. 110–121, 2015.
DOI: 10.1007/978-3-319-17142-5_11

For a bit $x \in \{0,1\}$, a pair of face-down cards $\boxed{?}\boxed{?}$ that has a value equaling x according to encoding rule (1) is called a *commitment* to x and is written as

$$\underbrace{\boxed{?}\boxed{?}}_{x}.$$

"Committed-format" protocols [3,6,10–13,15] produce the output as a commitment; for example, given commitments to bits a and b, we can obtain a commitment

$$\underbrace{\boxed{?}\boxed{?}}_{a \wedge b} \quad \text{or} \quad \underbrace{\boxed{?}\boxed{?}}_{a \oplus b}$$

as the output of an AND or XOR protocol.

Given a pair of bits (x,y), we define two operations, get and shift, as

$$\mathsf{get}^0(x,y) = x, \quad \mathsf{get}^1(x,y) = y;$$
$$\mathsf{shift}^0(x,y) = (x,y), \quad \mathsf{shift}^1(x,y) = (y,x).$$

Using these operations, the AND function can be written as

$$a \wedge b = \mathsf{get}^{a \oplus r}(\mathsf{shift}^r(0,b)) \tag{2}$$

for an arbitrary bit $r \in \{0,1\}$ [13]. Hereafter, for two bits x and y, the notation (i) below implies (ii).

$$\text{(i)} \underbrace{\boxed{?}\boxed{?}\boxed{?}\boxed{?}}_{(x,y)}, \quad \text{(ii)} \underbrace{\boxed{?}\boxed{?}}_{x}\underbrace{\boxed{?}\boxed{?}}_{y}.$$

1.2   AND Protocol

Next, we introduce the most efficient AND protocol [10] currently known. Given commitments to bits a and b together with two additional cards, it achieves a committed-format AND computation as follows.

1. Arrange three commitments to a, 0, and b:

$$\underbrace{\boxed{?}\boxed{?}}_{a}\boxed{\clubsuit}\boxed{\heartsuit}\underbrace{\boxed{?}\boxed{?}}_{b} \rightarrow \underbrace{\boxed{?}\boxed{?}}_{a}\underbrace{\boxed{?}\boxed{?}}_{0}\underbrace{\boxed{?}\boxed{?}}_{b}.$$

2. Rearrange the sequence of six cards as

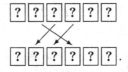

3. Bisect the sequence of six cards and switch the two portions (each of which consists of three cards) randomly; we call this a *random bisection cut* [10] and

denote it by $[\cdot\,|\,\cdot]$:

$$\left[\boxed{?}\,\boxed{?}\,\boxed{?}\,\Big\|\,\boxed{?}\,\boxed{?}\,\boxed{?}\,\right] \rightarrow \boxed{?}\,\boxed{?}\,\boxed{?}\,\boxed{?}\,\boxed{?}\,\boxed{?}\,.$$

4. Rearrange the sequence as

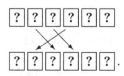

Then, we have

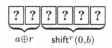

where r is a (uniformly distributed) random bit because of the random bisection cut.

5. Reveal the two left-most cards; then, the value of $a \oplus r$ along with Eq. (2) gives us the position of the desired commitment to $a \wedge b$:

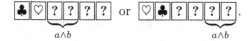

Since r is random, revealing the commitment to $a \oplus r$ does not cause any information about bit a to be leaked, and hence, this protocol achieves an information-theoretically secure computation[1]. Note that the two revealed cards can be used for another computation (we call such an available card a *free* card).

1.3 Copy Protocol

Given a commitment to bit a together with four additional cards, we can make two copied commitments to a [10] as follows.

1. Arrange three commitments to a, 0, and 0.

$$\underbrace{\boxed{?}\,\boxed{?}}_{a}\,\boxed{\clubsuit}\,\boxed{\heartsuit}\,\boxed{\clubsuit}\,\boxed{\heartsuit} \rightarrow \underbrace{\boxed{?}\,\boxed{?}}_{a}\,\underbrace{\boxed{?}\,\boxed{?}}_{0}\,\underbrace{\boxed{?}\,\boxed{?}}_{0}\,.$$

2. Rearrange the sequence, apply a random bisection cut, and rearrange it again:

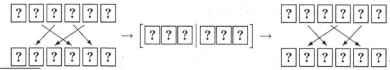

---

[1] Security is dependent on physical properties such as cards of the same color being indistinguishable and a random bisection cut being applied *truly randomly*. A formal treatment appears in [8], and the settings of this study are based on the formalization of card-based protocols. It is also known that one can practically assume a semi-honest model, i.e., a protocol is always executed properly [9].

Then, we have

$$a \oplus r \quad 0 \oplus r \quad 0 \oplus r$$

where r is a random bit.

3. Reveal the two left-most cards; then, we know whether $r = a$ or $r = \bar{a}$, and we have

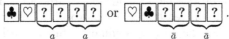

Hence we obtain two commitments to a.

Note that swapping the two cards that constitute a commitment to a bit results in a commitment to the negation of the bit (recall encoding rule (1)), i.e., the NOT computation is trivial. Therefore, hereafter, we omit detailed descriptions of how a commitment to negation $\bar{x}$ is transformed into a commitment to x.

If we start this protocol with commitments to a, b, and 0 in step 1 instead, commitments to $a \oplus b$ and a will be obtained [13].

Similarly, given commitments to bits a and b, we easily obtain

$$a \oplus r \quad b \oplus r$$

and hence the existing XOR protocol [10] produces a commitment to $a \oplus b$ without the use of any additional card.

1.4 Our Results

The existing AND, XOR, and NOT protocols introduced thus far immediately imply the following theorem.

Theorem 1 ([10]). *Given commitments to x_1 and x_2 together with two additional cards ♣♡, we can securely produce a commitment to the value of any 2-variable Boolean function $f(x_1, x_2)$.*

It is also known that the following holds.

Theorem 2 ([13]). *Given commitments to x_1, x_2, x_3 together with two additional cards ♣♡, we can securely produce a commitment to the value of any 3-variable Boolean function $f(x_1, x_2, x_3)$.*

These two results raise a natural question: what about the case of any general Boolean function having four or more variables? Of course, by combining the existing AND, XOR, NOT, and copy protocols, one can securely compute any (multivariable) Boolean function $f(x_1, x_2, \ldots, x_n)$ as long as there are plenty of additional cards. However, an explicit sufficient number of cards for computing any function has not been revealed thus far. We investigate this open problem and propose a general approach to constructing an efficient protocol, showing sufficient conditions on the numbers of additional cards.

The remainder of this paper is organized as follows. In Sect. 2, we improve the existing AND and half-adder protocols. In Sect. 3, using the improved AND protocol, we demonstrate the construction of a protocol that securely computes any given n-variable Boolean function with n input commitments and six additional cards, i.e., we prove that six additional cards are sufficient for this case. In Sect. 4, using our improved half-adder protocol, we show that two additional cards are sufficient for the case of symmetric functions. Finally, the paper is concluded in Sect. 5.

2   Building Blocks

In this section, we create two new protocols as building blocks for the main results (presented in Sects. 3 and 4) by modifying the known AND protocol [10] introduced in Sect. 1.2. The first new protocol produces a commitment to $a \wedge b$ as well as a commitment to b, as described in Sect. 2.1. The second one achieves half-adder computation with only two additional cards, as described in Sect. 2.2.

2.1   Improved AND Protocol

Recall the AND protocol [10] introduced in Sect. 1.2. When a commitment to $a \wedge b$ is obtained as the output of the protocol, the other two face-down cards will constitute a commitment to $\bar{a} \wedge b$, as known from Eq. (2):

$$a \wedge b \quad \bar{a} \wedge b \qquad \bar{a} \wedge b \quad a \wedge b$$

From this observation and the identity

$$ab \oplus \bar{a}b = (a \oplus \bar{a})\, b = b$$

(where we omit the conjunction symbol $\wedge$ hereafter), we can improve the AND protocol so that one of the input commitments will be retained, as follows.

1. Arrange three commitments to a, 0, and b.

$$\underbrace{\qquad}_{a} \qquad \underbrace{\qquad}_{b} \qquad \rightarrow \qquad \underbrace{\qquad}_{a} \quad \underbrace{\qquad}_{0} \quad \underbrace{\qquad}_{b}$$

2. Apply the known AND protocol [10]; then, we have

$$ab \quad \bar{a}b \qquad \bar{a}b \quad ab$$

3. Rearrange the sequence as

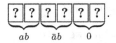

4. Apply steps 2 and 3 of the copy protocol [10] introduced in Sect. 1.3; then, we have

Since $ab \oplus \bar{a}b = b$, we have

Thus, this protocol allows us to retain a commitment to b. Therefore, the following lemma holds.

Lemma 3. *Given commitments to x_1 and x_2 together with two additional cards* ♣♡, *we can securely produce commitments to $x_1 x_2$ and x_2.*

2.2   Improved Half-Adder Protocol

It is known that half-adder computation can be achieved with eight cards [6], i.e., given commitments to a and b together with four additional cards, the existing protocol produces commitments to $a \oplus b$ and ab. In this subsection, we improve the half-adder protocol by applying the improved AND protocol described in the previous subsection. Our half-adder protocol requires only two additional cards and proceeds as follows.

1. Arrange three commitments to a, b, and 0.

2. Apply steps 2 and 3 of the copy protocol [10] introduced in Sect. 1.3; then, we have

3. Rearrange the sequence as

4. Apply the improved AND protocol described in the previous subsection; then, we have

Since $a\,(\overline{a \oplus b}) = a\bar{a} \oplus ab = ab$, we have

Thus, this protocol achieves half-adder computation using only two additional cards. Therefore, the following lemma holds.

Lemma 4. *Given commitments to x_1 and x_2 together with two additional cards* ♣♡, *we can securely produce commitments to $x_1 \oplus x_2$ and $x_1 x_2$.*

3 Computation of Any Multivariable Function

In this section, we present a general approach to constructing an efficient protocol for any given n-variable Boolean function by showing that any n-variable function can be securely computed with n input commitments and six additional cards.

3.1 Concepts and Sub-Protocol

Remember that XOR computation can be easily achieved [10] as described in Sect. 1.3. Hence, XOR computation should be employed to construct an efficient protocol. Therefore, we consider AND-XOR expressions of a given function. Indeed, it is well known that any n-variable function $f(x_1, x_2, \ldots, x_n)$ can be expressed as the Shannon expansion (or Boole's expansion) [14]:

$$f(x_1, x_2, \ldots, x_n) = \bar{x}_1 \bar{x}_2 \cdots \bar{x}_n f(0, 0, \ldots, 0) \oplus x_1 \bar{x}_2 \cdots \bar{x}_n f(1, 0, \ldots, 0)$$
$$\oplus\, \bar{x}_1 x_2 \cdots \bar{x}_n f(0, 1, \ldots, 0) \oplus x_1 x_2 \cdots \bar{x}_n f(1, 1, \ldots, 0)$$
$$\oplus \cdots \oplus x_1 x_2 \cdots x_n f(1, 1, \ldots, 1).$$

i.e., $f(x_1, x_2, \ldots, x_n)$ can be expressed uniquely by combining 2^n product terms with XORs, where a product term can be deleted if the corresponding value of f is 0.

Now, we want to handle product terms $v_1 v_2 \cdots v_n$, where v_i, $1 \leq i \leq n$, is a literal (either x_i or $\bar{x}_i$): given commitments to $v_1, v_2, \ldots, v_n$ together with four additional cards, the following sub-protocol securely generates a commitment to the product term $v_1 v_2 \cdots v_n$.

1. Make two copied commitments to v_1 using the copy protocol [10] introduced in Sect. 1.3:

2. Apply the AND protocol described in Sect. 2.1; then, by Lemma 3 we have

$$\underbrace{\boxed{?}\,\boxed{?}}_{v_1}\,\underbrace{\boxed{?}\,\boxed{?}}_{v_2}\,\boxed{\clubsuit}\,\boxed{\heartsuit}\,\underbrace{\boxed{?}\,\boxed{?}}_{v_1 v_2}\,\underbrace{\boxed{?}\,\boxed{?}}_{v_3}\cdots\underbrace{\boxed{?}\,\boxed{?}}_{v_n}.$$

3. Similarly, by Lemma 3 we have

$$\underbrace{\boxed{?}\,\boxed{?}}_{v_1}\,\underbrace{\boxed{?}\,\boxed{?}}_{v_2}\,\underbrace{\boxed{?}\,\boxed{?}}_{v_3}\,\boxed{\clubsuit}\,\boxed{\heartsuit}\,\underbrace{\boxed{?}\,\boxed{?}}_{v_1 v_2 v_3}\,\underbrace{\boxed{?}\,\boxed{?}}_{v_4}\cdots\underbrace{\boxed{?}\,\boxed{?}}_{v_n}.$$

4. Repeat this up to v_n so that we have

$$\underbrace{\boxed{?}\,\boxed{?}}_{v_1}\,\underbrace{\boxed{?}\,\boxed{?}}_{v_2}\cdots\underbrace{\boxed{?}\,\boxed{?}}_{v_n}\,\boxed{\clubsuit}\,\boxed{\heartsuit}\,\underbrace{\boxed{?}\,\boxed{?}}_{v_1 v_2\cdots v_n}.$$

Thus, four additional cards allow us to generate a commitment to the product term without losing the input commitments.

Lemma 5. *Given commitments to literals $v_1, v_2, \ldots, v_n$ together with four additional cards* $\boxed{\clubsuit}\,\boxed{\clubsuit}\,\boxed{\heartsuit}\,\boxed{\heartsuit}$, *we can securely produce commitments to $v_1, v_2, \ldots, v_n$, and $v_1 v_2 \cdots v_n$.*

3.2   Complete Description of Protocol

Now, we are ready to present our general protocol for securely computing any function.

Let f be an arbitrary n-variable function. Given n commitments

$$\underbrace{\boxed{?}\,\boxed{?}}_{x_1}\,\underbrace{\boxed{?}\,\boxed{?}}_{x_2}\cdots\underbrace{\boxed{?}\,\boxed{?}}_{x_n}$$

and six additional cards $\boxed{\clubsuit}\,\boxed{\clubsuit}\,\boxed{\clubsuit}\,\boxed{\heartsuit}\,\boxed{\heartsuit}\,\boxed{\heartsuit}$, the following protocol securely produces a commitment to $f(x_1, x_2, \ldots, x_n)$.

1. Let $T_1 \oplus T_2 \oplus \cdots \oplus T_\ell$ be the Shannon expansion of f after removing the constant-zero terms (where T_i, $1 \le i \le \ell$, is a product term). Generate a commitment to T_1 using the sub-protocol described in Sect. 3.1. Then, by Lemma 5 we have

$$\underbrace{\boxed{?}\,\boxed{?}}_{x_1}\,\underbrace{\boxed{?}\,\boxed{?}}_{x_2}\cdots\underbrace{\boxed{?}\,\boxed{?}}_{x_n}\,\boxed{\clubsuit}\,\boxed{\heartsuit}\,\boxed{\clubsuit}\,\boxed{\heartsuit}\,\underbrace{\boxed{?}\,\boxed{?}}_{T_1}.$$

2. Generate a commitment to T_2 by Lemma 5:

$$\underbrace{\boxed{?}\,\boxed{?}}_{x_1}\,\underbrace{\boxed{?}\,\boxed{?}}_{x_2}\cdots\underbrace{\boxed{?}\,\boxed{?}}_{x_n}\,\boxed{\clubsuit}\,\boxed{\heartsuit}\,\underbrace{\boxed{?}\,\boxed{?}}_{T_1}\,\underbrace{\boxed{?}\,\boxed{?}}_{T_2}.$$

3. Apply the XOR protocol [10] to the two right-most commitments:

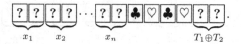

4. Generate a commitment to T_3 by Lemma 5:

$$\boxed{?}\boxed{?}\boxed{?}\boxed{?}\cdots\boxed{?}\boxed{?}\boxed{\clubsuit}\boxed{\heartsuit}\boxed{?}\boxed{?}\boxed{?}\boxed{?}.$$
$$\underbrace{\quad}_{x_1}\ \underbrace{\quad}_{x_2}\quad\underbrace{\quad}_{x_n}\quad\underbrace{\quad}_{T_1\oplus T_2}\ \underbrace{\quad}_{T_3}$$

5. Apply the XOR protocol [10] to the two right-most commitments:

$$\boxed{?}\boxed{?}\boxed{?}\boxed{?}\cdots\boxed{?}\boxed{?}\boxed{\clubsuit}\boxed{\heartsuit}\boxed{\clubsuit}\boxed{\heartsuit}\quad\boxed{?}\boxed{?}\ .$$
$$\underbrace{\quad}_{x_1}\ \underbrace{\quad}_{x_2}\quad\underbrace{\quad}_{x_n}\qquad\underbrace{\quad}_{T_1\oplus T_2\oplus T_3}$$

6. Repeat this until we get a commitment to $T_1 \oplus T_2 \oplus \cdots \oplus T_\ell$, which is equal to $f(x_1, x_2, \ldots, x_n)$:

$$\boxed{?}\boxed{?}\boxed{?}\boxed{?}\cdots\boxed{?}\boxed{?}\boxed{\clubsuit}\boxed{\heartsuit}\boxed{\clubsuit}\boxed{\heartsuit}\quad\boxed{?}\boxed{?}\ .$$
$$\underbrace{\quad}_{x_1}\ \underbrace{\quad}_{x_2}\quad\underbrace{\quad}_{x_n}\qquad\underbrace{\quad}_{f(x_1,x_2,\ldots,x_n)}$$

The remaining commitments to $x_1, x_2, \ldots, x_n$ as well as the four free cards $\boxed{\clubsuit}\boxed{\clubsuit}\boxed{\heartsuit}\boxed{\heartsuit}$ can be used for another computation. Thus, we have the following theorem.

Theorem 6. *Let f be an n-variable function. Given commitments to $x_1, x_2, \ldots,$ x_n together with six additional cards* $\boxed{\clubsuit}\boxed{\clubsuit}\boxed{\clubsuit}\boxed{\heartsuit}\boxed{\heartsuit}\boxed{\heartsuit}$, *we can securely produce commitments to $x_1, x_2, \ldots, x_n$, and the value $f(x_1, x_2, \ldots, x_n)$.*

Note that our improved AND protocol (Lemma 3) plays an important role in reducing the number of required cards; without it, two more cards would be required to run the sub-protocol, and consequently, the above protocol. Furthermore, although we used the Shannon expansion in step 1, one may use any AND-XOR expression instead, which can be obtained by applying some simplification algorithm [14].

4   Case of Symmetric Functions

The previous section described the construction of a protocol that securely produces a commitment to the value of any function using n input commitments and six additional cards. In this section, we focus our attention on symmetric functions because practically important functions in MPC are often symmetric. Specifically, we prove that two additional cards are sufficient for the case of symmetric functions.

Let f be an n-variable symmetric function. Then, the value $f(x_1, x_2, \ldots, x_n)$ depends on only the number of variables that take 1, namely $\sum_{i=1}^{n} x_i$. For instance, the 3-input majority MAJ_3, which is a 3-variable symmetric function, can be expressed using a function $g : \{0,1,2,3\} \to \{0,1\}$ as

$$\mathsf{MAJ}_3(x_1, x_2, x_3) = g\left(\sum x_i\right) = \begin{cases} 0 & \text{if } \sum x_i \leq 1 \\ 1 & \text{otherwise.} \end{cases}$$

Thus, for any n-variable symmetric function $f : \{0,1\}^n \to \{0,1\}$, there exists a unique function $g : \{0,1,\ldots,n\} \to \{0,1\}$ such that $f(x_1, x_2, \ldots, x_n) = g\left(\sum x_i\right)$.

Given commitments to $x_1, x_2, \ldots, x_n$, it is obvious that the half-adder computation described in Sect. 2.2 enables us to securely generate a sequence of commitments corresponding to the binary representation of $\sum x_i$; Lemma 4 implies that two additional cards are sufficient for this purpose.

Lemma 7. *Given commitments to $x_1, x_2, \ldots, x_n$ together with two additional cards* ♣♡*, we can securely produce a $(\lfloor \log_2 n \rfloor + 1)$-bit sequence of commitments corresponding to $\sum_{i=1}^{n} x_i$.*

Note that after generating a sequence of commitments to $\sum x_i$ from commitments to $x_1, x_2, \ldots, x_n$ (and two additional cards), some free cards will arise; more specifically, we will have a total of $2\left(n - \lfloor \log_2 n \rfloor\right)$ free cards.

Now, we are ready to present our main result of this section, i.e., only two additional cards are sufficient for the case of symmetric functions.

Theorem 8. *Let $n \geq 4$ and let f be an n-variable symmetric function. Given commitments to $x_1, x_2, \ldots, x_n$ together with two additional cards* ♣♡*, we can securely produce a commitment to the value $f(x_1, x_2, \ldots, x_n)$.*

Proof. Let $g : \{0,1,\ldots,n\} \to \{0,1\}$ be the function such that $g\left(\sum x_i\right) = f(x_1, x_2, \ldots, x_n)$. By Lemma 7, we obtain a $(\lfloor \log_2 n \rfloor + 1)$-bit sequence of commitments corresponding to $\sum x_i$ and $2\left(n - \lfloor \log_2 n \rfloor\right)$ free cards. If $n \geq 5$, then $2\left(n - \lfloor \log_2 n \rfloor\right) \geq 6$, and hence there are at least 6 free cards, and consequently, we can securely generate a commitment to the value $g\left(\sum x_i\right)$ by Theorem 6 (by regarding the domain of g as $\{0,1\}^{\lfloor \log_2 n \rfloor + 1}$). Similarly, if $n = 4$, then there are a 3-bit sequence of commitments and 4 free cards, and hence Theorem 2 completes the proof.   □

5   Conclusion

We proposed a general approach to designing an efficient card-based protocol for any given function. Specifically, using two-level AND-XOR representations, we can construct a protocol that requires only six additional cards to securely produce a commitment to the value of any n-variable function, regardless of how large n is (Theorem 6). Further, we showed that two additional cards are sufficient for the case of symmetric functions (Theorem 8).

As mentioned above, six additional cards are sufficient for general functions, and two additional cards are sufficient for symmetric functions. Determining whether they are necessary is an open problem; for example, is there a symmetric function that needs at least two additional cards? Note that to prove such a lower bound, one has to follow the formal computational model for card-based protocols [8].

Cryptography and playing cards share a deep connection (e.g., [2,4,5,16]). One benefit of considering such a connection is that it enables us to easily demonstrate the underlying concepts of MPC and cryptography to non-specialists. In addition, we have already confirmed that ordinary people such as high-school students can use card-based protocols in their daily activities.

Acknowledgments. This work was supported by JSPS KAKENHI Grant Number 26330001.

References

1. den Boer, B.: More efficient match-making and satisfiability: the five card trick. In: Quisquater, J.-J., Vandewalle, J. (eds.) Advances in Cryptology- EUROCRYPT 1989. LNCS, vol. 434, pp. 208–217. Springer, Heidelberg (1990)
2. Cordón-Franco, A., Van Ditmarsch, H., Fernández-Duque, D., Soler-Toscano, F.: A colouring protocol for the generalized Russian cards problem. Theor. Comput. Sci. **495**, 81–95 (2013)
3. Crépeau, C., Kilian, J.: Discreet solitary games. In: Stinson, D.R. (ed.) Advances in Cryptology-CRYPTO 1993. LNCS, vol. 773, pp. 319–330. Springer, Heidelberg (1994)
4. Duan, Z., Yang, C.: Unconditional secure communication: a Russian cards protocol. J. Comb. Optim. **19**(4), 501–530 (2010)
5. Fischer, M.J., Wright, R.N.: Bounds on secret key exchange using a random deal of cards. J. Cryptology **9**(2), 71–99 (1996)
6. Mizuki, T., Asiedu, I.K., Sone, H.: Voting with a logarithmic number of cards. In: Mauri, G., Dennunzio, A., Manzoni, L., Porreca, A.E. (eds.) UCNC 2013. LNCS, vol. 7956, pp. 162–173. Springer, Heidelberg (2013)
7. Mizuki, T., Kumamoto, M., Sone, H.: The five-card trick can be done with four cards. In: Wang, X., Sako, K. (eds.) ASIACRYPT 2012. LNCS, vol. 7658, pp. 598–606. Springer, Heidelberg (2012)
8. Mizuki, T., Shizuya, H.: A formalization of card-based cryptographic protocols via abstract machine. Int. J. Inf. Secur. **13**(1), 15–23 (2014)
9. Mizuki, T., Shizuya, H.: Practical card-based cryptography. In: Ferro, A., Luccio, F., Widmayer, P. (eds.) FUN 2014. LNCS, vol. 8496, pp. 313–324. Springer, Heidelberg (2014)
10. Mizuki, T., Sone, H.: Six-card secure and and four-card secure xor. In: Deng, X., Hopcroft, J.E., Xue, J. (eds.) FAW 2009. LNCS, vol. 5598, pp. 358–369. Springer, Heidelberg (2009)
11. Mizuki, T., Uchiike, F., Sone, H.: Securely computing XOR with 10 cards. Australas. J. Comb. **36**, 279–293 (2006)
12. Niemi, V., Renvall, A.: Secure multiparty computations without computers. Theor. Comput. Sci. **191**(1–2), 173–183 (1998)
13. Nishida, T., Hayashi, Y., Mizuki, T., Sone, H.: Securely computing three-input functions with eight cards. IEICE Trans. Fundam. Electron., Commun. Comput. Sci. **E98-A**(6) (2015, to appear)

14. Sasao, T.: Switching Theory for Logic Synthesis, 1st edn. Kluwer Academic Publishers, Norwell (1999)
15. Stiglic, A.: Computations with a deck of cards. Theor. Comput. Sci. **259**(1–2), 671–678 (2001)
16. Swanson, C.M., Stinson, D.R.: Combinatorial solutions providing improved security for the generalized Russian cards problem. Designs, Codes and Cryptography **72**(2), 345–367 (2014)

Size of Sets with Small Sensitivity: A Generalization of Simon's Lemma

Andris Ambainis and Jevgēnijs Vihrovs[⊠]

Faculty of Computing, University of Latvia, Raiņa bulv. 19, Riga LV-1586, Latvia
jevgenijs.vihrovs@lu.lv

Abstract. We study the structure of sets $S \subseteq \{0,1\}^n$ with small sensitivity. The well-known Simon's lemma says that any $S \subseteq \{0,1\}^n$ of sensitivity s must be of size at least 2^{n-s}. This result has been useful for proving lower bounds on the sensitivity of Boolean functions, with applications to the theory of parallel computing and the "sensitivity vs. block sensitivity" conjecture.

In this paper we take a deeper look at the size of such sets and their structure. We show an unexpected "gap theorem": if $S \subseteq \{0,1\}^n$ has sensitivity s, then we either have $|S| = 2^{n-s}$ or $|S| \geq \frac{3}{2} 2^{n-s}$.

This provides new insights into the structure of low sensitivity subsets of the Boolean hypercube $\{0,1\}^n$.

1 Introduction

The complexity of computing Boolean functions (for example, in the decision tree model of computation) is related to a number of combinatorial quantities, such as the sensitivity and block sensitivity of the function, its certificate complexity and the degree of polynomials that represent the function exactly or approximately [5]. Study of these quantities has resulted in both interesting results and longstanding open problems.

For example, it has been shown that decision tree complexity in either a deterministic, a probabilistic or a quantum model of computation is polynomially related to a number of these quantities: certificate complexity, block sensitivity and the minimum degree of polynomials that represent or approximate f [4,9]. This result, in turn, implies that deterministic, probabilistic and quantum decision tree complexities are polynomially related — which is very interesting because a

The research leading to these results has received funding from the European Union Seventh Framework Programme (FP7/2007–2013) under projects QALGO (Grant Agreement No. 600700) and RAQUEL (Grant Agreement No. 323970), ERC Advanced Grant MQC and Latvian State Research programme NexIT project No.1. Part of this work was done while Andris Ambainis was visiting Institute for Advanced Study, Princeton, supported by National Science Foundation under agreement No. DMS-1128155. Any opinions, findings and conclusions or recommendations expressed in this material are those of the author(s) and do not necessarily reflect the views of the National Science Foundation.

R. Jain et al. (Eds.): TAMC 2015, LNCS 9076, pp. 122–133, 2015.
DOI: 10.1007/978-3-319-17142-5_12

similar result is not known in the Turing machine world; and, for deterministic vs. quantum complexity, is most likely false because of Shor's factoring algorithm.

The question about the relation between the sensitivity of a function and the other quantities is, however, a longstanding open problem, known as the "sensitivity vs. block sensitivity" question. Since the other quantities are all polynomially related, showing a polynomial relation between sensitivity and any one of them would imply a polynomial relation between sensitivity and all of them. This question, since first being posed by Nisan in 1991 [8], has attracted much attention but there has been quite little progress and the gap between the best upper and lower bounds remains huge. The examples that achieve the asymptotically biggest separation between the two quantities give $bs(f) = \Omega(s^2(f))$ [3,10,12], while the best upper bound on $bs(f)$ in terms of $s(f)$ is exponential: $bs(f) \leq s(f)2^{s(f)-1}$ [1,7]. Here $bs(f)$ and $s(f)$ denote the block sensitivity and the sensitivity of f, respectively.

In this paper we study the following question: assume that a subset S of the Boolean hypercube $\{0,1\}^n$ has low sensitivity: that is, for every $x \in S$ there are at most s indices $i \in \{1,\ldots,n\}$ such that changing x_i to the opposite value results in $y \notin S$. What can we say about this set?

Most of the upper bounds on $bs(f)$ in terms of $s(f)$ are based on Simon's lemma [11]. We say that a subset S of the Boolean hypercube $\{0,1\}^n$ has sensitivity s if, for every $x \in S$, there are at most s indices $i \in \{1,\ldots,n\}$ such that changing x_i to the opposite value results in $y \notin S$. Simon's lemma [11] says that any $S \subset \{0,1\}^n$ with sensitivity s must contain at least 2^{n-s} input vectors $x \in S$.

Simon [11] then used this result to show that $s(f) \geq \frac{1}{2}\log_2 n - \frac{1}{2}\log_2\log_2 n + \frac{1}{2}$ for any Boolean function that depends on n variables. Since $bs(f) \leq n$, this implies $bs(f) \leq s(f)4^{s(f)}$. This was the first upper bound on $bs(f)$ in terms of $s(f)$. A more recent upper bound of $bs(f) \leq s(f)2^{s(f)-1}$ by Ambainis et al. [1] is also based on Simon's lemma. If it was possible to improve Simon's lemma, this would result in better bounds on $bs(f)$.

However, Simon's lemma is known to be exactly optimal. Let S be a subcube of the hypercube $\{0,1\}^n$ obtained by fixing s of variables x_i. That is, S is the set of all $x = (x_1,\ldots,x_n)$ that satisfy $x_{i_1} = a_1$, ..., $x_{i_s} = a_s$ for some choice of distinct $i_1,\ldots,i_k \in \{1,\ldots,n\}$ and $a_1,\ldots,a_s \in \{0,1\}$. Then every $x \in S$ is sensitive to changing s bits $x_{i_1},\ldots,x_{i_k}$ and $|S| = 2^{n-s}$.

In this paper, we discover a direction in which Simon's lemma can be improved! Namely, we show that any S with sensitivity s that is not a subcube must be substantially larger. To do that, we study the structure of sets S with sensitivity s by classifying them into two types:

1. sets S that are contained in a subcube $S' \subset \{0,1\}^n$ obtained by fixing one or more of values x_i;
2. sets S that are not contained in any such subcube.

There is one-to-one correspondence between the sets of the first type and low-sensitivity subsets of $\{0,1\}^{n-k}$ for $k \in \{1,\ldots,s\}$.[1] In contrast, the sets of the second type do not reduce to low-sensitivity subsets of $\{0,1\}^{n-k}$ for $k > 0$. Therefore, we call them *irreducible*.

Our main technical result (Theorem 2) is that any irreducible $S \subseteq \{0,1\}^n$ must be of size $|S| \geq 2^{n-s+1} - 2^{n-2s}$, almost twice as large as a subcube obtained by fixing s variables, and this bound is tight.

As a consequence, we obtain a surprising result: if $S \subseteq \{0,1\}^n$ has sensitivity s, then either $|S| = 2^{n-s}$ or $|S| \geq \frac{3}{2}2^{n-s}$. That is, such a set S cannot have a size between 2^{n-s} and $\frac{3}{2}2^{n-s}$ (Theorem 3).

In a following work [2], we have applied this theorem to obtain a new upper bound on block sensitivity in terms of sensitivity:

$$bs(f) \leq \max\left(2^{s(f)-1}\left(s(f) - \frac{1}{3}\right), s(f)\right). \tag{1}$$

Related Work. A gap theorem of a similar type is known for the spectral norm of Boolean functions [6]: the spectral norm of a Boolean function is either equal to 1 or is at least $\frac{3}{2}$. Both results have the constant $\frac{3}{2}$ appearing in them and there is some resemblance between the constructions of optimal sets/functions but the proof methods are quite different and it is not clear to us if there is a more direct connection between the results.

2   Preliminaries

In this section we give the basic definitions used in the paper. Let $f : \{0,1\}^n \to \{0,1\}$ be a Boolean function of n variables, where the i-th variable is denoted by x_i. We use $x = (x_1,\ldots,x_n)$ to denote a tuple consisting of all input variables x_i.

Definition 1. *The* sensitivity complexity $s(f,x)$ *of* f *on an input* x *is defined as* $|\{i \mid f(x) \neq f(x^{(i)})\}|$, *where* $x^{(i)}$ *is an input obtained from* x *by flipping the value of the* i-th *variable. The* sensitivity $s(f)$ *of* f *is defined as*

$$s(f) = \max\{s(f,x) \mid x \in \{0,1\}^n\}. \tag{2}$$

The c-sensitivity $s_c(f)$ *of* f *is defined as*

$$s_c(f) = \max\{s(f,x) \mid x \in \{0,1\}^n, f(x) = c\}. \tag{3}$$

In this paper we will look at $\{0,1\}^n$ as a set of vertices for a graph Q_n (called the *n-dimensional Boolean cube* or *hypercube*) in which we have an edge (x,y) whenever $x = (x_1,\ldots,x_n)$ and $y = (y_1,\ldots,y_n)$ differ in exactly one position. We look at subsets $S \subseteq \{0,1\}$ as subgraphs (induced by the subset of vertices S) in this graph.

---

[1] If a set S of sensitivity s is contained in a subcube S' obtained by fixing $x_{i_1},\ldots,x_{i_k}$, removing the variables that have been fixed gives us a set $S'' \subseteq \{0,1\}^{n-k}$ of sensitivity $s - k$.

Definition 2. *We define an m-dimensional* subcube *or m-subcube of Q_n to be a cube induced by the set of all vertices that have the same bit values on $n - m$ positions $x_{i_1}, \ldots, x_{i_{n-m}}$ where i_j are all different.*

We denote a subcube that can be obtained by fixing some continuous sequence b of starting bits by Q_b. For example, Q_0 and Q_1 can be obtained by fixing the first bit and Q_{01} can be obtained by fixing the first two bits to 01. We use a wildcard * symbol to indicate that the bit in the corresponding position is not fixed. For example, by Q_{*10} we denote a cube obtained by fixing the second and the third bit to 10.

Definition 3. *Two m-dimensional subcubes of Q_n are* adjacent *if the fixed $n-m$ positions of both subcubes are the same and their bit values differ in exactly one position.*

Each Boolean function f can be uniquely represented as a set of vertices $V(f) = \{x \mid f(x) = 1\}$, thus each function of n variables represents a single subgraph $G(f)$ of Q_n induced by $V(f)$. Note that for an input $x \in V(f)$, the sensitivity $s(f, x)$ is equal to the number of vertices not in $V(f)$ and connected to x with an edge in Q_n. Thus the sensitivity of $V(f)$ is equal to $s_1(f)$.

For a Boolean function f, the minimum degree $\delta(G(f))$ corresponds to $n - s_1(f)$, and the minimum degree of a graph induced by $\{0,1\}^n \setminus V$ corresponds to $n - s_0(f)$.

In the rest of this paper we phrase our results in terms of subgraphs of Q_n.

Definition 4. *Let X and Y be subgraphs of Q_n. By $X \cap Y$ we denote the intersection graph of X and Y that is the graph $(V(X) \cap V(Y), E(X) \cap E(Y))$. By $X \setminus Y$ denote the* complement *of Y in X that is the graph induced by the vertex set $V(X) \setminus V(Y)$ in X.*

We also denote the degree of a vertex v in a graph G by $\deg(v, G)$.

The main focus of the paper is on the *irreducible* class of subgraphs:

Definition 5. *We call a subgraph $G \subset Q_n$* reducible *if it is a subgraph of some graph $S \subset Q_n$ where $V(S)$ can be obtained by fixing one or more of values x_i. Conversely, other subgraphs we call* irreducible.

Another way to define the irreducible graphs is to say that each such graph contains at least one vertex in each of the $(n - 1)$-subcubes of Q_n.

3 Simon's Lemma

In this section we present a theorem proved by Simon [11].

Theorem 1 (Simon). *Let $G = (V, E)$ be a non-empty subgraph of Q_n $(n \geq 0)$ of minimum cardinality among the subgraphs with $\delta(G) = d$ $(d \geq 0)$. Then G is a d-dimensional subcube of Q_n and $|V| = 2^d$.*

This theorem implies:

Corollary 1. *Let $f(x)$ be a Boolean function on n variables. If $f(x)$ is not always 0, then*

$$|\{x \mid f(x) = 1\}| \geq 2^{n-s_1(f)}, \tag{4}$$

and the minimum is obtained iff some $s_1(f)$ positions hold the same bit values for all $x : f(x) = 1$.

Proof. Let G be a subgraph of Q_n induced by the set of vertices $V = \{x \mid f(x) = 1\}$. The minimum degree of G is $\delta(G) = n - s_1(f)$. Then by Theorem 1 $|V| \geq 2^{n-s_1(f)}$. The minimum is obtained iff G is an $(n - s_1(f))$-subcube of Q_n. This means that it is defined by some bits fixed in $s_1(f)$ positions. □

4   Smallest Irreducible Subgraphs

In this section we prove the main theorem.

Theorem 2. *Let $G = (V, E)$ be a non-empty irreducible subgraph of Q_n $(n \geq 1)$ with the minimum degree $d \geq 0$. Let the smallest possible cardinality of V be $S(n, d)$. Then*

$$S(n, d) = \lceil 2^{d+1} - 2^{2d-n} \rceil . \tag{5}$$

The proof of Theorem 2 is by induction on n and involves case analysis going as deep as considering $(n - 3)$-dimensional subcubes of Q_n.

In the language of Boolean functions, this theorem corresponds to:

Corollary 2. *Let $f(x)$ be a Boolean function on n variables. If $\forall i \in [n] \, \forall b \in \{0, 1\} \, \exists x \, (x_i = b, f(x) = 1)$, then*

$$|\{x \mid f(x) = 1\}| \geq 2^{n-s_1(f)+1} - 2^{n-2s_1(f)}. \tag{6}$$

Theorem 2 together with Lemma 1 imply the following generalization of Simon's lemma:

Theorem 3. *Let $G = (V, E)$ be a non-empty subgraph of Q_n $(n \geq 0)$ with $\delta(G) = d$. Then either $|V| = 2^d$ or $|V| \geq \frac{3}{2} \cdot 2^d$, with $V = |2^d|$ achieved if and only if G is a d-subcube.*

Equivalently, if G has sensitivity s, then either $|V| = 2^{n-s}$ or $|V| \geq \frac{3}{2}2^{n-s}$. Thus there is a gap between the possible values for $|V|$ — which we find quite surprising.

In the next two subsections we prove Theorem 2 and in the last two subsections we show how it implies Corollary 2 and Theorem 3.

4.1   Instances Achieving the Minimum

In this section we prove that the given number of vertices is sufficient. We distinguish three cases:

1. $n = 1$. The only valid graph satisfying the properties is $G = Q_n$ with $d = 1$. Then $|V| = 2$.
2. $n > 1$, $2d < n$. Since $2^{2d-n} < 1$, $|V|$ should be 2^{d+1}. We take

$$S_j = \{x \mid \forall i \in [n - d] \, (x_i = j)\} \tag{7}$$

for $j \in \{0, 1\}$ and $V = S_0 \cup S_1$. Let G be the graph induced by V in Q_n. Then G consists of two d-subcubes of Q_n with no common vertices. Since $n - d > 1$, no edge connects any two vertices between these subcubes, thus $\delta(G) = d$. For the irreducibility, suppose that some $(n - 1)$-subcube H is defined by fixing $x_i = j$. If $i \leq n - d$, then $H \cap S_j \neq \varnothing$. If $i > n - d$, then $H \cap S_j \neq \varnothing$ for any j. Then $|V| = 2 \cdot 2^d = 2^{d+1}$.
3. $n > 1$, $2d \geq n$. Then $|V|$ should be $2^{d+1} - 2^{2d-n}$. We take

$$S_l = \{x \mid \forall i \in [n - d] \, (x_i = 1)\}, \tag{8}$$
$$S_r = \{x \mid \forall i \in [n - d + 1; 2(n - d)] \, (x_i = 1)\} \tag{9}$$

and $V = S_l \cup S_r$. Let G be the graph induced by V in Q_n. Graphs induced by S_l and S_r are d-dimensional subcubes of Q_n. Since they are not adjacent, $\delta(G) = d$. For the irreducibility, observe that any bit position i is not fixed for at least one of S_l or S_r. Then the $(n-1)$-subcube H obtained by fixing x_i holds at least one of the vertices of G. Since $S_l \cap S_r = \{x \mid \forall i \in [2(n - d)](x_i = 1)\}$, it follows that

$$|V| = 2 \cdot 2^d - 2^{n-2(n-d)} = 2^{d+1} - 2^{2d-n}. \tag{10}$$

4.2   Optimality

In this section we prove that there are no such graphs with a number of vertices less than $\lceil 2^{d+1} - 2^{2d-n} \rceil$.

The proof is by induction on n. As the base case we take $n \leq 2$. From the fact that each $(n - 1)$-subcube contains at least one vertex of G it follows that $|V| \geq 2$. This proves the cases $n = 1$, $d = 1$ and $n = 2$, $d = 0$ (and the case $n = 1$, $d = 0$ is not possible). Suppose $n = 2$, $d = 1$: if there were 2 vertices in G, then either some of the 1-subcubes would contain no vertex of G or there would be a vertex of G with degree 0 (which contradicts $d = 1$). Thus, in this case $|V| \geq 3 = 2^{1+1} - 2^{2-2}$. Suppose $n = 2$, $d = 2$. Then $G = Q_n$ and $|V| = 4 = 2^{2+1} - 2^{4-2}$.

Inductive step. First suppose that each $(n - 2)$-subcube of Q_n contains at least one vertex of G, then $G \cap Q_0$ and $G \cap Q_1$ are irreducible. The minimum degrees of $G \cap Q_0$ and $G \cap Q_1$ are at least $d - 1$, since each vertex of $G \cap Q_0$ can have at most one neighbour in Q_1 (and conversely). By applying the inductive assumption to the cubes Q_0 and Q_1, we obtain that

$$|V| \geq 2 \cdot \left\lceil 2^{(d-1)+1} - 2^{2(d-1)-(n-1)} \right\rceil = \tag{11}$$
$$= 2 \cdot \left\lceil 2^d - 2^{2d-n-1} \right\rceil \geq \tag{12}$$
$$\geq \left\lceil 2^{d+1} - 2^{2d-n} \right\rceil. \tag{13}$$

Now suppose that there is some $(n-2)$-subcube without vertices of G. WLOG assume it is Q_{00}, i.e., $G \cap Q_{00} = \varnothing$. We prove two lemmas.

Lemma 1. *Let $G = (V, E)$ be a non-empty subgraph of Q_n ($n \geq 0$) with $\delta(G) = d$ ($d \geq 0$). Then either $|V| = 2^d$ or $|V| \geq \min_{i=d+1}^{n} S(i, d)$.*

Proof. The proof is by induction on n. Base case: $n = 0$. Then $G = Q_n$, $d = 0$ and $|V| = 1 = 2^{0-0}$. In the inductive step we prove the statement for $n > 0$. If $n = d$, then $G = Q_n$, and $|V| = 2^n = 2^d$. Otherwise $n > d$. If each $(n-1)$-subcube of Q_n contains vertices of G, then $|V| \geq S(n, d)$ by the definition of S. Otherwise there is an $(n-1)$-subcube of Q_n that does not contain any vertex of G. Then by induction the other $(n-1)$-subcube contains either 2^d or at least $\min_{i=d+1}^{n-1} S(i, d)$ vertices of G. Combining the two cases together gives us the result. □

Lemma 2. *Let $G = (V, E)$ be a subgraph of Q_n ($n \geq 1$). Let $G' = G \cap Q_0$. If G' is not empty and $\min_{v \in G'} \deg(v, G) \geq d$, then $|V| \geq 2^d$.*

Note that this lemma is also a stronger version of Simon's result. Here we require the lower bound for the minimum degree only for vertices of G in one of the $(n-1)$-subcubes of Q_n.

Proof. The proof is by induction on n.

(a) Base case, $n = 1$. Since G' is non-empty, $G' = Q_0$. If $d = 0$, $|V| \geq 1 = 2^0$. If $d = 1$, then $G = Q_n$ and $|V| = 2 = 2^1$.

(b) In the inductive step we prove the statement for $n > 1$. If $Q_{0j} \cap G'$ is empty for some $j \in \{0, 1\}$, then $G' \subseteq Q_{0(1-j)}$. Thus by the induction hypothesis $|V(Q_{*(1-j)})| \geq 2^d$. Otherwise both Q_{00} and Q_{01} contain some vertices of G. Since each vertex of $Q_{0j} \cap G$ has at most one neighbour in $Q_{0(1-j)} \cap G$, it follows that $\min_{v \in Q_{0j} \cap G} \deg(v, Q_{*j}) \geq d - 1$ for any $j \in \{0, 1\}$. By applying the induction hypothesis for $Q_{*j} \cap G$ in the cube Q_{*j} for each j, we obtain that $|V| \geq 2 \cdot 2^{d-1} = 2^d$.

□

We now have that $\delta(G \cap Q_{01}) \geq d - 1$ and $\delta(G \cap Q_{10}) \geq d - 1$ becase Q_{11} may contain vertices of G but on the other hand we are assuming $G \cap Q_{00} = \varnothing$. Now we distinguish two cases:

1. $|V(G \cap Q_{01})| \neq 2^{d-1}$ and $|V(G \cap Q_{10})| \neq 2^{d-1}$.
   Cube Q_{01} has $n - 2$ dimensions and $\delta(Q_{01} \cap G) \geq d - 1$. By Lemma 1

$$|V(Q_{01} \cap G)| \geq \min_{i=(d-1)+1}^{n-2} S(i, d-1) = \min_{i=d}^{n-2} S(i, d-1). \tag{14}$$

It follows by induction that

$$|V(Q_{01} \cap G)| \geq \min_{i=d}^{n-2} \left[2^{(d-1)+1} - 2^{2(d-1)-i} \right]. \tag{15}$$

The minimum is achieved when i is the smallest, $i = d$. Thus $|V(Q_{01} \cap G)| \geq \lceil 2^d - 2^{d-2} \rceil$. Similarly we prove that $|V(Q_{10} \cap G)| \geq \lceil 2^d - 2^{d-2} \rceil$.

It remains to estimate the number of vertices of G in Q_{11}. We deal with two cases:

1.1. Some $(n-3)$-subcube of Q_n in Q_{11} does not contain vertices of G. WLOG we assume it is Q_{110}, i.e., $G \cap Q_{110} = \varnothing$. We again distinguish two cases:

    1.1.1. One of the subcubes Q_{010} and Q_{100} does not contain vertices of G. WLOG assume it is Q_{010}, i.e., $G \cap Q_{010} = \varnothing$. Then for the subcube Q_{011} it holds that $\min_{v \in G \cap Q_{011}} \deg(v, G \cap Q_{*11}) \geq d$, since $G \cap Q_{001} = \varnothing$ (because $Q_{001} \subset Q_{00}$), $G \cap Q_{010} = \varnothing$ and Q_{111} may contain vertices of G. Applying Lemma 2 to $G \cap Q_{011}$ in Q_{*11}, we get $|V(G \cap Q_{*11})| \geq 2^d$. Similarly we prove that $|V(G \cap Q_{10*})| \geq 2^d$. That gives us

$$|V| \geq 2 \cdot 2^d = 2^{d+1} \geq \lceil 2^{d+1} - 2^{2d-n} \rceil \qquad (16)$$

    and the case is done.

    1.1.2. Both of the subcubes Q_{010} and Q_{100} contain vertices of G. Then for the subcube Q_{010} it holds that $\min_{v \in G \cap Q_{010}} \deg(v, G \cap Q_{01*}) \geq d$, since $G \cap Q_{000} = \varnothing$, $G \cap Q_{110} = \varnothing$, and Q_{011} may contain vertices of G. Applying Lemma 2 to $G \cap Q_{010}$ in Q_{01*}, we get $|V(G \cap Q_{01*})| \geq 2^d$. Similarly we prove that $|V(G \cap Q_{10*})| \geq 2^d$. That gives us

$$|V| \geq 2 \cdot 2^d = 2^{d+1} \geq \lceil 2^{d+1} - 2^{2d-n} \rceil \qquad (17)$$

    and this case also is done.

1.2. Each $(n-3)$-subcube of Q_n in Q_{11} contains vertices of G. Since Q_{11} is adjacent to Q_{01} and Q_{10}, $\delta(G \cap Q_{11}) \geq d - 2$. From the inductive assumption it follows that

$$|V(G \cap Q_{11})| \geq 2^{(d-2)+1} - 2^{2(d-2)-(n-2)} = 2^{d-1} - 2^{2d-n-2}. \qquad (18)$$

Thus

$$|V| = |V(G \cap Q_{01})| + |V(G \cap Q_{10})| + |V(G \cap Q_{11})| \geq \qquad (19)$$

$$\geq 2 \cdot \lceil 2^d - 2^{d-2} \rceil + \lceil 2^{d-1} - 2^{2d-n-2} \rceil \geq \qquad (20)$$

$$\geq \lceil 2 \cdot (2^d - 2^{d-2}) + 2^{d-1} - 2^{2d-n-2} \rceil = \qquad (21)$$

$$= \lceil 2^{d+1} - 2^{d-1} + 2^{d-1} - 2^{2d-n-2} \rceil = \qquad (22)$$

$$= \lceil 2^{d+1} - 2^{2d-n-2} \rceil \geq \qquad (23)$$

$$\geq \lceil 2^{d+1} - 2^{2d-n} \rceil. \qquad (24)$$

Hence this case is complete.

2. $|V(G \cap Q_{01})| = 2^{d-1}$ or $|V(G \cap Q_{10})| = 2^{d-1}$. WLOG assume that this holds for Q_{01}.

By Theorem 1 it follows that $G \cap Q_{01}$ is a $(d-1)$-dimensional subcube of Q_n, denote it by D_0. On the other hand, we are assuming $G \cap Q_{00} = \varnothing$. Thus WLOG we can assume that D_0 is induced on the set of vertices

$$\{x \mid x_1 = 0, \forall i \in [2; n - d + 1] (x_i = 1)\} = V(G \cap Q_0). \tag{25}$$

Observe that $\deg(v, G \cap Q_{01}) = d - 1$ for all $v \in G \cap Q_{01}$. Since $\delta(G) = d$, each $x \in V(G \cap Q_{01})$ has $x^{(1)}$ as a neighbour in G. Then $\{x^{(1)} \mid x \in V(G \cap Q_{01})\} \subseteq V(G \cap Q_{11})$, and $G \cap Q_{11}$ contains a $(d-1)$-subcube of Q_n adjacent to D_0. We denote it by D_1, with

$$\{x \mid x_1 = 1, \forall i \in [2; n - d + 1] (x_i = 1)\} \subseteq V(G \cap Q_1). \tag{26}$$

Then $D = D_0 \cup D_1$ is a d-dimensional subcube.

It remains to estimate the number of vertices of G in Q_1 that do not belong to D_1, denote it by $R = |V((G \cap Q_1) \setminus D_1)|$. We will prove the following claim:

Claim. By k denote the co-dimension of D_1 in Q_1, which is $(n-1) - (d-1) = n - d$. Then $R \geq 2^d - 2^{d-k}$.

Proof. We will denote the subcube of Q_1 obtained by restricting some t bits $x_{i_1} = b_1, \ldots, x_{i_t} = b_t$ by $Q_1(x_{i_1} = b_1, \ldots, x_{i_t} = b_t)$. Further note that $D_1 \subseteq Q_1(x_i = 1, x_j = 1)$ for $i, j \in [2; k + 1]$.

Since $G \cap Q_0 = D_0$, any vertex of $(G \cap Q_1) \setminus D_1$ can have a neighbour in G only in Q_1. Thus we have that

$$\min_{v \in (G \cap Q_1) \setminus D_1} \deg(v, G \cap Q_1) \geq d. \tag{27}$$

Pick any $i \in [2; k + 1]$. Examine the $(n-2)$-subcube $Q_1(x_i = 0)$. It does not overlap with D. But G is irreducible, so $G \cap Q_1(x_i = 0) \neq \varnothing$.

Assume $k = 1$. Then $D_1 = Q_{11}$ and $\delta(G \cap Q_{10}) = d - 1$. By Theorem 1, it follows that

$$R = |V(G \cap Q_{10})| \geq 2^{d-1} = 2^d - 2^{d-1}. \tag{28}$$

Otherwise $k \geq 2$. We will prove it can be assumed that for any $i, j \in [2; k+1]$, $i \neq j$ and $b \in \{0, 1\}$ we have $G \cap Q_1(x_i = 0, x_j = b) \neq \varnothing$.

- Let $G \cap Q_1(x_i = 0, x_j = 0) = \varnothing$. Then $\delta(G \cap Q_1(x_i = 0, x_j = 1)) \geq d - 1$ and $\delta(G \cap Q_1(x_i = 1, x_j = 0)) \geq d - 1$. By Theorem 1, we have $|V(G \cap Q_1(x_i = 0, x_j = 1))| \geq 2^{d-1}$ and $|V(G \cap Q_1(x_i = 1, x_j = 0))| \geq 2^{d-1}$. Thus in this case

$$R \geq 2 \cdot 2^{d-1} = 2^d > 2^d - 2^{d-k}. \tag{29}$$

- Let $G \cap Q_1(x_i = 0, x_j = 1) = \varnothing$. Then $\min_{v \in G \cap Q_1(x_i = 0, x_j = 0)} \deg(v, G \cap Q_1(x_j = 0)) \geq d$ and since $G \cap Q_1(x_j = 0) \neq \varnothing$, by Lemma 2 we have

$$R > |V(G \cap Q_1(x_j = 0))| \geq 2^d > 2^d - 2^{d-k}. \tag{30}$$

Now examine a subcube $G \cap Q_1(x_i = 0)$ for an $i \in [2; k + 1]$. Since $G \cap Q_0(x_i = 0) = \varnothing$, we have $\delta(G \cap Q_1(x_i = 0)) \geq d - 1$. By Lemma 1, either $|V(G \cap Q_1(x_i = 0))| = 2^{d-1}$ or $|V(G \cap Q_1(x_i = 0))| \geq \min_{t=d}^{n-1} S(t, d - 1)$.

- Assume it is the latter case; by the induction of this section, we have that the minimum is achieved by $t = d$ with $|V(G \cap Q_1(x_i = 0))| \geq 2^d - 2^{2(d-1)-d} = 2^d - 2^{d-2}$.

  We can now assume that $G \cap Q_1(x_i = 1, x_j = 0) \neq \varnothing$, for $j \in [2; k+1]$, $i \neq j$. Since $G \cap Q_0(x_i = 1, x_j = 0) = \varnothing$, we have $\delta(G \cap Q_1(x_i = 1, x_j = 0)) \geq d - 2$ and by Theorem 1 we have $|V(G \cap Q_1(x_i = 1, x_j = 0))| \geq 2^{d-2}$. Thus

$$R \geq |V(G \cap Q_1(x_i = 0))| + |V(G \cap Q_1(x_i = 1, x_j = 0))| \geq \qquad (31)$$
$$\geq (2^d - 2^{d-2}) + 2^{d-2} = 2^d > 2^d - 2^{d-k}. \qquad (32)$$

- Otherwise it is the former case in Lemma 1 for each i, $|V(G \cap Q_1(x_i = 0))| = 2^{d-1}$. By Theorem 1, $G \cap Q_1(x_i = 0)$ must be a $(d-1)$-subcube of Q_n.

  Pick $i_1, i_2 \in [2; k+1]$, $i_1 \neq i_2$. We have that $Q_a = G \cap Q_1(x_{i_1} = 0)$ and $Q_b = G \cap Q_1(x_{i_2} = 0)$ are both $(d-1)$-subcubes. We can now assume that $G \cap Q_1(x_{i_1} = 0, x_{i_2} = b) \neq \varnothing$, for $b \in \{0, 1\}$. This means that the i_2-th bit is not fixed for the subcube Q_a. Thus $G \cap Q_1(x_{i_1} = 0, x_{i_2} = 0)$ is a $(d-2)$-subcube. Hence Q_a and Q_b overlap exactly in a $(d-2)$-subcube.

  Examine $Q_a \cap Q_b$. Two of its fixed bits are the i_1-th and the i_2-th, which are distinct positions. Thus it has $n - (d-2) - 2 = n - d$ fixed positions not in $[2; k+1]$. Let the d-subcube defined by these restrictions be C. As Q_a and Q_b are both $(d-1)$-subcubes, they must share these $n - d$ fixed positions. As this applies for any $i_1 \neq i_2$, we have that $G \cap Q_1(x_i = 0) \subset C$ for any $i \in [2; k+1]$. We show that $C \subset G$. Pick $x \in C$. Suppose for some $i \in [2; k+1]$, we have $x_i = 0$. Then $x \in G \cap Q_1(x_i = 0)$. Otherwise we have $x_i = 1$ for each $i \in [2; k+1]$. But then $x \in D$.[2]

  Examine the intersection of D and C. Each position of $[2; k+1]$ is fixed in D (k positions). On the other hand, $n - d$ more positions not in $[2; k+1]$ are fixed in C. Thus their intersection is a $(d-k)$-subcube, and $R = 2^d - 2^{d-k}$. Since $k = n - d$, we have $R \geq 2^d - 2^{2d-n}$. Ultimately we get

$$|V| = |V(D)| + R \geq 2^d + (2^d - 2^{2d-n}) = 2^{d+1} - 2^{2d-n}. \qquad (33)$$

This completes the proof of Theorem 2.   □

4.3   Application for Boolean Functions

Theorem 2 implies:

Corollary 2. *Let $f(x)$ be a Boolean function on n variables. If $\forall i \in [n] \forall b \in \{0, 1\} \exists x\, (x_i = b, f(x) = 1)$, then*

$$|\{x \mid f(x) = 1\}| \geq 2^{n-s_1(f)+1} - 2^{n-2s_1(f)}. \qquad (34)$$

---

[2] In this case, we have obtained that G is a union of two d-dimensional subcubes D and C, such that each bit position is fixed in at most one of them. This is essentially the same construction as given in subsection 4.1.

Proof. Let G be a subgraph of Q_n induced by the set of vertices $V = \{x \mid f(x) = 1\}$. The minimum degree of G is $\delta(G) = n - s_1(f)$. The given constraint means that G is irreducible. Then, by Theorem 2, □

$$|V| \geq 2^{(n-s_1(f))+1} - 2^{2(n-s_1(f))-n} = 2^{n-s_1(f)+1} - 2^{n-2s_1(f)}. \tag{35}$$

4.4 Generalization of Simon's Lemma

We use Theorem 2 and Lemma 1 to prove Theorem 3, which is a stronger version of Simon's lemma (Theorem 1):

Theorem 3. *Let $G = (V, E)$ be a non-empty subgraph of Q_n $(n \geq 0)$ with $\delta(G) = d$. Then either $|V| = 2^d$ or $|V| \geq \frac{3}{2} \cdot 2^d$, with $V = |2^d|$ achieved if and only if G is a d-subcube.*

Proof. By Theorem 2 we may substitute $\lceil 2^{d+1} - 2^{2d-n} \rceil$ instead of $S(n, d)$ in Lemma 1. Then in

$$\min_{i=d+1}^{n} S(i, d) = \min_{i=d+1}^{n} \lceil 2^{d+1} - 2^{2d-i} \rceil \tag{36}$$

the minimum is obtained for $i = d + 1$. Thus either $|V| = 2^d$ or $|V| \geq 3 \cdot 2^{d-1}$. □

5 Conclusion

In this paper, we have shown two results on the structure of low sensitivity subsets of Boolean hypercube:

– Theorem 2: a tight lower bound on the size of irreducible low sensitivity sets $S \subseteq \{0,1\}^n$, that is, sets S that are not contained in any subcube of $\{0,1\}^n$ obtained by fixing one or more variables x_i;
– Theorem 3: a gap theorem that shows that $S \subseteq \{0,1\}^n$ of sensitivity s must either have $|S| = 2^{n-s}$ or $|S| \geq \frac{3}{2}2^{n-s}$.

The gap theorem follows from the first result by classifying $S \subseteq \{0,1\}^n$ into irreducible sets and sets that are constructed from irreducible subsets $S' \subseteq \{0,1\}^{n-k}$ for some $k \in \{1, 2, \ldots, s\}$ and then using the first result for each of those categories. We find this gap theorem quite surprising.

Both results contribute to understanding the structure of low-sensitivity subsets of the Boolean hypercube. After this paper was completed, we have used the gap theorem to obtain a new upper bound on block sensitivity in terms of sensitivity:

$$bs(f) \leq \max\left(2^{s(f)-1}\left(s(f) - \frac{1}{3}\right), s(f)\right). \tag{37}$$

We report this result in [2].

References

1. Ambainis, A., Bavarian, M., Gao, Y., Mao, J., Sun, X., Zuo, S.: Tighter relations between sensitivity and other complexity measures. In: Esparza, J., Fraigniaud, P., Husfeldt, T., Koutsoupias, E. (eds.) ICALP 2014. LNCS, vol. 8572, pp. 101–113. Springer, Heidelberg (2014)
2. Ambainis, A., Prūsis, K., Vihrovs, J.: Sensitivity versus Certificate Complexity of Boolean Functions. Preprint available at http://arxiv.org/abs/1503.07691
3. Ambainis, A., Sun, X.: New separation between $s(f)$ and $bs(f)$. CoRR, abs/1108.3494 (2011)
4. Beals, R., Buhrman, H., Cleve, R., Mosca, M., de Wolf, R.: Quantum lower bounds by polynomials. J. ACM **48**(4), 778–797 (2001)
5. Buhrman, H., de Wolf, R.: Complexity measures and decision tree complexity: a survey. Theor. Comput. Sci. **288**(1), 21–43 (2002). (Complexity and Logic)
6. Green, B., Sanders, T.: Boolean functions with small spectral norm. Geom. Funct. Anal. **18**(1), 144–162 (2008)
7. Kenyon, C., Kutin, S.: Sensitivity, block sensitivity, and ℓ-block sensitivity of boolean functions. Info. Comput. **189**(1), 43–53 (2004)
8. Nisan, N.: CREW PRAMS and decision trees. In: Proceedings of the Twenty-first Annual ACM Symposium on Theory of Computing, STOC 1989, pp. 327–335. ACM, New York (1989)
9. Nisan, N., Szegedy, M.: On the degree of boolean functions as real polynomials. Comput. Complex. **4**(4), 301–313 (1994)
10. Rubinstein, D.: Sensitivity vs. block sensitivity of boolean functions. combinatorica **15**(2), 297–299 (1995)
11. Simon, H.-U.: A tight $\Omega(\log \log N)$-bound on the time for parallel Ram's to compute nondegenerated boolean functions. In: Karpinski, M. (ed.) Foundations of Computation Theory. LNCS, vol. 158, pp. 439–444. Springer, Heidelberg (1983)
12. Virza, M.: Sensitivity versus block sensitivity of boolean functions. Inf. Process. Lett. **111**(9), 433–435 (2011)

Graph Theory

Star Shaped Orthogonal Drawing

Xin He[✉] and Dayu He

Department of Computer Science and Engineering,
State University of New York at Buffalo, Buffalo, NY 14260, USA
{xinhe,dayuhe}@buffalo.edu

Abstract. An *orthogonal drawing* of a plane graph G is a planar drawing, denoted by $D(G)$, of G such that each vertex of G is drawn as a point on the plane, and each edge is drawn as a sequence of horizontal and vertical line segments with no crossings. $D(G)$ is called *orthogonally convex* if each of its faces is an *orthogonally convex polygon* P. (Namely, for any horizontal or vertical line L, the intersection of L and P is a single line segment or empty). Recently, Chang et al. [1] gave a necessary and sufficient condition for a plane graph to have such a drawing.

$D(G)$ is called a *star-shaped orthogonal drawing* (SSOD) if each of its faces is a star-shaped polygon P. (Namely there is a point $p \in P$ such that the entire P is visible from p). Every SSOD is an orthogonally convex drawing, but the reverse is false. SSOD is visually more appealing than orthogonally convex drawings. In this paper, we show that if G satisfies the same conditions as in [1], it not only has an orthogonally convex drawing, but also a SSOD, which can be constructed in linear time.

1 Introduction

Among many graph drawing styles, *orthogonal drawing* has attracted much attention due to its various applications in circuit schematics, relationship diagrams, data flow diagrams etc. [2]. An *orthogonal drawing* of a plane graph G is a planar drawing, denoted by $D(G)$, of G such that each vertex of G is drawn as a point on the plane, and each edge is drawn as a sequence of horizontal and vertical line segments with no crossings. A *bend* is a point where an edge changes its direction. (See Fig. 1 (1) and (2). The point p is a bend).

Rahman et al. [8] gave a necessary and sufficient condition for a plane graph G of maximum degree 3 to have an orthogonal drawing without bends. A linear time algorithm to find such a drawing was also presented in [8]. In the drawing obtained in [8], the faces of $D(G)$ can be of complicated shapes. An orthogonal polygon P is *orthogonally convex* if, for any horizontal or vertical line L, the intersection of L and P is either empty or a single line segment. (Fig. 1 (3) shows an orthogonally convex polygon. The face marked by F in Fig. 1 (2) is not orthogonally convex). An orthogonal drawing $D(G)$ is *orthogonally convex* if all faces of $D(G)$ are orthogonally convex polygons. The orthogonally convex drawings are more visually appealing than arbitrary orthogonal drawings.

Research supported in part by NSF Grant CCR-1319732.

R. Jain et al. (Eds.): TAMC 2015, LNCS 9076, pp. 137–149, 2015.
DOI: 10.1007/978-3-319-17142-5_13

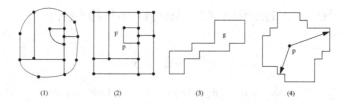

Fig. 1. (1) A plane graph G; (2) An orthogonal drawing of G; (3) An orthogonally convex polygon; (4) A star-shaped orthogonal polygon.

Chang et al. [1] gave a necessary and sufficient condition (which strengthens the conditions in [8]) for a plane graph G of maximum degree 3 to have an orthogonally convex drawing without bends. A linear time algorithm to find such a drawing was also obtained in [1].

An orthogonal polygon P is called *star-shaped* if there exists a point p in P such that the entire polygon P is *visible* from p. (See Fig. 1 (4)). It is easy to see that any star-shaped orthogonal polygon is always orthogonally convex. But the reverse is not true. An orthogonal drawing $D(G)$ is called a *star-shaped orthogonal drawing* (SSOD) if every inner face of $D(G)$ is a star-shaped orthogonal polygon. The star-shaped orthogonal drawings are more visually appealing than orthogonally convex drawings. In this paper, we show that if G satisfies the same conditions as in [1], then G has a SSOD without bends. In addition, such a drawing can be constructed in linear time.

To the best knowledge of the authors, SSOD is a new drawing style. Although star-shaped drawings have been studied before [5], the polygons in their drawings are required to be star-shaped but not orthogonal. In [7], the problem of covering orthogonal polygons by star-shaped orthogonal polygons is studied.

The paper is organized as follows. In Sect. 2, we present the definitions and preliminary results. Section 3 describes a special rectangular dual needed by our algorithm. In Sect. 4, we present our SSOD algorithm. Section 5 concludes the paper.

2   Preliminaries

Let $G = (V, E)$ be a graph with n vertices. The *degree* of a vertex v is the number of neighbors of v in G. A vertex of degree 2 is called a *2-vertex*. G is called a *d-graph* if the maximum degree of vertices of G is $\leq d$. A *planar graph* is a graph G that can be drawn on the plane without edge crossings. A *plane graph* is a planar graph with a fixed plane embedding. For the rest of this paper, as in [1,8], G always denotes a biconnected plane 3-graph.

The embedding of G divides the plane into a set of connected regions called *faces*. The unbounded face of G is called the *exterior face*. Other faces are called *interior faces*. The *contour* of a face is the cycle formed by the vertices and edges on the boundary of the face. The contour of the exterior face of G is denoted by $C_o(G)$. If a vertex a is on the contour of a face f, we say f is *incident to* a.

A cycle C of G with k edges is called a *k-cycle*. A *triangle* is a 3-cycle. G is called *internally triangulated* if all of its interior faces are triangles. A cycle

C divides the plane into its interior and exterior regions. A *separating cycle* of G is a cycle C such that there are vertices in both its interior and exterior. A separating cycle may be contained in other separating cycles. A separating cycle C is called *maximal* if it's not contained in other separating cycles.

Let $D(G)$ be an orthogonal drawing of G without bends. Each cycle C of G is drawn as an orthogonal polygon $D(C)$ in $D(G)$. Let a be a vertex of C. We will also use a to denote the point in $D(C)$ that corresponds to a. A vertex a of $D(C)$ is called a *corner* of $D(C)$ if the interior angle of $D(C)$ at a is 90° or 270°. A corner with 90° (270°, respectively) interior angle is called a *convex* (*concave*, respectively) corner. For an orthogonal drawing $D(G)$ without bends, any concave corner a of $D(G)$ must correspond to a 2-vertex in G.

In the definition of the orthogonal drawing of G, the exterior face $C_o(G)$ is not necessarily drawn as a rectangle. However, the algorithm **Bi-Orthogonal-Draw** in [8] (which finds an orthogonal drawing of G) produces an orthogonal drawing such that $C_o(G)$ is actually a rectangle. The first step of algorithm **Bi-Orthogonal-Draw** arbitrarily selects four degree-2 vertices on $C_o(G)$ as the four corners of the exterior rectangle of the drawing. Since the drawing in [1] is produced by a modified version of the algorithm **Bi-Orthogonal-Draw**, this is also true for the drawing in [1]. Thus, without loss of generality, we assume the input to our problem is a plane graph H with four specified degree-2 vertices a, b, c, d on $C_o(H)$ in clockwise order. Our goal is to produce an orthogonal drawing $D(G)$ of G such that $C_o(H)$ is drawn as a rectangle with a, b, c, d as the northwest, northeast, southeast and southwest corner of $D(H)$, respectively.

To simplify the presentation, we construct a graph G from H (Fig. 2 (1)):

1. Add eight new vertices $a'', a', b'', b', c'', c', d'', d'$ in the exterior face of G; connect them into a clockwise cycle;
2. Add four new edges $(a, a'), (b, b'), (c, c'), (d, d')$.

Clearly, H has an orthogonal drawing with no bends (with four corners a, b, c, d) if and only if G has an orthogonal drawing with no bends (with four corners a'', b'', c'', d'', see Fig. 2 (2)). Note that G satisfies the following properties:

Property 1.

- G is a biconnected plane 3-graph; On the exterior face $C_o(G)$, there are four degree-2 vertices and four degree-3 vertices; the degree-2 and degree-3 vertices alternate on $C_o(G)$;
- The four degree-2 vertices on $C_o(G)$ are specified as the northwest, northeast, southeast, southwest vertices.

In the rest of the paper, without loss of generality, we always assume G satisfies Property 1. Let C be a cycle of G. A *leg* of C is an edge e that is in the exterior of C and has exactly one vertex on C. The vertex of e that is on C is called a *leg vertex* of C. C is a *k-legged cycle* if it has exactly k legs. The k leg vertices divide C into k sub-paths. Each sub-path is called a *contour path* of C.

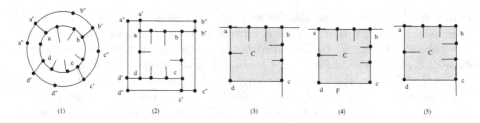

Fig. 2. (1) The construction of G from H; (2) drawings of H and G; (3) and (4) Conditions in Theorem 1; (5) Conditions in Theorem 2.

Theorem 1 [8]. *Let G be a plane graph that satisfies the conditions in Property 1. Then G has an orthogonal drawing without bends if and only if the following two conditions hold: (1) Every 3-legged cycle C has at least one 2-vertex and (2) Every 2-legged cycle C has at least two 2-vertices.*

Figure 2 (3) shows a 3-legged cycle $C = \{a, b, c, d\}$ and its orthogonal drawing. Figure 2 (4) shows a 2-legged cycle $C = \{a, b, c, d\}$ and its orthogonal drawing.

Theorem 2 [1]. *Let G be a plane graph that satisfies the conditions in Property 1. Then G has an orthogonally convex drawing without bends if and only if the following two conditions hold: (1) Every 3-legged cycle C has at least one 2-vertex and (2) Every 2-legged cycle C has at least two 2-vertices, at least one on each of its two contour paths.*

Figure 2 (5) shows a 2-legged cycle $C = \{a, b, c, d\}$ and an orthogonal drawing of C (b and d are two 2-vertices). Note that, in Fig. 2 (4), the 2-legged cycle C satisfies the condition 2 in Theorem 1, but not the condition 2 in Theorem 2. Hence there exists no orthogonally convex drawing: In any drawing, the face outside of C (marked by F) cannot be orthogonally convex. In Sect. 4, we will show that if G satisfies the conditions in Theorem 2, then G has a SSOD without bends.

Let $G^* = (V^*, E^*)$ be the dual graph of G. To avoid confusion, the members of V^* are called *nodes*. Each node in V^* corresponds to an interior face f of G, and two nodes in V^* are adjacent to each other if and only if their corresponding faces in G share an edge as common boundary. Note that G^* is an internally triangulated plane graph and the exterior face of G^* has four nodes. A *rectangular dual* of such a graph G^* is a rectangle R divided into smaller rectangles such that the following hold:

- No four smaller rectangles meet at the same point.
- Each smaller rectangle corresponds to a node of G^*.
- Two nodes of G^* are adjacent in G^* if and only if their corresponding small rectangles share a line segment as their common boundary.

See Fig. 3 (a) for an example. It's easy to see that a rectangular dual R of G^* is an orthogonal drawing $D(G)$ of the original graph G, and each face of $D(G)$ is a rectangle. Not every internally triangulated plane graph G^* has a rectangular dual. The following theorem characterizes such graphs.

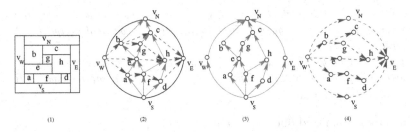

Fig. 3. (a) A rectangular dual of the graph shown in (b); (b) an **REL** $\mathcal{R} = \{T_1, T_2\}$; (c) the subgraph consisting of edges in T_1 and the 4 exterior edges oriented from v_S to v_N; (d) the subgraph consisting of edges in T_2 and the 4 exterior edges oriented from v_W to v_E (Color figure online).

Theorem 3 [6]. *A plane graph G^* has a rectangular dual with four rectangles on its boundary if and only if: (1) Every interior face of G^* is a triangle and the exterior face of G^* is a quadrangle; and (2) G^* has no separating triangles.*

G is called a *proper triangular plane (PTP) graph* if it satisfies the two conditions in Theorem 3. Our algorithm heavily depends on the following concept:

Definition 1. *A regular edge labeling **REL** $\mathcal{R} = \{T_1, T_2\}$ of a PTP graph G^* is a partition of the interior edges of G^* into two subsets T_1, T_2 of directed edges such that the following conditions hold:*

1. *For each interior node v, the edges incident to v appear in clockwise order around v as follows: a set of edges in T_1 leaving v; a set of edges in T_2 leaving v; a set of edges in T_1 entering v; a set of edges in T_2 entering v. (All four sets are not empty.)*
2. *Let v_N, v_E, v_S, v_W be the four exterior nodes of G^* in clockwise order. All interior edges incident to v_N are in T_1 entering v_N. All interior edges incident to v_E are in T_2 entering v_E. All interior edges incident to v_S are in T_1 leaving v_S. All interior edges incident to v_W are in T_2 leaving v_W.*

Figure 3 (b) shows an example of **REL** of a PTP graph. The red solid lines are edges in T_1. The green dashed lines are edges in T_2.

Theorem 4 [3,4]. *Every PTP graph G^* has an **REL** which can be constructed in linear time. From an **REL** of G^*, a rectangular dual of G^* can be constructed in linear time.*

3 A Special Rectangular Dual

A PTP graph G^* may have many different **REL**s. From the same **REL** of G^*, we may obtain different rectangular duals. In this section, we describe a rectangular dual of G^* with special properties, which is needed by our SSOD construction.

Lemma 1. *Any PTP graph G^* has a rectangular dual R such that the following properties hold for any node u in G^*.*

1. *Let $v_1 \rightarrow u$ be the first clockwise T_1 edge entering u and $u \rightarrow v_2$ the first clockwise T_1 edge leaving u. Then there exists a vertical stripe in R that intersects r_{v_1}, r_u, r_{v_2}.*
2. *Let $w_1 \rightarrow u$ be the first clockwise T_2 edge entering u and $u \rightarrow w_2$ the first clockwise T_2 edge leaving u. Then there exists a horizontal stripe in R that intersects r_{w_1}, r_u, r_{w_2}.*

The proof is omitted due to space limitation.

4   Star-Shaped Orthogonal Convex Drawing

Let G be a plane graph that satisfies the conditions in Theorem 2. In this section, we describe how to find a SSOD without bends for G.

Let v be a 2-vertex in G with two neighbors u, w. The operation *contracting* v is defined as follows: delete v and replace the two edges (u, v) and (v, w) by a single edge (u, w). First we modify G as follows. For every 3-legged cycle C in G with more than one 2-vertex on C, we arbitrarily choose one 2-vertex and contract every other 2-vertices on C. For every 2-legged cycle C in G with more than two 2-vertices on C, we arbitrarily choose one 2-vertex on each contour path of C and contract every other 2-vertices on C. After this modification, the resulting graph H has the following properties:

Property 2.

– Each 3-legged cycle C of H has exactly one 2-vertex on C.
– Each 2-legged cycle C of H has exactly one 2-vertex on each of the two contour paths of C.

After we construct a SSOD $D(H)$ of H, we can obtain a SSOD $D(G)$ of G as follows: Consider any 2-vertex v that was contracted from G. Let u, w be the two neighbors of v in G. In the drawing $D(H)$, the edge (u, w) is drawn as a line segment L. We simply draw v in the middle of L. After doing this for every contracted vertex v, we get a SSOD $D(G)$ for G. Thus, without loss of generality, we assume G satisfies the conditions in Property 2 from now on.

Let G^* be the dual graph of G. So G^* has exactly four nodes on its exterior face. Each 2-vertex of G corresponds to a pair of parallel edges in G^*. We only keep one of them in G^*. These edges in G^* are called *marked edges*.

Note that every 3-legged cycle C in G corresponds to a separating triangle C^* in G^*, and every 2-legged cycle C in G corresponds to a separating 2-cycle C^* in G^*. A 3-legged cycle C is shown in Fig. 4 (1). The edges in G are drawn as dashed lines, the edges in G^* are drawn as solid lines. The nodes in G^* are drawn as empty cycles. g is a 2-vertex in G. It corresponds to two parallel edges (w, x) in G^*. We keep only one of them in G^* and (w, x) is a *marked edge*. Figure 4 (2) shows a 2-legged cycle and its corresponding separating 2-cycle in G^*.

We first outline the main ideas of our algorithm. Basically, we want to construct a rectangular dual R of G^* which will be the "skeleton" of the drawing

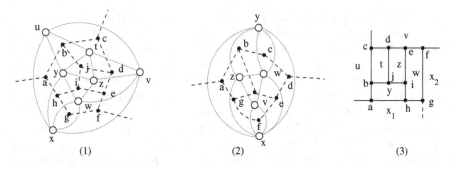

Fig. 4. (1) A 3-legged cycle $C = \{a, b, c, d, e, f, g, h\}$ and the dual separating triangle $C^* = \{u, v, x\}$; (2) A 2-legged cycle $C = \{a, b, c, d, e, f, g\}$ and the dual separating 2-cycle $C^* = \{x, y\}$. (3) The drawing of the graph in (1).

$D(G)$. However, because G^* has separating 2-cycles and 3-cycles, it is not a PTP graph and hence has no rectangular dual. We have to modify G^* to get a PTP graph $G^{*\prime}$ as follows. For each separating 2-cycle or 3-cycle C^* in G^* incident to a node x, we perform a *node split* operation on x as follows: This operation "splits" x into two nodes and "destroys" C^*. After all separating 2-cycles and 3-cycles in G^* are destroyed, the resulting graph $G^{*\prime}$ is a PTP graph. Each node x in G^* either corresponds to a node in $G^{*\prime}$ (if x is not split); or a set of nodes in $G^{*\prime}$ (since there may be multiple separating cycles incident to x, we may have to split x multiple times). We then find an **REL** $\mathcal{R}'$ of $G^{*\prime}$ and construct a rectangular dual $D(G^{*\prime})$ of $G^{*\prime}$ by Lemma 1. $D(G^{*\prime})$ is a "skeleton" of a SSOD $D(G)$ of G. Each face f of $D(G)$ corresponds to a node x in G^*, which either corresponds to a single rectangle in $D(G^{*\prime})$ (if x is not split), or an orthogonal polygon F that is the union of several rectangles in $D(G^{*\prime})$ (each rectangle corresponds to a split node of x). Figure 4 (3) illustrates the drawing $D(G)$ for the graph G in Fig. 4 (1) by using this process. We split the node x into two nodes x_1 and x_2 in order to destroy the separating triangle $C^* = \{u, v, x\}$. In Fig. 4 (3), each rectangle corresponds to a node in $G^{*\prime}$. The union of the two rectangles marked by x_1 and x_2 corresponds to the node x. The drawing in Fig. 4 (3) is an orthogonal drawing of the graph G in Fig. 4 (1). Note the location of the 2-vertex g in $D(G)$.

4.1 Node Split Operation

Since we want $D(G)$ to be a SSOD of G, we must make sure each face F in $D(G)$ is star-shaped. This is done by carefully constructing the **REL** $\mathcal{R}'$ so that certain properties are satisfied (to be defined later). Next we describe the details of our algorithm. Let G_1^* be the graph obtained from G^* as follows:

- For each maximal separating triangle C^*, delete all interior nodes of C^*.
- For each maximal separating 2-cycle C^*, delete all interior nodes of C^*, and replace the two edges of C^* by a single edge. We call these edges the *merged 2-cycle edges*.

Clearly G_1^* is a PTP graph. By Theorem 4, G_1^* has an **REL** $\mathcal{R}_1 = \{T_1, T_2\}$. We now need to add the deleted nodes back into G_1^*. We process the separating cycles of G^* one by one. Consider a maximal separating triangle C^* in G^*. Let $G^*(C^*)$ denote the induced subgraph of G^* consisting of the nodes on and in the interior of C^*. Let $G_1^* \cup G^*(C^*)$ be the graph obtained by adding the interior nodes of C^* back into G_1^*. We want to construct an **REL** for $G_1^* \cup G^*(C^*)$. However, $G_1^* \cup G^*(C^*)$ is not a PTP graph because C^* is a separating triangle. We must modify $G_1^* \cup G^*(C^*)$ so that C^* is not a separating triangle in it.

Let C be the 3-legged cycle in G corresponding to C^*. By Property 2, there is exactly one 2-vertex a in G on C. The vertex a corresponds to a marked edge e_a^* in G^*. e_a^* must be incident to a node on C^*. Let x be this node. We say the separating triangle C^* is *assigned to* x. (In Fig. 4 (1), the marked edge $e^* = (x, w)$ in G^* corresponds to the 2-vertex g in G. e^* is incident to the node x. So the separating triangle $C^* = \{u, v, x\}$ is assigned to x). The *node split* operation at x with respect to two specified edges (x, y_i) and (x, y_j) is illustrated in Fig. 5.

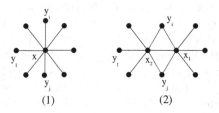

(1)   (2)

Fig. 5. Node split operation. (a) Before split; (b) After split.

Consider a separating triangle C^* assigned to x. After splitting x into two nodes, C^* becomes a quadrangle. Then we can add back the deleted interior nodes of C^*. Let e_1^*, e_2^* and e_3^* be the three edges of C^*. Two of them, say e_1^* and e_2^*, are incident to x. Depending on the pattern of these two edges in $\mathcal{R}_1$, there are eight cases (see Fig. 6). If both e_1^* and e_2^* are T_1 edges entering x, we call it the case south. If e_1^* is a T_2 edge entering x and e_2^* is a T_1 edge entering x, we call it the case southwest. The other six cases are shown in Fig. 6.

For example, consider the case south. We split x with respect to two edges: (z, x) is the marked edge in G^* that is in the interior of C^*; and (x, y) is a T_1 edge in the exterior of C^* leaving x (we will specify how to pick the edge (x, y) later). In Fig. 6, the left figure for the case south shows the edge pattern of C^* before the node split operation. The right figure shows the edge pattern of C^* after the node split operation. In Fig. 6, a blue dotted circle indicates the component inside C^* that was deleted. The blue dotted arrow (z, x) indicates the marked edge inside C^*.

Note that when looking from outside of C^*, the patterns of the involved edges are identical before and after the node split operation. After the node split operation, x is split into two nodes x_1 and x_2. Each of the two edges (z, x) and (x, y) is split into two edges. C^* becomes a quadrangle with four exterior nodes x_1, x_2, u, v in clockwise order. We recursively construct an **REL** $\mathcal{R}(C^*)$

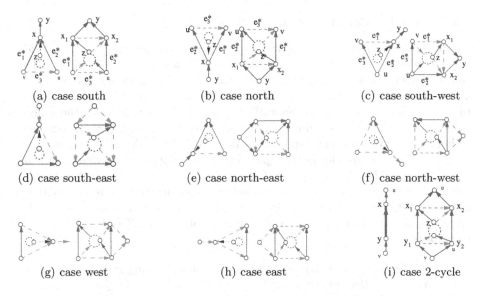

(a) case south   (b) case north   (c) case south-west

(d) case south-east   (e) case north-east   (f) case north-west

(g) case west   (h) case east   (i) case 2-cycle

Fig. 6. Cases of node split operation (Color figure online).

for $G^*(C^*)$ with x_1, x_2, u, v as the north, east, south and west node respectively. Now we put the nodes and the edges in the interior of the subgraph $G^*(C^*)$ back into G_1^*, together with the edge pattern specified in $\mathcal{R}(C^*)$. It is easy to see that after these operations, we get a valid **REL** of the graph $G_1^* \cup G^*(C^*)$.

The other cases are similar as shown in Fig. 6. For each of the eight cases, we get a valid **REL** of the graph $G_1^* \cup G^*(C^*)$ after the node split operation.

Now consider a separating 2-cycle C^* in G^*. We want to add the interior nodes of C^* back into G_1^*. C^* corresponds to a merged 2-cycle edge $e^* = (x, y)$ for some nodes x and y in G_1^*. Let C be the 2-legged cycle in G corresponding to C^*. By Property 2, C has two 2-vertices, a and b, one on each of its two contour paths. a and b correspond to two marked edges e_a^* and e_b^* in G^*. One of them, say e_a^*, is incident to the node x. The other (e_b^*) is incident to the node y. We say e^* is *assigned to* both x and y. Or equivalently, we say the separating 2-cycle C^* is assigned to both x and y. (In Fig. 4 (2), the edges (x, v) and (y, w) are two marked edges in G^*. They are incident to x and y, respectively. So the separating 2-cycle $C^* = \{x, y\}$ is assigned to both x and y). The processing of C^* is similar to a separating triangle. The only difference is that we need to split both x and y. Depending on the pattern of $e^* = (x, y)$ in $\mathcal{R}_1$, there are four cases. For example, if $e^* = y \to x$ is in T_1, then we split x according to the case south, and split y according to the case north. (See Fig. 6 (i), case 2-cycle). After performing these two node split operations, C^* becomes a quadrangle with four exterior nodes x_1, x_2, y_2, y_1 in clockwise order. We recursively construct an **REL** $\mathcal{R}(C^*)$ for $G^*(C^*)$ with x_1, x_2, y_2, y_1 as the north, east, south and west nodes respectively. Putting $\mathcal{R}_1$ and $\mathcal{R}(C^*)$ together, we get a valid **REL** of $G_1^* \cup G^*(C^*)$.

4.2   The Edge Pattern Around a Node

Although we can process the separating cycles of G^* in arbitrary order to add all deleted nodes back into G_1^*, doing so does not guarantee a SSOD of G at the end. Consider a node x in G_1^*. Let $\mathcal{C}$ be the set of all separating cycles of G^* assigned to x. If $\mathcal{C}$ contains several separating cycles, x must be split multiple times in order to destroy all separating cycles in $\mathcal{C}$. To make sure the union of the rectangles corresponding to these split nodes constitutes a star-shaped orthogonal polygon, we must split the node x carefully as described below.

Figure 7 (1) shows the general pattern of the edges in G_1^* around x with respect to the **REL** $\mathcal{R}_1 = \{T_1, T_2\}$. (In Fig. 7 (1), a blue dotted circle indicates the component inside a separating triangle C^* assigned to x. The blue dotted arrow indicates the marked edge inside C^*. A thick line indicates a merged 2-cycle edge assigned to x.) We partition $\mathcal{C}$ into four subsets (some subsets may be empty):

- $\mathcal{C}_S = \{C^* \in \mathcal{C} \mid C^* \text{ is a case south or southwest separating cycle}\}$.
  Let $m_S = |\mathcal{C}_S|$. Denote the separating cycles in $\mathcal{C}_S$ by C_{si}^* $(1 \leq i \leq m_S)$.
- $\mathcal{C}_E = \{C^* \in \mathcal{C} \mid C^* \text{ is a case east or southeast separating cycle}\}$.
  Let $m_E = |\mathcal{C}_E|$. Denote the separating cycles in $\mathcal{C}_E$ by C_{ei}^* $(1 \leq i \leq m_E)$.
- $\mathcal{C}_N = \{C^* \in \mathcal{C} \mid C^* \text{ is a case north or northeast separating cycle}\}$.
  Let $m_N = |\mathcal{C}_N|$. Denote the separating cycles in $\mathcal{C}_N$ by C_{ni}^* $(1 \leq i \leq m_N)$.
- $\mathcal{C}_W = \{C^* \in \mathcal{C} \mid C^* \text{ is a case west or northwest separating cycle}\}$.
  Let $m_S = |\mathcal{C}_S|$. Denote the separating cycles in $\mathcal{C}_S$ by C_{wi}^* $(1 \leq i \leq m_W)$.

We create a subgraph around x as follows (Fig. 7 (1) and (2)):

- Replace x by a new node x_0 and create a cycle K around x_0. K contains four corner nodes $x_{sw}, x_{se}, x_{ne}, x_{nw}$. The edge $x_{sw} \to x_0$ is in T_2. The edge $x_{se} \to x_0$ is in T_1. The edge $x_0 \to x_{ne}$ is in T_2. The edge $x_0 \to x_{nw}$ is in T_1.
- Between x_{sw} and x_{se}, K has a sub-path K_S containing $\max\{1, m_S\}$ edges. All edges in K_S are in T_2 directed counterclockwise. The nodes on K_S are named as x_{si} $(1 \leq i \leq m_S - 1)$ counterclockwise. For $1 \leq i < m_S$, the edge $x_{si} \to x_0$ is in T_1. For $1 \leq i \leq m_S$, the edge $(x_{s(i-1)}, x_{si})$ is used to destroy the separating cycle C_{si}^*. Namely, $(x_{s(i-1)}, x_{si})$ is an edge of the quadrangle obtained from C_{si}^*. Here $x_{s0} = x_{sw}$ and $x_{sm_S} = x_{se}$.
- The other sides of K are similar.

When some of $\mathcal{C}_S, \mathcal{C}_E, \mathcal{C}_N, \mathcal{C}_W$ are empty, they are treated as a special case. For example, when $\mathcal{C}_W = \emptyset$, K_W just contains one T_1 edge $x_{sw} \to x_{nw}$. Then we split the edge (w_1, x) into two edges (w_1, x_{sw}) and (w_1, x_{nw}). (See Fig. 7 (2)).

Note that, for each separating 2-cycle, both end nodes of e^* are split. For example, for the separating 2-cycle C_{s3}^* represented by $e^* = (s_3, x)$, C_{s3}^* becomes a quadrangle with nodes $x_{s2}, x_{se}, s_3', s_3$ (s_3' is a split node from s_3).

This construction deals with the most general case. If some of the sets $\mathcal{C}_S, \mathcal{C}_E, \mathcal{C}_N, \mathcal{C}_W$ are empty, the construction can be simplified.

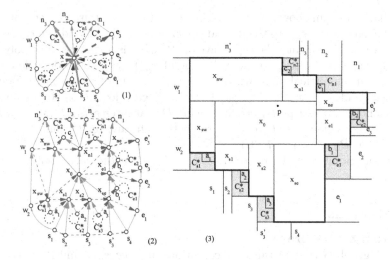

Fig. 7. (1) The edge pattern around a node x; (2) The subgraph created for x; (3) the orthogonal drawing of the subgraph in (2).

Figure 7 (3) shows an orthogonal drawing D of the nodes in the subgraph shown in Fig. 7 (2). Let r_x be the union of the rectangle x_0 and all rectangles $x_{\alpha i}$ ($\alpha \in \{s, e, n, w\}$ and $1 \leq i \leq m_\alpha$). This orthogonal polygon r_x is the face in the drawing $D(G)$ corresponding to the node x in G_1^*. In Fig. 7 (3), r_x is outlined by the thick line segments. A shaded rectangle indicates the region to draw interior nodes in a separating cycle $C_{\alpha i}^*$. Look at C_{s2}^*. The node a_2 is in the interior of C_{s2}^*. The edge (a_2, x) is a marked edge in G^*, and it corresponds to a 2-vertex in G. The northeast corner of the rectangle a_2 in Fig. 7 (3) is this 2-vertex.

Lemma 2. *For any node x in G_1^*, the orthogonal polygon r_x is star-shaped.*

Proof. r_x is obtained by adding the rectangles $x_{\alpha i}$ ($\alpha \in \{s, e, n, w\}$ and $1 \leq i \leq m_\alpha$) to the rectangle x_0. Let P_S be the lower envelop of r_x. P_S consists of the lower boundary of the rectangles $x_{s0}, x_{s1}, \ldots, x_{sm_S-1}, x_{sm_S}$ (where $x_{s0} = x_{sw}$ and $x_{sm_S} = x_{se}$). For $1 \leq i \leq m_S$, there is a marked edge (a_i, x) in the interior of the separating cycle C_{si}^*. Note that $a_i \rightarrow x_{s(i-1)}$ is a T_1 edge and $a_i \rightarrow x_{si}$ is a T_2 edge. So the rectangle x_{a_i} must touch the lower side of the rectangle $x_{s(i-1)}$ and touch the left side of the rectangle x_{si}. So the lower side of x_{si} must be below the lower side of $x_{s(i-1)}$. Since this is true for any $1 \leq i \leq m_S$, the lower envelop P_S of r_x must be a downward staircase-like poly-line, with the lower side of x_{se} as its lowest horizontal segment.

Similarly, we can show that the upper envelop P_N of r_x must be an upward staircase-like poly-line (from right to left, namely from x_{ne} to x_{nw}) with the upper side of x_{nw} as the highest horizontal segment. Because $x_{se} \rightarrow x_0$ is the first clockwise T_1 edge entering x_0 and $x_0 \rightarrow x_{nw}$ is the first clockwise T_1 edge leaving x_0, by Lemma 1, there is a vertical stripe L_v in the drawing D that intersects x_{se}, x_0, x_{nw}. Any point p in the region $x_0 \cap L_v$ can see the entire lower envelop P_S and the entire upper envelop P_N. (See Fig. 7 (3)).

Similarly, we can show the left envelop P_W of r_x is a staircase-like poly-line (from the left side of x_{nw} to the left side of x_{sw}), with the left side of x_{sw} as the leftmost vertical segment. The right envelop P_E of r_x is a staircase-like poly-line (from the right side of x_{se} to the right side of x_{ne}), with the right side of x_{ne} as the rightmost vertical segment. Because $x_{sw} \to x_0$ the first clockwise T_2 edge entering x_0 and $x_0 \to x_{ne}$ is the first clockwise T_2 edge leaving x_0, by Lemma 1, there is a horizontal stripe L_h in the drawing D that intersects x_{sw}, x_0, x_{ne}. Any point p in the region $x_0 \cap L_h$ can see the entire left envelop P_W and the entire right envelop P_E. (See Fig. 7 (3)).

Pick any point p in the region $x_0 \cap L_v \cap L_h$, then the entire polygon r_x is visible from p. $\square$

4.3   Algorithm

Algorithm SSOD-Draw:
Input: A graph G that satisfies the conditions in Theorem 2 and Property 2.

1. Construct the dual graph G^* of G.
2. Construct the graph G_1^*, by deleting all nodes in the interior of maximal separating cycles in G^*.
3. Construct a **REL** $\mathcal{R}_1$ of G_1^*.
4. By using the procedure described above, perform node split operation for all nodes x with at least one maximal separating cycle C^* assigned to it. When C^* is destroyed, make recursive call to construct a **REL** $\mathcal{R}(C^*)$ for $G^*(C^*)$. Let $G^{*\prime}$ be the PTP graph obtained from G_1^* by adding all deleted nodes back into G_1^*. Let $\mathcal{R}'$ be the **REL** of $G^{*\prime}$ obtained in this process.
5. Construct a rectangular dual R' of $G^{*\prime}$ by using $\mathcal{R}'$ as in Lemma 1.
6. Let $D(G)$ be the orthogonal drawing of G obtained from R' as above.

By Lemma 2, for any node x in G_1^*, the orthogonal polygon r_x corresponding to x is star-shaped. Any node y not in G_1^* is in the interior of a maximal separating cycle C^*. The orthogonal polygon r_y for y in $D(G)$ is contained in the drawing for $G^*(C^*)$. Our argument can be recursively applied to the drawing of $G^*(C^*)$ to show r_y is a star-shaped orthogonal polygon. Hence $D(G)$ is a SSOD of G. All steps in Algorithm **SSOD-Draw** can be done in linear time by Theorem 4 and basic algorithmic techniques for planar graphs. In summary:

Theorem 5. *Let G be a graph that satisfies the conditions in Theorem 2. Then G has a SSOD drawing, which can be constructed in linear time.*

5   Conclusion

In this paper, we strengthen the result in [1]. We show that if G satisfies the same conditions as in [1], it not only has an orthogonally convex drawing, but also a stronger star-shaped orthogonal drawing. The method we use is quite different from the methods used in [1,8]. It will be interesting to see if this method can be used to solve other orthogonal drawing problems.

References

1. Chang, Y.-J., Yen, H.-C.: On orthogonally convex drawings of plane graphs. In: Wismath, S., Wolff, A. (eds.) GD 2013. LNCS, vol. 8242, pp. 400–411. Springer, Heidelberg (2013)
2. Duncan, C.A., Goodrich, M.T.: Planar orthogonal and polyline drawing algorithms. In: Tammassia, R. (ed.) Handbook of Graph Drawings and Visualization, Chap. 7, pp. 223–246. CRC Press (2013)
3. He, X.: On finding the rectangular duals of planar triangular graphs. SIAM J. Comput. **22**, 1218–1226 (1993)
4. He, X.: On floor-plan of plane graphs. SIAM J. Comput. **28**, 2150–2167 (1999)
5. Hong, S.-H., Nagamochi, H.: A linear-time algorithm for star-shaped drawings of planar graphs with the minimum number of concave corners. Algorithmica **62**, 1122–1158 (2012)
6. Koźmiński, K., Kinnen, E.: Rectangular duals of planar graphs. Networks **5**, 145–157 (1985)
7. Lingas, A., Wasylewicz, A., Żyliński, P.: Note on covering monotone orthogonal polygons with star-shaped polygons. Info. Proc. Lett. **104**, 220–227 (2007)
8. Rahman, M., Nishizeki, T.: Orthogonal drawings of plane graphs without bends. J. Gr. Algorithms Appl. **7**, 335–362 (2003)

The Domination Number of On-line Social Networks and Random Geometric Graphs

Anthony Bonato[1]([⊠]), Marc Lozier[1], Dieter Mitsche[2], Xavier Pérez-Giménez[1], and Paweł Prałat[1]

[1] Ryerson University, Toronto, Canada
{abonato,marc.lozier,xperez,pralat}@ryerson.ca
[2] Université de Nice Sophia-Antipolis, Nice, France
dmitsche@unice.fr

Abstract. We consider the domination number for on-line social networks, both in a stochastic network model, and for real-world, networked data. Asymptotic sublinear bounds are rigorously derived for the domination number of graphs generated by the memoryless geometric protean random graph model. We establish sublinear bounds for the domination number of graphs in the Facebook 100 data set, and these bounds are well-correlated with those predicted by the stochastic model. In addition, we derive the asymptotic value of the domination number in classical random geometric graphs.

1 Introduction

On-line social networks (or OSNs) such as Facebook have emerged as a hot topic within the network science community. Several studies suggest OSNs satisfy many properties in common with other complex networks, such as: power-law degree distributions [2,13], high local clustering [36], constant [36] or even shrinking diameter with network size [23], densification [23], and localized information flow bottlenecks [12,24]. Several models were designed to simulate these properties [19,20], and one model that rigorously asymptotically captures all these properties is the geometric protean model (GEO-P) [5–7] (see [16,25,30,31] for models where various ranking schemes were first used, and which inspired the GEO-P model). For a survey of OSN models see [8], and for more general complex networks [3]. A fundamental difference with GEO-P versus other models [2,21–23] is that it posits an underlying feature or metric space. This metric space mirrors a construction in the social sciences called *Blau space* [26]. In Blau space, agents in the social network correspond to points in a metric space, and the relative position of nodes follows the principle of *homophily* [27]: nodes with similar socio-demographics are closer together in the space. We give the precise definition of the GEO-P model (actually, one of its variants, the so-called MGEO-P model) below. We focus on the MGEO-P model, since it is simpler than GEO-P and generates graphs with similar properties.

Supported by grants from NSERC.

R. Jain et al. (Eds.): TAMC 2015, LNCS 9076, pp. 150–163, 2015.
DOI: 10.1007/978-3-319-17142-5_14

The study of domination and dominating sets plays a prominent role in graph theory with a number of application to real-world networks. A *dominating set* in a graph G is a set of nodes S in G such that every node not in S is adjacent to at least one node in S. The *domination number* of G, written $\gamma(G)$, is the minimum cardinality of a dominating set in G. Computing $\gamma(G)$ is a well-known **NP**-complete problem, so typically heuristic algorithms are used to compute it for large-scale networks. Dominating sets appear in numerous applications such as: network controllability [11], as a centrality measure for efficient data routing [34], and detecting biologically significant proteins in protein-protein interaction network [28]. For more additional background on domination in graph theory, see [15].

In social networks, we consider the hypothesis that minimum order dominating sets contain agents with strong influence over the rest of the network. Our goal in the present paper is to consider the problem of finding bounds on dominating sets in stochastic models of OSNs, and also in real-world data derived from OSNs. We consider bounds on the domination number of a stochastic model (see next paragraph), and upper bounds for that model are well-correlated with real-world OSN data. We note that the domination number has been studied previously in complex network models, including preferential attachment [9], and recently in [29].

The OSN model we consider is called the *memoryless geometric protean model* (*MGEO-P*), first introduced in [4]. The MGEO-P model depends on five parameters which consist of: the number of nodes n, the dimension of the metric space m, the attachment parameter $0 < \alpha < 1$, the density parameter $0 < \beta < 1 - \alpha$, and the connection probability $0 < p \leq 1$.

The nodes and edges of the network arise from the following process. Initially, the network is empty. At each of n steps, a new node v arrives and is assigned both a random position q_v in $\mathbb{R}^m$ within the unit-hypercube $[0, 1)^m$ and a random rank r_v from those unused ranks remaining in the set 1 to n. The influence radius of any node v is computed based on the formula:

$$I(r_v) = \tfrac{1}{2}\left(r_v^{-\alpha} n^{-\beta}\right)^{1/m}.$$

With probability p, the node v is adjacent to each existing node u satisfying $\mathcal{D}(v, u) \leq I(r_u)$, where the distances are computed with respect to the following metric:

$$\mathcal{D}(v, u) = \min\left\{\|q_v - q_u - z\|_\infty : z \in \{-1, 0, 1\}^m\right\},$$

and where $\|\cdot\|_\infty$ is the infinity-norm. We note that this implies that the geometric space is symmetric in any point as the metric "wraps" around like on a torus. The volume of the space influenced by the node is $r_v^{-\alpha} n^{-\beta}$. Then the next node arrives and repeats the process until all n nodes have been placed. We refer to this model by MGEO-P(n, m, α, β, p).

We give rigorous bounds on the domination number of a typical graph generated by the MGEO-P. An event A_n holds *asymptotically almost surely* (*a.a.s.*) if it holds with probability tending to 1 as n tends to infinity. Our main result on MGEO-P is the following.

Theorem 1. *If $m = o(\log n)$, then a.a.s. the domination number of a graph G sampled from the MGEO-P(n, m, α, β, p) model satisfies*

$$\gamma(G) = \Omega(C^{-m/(1-\alpha)}n^{\alpha+\beta}) \quad and \quad \gamma(G) = O(n^{\alpha+\beta}\log n),$$

where C is any constant greater than 6. In particular, a.a.s. $\gamma(G) = n^{\alpha+\beta+o(1)}$.

We defer the proof of Theorem 1 to Sect. 2. It is noteworthy that the domination number of the preferential attachment model is linear in the order of the graphs sampled; see [9]; this fact further demonstrates the differences between MGEO-P and other complex graph models.

Theorem 1 suggests a sublinear bound on the domination number for OSNs, and we evidence for this in real-world data. In Sect. 3, we find bounds for graphs in the Facebook 100 data set, and compare these results to those for the stochastic models. We chose to work with the so-called *Facebook 100* (or FB100) data set, as it provides representative samples from the network of increasing orders. Hence, we may consider trends for the domination number in the data. While the data presented is our first and initial study, the bounds we find for the domination number of FB100 are of sublinear order, and these bounds are well-correlated with those from MGEO-P. Sublinear domination results for other complex networks were also reported in [29]; our approach is distinct as we consider social networks of increasing orders.

In addition to the results above, we find rigorous bounds on the domination number for classical random geometric graphs. Given a positive integer n, and a non-negative real r, we consider a *random geometric graph* $G = (V, E) \in \mathscr{G}(n, r)$ defined as follows. The node set V of G is obtained by choosing n points independently and uniformly at random in the square $\mathcal{S} = [0, 1]^2$. (Note that, with probability 1, no point in $\mathcal{S}$ is chosen more than once, and hence, we may assume that $|V| = n$.) For notational purposes, we identify each node $v \in V$ with its corresponding geometric position $v = (v_x, v_y) \in \mathcal{S}$, where v_x and v_y denote the usual x- and y-coordinates in $\mathcal{S}$, respectively. Finally, the edge set E is constructed by connecting each pair of nodes u and v by an edge if and only if $d_E(u, v) \leq r$, where d_E denotes the Euclidean distance in $\mathcal{S}$.

Random geometric graphs were first introduced in a slightly different setting by Gilbert [14] to model the communications between radio stations. Since then several closely related variants on these graphs have been widely used as a model for wireless communication, and have also been extensively studied from a mathematical point of view. The basic reference on random geometric graphs is the monograph by Penrose [32].

We note that our study is the first to explicitly provide provable bounds on the domination number of random geometric graphs. In particular, we derive the following result.

Theorem 2. *Let $G \in \mathscr{G}(n, r)$ and let $\omega = \omega(n)$ be any function tending to infinity as $n \to \infty$. Then a.a.s. the following holds:*

(a)Denote by $N(x)$ the minimal number of balls of radius x needed to cover S. If $r = \Theta(1)$, then

$$\Omega(1) = N(r + \sqrt{\omega \log n / n}) \le \gamma(G) \le N(r - \omega / \sqrt{n}).$$

(b) Define $C = 2\pi\sqrt{3}/9 \approx 1.209$. If $\omega\sqrt{\log n / n} \le r = o(1)$, then

$$\gamma(G) = (C / \pi + o(1))r^{-2}.$$

(c) If $1/\sqrt{n} \le r < \omega\sqrt{\log n / n}$, then

$$\gamma(G) = \Theta(r^{-2}).$$

(d) If $r < 1/\sqrt{n}$, then

$$\gamma(G) = \Theta(n).$$

It is straightforward to verify that the bounds on $\gamma(G)$ in part (a) differ by at most 1 if ω is sufficiently small, but in general we do not give accurate estimations of $N(r)$ for $r = \Theta(1)$. The proof of Theorem 2 is deferred to Sect. 4. The final section summarizes our results and presents open problems.

2   Proof of Theorem 1

For each node $v \in [n]$, we consider the ball $B_v = \{x \in [0,1)^m : D(x, v) \le I(r_v)\}$, which has volume $b_v = r_v^{-\alpha}n^{-\beta}$. The next lemma will be useful to estimate the sum of volumes of the balls corresponding to a set of nodes.

Lemma 1. *Let T be a set of t nodes (fixed before ranks are chosen) with $\omega n^\alpha \log n \le t \le n$, for a function ω going to infinity with n arbitrarily slowly.*

(a) Then a.a.s.

$$\sum_{i \in T} r_i^{-\alpha} = (1 + o(1))\frac{tn^{-\alpha}}{1 - \alpha}. \tag{1}$$

(b) Furthermore, given any integer s such that $1 \le s \le t$ and $s^{1-\alpha} \ge \omega(n/t)^\alpha \log n$, a.a.s. all subsets $S \subseteq T$ of s nodes satisfy

$$\sum_{i \in S} r_i^{-\alpha} \le (1 + o(1))\frac{s^{1-\alpha}(t/n)^\alpha}{1 - \alpha}. \tag{2}$$

Observe that the sum in (1) is asymptotic to what one would expect. Indeed, if the ranks of the nodes in T are distributed evenly, then one would obtain $\sum_{i \in T}^t r_i^{-\alpha} = \sum_{i=1}^t (in/t)^{-\alpha} = tn^{-\alpha}/(1 - \alpha) + O(1)$.

Proof. Let s and t be integers satisfying all the conditions of the statement in part (b). Set $\hat{\omega} = \omega^{1/4} \to \infty$, so we have $t \ge \hat{\omega}^4(t/s)^{1-\alpha}n^\alpha \log n$. This also implies $\hat{\omega} = o((sn/t)^{1-\alpha})$. Let Y_j be the number of elements in T with rank at most j. Observe that Y_j has expectation jt/n, and follows a hypergeometric distribution. For $(sn/t)^{1-\alpha}/\hat\omega \le j \le n$, a Chernoff bound (see e.g. [17]) gives that

$$\mathbf{Pr}\left(\left|Y_j - \frac{jt}{n}\right| \ge (1/\hat\omega)\frac{jt}{n}\right) \le 2\exp\left(-\frac{jt}{3\hat\omega^2 n}\right) \le 2e^{-\hat\omega \log n/3} = o(1/n^2).$$

We apply a union bound over all j, and conclude that a.a.s., for every $(sn/t)^{1-\alpha}/\hat\omega \le j \le n$,

$$(1 - 1/\hat\omega)jt/n < Y_j < (1 + 1/\hat\omega)jt/n.$$

In order to estimate the sums in the statement, we assume w.l.o.g. that $T = [t]$ and $r_1 < r_2 < \cdots < r_t$ (otherwise we permute the indices of the vertices in T). It follows that a.a.s., for every $(sn/t)^{1-\alpha}/\hat\omega \le j \le n$,

$$r_{\lfloor (1-1/\hat\omega)jt/n \rfloor} \le j \le r_{\lceil (1+1/\hat\omega)jt/n \rceil}.$$

Therefore, setting $\ell = 2s^{1-\alpha}t^\alpha n^{-\alpha}/\hat\omega$, we have that a.a.s., for every $\ell \le i \le t$,

$$\left\lfloor \frac{1}{1+1/\hat\omega}in/t \right\rfloor \le r_i \le \left\lceil \frac{1}{1-1/\hat\omega}in/t \right\rceil.$$

For the lower bound on r_i below, we need to use the fact that $\left\lfloor \frac{1}{1+1/\hat\omega}in/t \right\rfloor \ge (sn/t)^{1-\alpha}/\hat\omega$, which is easily verified to be true since $\hat\omega = o\left((sn/t)^{1-\alpha}\right)$. Finally, we infer that a.a.s., for any choice of S,

$$\sum_{i\in S} r_i^{-\alpha} \le \sum_{i=1}^{s} r_i^{-\alpha} = (1 + o(1))\left(\frac{n}{t}\right)^{-\alpha}\sum_{i=\ell}^{s} i^{-\alpha} + O(\ell) = (1 + o(1))\frac{1}{1-\alpha}s^{1-\alpha}t^\alpha n^{-\alpha}.$$

This proves statement (b). For statement (a), take $s = t$ and note that for this choice of s, for any $wn^\alpha \log n \le t \le n$, the condition $s^{1-\alpha} \ge w(n/t)^\alpha \log n$ is satisfied. Observe that then $S = T = [t]$, so the first inequality in the above equation is an equality. $\qquad\square$

Upper Bound: Fix a constant $K > \frac{1-\alpha}{p}$, and let D be the set containing the first $t = \lfloor Kn^{\alpha+\beta}\log n \rfloor$ nodes added in the process. We will show that a.a.s. D is a dominating set. By Lemma 1, we may condition on the event that (1) holds for $t = |D| = \lfloor Kn^{\alpha+\beta}\log n \rfloor$. Note that this assumption on the ranks does not affect the distribution of the location of the nodes in $[0,1)^m$. Therefore, given a node $u > t$ (appearing in the process later than nodes in D), the probability that u is not dominated by D is

$$\prod_{i=1}^{t}(1 - pr_i^{-\alpha}n^{-\beta}) \le \exp\left(-pn^{-\beta}\sum_{i=1}^{t} r_i^{-\alpha}\right) = \exp\left(-(1+o(1))\frac{p}{1-\alpha}tn^{-\alpha-\beta}\right)$$

$$= \exp\left(-(1+o(1))\frac{Kp}{1-\alpha}\log n\right) = o(1/n).$$

Taking a union bound over all nodes not in D, we can guarantee that a.a.s. all nodes are dominated.

As an alternative and relatively simple approach, one may prove the same upper bound on the domination number as follows. First, show that a.a.s. the minimum degree δ is at least $(1 + o(1))pn^{1-\alpha-\beta}$. Then we may use Theorem 1.2.2 in [1], which states that for every graph G with minimum degree δ,

$$\gamma(G) \leq n\frac{1 + \log(\delta + 1)}{\delta + 1}. \tag{3}$$

Lower Bound: We consider for convenience a natural directed version of MGEO-P(n, m, α, β, p), by orienting each edge from its "younger" end node (that is, appearing later in the process) to its "older" end node. For a set of nodes $D \subseteq [n]$, $N_{\text{in}}(D)$ denotes the set of nodes $u \in [n] \setminus D$ such that there is a directed edge from u to some node in D or, equivalently, such that there is an edge from u to some node in D that is older than u. $N_{\text{out}}(D)$ is defined analogously, replacing older by younger.

Define $t = \lceil n^{\alpha+\beta} \rceil$, and let $T = [t]$ be the set of the oldest t nodes in the process. We want to show that a.a.s. there is no dominating set of order at most $\xi\mu^{-m}n^{\alpha+\beta}$, where ξ and μ are specified later. We give more power to our adversary by allowing her to pick (deterministically after the graph has been revealed) two sets of nodes D_1 and D_2 of order $\lfloor \xi\mu^{-m}n^{\alpha+\beta} \rfloor$ each (not necessarily disjoint). Her goal is also easier than the original one; she needs to achieve that, for every node $v \in T$, either v is in-dominated by D_1 (that is, $v \in D_1 \cup N_{\text{in}}(D_1)$) or v is out-dominated by D_2 (that is, $v \in D_2 \cup N_{\text{out}}(D_2)$); nothing is required for young nodes in $[n] \setminus T$. We show that a.a.s. the adversary cannot succeed, that is, regardless of her choice of D_1, D_2 we have always some node in T not in $D_1 \cup D_2 \cup N_{\text{in}}(D_1) \cup N_{\text{out}}(D_2)$.

Out-domination: Given a constant $0 < \varepsilon < 1$, we define T' to be the set of nodes in T with rank greater than $(1 - \varepsilon)n$. Note that $|T'|$ has a hypergeometric distribution, so it follows easily from Chernoff's bound (see [17]) that a.a.s. $|T'| \geq (\varepsilon/2)n^{\alpha+\beta}$. For convenience, we choose $\varepsilon = \varepsilon(\alpha)$ to be the only real in $(0, 1)$ satisfying $\varepsilon = 2(1 - \varepsilon)^{\alpha}$. For every node $i \in T'$, the corresponding ball B_i has length at most

$$((1 - \varepsilon)^{-\alpha}n^{-\alpha-\beta})^{1/m} = ((2/\varepsilon)n^{-\alpha-\beta})^{1/m}.$$

We consider a tessellation of $[0, 1)^2$ into large cells. At the centre of each large cell we consider a smaller cell. Small cells have side length $((2/\varepsilon)n^{-\alpha-\beta})^{1/m}$ and large ones have side length $2((2/\varepsilon)n^{-\alpha-\beta})^{1/m}$. There are

$$N = \left\lfloor \frac{1}{2}((\varepsilon/2)n^{\alpha+\beta})^{1/m} \right\rfloor^m = \frac{\varepsilon}{2}(2 + o(1))^{-m}n^{\alpha+\beta} \to \infty$$

large cells fully contained in $[0, 1)^m$ (we discard the rest), and thus N small cells inside of those. By construction, if a node in T' falls into a small cell, then its ball is contained into in the corresponding large cell. Let $\mathcal{X}$ be the set of small cells that contain at least one node in T', and let $T'' \subseteq T'$ be a set of $X = |\mathcal{X}|$ nodes such that each cell in $\mathcal{X}$ contains precisely one node in T'' (if a given small cell contains at least two nodes in T', then a node is selected arbitrarily to be placed in T''). Vertices in T'' are potentially dangerous for the adversary since, one node in D_2 can "out-dominate" at most one single node in T''. However, she may in theory get lucky and in-dominate many of these nodes (in the next section we will show that this will not happen a.a.s.).

We want to show that a.a.s. $X \geq N/4$. The probability that there are at least $3N/4$ small cells containing no nodes in T' is at most

$$\binom{N}{\lceil 3N/4 \rceil} \left(1 - (3N/4)(2/\varepsilon)n^{-\alpha-\beta}\right)^{(\varepsilon/2)n^{\alpha+\beta}} \leq 2^N \exp(-3N/4) = o(1).$$

Therefore,

$$X \geq N/4 = \frac{\varepsilon}{8}\lambda^{-m}n^{\alpha+\beta}, \qquad \text{for some} \quad \lambda = 2 + o(1). \tag{4}$$

In-domination: Let $\xi = \xi(\alpha)$ be a sufficiently small positive constant, and define

$$\mu = \left(\lambda\left(1 + 2(2/\varepsilon)^{1/m}\right)\right)^{1/(1-\alpha)} > 3\lambda.$$

The adversary chooses a set $D_1 \subseteq T$ of $s = \lfloor \xi\mu^{-m}n^{\alpha+\beta} \rfloor$ nodes in her attempt to in-dominate T'. By Lemma 1(b), a.a.s. regardless of her choice,

$$\sum_{i \in D_1} r_i^{-\alpha} \leq \frac{(1+o(1))}{1-\alpha}s^{-\alpha+1}t^{\alpha}n^{-\alpha} = (1+o(1))\frac{\xi^{1-\alpha}}{1-\alpha}\mu^{-(1-\alpha)m}n^{\beta}. \tag{5}$$

We tessellate the space into cells of volume $(2/\varepsilon)n^{-\alpha-\beta}$ (same size as the small cells in the out-domination part, but now we have the whole space partitioned into cells of that size). Recall that, for each node $i \in D_1$, the ball B_i has length $b_i^{1/m} \geq n^{-(\alpha+\beta)/m}$. Therefore, the volume of the set of cells intersected by B_i is at most

$$\left(b_i^{1/m} + 2((2/\varepsilon)n^{-\alpha-\beta})^{1/m}\right)^m \leq \left(1 + 2(2/\varepsilon)^{1/m}\right)^m b_i = \left(\mu^{1-\alpha}/\lambda\right)^m b_i.$$

Combining this and (5), a.a.s. and regardless of the adversary's choice, the total volume of the cells intersected by the balls of the nodes in D_1 is at most

$$\sum_{i \in D_1} \left(\mu^{1-\alpha}/\lambda\right)^m b_i = \left(\mu^{1-\alpha}/\lambda\right)^m n^{-\beta} \sum_{i \in D_1} r_i^{-\alpha} \leq (1+o(1))\frac{\xi^{1-\alpha}}{1-\alpha}\lambda^{-m}. \tag{6}$$

Let $\mathcal{Y}$ be the set of cells intersected by the balls of the nodes in D_1, and put $Y = |\mathcal{Y}|$. By (6), a.a.s.

$$Y \leq \frac{\varepsilon\xi^{1-\alpha}}{2(1-\alpha)}\lambda^{-m}n^{\alpha+\beta}.$$

Thus, in view of (4) we just need to make ξ small enough so that $|\mathcal{X} \setminus \mathcal{Y}|$ is larger than $|D_2| = \lfloor \xi \mu^{-m} n^{\alpha+\beta} \rfloor$. That is because dangerous cells in $\mathcal{X} \setminus \mathcal{Y}$ contain nodes in T' that are not in-dominated by D_1, and each one of these cells requires one different node in D_2 to out-dominate its nodes. Recall that our choice of $\varepsilon \in (0,1)$ depends only on α. Then picking ξ sufficiently small so that $\frac{\varepsilon \xi^{1-\alpha}}{2(1-\alpha)} + \xi < \frac{\varepsilon}{8}$, we get

$$|\mathcal{X} \setminus \mathcal{Y}| \geq X - Y \geq \left(\frac{\varepsilon}{8} - \frac{\varepsilon \xi^{1-\alpha}}{2(1-\alpha)} \right) \lambda^{-m} n^{\alpha+\beta} > \xi \mu^{-m} n^{\alpha+\beta} \geq \lfloor \xi \mu^{-m} n^{\alpha+\beta} \rfloor = |D_2|,$$

where we also used that $\mu > 3\lambda > \lambda$. Finally, distinguishing the cases $m = O(1)$ and $m \to \infty$, we observe that

$$\mu^{-m} = \left(\lambda (1 + 2(2/\varepsilon)^{1/m}) \right)^{-m/(1-\alpha)} = \begin{cases} \Theta(1) & \text{for} \quad m = O(1), \\ (6 + o(1))^{-m/(1-\alpha)} & \text{for} \quad m \to \infty, \end{cases}$$

so the claimed lower bound follows.

3   Domination in Facebook 100 Graphs

Facebook distributed 100 samples of social networks from universities within the United States measured as of September 2005 [35], which range in size from 700 nodes to 42,000 nodes. We call these networks the *Facebook 100* (or simply FB100) graphs. As the domination number is sensitive to nodes of low degree, we used the k-core of the network, where $1 \leq k \leq 5$; see [33]. For $k \in \mathbb{N}$, the *k-core* of a graph is the largest induced subgraph of minimum degree at least k. The k-core can be found by a simple node deletion algorithm that repeatedly deletes nodes with degree less than k. This algorithm always terminates with the k-core of the graph, which is possibly empty.

Several algorithms were used to bound the domination number of the FB100 graphs, but one providing the smallest dominating sets is an adaptation of the *DS-DC* algorithm [28]. In the algorithm, initially all nodes V are in the dominating set S. It then selects a node u of minimum degree in S, and deletes it only if the set $S \setminus \{u\}$ remains dominating. The algorithm then repeats these steps for all nodes in S in order of their increasing degrees. We considered other algorithms, such as greedy algorithms where high degree nodes are added to an empty dominating set sequentially, or by choosing a random dominating set, but DS-DC outperformed these algorithms. We omit a detailed discussion of the performance of other algorithms owing to space.

Figure 1 presents the DS-DC predicted upper bounds on $\gamma(G)$, where G is a graph in the FB100 data set. We plotted the upper bound predicted by the MGEO-P model in Theorem 1, and we note the close similarity between that bound and the ones for FB100. Note that we ignore constants in the big Oh term in the upper bound from the model, and simply plot the bound generated by $n^{\alpha+\beta} \log n$. The values for α, β, and the dimension parameter m for each of the FB100 graphs are taken from tables provided in [4]. (For example, in order

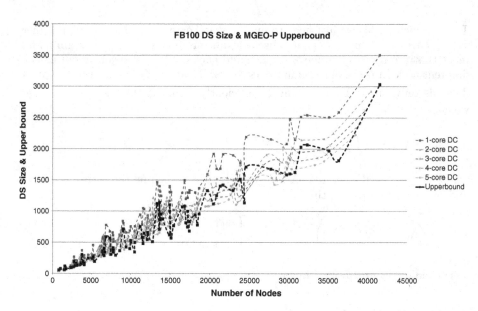

Fig. 1. Upper bounds on the domination number of the FB100 networks vs MGEO-P.

to determine the power-law exponent, the Clauset-Shalizi-Newman power law exponent estimator was used; see [4] for more details.) The MGEO-P bound seems well-correlated with the bounds provided in the k-core, especially where $k = 3, 4, 5$. See Table 1, which fits the domination number of the FB100 graphs to the curve $y = n^x \log n$.

Table 1. Fitting the domination number of the k-cores of FB100 to $y = n^x \log n$, suggesting a sub-linear trend.

k	x	R^2
1	0.509	0.8472
2	0.492	0.8292
3	0.4818	0.8179
4	0.4741	0.8093
5	0.4677	0.803

To contrast the bounds provided in Fig. 1 with the bound in (3), we plot them in Fig. 2. We plotted the theoretical bound using $\delta = 5$ (that is, the minimum degree of the 5-core). The figure shows a significant over-estimate of the domination number of the bound in (3), further corroborating the claim that the domination numbers of the FB100 graphs are sublinear with respect to the order of the graph.

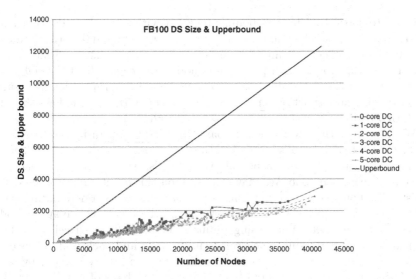

Fig. 2. Upper bounds on the domination number of the FB100 networks vs the bound in (3), showing a substantial overestimation

4   Proof of Theorem 2

We will relate the domination number to the problem of covering the plane with circles. Given $x \in \mathbb{R}^2$ and $\rho > 0$, we denote by $\mathcal{B}(x, \rho)$ the ball with centre x and of radius ρ. The following theorem is well known [18].

Theorem 3 ([18]). *Given a bounded subset of the plane M, for $\varepsilon > 0$ let $N(\varepsilon)$ be the minimum number of balls of radius ε that can cover M. Then we have that*

$$\lim_{\varepsilon \to 0} \pi \varepsilon^2 N(\varepsilon) = C \ \text{Area}(\overline{M}),$$

where $\overline{M}$ denotes the closure of M.

Observe that $(C - 1)$ can therefore, be seen as measuring the proportion of unavoidable overlapping. Moreover, [18] shows that an optimal covering of the square S using balls of radius ε corresponds to arranging the balls in such a way that their centers are the centers of the cells of a hexagonal tiling of length ε. More precisely, consider the lattice

$$\mathcal{L}_\varepsilon = \{i\varepsilon(\sqrt{3}, 0) + j\varepsilon(\sqrt{3}/2, 3/2) : i, j \in \mathbb{Z}\}. \tag{7}$$

Then the set of balls of radius ε and centre in $\mathcal{L}_\varepsilon$ that intersect S form a covering of S that gives the limit in Theorem 3.

Note that for all G with maximum degree Δ, we trivially have $\gamma(G) \geq n/(1 + \Delta(G))$ (for further relations between $\gamma(G)$ and other graph parameters, see, for example, [15]). Given any constant $c > 0$, for $G \in \mathscr{G}(n, r)$ with $r \geq c\sqrt{\log n / n}$, it is easy to show, by Chernoff bounds together with union bounds, that a.a.s. $\Delta(G) = O(r^2 n)$. Therefore, a.a.s. we have $\gamma(G) = \Omega(r^{-2})$. On the other hand, we can trivially construct a dominating set of $\mathscr{G}(n, r)$ by tessellating $\mathcal{S}$ into square cells of side length $r/\sqrt{2}$ and picking one node from each cell (if the cell is not empty). This holds deterministically for any geometric graph (not necessarily random), with no restriction on r, and gives $\gamma(G) = O(r^{-2})$. It follows that, for $G \in \mathscr{G}(n, r)$ with $r \geq c\sqrt{\log n / n}$, a.a.s. $\gamma(G) = \Theta(r^{-2})$.

We first prove the lower bound in part (b). Fix an arbitrarily small constant $\delta > 0$. Tessellate $\mathcal{S}$ into cells of side length $\alpha = \sqrt{\omega \log n / n} = o(r)$. By Chernoff bounds together with a union bound over all cells, we get that a.a.s. each cell contains at least one node. We may condition on this event, and proceed deterministically. For a contradiction, suppose that there exists a dominating set of size $s = \lfloor (C/\pi - \delta)r^{-2} \rfloor$. Consider then s balls whose centers are at the nodes of the dominating set. When using radius r for all balls, each cell is at least touched by some ball, since each cell is non-empty and each node is covered. Hence, by using radius $r' = r + \alpha\sqrt{2}$, each square is totally covered by some ball. Therefore, $\mathcal{S}$ can be covered by $\lfloor (C/\pi - \delta)r^{-2} \rfloor$ balls of radius r'. On the other hand,

$$\pi r'^2 \lfloor (C/\pi - \delta)r^{-2} \rfloor \leq (r^2 + 2\sqrt{2}r\alpha + 2\alpha^2)(C - \pi\delta)r^{-2} = (1 + o(1))(C - \pi\delta),$$

since $\alpha = o(r)$. This contradicts Theorem 3 and so $\gamma(G) > s$. Since the argument holds for any $\delta > 0$, we get the desired lower bound.

For the upper bound, we will show that we can find a covering of $\mathcal{S}$ with $(C/\pi + o(1))r^{-2}$ balls of radius r that are centered at some nodes of G. Again, fix some arbitrarily small constant $\delta > 0$. Let $r' = (1 - \delta)r$, and consider the lattice $\mathcal{L}_{r'}$, as defined in (7). Let $\mathcal{L}'_{r'}$ be the set of all points $x \in \mathcal{L}_{r'}$ such that the ball with centre x and radius r' intersects $\mathcal{S}$. Recall that $\mathcal{L}'_{r'}$ gives the optimal covering of $\mathcal{S}$ with balls of radius r', and therefore, attains the bound given by Theorem 3

$$s = |\mathcal{L}'_{r'}| = \left(\frac{C}{\pi r'^2}\right)(1 + o(1)) = \left(\frac{C}{\pi}\right)r^{-2}(1 - \delta)^{-2}(1 + o(1)).$$

It might happen that some point $x \in \mathcal{L}'_{r'}$ does not belong to $\mathcal{S}$. In this case, we replace x by the closest point $\hat{x}$ on the boundary of $\mathcal{S}$ (this can be uniquely done, since $\mathcal{S}$ is closed and convex). Note that $\mathcal{B}(x, r') \cap \mathcal{S} \subseteq \mathcal{B}(\hat{x}, r') \cap \mathcal{S}$. We denote $\hat{\mathcal{L}}_{r'}$ the modified set of points that we obtained. By construction, $\hat{\mathcal{L}}_{r'} \subseteq \mathcal{S}$, and we can cover $\mathcal{S}$ using balls with centre in $\hat{\mathcal{L}}_{r'}$ and radius r' or larger. Moreover, $|\hat{\mathcal{L}}_{r'}| = s$. Clearly, if we can guarantee that for each $x \in \hat{\mathcal{L}}_{r'}$ there exists a node of G inside $\mathcal{B}(x, \delta r) \cap \mathcal{S}$, then G is dominated by these nodes, and hence, s yields an upper bound for $\gamma(G)$.

Observe that for any point $x \in \hat{\mathcal{L}}_{r'}$ (and therefore, in $\mathcal{S}$), the area of $\mathcal{B}(x, \delta r) \cap \mathcal{S}$ is at least $(\delta r)^2 \pi / 4$, since at least a quarter of a ball must be inside $\mathcal{S}$. The probability that there is no node of G in $\mathcal{B}(x, \delta r) \cap \mathcal{S}$ is at most

$$\left(1 - \frac{(\delta r)^2 \pi}{4}\right)^n \leq \exp\left(-\frac{n(\delta r)^2 \pi}{4}\right) \leq \exp\left(-\frac{\omega^2 \delta^2 \pi \log n}{4}\right) = o(n^{-2}).$$

Since there are s events that we need to investigate and clearly $s \leq n$, by a union bound, a.a.s., for every $x \in \hat{\mathcal{L}}_{r'}$, the region $\mathcal{B}(x, \delta r) \cap \mathcal{S}$ contains at least one node of G. It follows that a.a.s. $\gamma(G) \leq s$ and since the argument holds for any $\delta > 0$, we derive the desired upper bound.

For the proof of part (a), note that $N(x)$ is non-decreasing function of x, and $N(x) = 1$ for $x \geq 1/\sqrt{2}$. Fix $r = \Theta(1)$. Tessellate $\mathcal{S}$ into cells of side length $\alpha = \sqrt{(\omega/2) \log n / n}$. For the lower bound, suppose for contradiction that $\gamma(G) \leq N(r + \alpha\sqrt{2}) - 1$. By Chernoff bounds together with a union bound over all cells, a.a.s. there is at least one node in each such cell. Now place $N(r + \alpha\sqrt{2}) - 1$ many balls with centers at the nodes of the dominating set. Since by using radius r each cell is at least touched by some ball, by using radius $r + \alpha\sqrt{2}$ each cell is totally covered by a ball, and therefore $\mathcal{S}$ is covered by $N(r + \alpha\sqrt{2}) - 1$ balls of radius $r + \alpha\sqrt{2}$, contradicting the definition of $N(x)$. Therefore, a.a.s. $\gamma(G) \geq N(r + \alpha\sqrt{2})$.

For the upper bound, consider an optimal arrangement of $N(r - \beta)$ balls of radius $r - \beta$, where $\beta = \omega/\sqrt{n}$. As before, if the centre p of a ball is outside $\mathcal{S}$, but $\mathcal{B}(p, r - \beta) \cap \mathcal{S} \neq \emptyset$, we may shift the centre of the ball towards its closest point p' on the boundary of $\mathcal{S}$. Since $\mathcal{B}(p, r - \beta) \cap \mathcal{S} \subseteq \mathcal{B}(p', r - \beta) \cap \mathcal{S}$, we still preserve the covering property, and therefore, we can obtain an optimal covering of $\mathcal{S}$ with balls of radius $r - \beta$ and centered at points inside of $\mathcal{S}$. As in part (a), it suffices to show the existence of a node $v \in V$ inside $\mathcal{B}(c, \beta) \cap \mathcal{S}$ for any centre c in this optimal arrangement of balls. Since $N(r - \beta) = O(1)$, the probability that there exists a centre c such that $(\mathcal{B}(c, \beta) \cap \mathcal{S}) \cap V = \emptyset$ is at most

$$O(1)(1 - \beta^2\pi / 4)^n = O(\exp(-n\beta^2\pi / 4)) = o(1),$$

and hence, a.a.s. for all centers c, we have that $\mathcal{B}(c, \beta) \cap \mathcal{S}$ contains at least one node of G. These nodes form a dominating set, and so a.a.s. $\gamma(G) \leq N(r - \beta)$.

Finally, note that the lower bound in part (b) can be easily adopted to show that a.a.s. $\gamma(G) = \Omega(r^{-2})$ as a.a.s. a positive fraction of cells contain at least one node for the range of r considered in part (c). As already mentioned, the upper bound of $O(r^{-2})$ holds for any (deterministic) geometric graph and any r. Hence, part (c) follows. For part (d), the upper bound is trivial. The lower bound comes from the fact that a.a.s. there will be $\Theta(n)$ isolated nodes, and a dominating set has to contain all of them. The proof of the theorem is finished. □

5 Conclusions and Open Problems

We considered the domination number of a stochastic model for OSNs, the MGEO-P model. Theorem 1 shows a sublinear bound on the domination number of OSNs, which is well correlated with estimates for the domination number taken for the Facebook 100 data set. In addition, we provided bounds for the domination number of random geometric graphs.

In future work, we would like to broaden our analysis of the domination number to other data sets, and to test larger samples of OSNs. We will contrast the estimates provided by other heuristic algorithms for computing minimum order dominating sets, and provide a fitting of the data to bounds provided by the model.

So-called "elites", those who exert strong influence on the ambient network, are studied extensively in the sociology literature (see [10] for an overview of the literature on this topic). One approach to detecting elites is via their relatively high degree; hence, the use of k-cores in [10]. A different approach to detecting elites is to search for them within a minimum order dominating set, as these sets reach the entire network. Further, if minimum order dominating sets have much smaller order than the network (as we postulate), then that reduces the computational costs of finding elites. We plan on considering this approach to finding elites via dominating sets in future work.

References

1. Alon, N., Spencer, J.: The Probabilistic Method. Wiley, New York (2000)
2. Barabási, A.L., Albert, R.: Emergence of scaling in random networks. Science **286**, 509–512 (1999)
3. Bonato, A.: A Course on the Web Graph. Graduate Studies Series in Mathematics. American Mathematical Society, Providence (2008)
4. Bonato, A., Gleich, D.F., Kim, M., Mitsche, D., Prałat, P., Tian, A., Young, S.J.: Dimensionality matching of social networks using motifs and eigenvalues. PLOS ONE **9**, e106052 (2014)
5. Bonato, A., Janssen, J., Prałat, P.: Geometric protean graphs. Internet Math. **8**, 2–28 (2012)
6. Bonato, A., Janssen, J., Prałat, P.: The geometric protean model for on-line social networks. In: Kumar, R., Sivakumar, D. (eds.) WAW 2010. LNCS, vol. 6516, pp. 110–121. Springer, Heidelberg (2010)
7. Bonato, A., Janssen, J., Prałat, P.: A geometric model for on-line social networks. In: Proceedings of 3rd Workshop on Online Social Networks (WOSN 2010) (2010)
8. Bonato, A., Tian, A.: Complex networks and social networks, invited book chapter. In: Kranakis, E. (ed.) Social Networks. Mathematics in Industry Series, pp. 269–285. Springer, New York (2013)
9. Cooper, C., Klasing, R., Zito, M.: Lower bounds and algorithms for dominating sets in web graphs. internet math. **2**, 275–300 (2005)
10. Corominas-Murtra, B., Fuchs, B., Thurner, S.: Detection of the elite structure in a virtual multiplex social system by means of a generalized k-core, Preprint (2014)
11. Cowan, N.J., Chastain, E.J., Vilhena, D.A., Freudenberg, J.S., Bergstrom, C.T.: Nodal dynamics, not degree distributions, determine the structural controllability of complex networks. PLOS ONE **7**, e38398 (2012)
12. Estrada, E.: Spectral scaling and good expansion properties in complex networks. Europhys. Lett. **73**, 649 (2006)
13. Faloutsos, M., Faloutsos, P., Faloutsos, C.: On power-law relationships of the internet topology. SIGCOMM Comput. Commun. Rev. **29**, 251–262 (1999)
14. Gilbert, E.N.: Random plane networks. J. Soc. Ind. Appl. Math. **9**, 533–543 (1961)
15. Haynes, T.W., Hedetniemi, S.T., Slater, P.J.: Fundamentals of Domination in Graphs. CRC Press, Boca Raton (1998)
16. Janssen, J., Prałat, P.: Protean graphs with a variety of ranking schemes. Theoret. Comput. Sci. **410**, 5491–5504 (2009)

17. Janson, S., Łuczak, T., Rucinski, A.: Random Graphs. Wiley-Interscience Series in Discrete Mathematics and Optimzation. John Wiley & Sons, New York (2000)
18. Kershner, R.: The number of circles covering a set. Am. J. Math. **61**, 665–671 (1939)
19. Kim, M., Leskovec, J.: Multiplicative attribute graph model of real-world networks. Internet Math. **8**, 113–160 (2012)
20. Kolda, T.G., Pinar, A., Plantenga, T., Seshadhri, C.: A scalable generative graph model with community structure, Preprint (2014)
21. Kumar, R., Raghavan, P., Rajagopalan, S., Sivakumar, S., Tomkins, A.: Stochastic models for the web graph. In: Proceedings of the 41st Annual Symposium on Foundations of Computer Science (2000)
22. Leskovec, J., Chakrabarti, D., Kleinberg, J., Faloutsos, C., Ghahramani, Z.: Kronecker graphs: an approach to modeling networks. J. Mach. Learn. Res. **11**, 985–1042 (2010)
23. Leskovec, J., Kleinberg, J., Faloutsos, C.: Graph evolution: densification and shrinking diameters. ACM Trans. Knowl. Discov. Data **1**, 1–41 (2007)
24. Leskovec, J., Lang, K.J., Dasgupta, A., Mahoney, M.W.: Community structure in large networks: natural cluster sizes and the absence of large well-defined clusters. Internet Math. **6**, 29–123 (2009)
25. Łuczak, T., Prałat, P.: Protean graphs. Internet Math. **3**, 21–40 (2006)
26. McPherson, J.M., Ranger-Moore, J.R.: Evolution on a dancing landscape: organizations and networks in dynamic blau space. Soc. Forces **70**, 19–42 (1991)
27. McPherson, M., Smith-Lovin, L., Cook, J.M.: Birds of a feather: homophily in social networks. Annu. Rev. Sociol. **27**, 415–444 (2001)
28. Milenković, T., Memišević, V., Bonato, A., Pržulj, N.: Dominating biological networks. PLOS ONE **6**(8), e23016 (2013)
29. Molnár Jr., F., Derzsy, N., Czabarka, É., Székely, L., Szymanski, B.K., Korniss, G.: Dominating scale-free networks using generalized probabilistic methods, Preprint (2014)
30. Prałat, P.: A note on the diameter of protean graphs. Discrete Math. **308**, 3399–3406 (2008)
31. Prałat, P., Wormald, N.: Growing protean graphs. Internet Math. **4**, 1–16 (2009)
32. Penrose, M.: Random Geometric Graphs. Oxford Studies in Probability. Oxford University Press, Oxford (2003)
33. Seidman, S.B.: Network structure and minimum degree. Soc. Netw. **5**, 269–287 (1983)
34. Stojmenovic, I., Seddigh, M., Zunic, J.: Dominating sets and neighbor elimination-based broadcasting algorithms in wireless networks. IEEE Trans. Parallel Distrib. Syst. **13**, 14–25 (2002)
35. Traud, A.L., Mucha, P.J., Porter, M.A.: Social structure of facebook networks, Preprint (2014)
36. Watts, D.J., Strogatz, S.H.: Collective dynamics of "small-world" networks. Nature **393**, 440–442 (1998)

A Linear Time Algorithm for Determining Almost Bipartite Graphs

Dayu He and Xin He$^{(\boxtimes)}$

Department of Computer Science and Engineering,
University at Buffalo, Buffalo, NY 14260, USA
`xinhe@buffalo.edu`

Abstract. A graph $G = (V, E)$ is called *almost bipartite* if G is not bipartite, but there exists a vertex $v \in V$ such that $G - \{v\}$ is bipartite. We consider the problem of testing if G is almost bipartite or not.

This problem arises from the study on the k-arch layout problem. It is known that, given a graph G and an integer $k \geq 2$, it is NP-complete to determine if G has a k-arch layout. On the other hand, G has a 1-arch layout if and only if G is almost bipartite [3]. It is straightforward to test if G is almost bipartite in $O(n(n+m))$ time by using depth first search.

In this paper, we present a simple linear time algorithm for solving this problem. The efficiency of the algorithm is achieved by sophisticated applications of depth first search tree and the study of the structure of such graphs.

1 Introduction

Let $G = (V, E)$ be an undirected graph with $|V| = n$ vertices and $|E| = m$ edges. G is called k-*colorable* if the vertices of G can be colored by k colors such that no two adjacent vertices have the same color. G is called *almost k-colorable* if G is not k-colorable, but there exists a vertex $v \in V$ such that $G - \{v\}$ is k-colorable. In particular, a 2-colorable graph is also called a *bipartite graph*, and an almost 2-colorable graph is called an *almost bipartite graph*. In this paper, we present a linear time algorithm for testing if G is almost bipartite.

This problem arises from the study on *linear layouts of graphs*. Such layouts have many applications and have been extensively studied in the literature (see [3] for a survey). There are several different versions of linear layouts. The version related to this paper is the k-*arch layout*. A k-arch layout of a graph $G = (V, E)$ consists of a total order σ of the vertices in V, and a partition of the edges in E into k subsets $E_1, \ldots, E_k$ such that any two edges $e_1 = (u_1, v_1)$ and $e_2 = (u_2, v_2)$ within each subset E_i must overlap with respect to the total order σ. The *arch-number* of G, denoted by $an(G)$, is the minimum k such that G has a k-arch layout. A basic question in this area is to determine $an(G)$ for an input graph G. Two results related to k-arch layouts were obtained in [3]:

1. For any graph G, $an(G) \leq k$ if and only if G is almost $(k+1)$-colorable.

Research supported in part by NSF Grant CCR-1319732.

R. Jain et al. (Eds.): TAMC 2015, LNCS 9076, pp. 164–176, 2015.
DOI: 10.1007/978-3-319-17142-5_15

2. Given G and an integer $k \geq 2$, it is NP-complete to determine if $an(G) \leq k$.

By result 1 above, $an(G) = 1$ if and only if G is almost bipartite. It is easy to check if a given graph G is bipartite in $O(n+m)$ time by using depth first search (DFS) tree (for example see [2]). To test if G is almost bipartite, we can check, for each vertex $v \in V$, if $G - \{v\}$ is bipartite or not. The whole process takes $O(n(n+m))$ time. It was posed as an open question in [3] whether there is a sub-quadratic algorithm for determining if G is almost bipartite. The properties of almost bipartite graphs are also studied in [5].

This problem is a special case of the *odd cycle transversal* (OCT) problem which, given a graph $G = (V, E)$, asks does there exist a set $S \subseteq V$ of at most k vertices such that $G \setminus S$ is bipartite. Two recent papers [4,6] present linear time algorithms for solving the OCT problem. These results imply linear time algorithms for testing almost bipartite graphs. However, the algorithms in these two papers use sophisticated (integer programming, max-flow, skew-symmetric multicuts) concepts and their linear runtime are achieved by using a sequence of reductions, and/or solving a complex related problem (such as finding primal/dual solution of an IP problem, finding max flows). So their implementations are complicated and the constants involved in the runtime are large.

In this paper, we present a simple $O(n+m)$ time algorithm for solving this problem. Our algorithm relies only on elementary graph algorithm techniques and depth first trees. Hence our algorithm is easy to implement, and the constant in its runtime is small. The efficiency of the algorithm is achieved by sophisticated applications of the DFS tree and the study of the structure of almost bipartite graphs. The paper is organized as follows. Section 2 introduces definitions and preliminary results. In Sect. 3, we prove a theorem that is essential for our algorithm. Section 4 discusses the details of the testing algorithm.

2   Preliminaries

In this paper, $G = (V, E)$ always denotes an undirected graph. It is clear that G is almost bipartite if and only if one connected component of G is almost bipartite and all other connected components of G are bipartite. To test if G is almost bipartite, we can test whether its connected components are bipartite or almost bipartite. Thus, without loss of generality, we assume G is connected.

A *path* P of G is a sequence of vertices $(v_1, ..., v_k)$ with $(v_i, v_{i+1}) \in E$ ($1 \leq i < k$). A *cycle* C of G is a sequence of vertices $(v_1, ..., v_k)$ with $(v_i, v_{i+1}) \in E$ ($1 \leq i < k$) and $(v_k, v_1) \in E$. The *length* of a path P (or a cycle C), denoted by $|P|$ (or $|C|$), is the number of edges in it. A cycle C is called an odd (or even) cycle if $|C|$ is odd (or even). The following result is well known.

Lemma 1. *[1] G is bipartite if and only if G contains no odd cycles.*

A *depth first search* (DFS) tree of G is a spanning tree T of G produced by the depth-first search on G [2]. Let T be a DFS tree of G with root r. For any vertex v, let $T(v)$ denote the sub-tree of T rooted at v. A vertex w is called an *ancestor* of another vertex u if the path in T from u to r contains w. If w is an ancestor of u, $T[w, u]$ denotes the path in T between w and u including both w and u. $T[w, u)$ denotes the path in T between w and u including w, but excluding u. The meanings of the notations $T(w, u]$ and $T(w, u)$ are similar. The *lowest common ancestor* of two vertices w and u, denoted by lca(w, u), is the lowest (the farthest from r) vertex in $T[r, w] \cap T[r, u]$.

Let level(v) denote the *level* of a vertex v in T which is defined to be the number of edges in the path $T[r, v]$. If level(v) is odd, v is called an *odd vertex* and we say the *parity* of v is 1. If level(v) is even, v is called an *even vertex* and we say the *parity* of v is 0. Let parity_diff$(w, v) = |\text{parity}(w) - \text{parity}(v)|$ denote the parity difference between two vertices w and v, which is either 0 or 1.

For a given DFS tree T of $G = (V, E)$, the edges in E can be partitioned into two subsets: the tree edges and the non-tree edges. A non-tree edge $e = (w, u)$ is called a *back edge* (with respect to T) if w is an ancestor of u. Since we only consider undirected graphs in this paper, we have the following property [2]:

Property 1. For an undirected graph G and a DFS tree T of G, all non-tree edges are back edges.

In the remainder of this paper, for a non-tree edge $e = (w, u)$, the first end vertex w is always the ancestor of the second end vertex u. Hence $w \in T[r, u)$. A non-tree edge $e = (w, u)$ in G together with the tree path $T[w, u]$ form a cycle in G. We call this cycle the *cycle induced by* e and denote it by $C(e)$. The non-tree edges of G can be divided into two subsets:

Definition 1. A non-tree edge $e = (w, u)$ is called an *odd-non-tree* (ONT) edge if $C(e)$ is an odd cycle. Note that e is an ONT edge if and only if parity_diff$(w, u) = 0$. A non-tree edge $e = (w, u)$ is called an *even-non-tree* (ENT) edge if $C(e)$ is an even cycle. Note that e is an ENT edge if and only if parity_diff$(w, u) = 1$.

Fig. 1. An example of a DFS-tree with two ONT edges and one ENT edge. The blue (red, resp.) dots represent even (odd, resp.) vertices. Thin blue lines represent ONT edges, dashed red lines represent ENT edges. Bold black lines represent the tree edges (Color figure online).

Figure 1 shows an example of ONT and ENT edges. The following is well-known:

Lemma 2. *[1] Let G be a connected graph and T be a DFS tree of G. Then G is bipartite if and only if it has no ONT edges.*

Based on Lemma 2, we have the following algorithm for testing the bipartiteness of G:

– Perform the depth first search on G and construct a DFS tree T of G. Calculate the parity of the vertices with respect to T.
– Check the non-tree edges one by one. If the two end vertices of any non-tree edge have the same parity, then G is not bipartite. Otherwise G is bipartite.

Clearly this algorithm takes $O(n + m)$ time [2]. In this paper, we want to test if a non-bipartite graph G is almost bipartite. From now on, we assume G is not bipartite which means G has at least one ONT edge by Lemma 2.

3   Main Theorem

First, we prove a lemma needed by our algorithm. Let T be a DFS tree of G.

Lemma 3. *A cycle C is odd if and only if C has an odd number of ONT edges.*

Proof. Let C be a cycle in G. Suppose C has k non-tree edges. Let $e_i = (x_i, y_i)$ $(1 \le i \le k)$ denote the k non-tree edges in the order they appear in C. Let t_i $(1 \le i \le k)$ be the tree path between y_i and x_{i+1} in C (t_i could be empty). See Fig. 2 for an example. Let $P_i = \{e_i\} \cup t_i$ denote the sub-path of C from the vertex x_i to the vertex x_{i+1}. The following properties are clear:

1. If e_i is an ONT edge, then $|P_i| \equiv$ parity_diff$(x_i, x_{i+1}) + 1$ (mod 2).
2. If e_i is an ENT edge, then $|P_i| \equiv$ parity_diff$(x_i, x_{i+1}) + 0$ (mod 2).

Suppose that C contains k_1 ONT edges and k_2 ENT edges. Then we have:

$$|C| = \sum_{i=1}^{k} |P_i| = \sum_{e_i \text{ is an ONT edge}} |P_i| + \sum_{e_j \text{ is an ENT edge}} |P_j|$$

$$\equiv \{ \sum_{e_i \text{ is an ONT edge}} (\text{parity\_diff}(x_i, x_{i+1}) + 1) +$$

$$\sum_{e_j \text{ is an ENT edge}} (\text{parity\_diff}(x_j, x_{j+1}) + 0)\} \, (\text{mod } 2)$$

$$\equiv \{ \sum_{i=1}^{k} (\text{parity\_diff}(x_i, x_{i+1})) + k_1 \} \, (\text{mod } 2) \equiv k_1 \, (\text{mod } 2)$$

(Note that $\sum_{i=1}^{k}$ parity_diff$(x_i, x_{i+1}) \equiv 0 \, (\text{mod } 2)$, since C is a cycle.) Hence, C is an odd cycle if and only if k_1 is odd. $\qquad \square$

Definition 2. A vertex v in G is a *witness vertex* if $G - \{v\}$ is a bipartite graph.

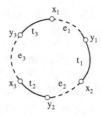

Fig. 2. An example of a cycle C in G with non-tree edges and tree paths. Dashed lines represent non-tree edges. Bold lines represent tree paths.

By this definition, a non-bipartite graph G is almost bipartite if and only if it has a witness vertex. So testing if G is almost bipartite is the same as finding witness vertices in G. However, it is not easy to search for witness vertices by using this definition. In the following, we provide an alternative definition of the witness vertex. The advantage is that we are able to search for such vertices according to the alternative definition. For any vertex v in G, we partition the vertices of G into three subsets with respect to the position of v in T:

- $\{v\}$;
- $\mathrm{Lower}(v) = T(v) - \{v\}$ is the subset of the vertices that are "below" v;
- $\mathrm{Upper}(v) = T - T(v)$ is the subset of the vertices that are "above" v.

Definition 3. A vertex v is called a *candidate vertex* if and only if both of the following conditions hold:

1. For any ONT edge $e_i = (w_i, u_i)$, either $v \in \{w_i, u_i\}$; or $w_i \in \mathrm{Upper}(v)$ and $u_i \in \mathrm{Lower}(v)$.
2. For any ENT edge $e_j = (w_j, u_j)$, one of the following holds:
   (a) Both w_j and u_j are in $\mathrm{Upper}(v) \cup \{v\}$.
   (b) u_j is in $T(v_{child})$ for a child v_{child} of v. If there exists an ONT edge $e_i = (w_i, u_i)$ such that $w_i \in \mathrm{Upper}(v)$ and $u_i \in T(v_{child})$, then $w_j \in T[v, u_j]$. (If no such ONT edge e_i exists, w_j can be anywhere in $T[r, u_j]$.)

Note the conditions 2 (a) and 2 (b) are mutually exclusive. Figure 3 (a) shows an example of a candidate vertex v with ONT and ENT edges that satisfy the conditions in Definition 3. The following theorem shows that the definitions of the witness vertex and the candidate vertex are equivalent.

Theorem 1. *Let G be a non-bipartite graph. A vertex v of G is a witness vertex if and only if v is a candidate vertex.*

Proof. First, we prove the if part of Theorem 1. Let v be a candidate vertex. We will show that any cycle C in $G - \{v\}$ must be even (which implies v is a witness vertex). C contains three types of edges: ONT edges, ENT edges and tree edges.

In $G - \{v\}$, any ONT edge $e_i = (w_i, u_i)$ of G must satisfy $w_i \in \mathrm{Upper}(v)$ and $u_i \in \mathrm{Lower}(v)$ by the condition 1 in Definition 3. If C contains no ONT

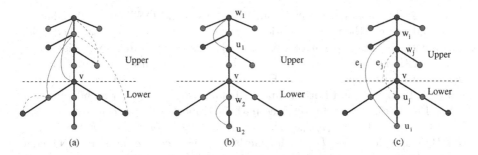

Fig. 3. (a) A candidate vertex v with two ONT edges and three ENT edges that satisfy the conditions in Definition 3. (b) Two ONT edges that violate the condition 1 in Definition 3. (c) An ENT edge that violates the condition 2 (b) in Definition 3.

edges, then C is even by Lemma 3. Suppose C contains at least one ONT edge $e_i = (w_i, u_i)$. Let v_{child} be the child of v such that $u_i \in T(v_{child})$. Imagine we travel along C starting from w_i. When passing through e_i, we "jump" from Upper(v) into $T(v_{child}) \subseteq$ Lower(v). In order to complete our travel along C, we must go back from $T(v_{child})$ to Upper(v). We consider the cases this can happen:

- Since v is deleted, we cannot travel from $T(v_{child})$ to Upper(v) or $T(v_{child'})$ (where $v_{child'} \neq v_{child}$ is any other child of v) by using tree paths.
- Consider any non-tree (either ONT or ENT) edge $e_k = (w_k, u_k)$ with $u_k \in T(v_{child})$. Since e_k is a back edge, w_k must be in the tree path $T[r, u_k]$ (where r is the root of T). So we cannot travel from $T(v_{child})$ to $T(v_{child'})$ (where $v_{child'} \neq v_{child}$ is any other child of v) by using e_k.
- Consider any ENT edge $e_j = (w_j, u_j)$ with $u_j \in T(v_{child})$. Since e_j satisfies the condition 2 (b) in Definition 3, w_j must be a vertex in the tree path $T[v, u_j]$. Thus, we cannot travel from $T(v_{child})$ to Upper(v) by using e_j.

So, once within $T(v_{child})$, we are "stuck" in $T(v_{child})$. The only way to get out of $T(v_{child})$ is to go back to Upper(v) by using another ONT edge. By repeating this argument, we see that C must contain an even number of ONT edges. By Lemma 3, C is even. So v is a witness vertex.

Next, we prove the only if part by contraposition. Suppose v is not a candidate vertex. Then at least one of the two conditions in Definition 3 fails.

Case 1: The condition 1 in Definition 3 fails. Then there exists at least one ONT edge $e_i = (w_i, u_i)$ such that either both w_i and u_i are in Upper(v); or both in Lower(v). See Fig. 3 (b). Suppose both w_i and u_i are in Upper(v). Then $C(e_i)$ is a cycle in $G - \{v\}$ containing exactly one ONT edge e_i. Thus $C(e_i)$ is odd and $G - \{v\}$ is not bipartite. So v is not a witness vertex. Suppose both w_i and u_i are in Lower(v). Let v_{child} be the child of v such that $u_i \in T(v_{child})$. Because e_i is a back edge, we must have $w_i \in T(v_{child})$ also. Thus $C(e_i)$ is a cycle in $T(v_{child})$ containing exactly one ONT edge e_i. So $C(e_i)$ is an odd cycle in $G - \{v\}$, and v is not a witness vertex.

Case 2: The condition 2 in Definition 3 fails. If all ONT edges of G have v as an end vertex, then the condition 2 in Definition 3 is always satisfied. This contradicts the assumption of this case. So there exists an ONT edge $w_i = (w_i, u_i)$ with $w_i \in \text{Upper}(v)$ and $u_i \in \text{Lower}(v)$. Because the condition 2 of Definition 3 fails, there exists an ENT edge $e_j = (w_j, u_j)$ with $w_j \in \text{Upper}(v)$ and $u_j \in \text{Lower}(v)$, and both u_i and u_j belong to $T(v_{child})$ for a child v_{child} of v. See Fig. 3 (c). Let P_1 be the tree path in $\text{Upper}(v)$ between w_i and w_j. Let P_2 be the tree path in $T(v_{child})$ between u_i and u_j. Then the union of e_i, e_j, P_1 and P_2 is a cycle in $G - \{v\}$ with exactly one ONT edge e_i and one ENT edge e_j. By Lemma 3, C is odd. So v is not a witness vertex.   □

4   Algorithm

We describe our algorithm for finding candidate vertices in this section. We first outline the main idea of the algorithm. Basically, we want to identify the subgraph G_s of G consisting of candidate vertices. Initially, G_s consists of all vertices of G. We consider the non-tree edges one by one. For a non-tree edge $e = (w, u)$, the conditions in Definition 3 fails for e with respect to some vertices v in G_s. Such vertices v are removed from G_s. After processing all non-tree edges, the conditions in Definition 3 hold for all non-tree edges with respect to any vertex v in the remaining G_s. In other words, these remaining vertices in G_s are candidate vertices. If G_s becomes empty during the process, then there exist no candidate vertices in G.

The processing is divided into two stages. The first stage processes the ONT edges. The second stage processes ENT edges. We discuss them separately.

4.1   Processing ONT Edges

Let T be a DFS tree of G with root r. Let $e_i = (w_i, u_i)$ $(1 \leq i \leq p)$ be all ONT edges in G. Since G is non-bipartite, it has at least one ONT edge. The following observation is clear:

Observation 1. *An ONT edge $e_i = (w_i, u_i)$ satisfies the condition 1 in Definition 3 with respect to a vertex v if and only if $v \in T[w_i, u_i]$.*

The following lemma immediately follows from Observation 1:

Lemma 4. *Let $e_i = (w_i, u_i)(1 \leq i \leq p)$ be all ONT edges in G. Define: $T[x, y] = \bigcap_{i=1}^{p} T[w_i, u_i]$. Then the condition 1 in Definition 3 is satisfied for all ONT edges $e_i = (w_i, u_i)(1 \leq i \leq p)$ with respect to a vertex v if and only if $v \in T[x, y]$.*

Note that $T[x, y]$ is a tree path in T (which may be empty). If $T[x, y] = \emptyset$, then there exists no vertex v for which the condition 1 in Definition 3 is satisfied for all ONT edges in G. In this case, there are no candidate vertices in G.

Next we describe the details of the processing of the ONT edges. The first step of the algorithm is to construct a DFS tree T of G. When performing

DFS on G, we also calculate two integers (between 1 and $2n$) for each vertex $v \in V$: the *start time* of v, denoted by $s(v)$, and the *finish time*, denoted by $f(v)$. We associate each v with an interval $I(v) = [s(v), f(v)]$. These values can be computed in $O(n + m)$ time as a by-product of DFS [2]. The following facts are well known [2]:

Fact 1.

- *A vertex w is an ancestor of another vertex u in T if and only if $I(u) \subset I(w)$. It takes $O(1)$ time to determine if w is an ancestor of u or not (by checking if $I(u) \subset I(w)$).*
- *For any vertex w with children $w_1, \ldots, w_d$, the intervals $I(w_1), \ldots, I(w_d)$ are pairwise disjoint. Moreover, for each $1 \le j < d$, $I(w_j)$ is located to the left of $I(w_{j+1})$. (Namely, $s(w_j) < f(w_j) < s(w_{j+1}) < f(w_{j+1})$).*

Initially, we set $T[x, y] \leftarrow T[w_1, u_1]$. Then the algorithm performs a sequence of steps. The step i ($2 \le i \le p$) processes the ONT edge $e_i = (w_i, u_i)$ and updates $T[x, y] \leftarrow T[x, y] \cap T[w_i, u_i]$.

In order to compute $T[x, y] \cap T[w_i, u_i]$ in $O(1)$ time, we need a data structure, after initializing $T[x, y] \leftarrow T[w_1, u_1]$, so that the following holds:

Fact 2. *For a vertex $u \in V$, it takes $O(1)$ time to determine the lowest common ancestor $z = lca(u_1, u)$ and the child z_{child} of z such that $u \in T(z_{child})$.*

The implementation of the operation in Fact 2 is outlined below. We number the vertices of G in the left-to-right post-order with respect to T. Let $r = a_1, a_2, \ldots, a_t = u_1$ be the vertices in the tree path $T[r, u_1]$. For each a_i ($1 \le k < t$), let $l_i^1, l_i^2 \ldots, l_i^{p_i}, a_{i+1}, r_i^1, r_i^2, \ldots, r_i^{q_i}$ be the children of a_i in T ordered from left to right. For the vertex a_t, let $l_t^1, l_t^2 \ldots, l_t^{p_t}, r_t^1, r_t^2, \ldots, r_t^{q_t}$ be the children of a_t in T ordered from left to right. Then the vertices in $V - T[r, u_1]$ appear, in the order, in the subtrees in the following list:

$$T(l_1^1), \ldots, T(l_1^{p_1}), T(l_2^1), \ldots, T(l_2^{p_2}), \ldots, T(l_{t-1}^1), \ldots, T(l_{t-1}^{p_{t-1}}), T(l_t^1), \ldots, T(l_t^{p_t}),$$

$$T(r_t^1), \ldots, T(r_t^{q_t}), T(r_{t-1}^1), \ldots, T(r_{t-1}^{q_{t-1}}), \ldots, T(r_1^1), \ldots, T(r_1^{q_1})$$

In $O(n)$ time, we can set up a look-up table $D[1..n]$ indexed by the vertices of G. For a vertex v, the entry $D[v]$ stores the identity of the vertex z such that $v \in T(z)$ in the above list . Then, given a vertex v, the information required in the operation in Fact 2 can be retrieved in $O(1)$ time from $D[v]$.

After processing all ONT edges, we get $T[x, y]$ at the end. In addition, we also need to label the children $y_1 \ldots, y_d$ of y as *marked* or *unmarked* with the following property. (These labels are needed for the processing of ENT edges discussed later).

Property 2.

- If there is an ONT edge $e_i = (w_i, u_i)$ such that $u_i \in T(y_l)$, then y_l is labeled "marked".

Algorithm 1. Processing ONT Edges

---

1. Initialize $T[x, y] \leftarrow T[w_1, u_1]$;
2. Set up the data structures so that the operations in Fact 1 and Fact 2 can be performed in $O(1)$ time;
3. For $i = 2$ to p do:

    3a. Process the ONT edge $e_i = (w_i, u_i)$; calculate $T[x', y'] = T[x, y] \cap T[w_i, u_i]$;
    3b. If $T[x', y'] = \emptyset$, stop and reports "no candidate vertex exists";
    3c. If $T[x', y'] \neq \emptyset$, update $T[x, y] \leftarrow T[x', y']$;
4. Label all children $y_1, \ldots, y_d$ of y as "unmarked".
5. For $i = 1$ to p do:

    5a. Process the ONT edge $e_i = (w_i, u_i)$. Find the child y_l of y such that $u_i \in T(y_l)$ and label y_l as "marked".

– Otherwise y_l is labeled "unmarked".

The algorithm for processing ONT edges is given in Algorithm 1.

Figure 4 shows an example of the processing of ONT edges and the labeling of the children of the vertex y. Tree edges are marked as black bold lines and ONT edges are marked as blue curve lines. Dashed line represents $T[x, y]$.

Lemma 5. *Algorithm 1 takes $O(m + n)$ time.*

Proof. The pre-processing (Steps 1 and 2) can be done in $O(n + m)$ time. The step 4 clearly takes $O(n)$ time. By using the operations described in Fact 1 and Fact 2, each iteration of the loop body in step 3 and step 5 can be done in $O(1)$ time. So Algorithm 5 takes $O(n + m)$ time. □

After processing all ONT edges, if the remaining $T[x, y] = \emptyset$, then G has no candidate vertex. Otherwise, we get a non-empty tree path $T[x, y]$. At this point, the condition 1 in Definition 3 is satisfied with respect to any vertex $v \in T[x, y]$. If x is an end vertex of every ONT edge $e_i = (w_i, u_i)$ of G, then x is clearly a candidate vertex. Then the algorithm can stop and report so. So we assume this is not the case in the rest of the paper. Hence the following properties hold:

Property 3. For any ONT edge $e_i = (w_i, u_i)$ $(1 \leq i \leq p)$ of G, $w_i \in T[r, x]$ and $u_i \in T(y)$. In addition, there exists at least one ONT edge $e_i = (w_i, u_i)$ such that $w_i \in T[r, x)$.

4.2 Processing ENT Edges

In this subsection, we discuss the second stage of the algorithm, which processes the ENT edges of G. At the beginning of this stage, we have a tree path $T[x, y]$ as G_s and the children $y_1, \ldots, y_d$ of y are labeled as "marked" and "unmarked" as in Property 2.

Let e_j $(1 \leq j \leq q)$ be all ENT edges of G. When an ENT edge $e_j = (w_j, u_j)$ is processed, the condition 2 in Definition 3 fails for e_j with respect to some vertices

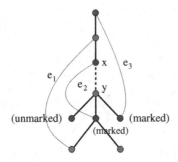

Fig. 4. An example of ONT edge processing and the labeling of the children of y.

$v \in G_s$. Such vertices v are removed from G_s. If $G_s \neq \emptyset$ after all ENT edges have been processed, then every vertex v in the remaining G_s is a candidate vertex. First, we need a lemma.

Lemma 6. *Let $e_j = (w_j, u_j)$ be an ENT edge in G. Let $z = lca(y, u_j)$. Then the condition 2 in Definition 3 is satisfied for e_j with respect to a vertex $v \in T[x, y]$ if and only if one of the following two conditions holds:*

1. *If $z = y$ and $u_j \in T(y_{child})$ for a marked child y_{child} of y, then $v \in T[x, y] - T(w_j, z]$.*
2. *Otherwise, $v \in T[x, y] - T(w_j, z)$.*

(When $z = w_j$ or z is an ancestor of w_j, $T(w_j, z)$ and $T(w_j, z]$ are empty sets).

Proof. Partition the vertices of $G = (V, E)$ into three subsets as follows:

- Top $= V - T(x)$.
- Mid $= (T(x) - T(y)) \cup \{y\}$. (Note that Mid contains both x and y).
- Bottom $= T(y) - \{y\}$.

The proof is divided into cases depending on the position of w_j and u_j.

Case 1: Both w_j and u_j are in Top (see the edge e_1 in Fig. 5 (a)). We have $w_j \in T[r, x)$ and $z \in T[r, x)$. So $T[x, y] - T(w_j, z) = T[x, y]$. The condition 2 in the lemma holds in this case. It is clear that, for any $v \in T[x, y]$, the condition 2 (a) in Definition 3 holds for $e_j = (w_j, u_j)$ with respect to v.

Case 2: both w_j and u_j are in Bottom (see the edge e_2 in Fig. 5 (a)). We have $z = lca(y, u_j) = y$. So z is an ancestor of w_j, and $T(w_j, z) = T(w_j, z] = \emptyset$. Hence $T[x, y] - T(w_j, z) = T[x, y] - T(w_j, z] = T[x, y]$. In this case, either the condition 1 or the condition 2 in the lemma holds. In both cases, the condition 2 (b) in Definition 3 holds for $e_j = (w_j, u_j)$ with respect to any $v \in T[x, y]$.

Case 3: $w_j \in$ Top and $u_j \in$ Mid (Fig. 5 (b)). The condition 2 in the lemma holds in this case. We have $w_j \in T[r, x)$ and $z \in T[x, y)$ (when $u_j \neq y$) or $z = y$ (when $u_j = y$). In either case, $T[x, y] - T(w_j, z) = T[z, y]$.

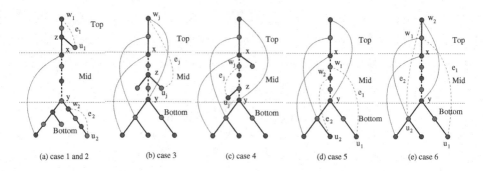

Fig. 5. Cases of ENT edges with respect to their location in $T[x, y]$.

Consider any vertex $v \in T[x, z)$. Recall that there exists at least one ONT edge $e_i = (w_i, u_i)$ such that $w_i \in T[r, x)$ and $u_i \in T(y)$. So both u_j and u_i are in $T(v_{child})$ where v_{child} is the child of v in $T[x, y]$. Thus the condition 2 (b) in Definition 3 fails for $e_j = (w_j, u_j)$ with respect to v.

Consider any vertex $v \in T[z, y]$. If $v \neq z$, both end vertices of $e_j = (w_j, u_j)$ are in Upper(v). So the condition 2 (a) in Definition 3 holds for e_j. If $v = z$, then $u_j \in T(v_{child})$ for a child v_{child} of v. Note that there exists no ONT edges $e_i = (w_i, u_i)$ such that $u_i \in T(v_{child})$. (This is because for any ONT edge $e_i = (w_i, u_i)$, $u_i \in T(y)$). So the condition 2 (b) in Definition 3 holds for e_j. Thus $T[x, y] - T(w_j, z) = T[z, y]$ contains exactly those vertices v for which the condition 2 in Definition 3 holds for e_j.

Case 4: w_j and u_j both $\in$ Mid (Fig. 5 (c).) The condition 2 in the lemma holds in this case. We have $w_j \in T[x, y]$ and $z \in T[x, y]$.

For any vertex $v \in T(w_j, z)$, the condition 2 (b) in Definition 3 fails for $e_j = (w_j, u_j)$ with respect to v. Note that $T[x, y]$ is "cut" by $T(w_j, z)$ into two parts: $T[x, w_j]$ and $T[z, y]$. For any vertex $v \in T[x, , w_j]$, the condition 2 (b) in Definition 3 holds for e_j with respect to v. For any vertex $v \in T[z, y]$, by using the same argument as in Case 3, either the condition 2 (a) (when $v \neq z$) or the condition 2 (b) (when $v = z$) in Definition 3 holds for e_j with respect to v. Thus $T[x, y] - T(w_j, z)$ contains exactly those vertices v for which the condition 2 in Definition 3 holds for e_j.

Case 5: $w_j \in$ Mid and $u_j \in$ Bottom (Fig. 5 (d)). We have $w_j \in T[x, y]$ and $z = lca(y, u_j) = y$. Let y_{child} be the child of y such that $u_j \in T(y_{child})$.

Case 5a: y_{child} is unmarked (see the edge e_1 in Fig. 5 (d)). The condition 2 in the lemma holds in this case.

For any vertex $v \in T(w_j, y)$, the condition 2 (b) in Definition 3 fails for $e_j = (w_j, u_j)$ with respect to v. (Recall that there exists at least one ONT edge $e_i = (w_i, u_i)$ such that $w_i \in T[r, x)$ and $u_i \in T(y)$.) Consider any vertex $v \in T[x, y] - T(w_j, y) = T[x, w_j] \cup \{y\}$. Regardless of whether $v = y$ or $v \in T[x, w_j]$, the condition 2 (b) in Definition 3 holds for e_j with respect to v. Thus $T[x, y] - T(w_j, y)$ contains exactly those vertices v for which the condition 2 in Definition 3 holds for e_j.

Case 5b: y_{child} is marked (see the edge e_2 in Fig. 5 (d)). The condition 1 in the lemma holds in this case.

For any vertex $v \in T(w_j, y]$, the condition 2 (b) in Definition 3 fails for $e_j = (w_j, u_j)$ with respect to v. For any vertex $v \in T[x, y] - T(w_j, y] = T[x, w_j]$, the condition 2 (b) in Definition 3 holds for e_j with respect to v. Thus $T[x, y] - T(w_j, y]$ contains exactly those vertices v for which the condition 2 in Definition 3 holds for e_j.

Case 6: $w_j \in$ Top and $u_j \in$ Bottom. (Fig. 5 (e)). We have $w_j \in T[r, x)$ and $z = lca(y, u_j) = y$. Let y_{child} be the child of y such that $u_j \in T(y_{child})$.

  Case 6a: y_{child} is unmarked (see the edge e_1 in Fig. 5 (e)).
  Case 6b: y_{child} is marked (see the edge e_2 in Fig. 5 (e)).

The proofs for Cases 6a and 6b are similar to the proofs for Cases 5a and 5b. The only difference is that, because $w_j \in T[r, x)$, $T[x, y] - T(w_j, y) = \{y\}$, and $T[x, y] - T(w_j, y] = \emptyset$. $\qquad \square$

Definition 4. Let $e_j = (w_j, u_j)$ be an ENT edges in G and let $z = lca(y, u_j)$. Define w'_j, v'_j to be the vertices in $T[x, y]$ such that:

1. $T[x, y] - T[w'_j, u'_j] = T[x, y] - T(w_j, z]$ if the condition 1 in Lemma 6 holds.
2. $T[x, y] - T[w'_j, u'_j] = T[x, y] - T(w_j, z)$ if the condition 2 in Lemma 6 holds.

The following lemma immediately follows from Lemma 6 and Definition 4:

Lemma 7. *Let* $e_j = (w_j, u_j)$ $(1 \leq j \leq q)$ *be all ENT edges in G. Then v is a candidate vertex of G if and only if*

$$v \in G_s = \bigcap_{j=1}^{q} (T[x, y] - T[w'_j, u'_j]) = T[x, y] - \bigcup_{j=1}^{q} T[w'_j, u'_j]$$

Note that the final set G_s could be a series of tree paths contained in $T[x, y]$. If $G_s = \emptyset$, then there are no candidate vertices in G. If $G_s \neq \emptyset$, then every vertex in G_s is a candidate vertex. The following algorithm calculates the final set G_s of candidate vertices.

Lemma 8. *Algorithm 2 takes $O(m + n)$ time.*

Proof. Let $m_1 < m$ be the number of ENT edges in G. Calculating (w'_j, u'_j) for each ENT edge $e_j = (w_j, u_j)$ can be done in $O(1)$ time by using the operations in Fact 1 and Fact 2. So the step 1 takes $O(m_1)$ time.

Algorithm 2. Processing ENT Edges

Input: The tree path $T[x, y]$. Label the vertices in $T[x, y]$ as $x = x_1, x_2, \ldots, x_t = y$.
1. For each ENT edge $e_j = (w_j, u_j)$ $(1 \leq j \leq q)$, calculate w'_j, u'_j as in Definition 4.
2. Sort the intervals $[w'_j, u'_j]$ $(1 \leq j \leq q)$ using bucket sort with w'_j as the key.
3. Scan the sorted intervals $[w'_j, v'_j]$ and calculate the union $\bigcup_{j=1}^{q} T[w'_j, v'_j]$.
4. Return $G_s = T[x, y] - \bigcup_{j=1}^{q} T[w'_j, v'_j]$.

Because w_j', v_j' are in the range $x_1, \ldots, x_t$ with $t \leq n$, the bucked sorting operation takes $O(m_1 + n)$ time. Once the intervals $[w_j', v_j']$ are sorted, their union (step 3) can be calculated in $O(m_1 + n)$ time. Thus, Algorithm 2 takes $O(n + m)$ time in total. □

In summary we have:

Theorem 2. *Let G be a non-bipartite graph. We can determine if G is an almost bipartite graph, and find the candidate vertices in G if it is, in $O(n + m)$ time.*

Proof. The preprocessing (constructing a DFS tree) takes $O(n + m)$ time. By Lemma 5, the processing of all ONT edges takes $O(m + n)$ time. By Lemma 8, the processing of all ENT edges takes $O(m + n)$ time. The correctness of the algorithm follows from the discussions in previous sections. □

References

1. Boundy, J.A., Murty, U.S.R.: Graph Theory with Applications. Elsevier, North Holland (1976)
2. Cormen, T.H., Leiserson, C.E., Rivest, R.L., Stein, C.: Introduction to Algorithms, 3rd edn. The MIT Press, Cambridge (2009)
3. Dujmović, M., Wood, D.R.: On linear layout of graphs. Discrete Math. Theoret. Comput. Sci. **6**, 339–358 (2004)
4. Iwata, Y., Oka, K., Yoshida, Y.: Linear-time FPT algorithms via network flow. In: Proceedings of the SODA, vol. 2014, pp. 1749–1761 (2014)
5. Proömel, H.J., Schickinger, T., Steger, A.: A note on triangle-free and bipartite graphs. Discrete Math. **257**(2–3), 531–540 (2002)
6. Ramanujan, M.S., Saurabh, S.: Linear time parameterized algorithms via skew-symmetric multicuts. In: Proceedings of the SODA, vol. 2014, pp. 1739–1748 (2014)

The First-Order Contiguity of Sparse Random Graphs with Prescribed Degrees

Nans Lefebvre[(✉)]

LIAFA, Université Paris VII, Paris, France
nans.lefebvre@liafa.univ-paris-diderot.fr

Abstract. Two models are first-order contiguous if they satisfy asymptotically almost surely the same sets of first-order formulas, a notion introduced to classify random structures from a logical point of view. We study in particular the random graph defined as the uniform distribution on graphs with a given degree sequence. We characterise degree sequences that define contiguous random graph sequences, and in particular contiguous to an Erdős-Rényi random graph. The method allows to extend a result of Lynch showing that a large class of degree sequences define random graphs that have a convergence law.

Keywords: Logic · Finite model theory · Random graphs · Convergence laws

1 Introduction

The logical analysis of random graphs began in [5] and independently in [4], and throughout the years successfully charted all of the so-called Erdős-Rényi model, as summarised in [12]. The ubiquity of graphs in computer science and other fields, from biology to physics, introduced the need for more complex models of graphs in order to model various phenomena. One such model is the random graph with prescribed degrees, where a graph is taken uniformly among graphs having a given degree sequence; the rationale is that various processes lead to graphs having degree sequences very close to some probability distribution (often power laws). Understanding these graphs may then give an insight into many areas such as social networks or protein interactions. An account of the relevance of this model for real-world networks can be found in [3].

The notion of first-order contiguity is the main concept introduced to characterise sequences of graphs that have a similar behaviour with regard to first-order logic. Two sequences of random graphs are (first-order) contiguous if they have the same set of almost sure formulas. It is analogous to the statistical notion of contiguity from [7]. This is a tool to compare various models of random graphs; in recent years many random processes generating random graphs were introduced, such as growing graphs, preferential attachment, etc., and some exhibit new behaviours while some are contiguous to simple random graph models. Contiguity also allows to determine how resilient a random process is, i.e. how much it must it be modified to give non-contiguous random graphs.

© Springer International Publishing Switzerland 2015
R. Jain et al. (Eds.): TAMC 2015, LNCS 9076, pp. 177–188, 2015.
DOI: 10.1007/978-3-319-17142-5_16

In [10] Lynch proved that a large class of asymptotic degree sequences define random graphs that have a convergence law, i.e. the probability that a formula of first-order logic is satisfied converges to a limiting value for any formula. This is the first result extending the logical convergence to a model studied for its likeness to real-world networks, including power law graphs. This paper builds upon [10], and extend it by giving a necessary and sufficient condition for which the theorem applies. Besides this extension, the main results are:

- A criterion to determine if two random graphs with given degree sequences are contiguous (Theorem 4), and in particular contiguous to the Erdős-Rényi graph $\mathcal{G}(n, \frac{1}{n})$ (Theorem 5).
- A complete taxonomy of the ordered continuum of theories generated by these degree sequences (Theorem 6).

The proof also gives a first-order axiomatisation of the first-order theories generated, as well as the limiting values of the formula probabilities.

In the second section the notations from random graphs, logic and probability are introduced. In the third section the convergence law result is reproven, and extended with a necessary and sufficient condition. The main results are stated and proved in the fourth section, and some remarks on future work conclude.

2   Preliminaries

This section introduces the prerequisite from graph theory, logic, and probability, to state the main results. Let $D \sim d$ denote a random variable sampled from distribution d, and $\mathcal{P}(\mathbb{N})$ denote the set of distributions on $\mathbb{N}$. Let $(n)_k$ denote the k-falling factorial of n, $(n)_k = \prod_{i=0}^{k-1}(n - i)$.

2.1   Graph Theory

A graph $G_n = (V, E) = (\{1, \ldots, n\}, \frown)$ is a set of vertices equipped with an edge relation, which is symmetric and irreflexive. A random graph, denoted by $\mathcal{G}_n$, is a probability distribution on the $2^{\binom{n}{2}}$ possible graphs; sequences of random graphs $(\mathcal{G}_n)_{n \in \mathbb{N}}$ are usually denoted by $\mathcal{G}$. A *multigraph* is a graph that can have multiple edges between pairs of vertices as well as loops on a single vertex, and the set of multigraphs is denoted by $\mathbb{G}$.

Let $\mathcal{C}$ be the class of all connected graphs, and $\mathcal{T} \subset \mathcal{C}$ be the class of all connected trees. Let A_k, P_k, C_k and K_k denote respectively the anticlique, the path, the cycle, and the clique on k vertices. The *excess* of a graph is $|E| - |V|$, and is denoted $\mathrm{exc}(H)$. If $H \in \mathcal{C}$ then $\mathrm{exc}(H) \geqslant -1$, with $\mathrm{exc}(H) = -1$ exactly for trees, while $\mathrm{exc}(H) = 0$ for unicyclic graphs (graphs that contain exactly one cycle), and $\mathrm{exc}(H) > 0$ for graphs that contain at least two cycles. The degree of vertex i in a subgraph H is denoted $\deg(i, H)$.

For any subgraph H of a graph G, $\Phi(H)$ denote the formula asserting that there is an embedding of H in G, and $\Psi(H)$ that there is an isolated component isomorphic to H. For example, $\Phi(K_2) = \exists x_1 x_2. (x_1 \frown x_2)$, and $\Psi(K_2) =$

$\exists x_1 x_2 \forall y. (x_1 \overset{\frown}{} x_2) \wedge \bigwedge_{i \in \{1,2\}} ((y = x_i) \vee \neg (y \overset{\frown}{} x_{3-i}))$. Similarly, let $\Phi(H)$ denote the formula with a set of $|V_H|$ free variables asserting that the subgraph restricted to the free variables is homomorphic to H (so $\Phi(K_2)(x_1, x_2) = (x_1 \overset{\frown}{} x_2)$), so that $\mathbb{E}[\Phi(H)]$ is the expected number of embeddings of H in G. A set of vertices $v_1, v_2, \ldots$ is a *realisation* of H if $\Phi(H)(v_1, \ldots)$ holds, and if the asymptotic probability that a subgraph is embedded is nonzero, the graph is *realisable*.

A *degree sequence* is just a sequence of natural numbers $d = (d(0), d(1), \ldots)$, where $d(i)$ is the number of degree i vertices, and a sequence is *feasible* if there is a graph with exactly that degree sequence; it implies that there exists a n such that $d(i) = 0$ for all $i \geqslant n$, $\sum_{i \in \mathbb{N}} d(i) = n$ and that $\sum_{i \in \mathbb{N}} i \cdot d(i)$ is even. Sequences of feasible degree sequences are denoted by $\mathcal{D} = (d_0, d_1, \ldots)$ and called *asymptotic degree sequences* (short a.d.s.) if there is a probability distribution $d_{\mathcal{D}} \in \mathcal{P}(\mathbb{N})$ such that $\lim_{n \to \infty} d_n(i)/n = d_{\mathcal{D}}(i)$ (usually the subscript is clear from the context and therefore omitted). Given d define d^* to be the degree distribution d stripped of its degree 0 vertices, $d^*(i) = d(i)/(1 - d(0))$ for all $i \neq 0$ and $d^*(0) = 0$. The random graph $\mathcal{G}(n, \mathcal{D})$ is defined as the uniform distribution on n vertices graphs having degree sequence $d_n = (d_n(0), d_n(1), \ldots, d_n(n-1))$, and defines naturally a random graph sequence $\mathcal{G}(\mathcal{D})$.

Definition 1. *Let $\lfloor \mathcal{D} \rfloor$ be the set of indices of $\mathcal{D}$ that are asymptotically 0. Similarly $\lceil \mathcal{D} \rceil$ is the complementary set, i.e. the set of degrees that are asymptotically represented with nonzero probability. The set $\lceil \mathcal{D} \rceil$ is called the* support *of $\mathcal{D}$.*

The degree sequences considered here satisfy condition $(\mathcal{H}_2)$ (they have a bounded second moment) given by:

Definition 2. *Let the condition $(\mathcal{H}_p)$ be:*

$$\lim_{n \to \infty} \frac{1}{n} \sum_{i=0}^{n} i^p d_n(i) = \sum_{i=0}^{\infty} i^p d(i)$$

The configuration model is a way to generate graphs with a given degree sequence, using the idea that it is much simpler to sample a multigraph having the degree sequence than a simple graph. A configuration on degree sequence d on n vertices is a set C of $\sum_{i=0}^{n} i \cdot d(i)$ elements provided with a partition in n equivalence classes and a random matching of its elements. A multigraph is obtained from this matching by collapsing each equivalence set of nodes into a vertex, and set $i \overset{\frown}{} j$ for every node in the equivalence class of i matched with a node in the class of j. This gives a distribution on multigraphs with degree sequence $\mathcal{D}$, such that, conditioned on the multigraph being simple, is the uniform distribution on graphs with degree sequence $\mathcal{D}$, provided that d satisfies $(\mathcal{H}_2)$ (see [8]). Therefore, the probability of a graph property can be obtained from the configuration model, using for a formula φ

$$\mu(\varphi, \mathcal{G}(n, \mathcal{D})) = \frac{\mu(\varphi \wedge \Phi, \mathbb{G}(n, \mathcal{D}))}{\mu(\Phi, \mathbb{G}(n, \mathcal{D}))}$$

with Φ being the property that the multigraph is simple, i.e. has no loops nor multiple edges (μ is defined in the next subsection). This allows to transfer results on the configuration model to graphs with given degree sequence.

2.2 Logic

The language of first-order (FO) logic on graphs is defined as the closure under logical connectives and quantifiers of atomic formulas, where variables stand for vertices, with equality and one binary predicate interpreted by the edge relation. The probability that a FO-formula φ is satisfied by $\mathcal{G}_n$ is denoted by $\mu(\varphi, \mathcal{G}_n)$, and, when it exists, $\mu(\varphi, \mathcal{G}) = \lim_{n \to \infty} \mu(\varphi, \mathcal{G}_n)$ denotes the limiting probability. A convergence law holds on $\mathcal{G}$ for the language $\mathcal{L}$ if any sentence of $\mathcal{L}$ has a limiting probability, and a 0-1 law holds if the limiting probabilities are all either 0 or 1. Here $\mathcal{L}$ is always FO-logic. If Φ is a set of sentences, called a set of *axioms*, the set of implied formulas $T_\Phi = \{\phi \mid \Phi \models \phi\}$ is called the *theory* of Φ. If every FO-sentence (or its negation) belongs to the theory, it is *complete*. It is called an *almost sure theory* if Φ is a set of (asymptotically) almost sure axioms. Similarly the theory T_D of a class of (random) structures D is the set of formulas that are (almost surely) true on this structure. Since every formula in a theory may be deduced from a finite set of axioms, an almost sure theory is consistent (it does not prove a formula and its negation). Now the notion of contiguity from [7] is adapted to first-order logic to compare structures according to their limit theories.

Definition 3. *Two random graph sequences $\mathcal{G}$ and $\mathcal{G}'$ are weakly FO-contiguous if for all FO-sentences φ, $\mu(\mathcal{G}, \varphi)$ exists and $\mu(\mathcal{G}, \varphi) = 1$ iff $\mu(\varphi, \mathcal{G}')$ exists and $\mu(\varphi, \mathcal{G}') = 1$ holds. If additionally $\mu(\varphi, \mathcal{G})$ exists iff $\mu(\varphi, \mathcal{G}')$ exists, then $\mathcal{G}$ and $\mathcal{G}'$ are FO-contiguous. Furthermore $\mathcal{G}$ and $\mathcal{G}'$ are strongly FO-contiguous if for all FO-sentences φ, $\mu(\varphi, \mathcal{G}) = \mu(\varphi, \mathcal{G}')$.*

2.3 Galton-Watson Branching Processes and Local Weak Convergence

In the random graphs studied here, the typical neighbourhood of a vertex converges towards a simple object, a branching process. This is called the local weak convergence [11]. For a distribution $\boldsymbol{d} \in \mathcal{P}(\mathbb{N})$ with positive first moment, define $\boldsymbol{d}^{**}$ as

$$\boldsymbol{d}^{**}(k-1) = \frac{k\boldsymbol{d}(k)}{\sum_i i\boldsymbol{d}(i)}.$$

The Galton-Watson tree with distribution $\boldsymbol{d}$ is defined by a unique root vertex in generation 0, which has $D \sim \boldsymbol{d}$ children in generation 1. Then each vertex in generation $i > 0$ has independently $D \sim \boldsymbol{d}^{**}$ children in generation $i + 1$. Let GW($\boldsymbol{d}$) denote the distribution on rooted trees induced by this process.

A graph G induces a distance between vertices $d_G(u, v)$ defined as the minimum length of a path between u and v. A graph G induces for every vertex v a rooted graph $G[v]$, where v is called the root; if G is not connected, then $G[v]$ is restricted to the connected component containing v. An isomorphism of rooted graphs $G_1[v_1]$ and $G_2[v_2]$ is an isomorphism of graphs that maps the root v_1 to the root v_2. For a graph G, let $\mathcal{B}_G[v, i]$ be the rooted graph that is the ball of diameter i (for the graph distance) centered on vertex v. This allows to define a distance between two rooted graphs as $d(G_1[v_1], G_2[v_2]) = 1/(1 + T)$ where

$$T = \sup\{t \mid \text{there exists an isomorphism from } \mathcal{B}_{G_1}[v_1, t] \text{ to } \mathcal{B}_{G_2}[v_2, t]\}$$

To a graph G can therefore be associated a probability measure on the set of all rooted graphs defined by

$$U(G) = \frac{1}{|V|} \sum_{v \in V} \delta[G[v]]$$

where δ is the Dirac function. Let $\rightsquigarrow$ denote the weak convergence of probability measures (see the textbook [1]).

Definition 4. *A sequence $(G_n)_{n \in \mathbb{N}}$ has local weak limit ρ if $U(G_n) \rightsquigarrow \rho$.*

This machinery is applied in the next section to show that with some hypotheses on the degree sequence, the local weak limit is a Galton-Watson tree.

3   Convergence of Graphs with Prescribed Degrees

To obtain a convergence law, it is necessary to put some restrictions on the degree sequences considered. Throughout the literature the sequences satisfying some conditions are called *smooth*, *well-behaved*, etc. Four conditions describe a class of a.d.s., as in [10].

Definition 5. *An asymptotic degree sequence $\mathcal{D}$ is an l-a.d.s., if the following conditions hold:*

(1) $\lim_{n \to \infty} d_n(i)/n = \boldsymbol{d}(i)$ uniformly in i, and there exists an N such that for all i with $\boldsymbol{d}(i) = 0$, $d_n(i) = 0$ for all $n > N$.
(2) $\sum_{i=0}^{\infty} i \cdot \boldsymbol{d}(i) = \Delta < \infty$.
(3) $\lim_{n \to \infty} \sum_{i=1}^{n-1} i \cdot d_n(i)/n = \Delta$.
(4) There exists a constant $a < 1/4$ such that for all n and $i > n^a$, $d_n(i) = 0$.

These conditions are sufficient to derive the following theorem:

Theorem 1 *[10]. Let $\mathcal{D}$ be an l-a.d.s., then $\mathcal{G}(n, \mathcal{D})$ has a convergence law.*

Note that the theorem is stated in [10] for *l*-a.d.s. satisfying an additional condition, which is that $\boldsymbol{d}(0) = 0$, which is not a loss of generality by Lemma 2.1 of the original paper [10]. To show this, consider $\mathcal{G}(n, d)$ a random graph with degree sequence d with k isolated vertices, then $G \sim \mathcal{G}(n, d)$ has the same law as $G' + A_k$, with $G' \sim \mathcal{G}(n - k, d^*)$. Since the explicit values of formulas are considered, it is easier to directly consider a.d.s. for which $\boldsymbol{d}(0) \neq 0$, while the subgraph counts are computed on $\boldsymbol{d}^*$.

The proof is in fact carried on configurations, using the fact that l-a.d.s. satisfy $(\mathcal{H}_2)$, so the distribution induced by configurations on simple graphs is uniform. Therefore the convergence of first-order formulas on the configurations is sufficient to deduce a convergence law for $\mathcal{G}(n, \mathcal{D})$. The main idea is to use the Gaifman Locality Lemma, and Ehrenfeucht-Fraïssé games to reduce the equivalence with regard to first-order logic to a game played on a finite set of structures. If the duplicator has a winning strategy, then no first-order formula can differentiate the two structures. Then it is shown that only certain classes of neighbourhoods (defined by their excess) need to be considered, and finally the theorem follows by counting the number of neighbourhoods in each class.

In some sense, a first-order formula with q quantifiers cannot count to more than q, anything more is considered to be 'infinite'. The notion of a q-sphere is introduced to formalise this notion: a q sphere is a ball with root r of diameter q where every vertex at distance i from r has at most $q + 1 - i$ neighbours of each type of $(q - i)$-sphere. The type of q-sphere associated to a ball $\mathcal{B}_G(v, q)$ is obtained by setting the number n of neighbours of a given type to $q + 1$ if it is greater than $q + 1$. Fix $q \in \mathbb{N}$, and for any $i < 3^q$ and B an isomorphism class of i-spheres, let $\tau_{j,B}$ denote the number of vertices v such that the ball $\mathcal{B}_{G_j}(v, i)$ is in class B, for $j \in \{1, 2\}$. Then Hanf's lemma states that G_1 is equivalent to G_2 up to q-quantifiers formulas if and only if $\tau_{1,B} = \tau_{2,B}$ or both $\tau_{1,B} > q$ and $\tau_{2,B} > q$, for all i-sphere B. Note that since the number of isomorphism classes of q-spheres is finite, only a finite quantity of information is needed to determine the truth of q-quantifiers formulas. Then the proof follows from the three following lemmas:

(i) If $\text{exc}(H) < 0$ then there are either zero or an unbounded number of realisations of H in G.

(ii) If $\text{exc}(H) = 0$ then the number of realisations of H in G converges to a finite value.

(iii) If $\text{exc}(H) > 0$ then there are almost surely no realisations of H in G.

The condition (i) follows from the convergence to the local weak limit:

Theorem 2 *[11]. Let $\mathcal{D}$ be an a.d.s. such that d satisfies $(\mathcal{H}_2)$. Then, as n goes to infinity, $\mathbb{E}[U(G_n)] \rightsquigarrow \text{GW}(d)$.*

From which follows that all trees that appear as embedded or induced subgraphs have an unbounded number of realisations. Furthermore a tree T appears as an induced component if and only if $d(\deg(i)) > 0$ for all $i \in T$.

Lemma 1 *[8](Subgraph count). Let $\mathcal{D}$ be an a.d.s., $H \in \mathcal{C}$ be a graph on k vertices with m edges, maximal degree $p \geqslant 1$ and number of automorphisms c. Let $D \sim d^*$. Suppose $\mathcal{D}$ respects condition $(\mathcal{H}_p)$, then the expected number of realisations of H in $\mathcal{G}(n, \mathcal{D})$ is*

$$\lim_{n \to \infty} \mathbb{E}[\Phi(H)] = n^{-\text{exc}(H)} \frac{\prod_{i=1}^k \mathbb{E}[(D)_{\deg(i,H)}]}{c(\mathbb{E}[D])^m}$$

and converges in distribution to a Poisson variable.

A detailed proof can be found in [2]. Since l-a.d.s. satisfy $(\mathcal{H}_4)$ by *(4)*, conditions (ii) and (iii) can be deduced. However a weaker (and sufficient condition) is required, as shown next.

Definition 6. *An asymptotic degree sequence $\mathcal{D}$ is an L-a.d.s., if it satisfies conditions (1)–(3) and (4'), which is that $(\mathcal{H}_2)$ holds and $\mathbb{E}[(D)_4] = o(n)$.*

Theorem 3. *The graph sequence $\mathcal{G}(n, \mathcal{D})$ has a convergence law and satisfies (i)–(iii) if and only if $\mathcal{D}$ is an L-a.d.s.*

Proof. First, note that (i) is implied by the local convergence, and (ii) holds if and only if $(\mathcal{H}_2)$ holds, as it is sufficient to consider the C_n subgraphs (more details on the number of cycles are given in the next section, see corollary 1). For condition (iii), any subgraph with positive excess contains a minimal subgraph of excess exactly 1 with maximal degree 4 (in the case of two cycles joined by one vertex) or two vertices of degree 3 and all the other vertices having degree 2. So by plugging these parameters into lemma 1, for any subgraph H with $\mathrm{exc}(H) = 1$ and maximal degree p, $\mathbb{E}[\boldsymbol{\Phi}(H)] = \Theta(\mathbb{E}[(D)_p]/n)$, so this goes to zero exactly when $\mathbb{E}[(D)_p] = o(n)$, as given by condition *(4')*. To show that this is a necessary condition, we consider a counterexample. Take $\boldsymbol{d}$ a power law of exponent 3, i.e. $\boldsymbol{d} \sim k^{-3}$. Then $\mathbb{E}[(D)_4] = \Theta(n)$, so $\mathbb{E}[(D)_4]/n \to l > 0$. So in the case of power law graphs, conditions (i)–(iii) hold exactly as long as $(\mathcal{H}_3)$ is satisfied, since $(\mathcal{H}_p)$ implies that the exponent is strictly greater than p. The reader is referred to the original paper [10] for a formal proof of the combinatorial lemmas and the winning strategy for Ehrenfeucht-Fraïssé used to deduce the convergence law. $\square$

4   Main Results

The subgraph count lemma and the local weak convergence give explicit formulas to compute the sets of realised subgraphs. This allows to describe the possible limit theories in detail and to give first-order axiom schemes.

4.1   Asymptotic Degree Sequences and 0-1 Laws

The probabilities of first-order formulas therefore depend on $\boldsymbol{d}$. One can ask whenever 0-1 laws hold, so some notation is introduced to answer this question.

Let $\mathcal{D}_0$ be the set of a.d.s. that with $\boldsymbol{d}(0) = 1$, $\mathcal{D}_1$ be the set of a.d.s. with $\boldsymbol{d}(1) = 1$, and $\mathcal{D}_{0,1}$ be the class of all a.d.s. such that $\lceil \mathcal{D} \rceil = \{0, 1\}$. By condition 1, the proportion of vertices of degree i is asymptotically 0 if and only if it is always 0 after some N, so all a.d.s. in $\mathcal{D}_0$ can be truncated to an a.d.s. with $d_n = (n, \overline{0})$ for all n. Therefore any d-regular sequence can be considered to have a unique L-a.d.s. representative.

Definition 7. *An asymptotic degree sequence $\mathcal{D}$ is an L^*-a.d.s., if $\mathcal{D}$ is an L-a.d.s. and $\mathcal{D}$ is not in $\mathcal{D}_0$, in $\mathcal{D}_1$, nor in $\mathcal{D}_{0,1}$.*

Lemma 2. *Let $\mathcal{D}$ be an L-a.d.s., then $\mathcal{G}(n, \mathcal{D})$ has a 0-1 law if and only if $\mathcal{D}$ belongs to $\mathcal{D}_0$, to $\mathcal{D}_1$, or to $\mathcal{D}_{0,1}$.*

Therefore only the trivial L-a.d.s. have 0-1 laws.

Proof. The if part is direct, and proven first. The sequence $\mathcal{G}(n, \mathcal{D}_0)$ has for every n only one support graph, the empty graph. This sequence converges to the infinite empty graph, which has an ω-categorical theory (all infinite countable models of the theory are isomorphic), axiomatised by the following formulas:

- No two vertices are adjacent: $\forall xy.\,(x \asymp y)$
- There are at least k distinct vertices: $A_k := \exists x_1 \ldots x_k.\,(\bigwedge_{i \neq j} x_i \neq x_j)$

The sequence $\mathcal{G}(n, \mathcal{D}_1)$ converges to an infinite collection of independent edges, which has an ω-categorical theory, axiomatised by the A_k axiom scheme and the formula

$$\forall x \exists y \forall z.\,(x \overset{\frown}{} y \wedge (x \overset{\frown}{} z \implies z = y)).$$

Finally, for any $\mathcal{D} \in \mathcal{D}_{0,1}$, the sequence converges to an infinite collection of isolated vertices and isolated edges, which has an ω-categorical theory, axiomatised by the A_k axiom scheme and the formula

$$\forall x \exists y \forall z.\,((x \overset{\frown}{} y \wedge (x \overset{\frown}{} z \implies z = y)) \vee x \asymp z).$$

The other direction is proved using the following lemma:

Lemma 3. *Let $\mathcal{D}$ be an L^*-a.d.s., then for all cycles C_n, $\mu(\Phi(C_n), \mathcal{G}(n, \mathcal{D})) \notin \{0, 1\}$, the probability that there is an embedding of C_n is neither 0 nor 1.*

Which follows from the subgraph count lemma specialised to cycles:

Corollary 1. *(Cycles count) Using the same hypothesis as in 1, for all $k \geqslant 3$, the expected number of cycles of length k in $\mathcal{G}(n, \mathcal{D})$ with $\mathcal{D}$ an L-a.d.s. is*

$$\lim_{n \to \infty} \mathbb{E}[\Phi(C_k)] = \frac{(\mathbb{E}[(D)_2])^k}{2k(\mathbb{E}[D])^k}$$

Now, this last corollary gives us that the number of expected cycles is finite, and it is furthermore possible to show using Stein's method that it converges in distribution to a Poisson variable. This expression can be rewritten as $(2k\Delta^k)^{-1} \times \sum_i i(i-1)d(i)$, so this number is positive if and only if $\lceil \mathcal{D} \rceil$ contains values greater than 1. Therefore for any L^*-a.d.s. $\mathcal{D}$, $0 < \mu(\Phi(C_n), \mathcal{G}(n, \mathcal{D})) < 1$, so $\mathcal{G}(n, \mathcal{D})$ has no 0-1 law, which concludes the proof. $\qquad\square$

4.2   First-Order Contiguity of $\mathcal{G}(n, \mathcal{D})$ Graphs

The previous results show that the limit theory depends on d. This intuition is formalised by the next theorem, which shows that the limit theory only depends on the support of d, and the actual probabilities do not matter.

Theorem 4. *Let* $\mathcal{D}$, $\mathcal{D}'$ *be L-a.d.s., then* $\mathcal{G}(n, \mathcal{D})$ *and* $\mathcal{G}(n, \mathcal{D}')$ *are contiguous if and only if* $\lfloor \mathcal{D} \rfloor = \lfloor \mathcal{D}' \rfloor$.

Proof. The idea is to show, given a $\mathcal{D}$, how to axiomatise the theory $T_{\lceil \mathcal{D} \rceil}$ from $\lceil \mathcal{D} \rceil$. From condition (i)–(iii), only tree components appear in the almost sure theory. The almost sure theory also contains the formulas stating that every subgraph with $\text{exc}(H) > 0$ is not realised. From Corollary 1, sentences about unicyclic components (or their negation) do not appear in the almost sure theory. Let the axiom scheme B_k be 'there is no set of k vertices with at least $k + 1$ edges'; $B_k \in T_{\lceil \mathcal{D} \rceil}$ for every k and $\lceil \mathcal{D} \rceil$. The axiom schemes corresponding to condition (i) are:

- $\Upsilon_{T,k}$: There are at least k different exact realisations of T for all $k \in \mathbb{N}$ and every T such that $\text{GW}(\boldsymbol{d})[T] > 0$. Let $\Upsilon_{\mathcal{D}}$ denote the set of $\Upsilon_{T,k}$ axioms of $\mathcal{D}$.
- $\deg_k$: There are no vertices of degree exactly k for all $k \in \lfloor \mathcal{D} \rfloor$.

Where an exact realisation of a tree T is an embedding of T where $\deg(i, T) = \deg(i, G)$ for every vertex *that is not a leaf* in T if GW cannot die out isomorphic to T, and including the degree of the leaves otherwise (for example if $\mathcal{D}$ is 3-regular, then the corresponding GW cannot die out and induced components are not considered). Note that the $\deg_k$ axiom scheme can be replaced by a single sentence if and only if the set $\lceil \mathcal{D} \rceil$ is finite or cofinite.

From the Gaifman locality lemma, these sets of axioms are sufficient to determine the truth of first-order formulas. Since they depend only on $\text{GW}(\boldsymbol{d})$, and $\text{GW}(\boldsymbol{d})[T] > 0$ depends only on $\lceil \mathcal{D} \rceil$, the claim follows. □

Corollary 2. *Let* $\mathcal{D}$, $\mathcal{D}'$ *be* L^*-*a.d.s., then* $\mathcal{G}(n, \mathcal{D})$ *and* $\mathcal{G}(n, \mathcal{D}')$ *are strongly contiguous if and only if* $\boldsymbol{d}_{\mathcal{D}} = \boldsymbol{d}_{\mathcal{D}'}$.

Note that this statement is false for L-a.d.s., any two $\mathcal{D}$, $\mathcal{D}'$ in $\mathcal{D}_{0,1}$ are strongly contiguous since a 0-1 law holds.

Proof. Let $i \geqslant 2$ be such that $\boldsymbol{d}(i) \neq \boldsymbol{d}'(i)$. Then consider the formula φ being "there is a vertex of degree exactly i in a triangle"; the limiting probability of this formula is a binomial random variable where the parameter n is three times the number of triangles in the graph and the parameter p is $\text{Pr}[\deg(v) = i]$. Since by assumption $\text{Pr}[\deg(v) = i]$ is different in $\mathcal{G}(n, \mathcal{D})$ and $\mathcal{G}(n, \mathcal{D}')$, either $\mu(\varphi, \mathcal{G}(n, \mathcal{D})) \neq \mu(\varphi, \mathcal{G}(n, \mathcal{D}'))$ or $\mu(\Phi(C_3), \mathcal{G}(n, \mathcal{D})) \neq \mu(\Phi(C_3), \mathcal{G}(n, \mathcal{D}'))$.

In the other direction, from previous lemmas it follows that the local geometry of the graph can be asymptotically computed from $\boldsymbol{d}$ alone. In particular it is sufficient to consider the probability of unicyclic neighbourhoods since the formulas that have a 0 or 1 limiting probability are the same. The claim follows from the fact that these probabilities only depend on $\boldsymbol{d}$. □

4.3   Contiguity with Erdős-Rényi Random Graphs

The Erdős-Rényi model $\mathcal{G}(n, p)$ is the simplest model of random graph, where each edge is independently present with probability $p(n)$. The sparse regime is when $p(n) = c/n$, so that the average degree is c. This regime is well-studied, $\mathcal{G}(n, c/n)$ has a convergence law and furthermore $\mathcal{G}(n, c_1/n)$ and $\mathcal{G}(n, c_2/n)$ are contiguous for any $c_1, c_2 \in \mathbb{N}$.

Theorem 5. *A random graph sequence $\mathcal{G}(n, \mathcal{D})$ is FO-contiguous to the Erdős-Rényi graph $\mathcal{G}(n, c/n)$ for any c if and only if $\mathcal{D}$ is such that $\lceil \mathcal{D} \rceil = \mathbb{N}$.*

Proof. To show that the two models produce contiguous random graphs, observe that conditions (i)–(iii) hold on $\mathcal{G}(n, c/n)$, as was proven in [9]. Condition (iii) is immediate, since the probability that there is a graph with positive excess is given by $\binom{n}{k} \cdot (\frac{c}{n})^{k+1} = \Theta(\frac{1}{n})$ since there are $\binom{n}{k}$ possible sets of k vertices and each of the $k+1$ edges is present independently with probability c/n. Condition (ii) can be proved, as on the configuration model, by using the method of moments to show that the number of cycles converges to a Poisson variable. Finally condition (i) follows from the convergence to the local weak limit $\mathrm{GW}(\mathrm{Poi}_c)$. Therefore every $T \in \mathcal{T}$ is embedded in the graph and appears as an induced component, so $\Upsilon_{\mathcal{G}(n,c/n)} = \Upsilon_{\mathbb{N}}$, where $\Upsilon_{\mathbb{N}}$ is the set of axioms for any $\mathcal{G}(n, \mathcal{D})$ with $\lceil \mathcal{D} \rceil = \mathbb{N}$. From this follows that for $\mathcal{G}(n, c/n)$, $\Phi = \Psi = \{T \mid T \in \mathcal{T}\}$. The other direction follows, since the first-order contiguity is an equivalence relation, from theorem 4, two sequences $\mathcal{G}(n, \mathcal{D})$ and $\mathcal{G}(n, \mathcal{D}')$ are contiguous if and only if $\lceil \mathcal{D} \rceil = \lceil \mathcal{D}' \rceil$. □

A consequence in the Erdős-Rényi case is that the limiting probability of every formula can be written using the constant 1, the values of $\mathrm{Poi}(c)$, subtraction and multiplication. This result has an analogous for L-a.d.s. depending on $\boldsymbol{d}$.

Corollary 3. *For any formula of first-order logic, its limiting value can be written using the constant 1, the values from $\mathrm{GW}(\boldsymbol{d})$ and $\mathbb{E}[\Phi(C_k)] \sim \mathrm{Poi}(f(k))$, subtraction and multiplication.*

Sketch of the proof. Since the only formulas that are not either almost surely true nor almost surely false are about unicyclic neighbourhoods, they can be written using the probabilities that any type of cycle is realised, and the probabilities that the involved vertices branch in a certain way.

4.4   A Taxonomy of the Limit Theories

The previous subsections associate axiom schemes to every possible support, which allows to order the limit theories. Two other natural orders can be defined. Let $\Phi_{\mathcal{D}} = \{\Phi(H) \mid \mu(\Phi(H), \mathcal{G}(n, \mathcal{D})) = 1 \wedge H \in \mathcal{C}\}$ denote the set of axioms defined by the set of connected subgraphs almost surely embedded in $\mathcal{G}(n, \mathcal{D})$. Let $\Psi_{\mathcal{D}} = \{\Psi(H) \mid \mu(\Psi(H), \mathcal{G}(n, \mathcal{D})) = 1 \wedge H \in \mathcal{C}\}$ denote the set of axioms defined by the set of subgraphs that almost surely appear as induced components. Note that $\Phi_{\mathcal{D}}$ is never empty since A_1 always belongs to it, while $\Psi_{\mathcal{D}}$ can be empty, if there is no small isolated component occurring with high probability. For example $\Phi_{\mathcal{D}_0} = \Psi_{\mathcal{D}_0} = \{A_1\}$, while $\Psi_{\mathcal{D}_2} = \emptyset$ and $\Phi_{\mathcal{D}_2} = \{P_n \mid n \in \mathbb{N}\}$. Let $\mathcal{D}$ be such that $\lceil \mathcal{D} \rceil = \mathbb{N}$, then $\Phi_{\mathcal{D}} = \Psi_{\mathcal{D}} = \{T \mid T \in \mathcal{T}\}$ (Fig. 1).

Theorem 6. *There is a continuum of different limit theories defined by $\mathcal{G}(n,\mathcal{D})$ graphs with $\mathcal{D}$ an L-a.d.s., naturally ordered by their supports $\lceil\mathcal{D}\rceil$. The natural orders correspond to natural orders on the Φ, Ψ and Υ axiom sets, and the theory of $\mathcal{G}(n,1/n)$ is maximal for all these orders.*

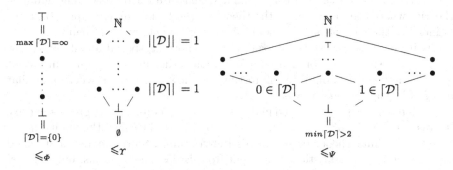

Fig. 1. The orders defined by the limit theories

Proof. The $\leqslant_\Upsilon$ order is defined by the inclusion order on supports, $T_\mathcal{D} \leqslant_\Upsilon T_{\mathcal{D}'}$ is true if $\lceil\mathcal{D}\rceil \subseteq \lceil\mathcal{D}'\rceil$. The maximum is $\top = \mathbb{N}$ but there is no minimum since $\bot = \emptyset$ which is not a possible support, so this partial order is only a semi-lattice. By definition if $\lceil\mathcal{D}\rceil \subseteq \lceil\mathcal{D}'\rceil$ then $\Upsilon_\mathcal{D} \subseteq \Upsilon_{\mathcal{D}'}$. These theories are also ordered by the Φ and Ψ sets of axioms corresponding respectively to the $\leqslant_\Phi$ and $\leqslant_\Psi$ order on supports. The order $T_\mathcal{D} \leqslant_\Psi T_{\mathcal{D}'}$ is defined by $\max\lceil\mathcal{D}\rceil \leqslant \max\lceil\mathcal{D}'\rceil$, so this is a linear order. Let the order $\leqslant_\Psi$ be the lattice defined by:

- $\bot = \emptyset$ for all $\mathcal{D}$ such that $\min\lceil\mathcal{D}\rceil \geqslant 2$
  - if $0 \in \mathcal{D}$ and $0 \in \mathcal{D}'$, then $\mathcal{D} \leqslant_\Psi \mathcal{D}'$ if and only if $\lceil\mathcal{D}\rceil \subseteq \lceil\mathcal{D}'\rceil$
  - if $1 \in \mathcal{D}$ and $1 \in \mathcal{D}'$, then $\mathcal{D} \leqslant_\Psi \mathcal{D}'$ if and only if $\lceil\mathcal{D}\rceil \subseteq \lceil\mathcal{D}'\rceil$
  - otherwise $\mathcal{D}$ and $\mathcal{D}'$ are incomparable.
- $\top = \{T \mid \forall T \in \mathcal{T}\}$

Each different limit theory is defined by a set of contiguous random graph sequences, each depending on the support of $\mathcal{D}$. There is an uncountable number of such supports since every nonempty subset of $\mathbb{N}$ can be the support of some L-a.d.s. $\mathcal{D}$. To see this, let S be a subset of $\mathbb{N}$, take any $\boldsymbol{d} \in \mathcal{P}(\mathbb{N})$ satisfying $(\mathcal{H}_3)$ with support S. One gets $\mathcal{D}$ by sampling n integers from $\boldsymbol{d}$ for every n; it is known that every $\boldsymbol{d}$ satisfying $(\mathcal{H}_2)$ is feasible for all $n > N$, for some N depending only on $\boldsymbol{d}$. Some care has to be taken to make the sum of the degrees even, which can be done by modifying only one vertex (unless the support contains only odd numbers, in which case $\mathcal{G}(n,\mathcal{D})$ is only defined on odd n). This process therefore generates a sequence of feasible degree sequences, and it converges to $\boldsymbol{d}$ with support S. $\qquad\square$

Sequences with a sublinear number of edges behave similarly, provided that $\mathcal{D}^* \rightsquigarrow \boldsymbol{d}^*$. The limit theories are exactly the same, only the rate of convergence would be modified (in the obvious way). However no degree sequence can be contiguous to an Erdős-Rényi graphs with a sublinear number of edges, e.g. $\mathcal{G}(n,c/n^{1+1/k})$, since the realised subgraphs are exactly the trees on at most $k-1$ vertices, whereas no branching process can generate exactly these trees.

5 Conclusion

In the light of first-order contiguity, the statistics of a degree sequence, important for many applications, are lost; the only information needed is the support of the distribution. Furthermore the L-a.d.s. generate an ordered continuum of theories where the maximum is the theory of the Erdős-Rényi graph $\mathcal{G}(n, 1/n)$. It includes the case of power law graphs for exponents greater than 3, so from the point of view of first-order logic, there is nothing special about power law graphs. The same question can be asked about other models of random graphs with parameters, it would be interesting to know other cases where a similar order of logical theories appear.

An interesting question asked in [10] is to consider non-sparse degree sequences. It was settled for n^α-regular graphs in [6] by sandwiching the limit theory between the limit theory of two Erdős-Rényi graphs. So the model of specified degree sequences is statistically different from Erdős-Rényi graphs, but in many cases it is similar from the point of view of logic. More exotic degree sequences need to be investigated to exhibit new logical behaviours. Such sequences are for example a.d.s. for which $(\mathcal{H}_1)$ holds but not $(\mathcal{H}_2)$. The diameter of these random graphs falls to $\Theta(\ln \ln n)$ because there is a core of highly connected vertices. This behaviour cannot be obtained from Erdős-Rényi graphs, so it would be interesting to know the logical theory of these graphs. In between, i.e. the degree sequences for which $(\mathcal{H}_2)$ holds but not $(\mathcal{H}_3)$, there may be an analogous order of theories as complex subgraphs start to appear.

References

1. Billingsley, P.: Convergence of Probability Measures. Wiley, New York (2009)
2. Bordenave, C.: Notes on random graphs and combinatorial optimization. Lecture notes. http://www.math.univ-toulouse.fr/bordenave/coursRG.pdf
3. Chung, F.R.K., Lu, L.: Complex Graphs and Networks, vol. 107. American mathematical society, Providence (2006)
4. Fagin, R.: Probabilities on finite models. J. Symbolic Logic **41**(01), 50–58 (1976)
5. Glebskii, Y.V., Kogan, D.I., Liogonki, M.I., Talanov, V.A.: Range and degree of realizability of formulas in the restricted predicate calculus. Kibernetika **5**, 17–27 (1969)
6. Haber, S., Krivelevich, M.: The logic of random regular graphs. J. Comb. **1**(3–4), 389–440 (2010)
7. Janson, S.: Random regular graphs: asymptotic distributions and contiguity. Comb. Probab. Comput. **4**(04), 369–405 (1995)
8. Janson, S.: The probability that a random multigraph is simple, II. preprint (2013). arXiv:1307.6344
9. Lynch, J.F.: Probabilities of sentences about very sparse random graphs. Random Struct. Algorithms **3**(1), 33–53 (1992)
10. Lynch, J.F.: Convergence law for random graphs with specified degree sequence. ACM Trans. Comput. Log. (TOCL) **6**(4), 727–748 (2005)
11. Mezard, M., Montanari, A.: Information, Physics, and Computation. Oxford University Press, Oxford (2009)
12. Spencer, J.: The Strange Logic of Random Graphs, vol. 22. Springer, Berlin (2001)

Streaming Algorithms for Smallest Intersecting Ball of Disjoint Balls

Wanbin Son[(✉)] and Peyman Afshani

MADALGO, Department of Computer Science, Aarhus University, Aarhus, Denmark
{wson,peyman}@cs.au.dk

Abstract. In this paper, we propose streaming algorithms for approximating the smallest intersecting ball of a set of disjoint balls in $\mathbb{R}^d$. This problem is a generalization of the 1-center problem, one of the most fundamental problems in computational geometry. We consider the single-pass streaming model; only one-pass over the input stream is allowed and a limited amount of information can be stored in memory. We introduce three approximation algorithms: one is an algorithm for the problem in arbitrarily dimensions, but in the other two we assume d is a constant. The first algorithm guarantees a $(2+\sqrt{2}+\varepsilon^*)$-factor approximation using $O(d^2)$ space and $O(d)$ update time where ε^* is an arbitrarily small positive constant. The second algorithm guarantees an approximation factor 3 using $O(1)$ space and $O(1)$ update time (assuming constant d). The third one is a $(1+\varepsilon)$-approximation algorithm that uses $O(1/\varepsilon^d)$ space and $O(1/\varepsilon^{(d-1)/2})$ amortized update time. They are the first approximation algorithms for the problem, and also the first results in the streaming model.

1 Introduction

Given a set D of n pairwise interior-disjoint balls in $\mathbb{R}^d$, $n > d$, we consider the problem of finding a center that minimizes the maximum distance between the center and a ball in D. The distance between a point p and a ball b centered at c of radius r is defined by $dist(p,b) = \max\{0, |pc| - r\}$ where $|\cdot|$ is a distance between two points. The problem can be also formulated as finding the smallest ball that intersects all the input balls, and hence we call it "the smallest intersecting ball of disjoint balls (SIBB) problem". Observe that if the input balls are points, which we call "the smallest enclosing ball of points (SEBP) problem", then we have an instance of the 1-center problem. The 1-center problem is a fundamental problem in computational geometry, specially in the area of facility location problems [1]. So in our problem we model facilities that can be "relocated" up to a fixed distance. Finally, as an additional motivation, we can view each ball as an uncertain point [2] and thus a solution for the SIBB problem would imply a lower bound for the 1-center problem in a set of uncertain points.

MADALGO—Center for Massive Data Algorithmics, a center of the Danish National Research Foundation.

© Springer International Publishing Switzerland 2015
R. Jain et al. (Eds.): TAMC 2015, LNCS 9076, pp. 189–199, 2015.
DOI: 10.1007/978-3-319-17142-5_17

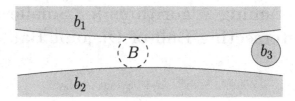

Fig. 1. Three input balls b_1 b_2 and b_3, and the smallest ball B that intersects b_1 and b_2.

Several properties of the SIBB problem are same as those of the SEBP problem. Both problems are LP-type problems with combinatorial dimension $d + 1$ [3], so the problem can be solved by a generic algorithm for an LP-type problem in the static setting. Because of the similarity between the SEBP problem and the SIBB problem, few studies [1,2,4] consider the SIBB problem.

In this paper, we are interested in the problem in the *single-pass streaming model*; only one-pass over the input stream is allowed and only a limited amount of information can be stored. The streaming model is attractive both in theory and in practice due to massive increase in the volume of data over the last decades, a trend that is most likely to continue. We assume that the memory size is much smaller than the size of input data in the streaming model, so it is important to develop an algorithm in which the space complexity does not depend on the size of input data.

In the streaming model, the similarities between the SIBB and SEBP problems break down as it turns out that a set of balls is much more difficult to process than a set of points. For example, let us consider a factor 1.5-approximation algorithm for the SEBP problem [5] which works as follows: For the first two input points, the algorithm computes the smallest enclosing ball B for them. For each next input point p_i, the algorithm updates B to be the smallest ball that contains B and p_i. For the SEBP problem, the algorithm gives the correct approximation factor. The obvious extension of this algorithm to the SIBB problem would be as follows: For the first two input balls b_1 and b_2, compute the smallest intersecting ball B for them. For each next input ball b_i, compute the smallest ball that contains B and intersects b_i. This algorithm gives a solution that intersects all the input balls, but it does not guarantee any approximation factor. In fact, as Fig. 1 shows, the approximation factor could be arbitrarily large: the first two input balls are b_1 and b_2, and B is the smallest intersecting ball for them. The third input ball on the stream is b_3 and the smallest ball that contains B and intersects b_3 can be arbitrarily larger than the optimal solution if b_3 is located far way from B. The reason why the algorithm does not work properly is that B does not "keep" enough information about b_1 and b_2. Because of similar reasons, the other approximation algorithms [5–7] for the SEBP problem also do not work for the SIBB problem.

Previous work in the static setting. Matoušek et al. [3] showed that the smallest intersecting ball of convex objects problem is an LP-type problem, so it can be solved in $O(n)$ time in fixed dimensions. Löffler and Kreveld [2] considered the smallest intersecting ball of balls problem as the 1-center problem for imprecise points. They mentioned that the problem is an LP-type problem.

Mordukhovich et al. [1] described sufficient conditions for the existence and uniqueness of a solution for the problem. In the plane, Ahn et al. [4] proposed an algorithm to compute the smallest two congruent disks that intersect all the input disks in $O(n^2 \log^4 n \log \log n)$ time.

For the SEBP problem, it is known that the problem is an LP-type problem, so it can be solved by an LP-type framework in linear time in fixed constant dimensions [3]. While the LP-type framework gives an exact solution, it is not attractive when d may be large because a hidden constant in the time complexity of the LP-type framework has exponential dependency on d. In high dimensions, Bâdoiu and Clarkson [8] presented a $(1 + \varepsilon)$-approximation algorithm that computes a solution in $O(nd/\varepsilon + (1/\varepsilon)^5)$ time.

The k-center problem is NP-hard if k is a part of input [9], so studies have been focused on the problem for small k [10,11] or developing approximation algorithms [12,13].

Previous work on data streams. To the best of our knowledge, our work contains the first approximation algorithms for the SIBB problem, and also the first results in the streaming model.

The SEBP problem, however, has been studied extensively in the streaming model. Zarrabi-Zadeh and Chan [5] showed a 1.5 approximation algorithm which uses the minimum amount of storage. Agarwal and Sharathkumar [6] presented a $((1 + \sqrt{3})/2 + \varepsilon)$-approximation algorithm using $O(d/\varepsilon^3 \log (1/\varepsilon))$ space, and Chan and Pathak [7] proved that the algorithm has approximation factor 1.22. Agarwal and Sharathkumar [6] also showed that any algorithm in the single-pass stream model that uses space polynomially bounded in d cannot achieve an approximation factor less than $(1 + \sqrt{2})/2 > 1.207$. In fixed dimensions, a $(1 + \varepsilon)$-approximation algorithm can be derived using $O(1/\varepsilon^{(d-1)/2})$ space and $O(1/\varepsilon^{(d-1)/2})$ update time [7]. For the k-center problem, several approximation algorithms [14–16] also have been proposed.

Our results. We describe a $(2+\sqrt{2}+\varepsilon^*)$-approximation algorithm that uses $O(d^2)$ space and $O(d)$ update time for arbitrary dimension d where ε^* is an arbitrarily small positive constant. After that we present two approximation algorithms for fixed constant dimension d. The first approximation algorithm guarantees a 3-approximation using $O(1)$ space and $O(1)$ update time, and the next one guarantees a $(1+\varepsilon)$-approximation using $O(1/\varepsilon^d)$ space and $O(1/\varepsilon^{(d-1)/2})$ update time. One may think the last two approximation algorithms have the same complexity

Table 1. Results for the smallest intersecting ball of disjoint balls problem over the single-pass streaming model. $O^*(x)$ denotes $O(x)$ amortized time, and ε^* denotes an arbitrarily small positive constant.

Dimension d	Factor	Space	Update time
Arbitrary dim. d	$(2 + \sqrt{2} + \varepsilon^*)$	$O(d^2)$	$O(d)$
Constant dim. d	3	$O(1)$	$O(1)$
	$(1 + \varepsilon)$	$O(1/\varepsilon^d)$	$O^*(1/\varepsilon^{(d-1)/2})$

for $\varepsilon = 2$, but the 3-approximation algorithm only uses space polynomial in d, so it is more valuable than the $(1 + \varepsilon)$-approximation algorithm in the streaming model. Table 1 shows a summary of our results.

2 Preliminaries

Let D be a set of n pairwise interior-disjoint balls in $\mathbb{R}^d$. The balls in D arrive one by one over the single-pass stream. They are labeled in order, so b_i is a ball in D that has arrived at the i-th step, that is, $D = \{b_1, b_2, ..., b_n\}$.

Let $b(c, r)$ denote a ball centered at c of radius r, and let $c(b)$ and $r(b)$ denote the center and the radius of a ball b, respectively. We denote B^* *the optimal solution*, and c^* and r^* denote $c(B^*)$ and $r(B^*)$, respectively.

The distance between any two points p and q is denoted by $|pq|$, and the distance between any two balls b and b' is denoted by $dist(b, b') = \max\{|c(b)c(b')| - (r(b) + r(b')), 0\}$. We use $dist(p, b)$ to denote the distance between a point p and a ball b, that is, $\max\{|pc(b)| - r(b), 0\}$.

Our goal is to approximate the smallest ball B^* that intersects all the input balls.

3 $(2 + \sqrt{2} + \varepsilon^*)$-Approximation Algorithm in Any Dimensions d

We introduce an algorithm that guarantees $(2 + \sqrt{2} + \varepsilon^*)$-approximation factor for any d where ε^* is an arbitrarily small positive constant. It is trivial to solve the problem for $d = 1$, and the algorithm in Sect. 4 gives a better result for $d = 2$, so we assume that $d \geq 3$ in this section.

Lemma 1. *The radius of d concurrent interior-disjoint balls is at most $\dfrac{\sqrt{d}}{\sqrt{2(d-1)}} r$ where r is the radius of the smallest enclosing ball of the centers of the d balls.*

Proof. Let us consider d points in a ball b with radius r. To maximize the distance of the closest pair of the points, they should satisfy the following conditions.

- they should lie on the boundary of the ball,
- the distances between all the pairs should be same, and
- the hyper-plane h defined by them should contain $c(b)$.

The above imply that they are vertices of a $(d - 1)$-dimensional regular simplex on h. The side length of a $(d-1)$-dimensional regular simplex is $2\dfrac{\sqrt{d}}{\sqrt{2(d-1)}} r$ where r is the radius of the circumscribed ball of it. Therefore the lemma holds. $\square$

We can derive the following lemma from Lemma 1. Let $c_d = \dfrac{\sqrt{2(d-1)}\sqrt{d}+d}{d-2}$.

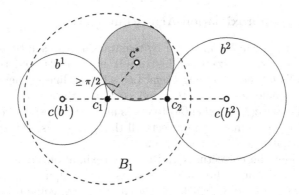

Fig. 2. The two input balls b^1 and b^2, B^* (the gray ball), and one of our solution B_1 (the dashed ball).

Lemma 2. *There are at most $d-1$ input balls such that radius of each of them is greater than $c_d r^*$.*

Proof. Assume to the contrary that there are d input balls such that radii of all of them are $c_d r^* + \varepsilon^*$ where ε^* is an arbitrarily small positive constant. Their centers should lie in $b(c^*, (1 + c_d)r^* + \varepsilon^*)$ by the problem definition. Let $r = (1 + c_d)r^* + \varepsilon^*$, then for $d \geq 3$

$$\frac{\sqrt{d}}{\sqrt{2(d-1)}}r = c_d r^* + \frac{\sqrt{d}}{\sqrt{2(d-1)}}\varepsilon^* < c_d r^* + \varepsilon^*$$

By Lemma 1, at least one pair of the balls should intersect each other, a contradiction. □

We propose a simple approximation algorithm by using Lemma 2 as follows. We keep the first d input balls, and then find the smallest ball b_{min} among them. We set the center of our solution to $c(b_{min})$, and then expand radius of our solution whenever a new input ball b arrives that does not intersect our solution.

By Lemma 2, $r(b_{min})$ is at most $c_d r^*$, so our solution guarantees an approximation factor $2 + c_d$. Because $\sqrt{2(d-1)}\sqrt{d} < \sqrt{2}(d-1) + \sqrt{d-1}$ for any $d \geq 3$ the following equation holds.

$$c_d = \frac{\sqrt{2(d-1)}\sqrt{d} + d}{d-2} < (1 + \sqrt{2}) + \frac{\sqrt{d-1} + \sqrt{2} + 2}{d-2}$$

We can use the algorithm in Sect. 4 for a small constant dimension, so following theorem holds.

Theorem 1. *For streaming balls in arbitrary dimensions d, there is an algorithm that guarantees a $(3 + \sqrt{2} + \varepsilon^*)$-approximation to the smallest intersecting ball of disjoint balls problem using $O(d^2)$ space and $O(d)$ update time where ε^* is an arbitrarily small positive constant.*

3.1 Improved Approximation Algorithm

The above algorithm can be improved by maintaining two solutions B_1 and B_2 as follows. See Fig. 2. We keep the first $d + 1$ input balls, and then find the two smallest balls b^1 and b^2 among them. Let s be the line segment connecting $c(b^1)$ and $c(b^2)$ (remember that $c(b^1)$ and $c(b^2)$ are the centers of b^1 and b^2, respectively). We set $c(B_1)$ and $c(B_2)$ to $b^1 \cap s$ and $b^2 \cap s$, respectively, and then expand each of them to make it intersects all the input balls. Our solution B at the end is the smaller one between B_1 and B_2.

Let us consider the correctness and the approximation factor of the above algorithm. Obviously, our solution B intersects all the input balls. As shown in Fig. 2, one of $\angle c(b^1)c_1 c^*$ and $\angle c(b^2)c_2 c^*$ is greater than or equal to $\pi/2$. Without loss of generality, let us assume that $\angle c(b^1)c_1 c^* \geq \pi/2$. The approximation factor of our solution is $(|c_1 c^*| + r^*)/r^*$, and

$$|c_1 c^*|^2 \leq |c(b^1)c^*|^2 - r(b^1)^2 \leq (r^* + r(b^1))^2 - r(b^1)^2 = (r^*)^2 + 2r^* r(b^1)$$

by the Pythagorean theorem. By Lemma 2, $r(b^1) \leq (1 + \sqrt{2} + \varepsilon^*)r^*$, so

$$|c_1 c^*| \leq \sqrt{1 + 2(1 + \sqrt{2})}r^* + 2\varepsilon^* r^* = (1 + \sqrt{2})r^* + 2\varepsilon^* r^*$$

which proves the following theorem.

Theorem 2. *For streaming balls in arbitrary dimensions d, there is an algorithm that guarantees a $(2 + \sqrt{2} + \varepsilon^*)$-approximation to the smallest intersecting ball of disjoint balls problem using $O(d^2)$ space and $O(d)$ update time.*

Because of the curse of dimensionality, there can be $d + 1$ balls each of radius is slightly smaller than $c_d r^*$ in high dimensions, and both of $\angle c(b^1)c_1 c^*$ and $\angle c(b^2)c_2 c^*$ can be $\pi/2$, so the analysis of our algorithm is tight. Next two sections introduce approximation algorithms in fixed constant dimensions d.

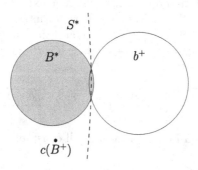

Fig. 3. Proof of Lemma 3

4   3-Approximation Algorithm in Fixed Dimensions d

In this section, we introduce a 3-approximation algorithm in fixed constant dimensions d. The following lemma is the heart of the algorithm.

Lemma 3. *Any set D' of $d + 2$ input balls satisfies the one of the following conditions.*

1. *there is a ball b in D' such that $r(b) \leq r^*$, or*
2. *the smallest ball B^+ that intersects all the balls in D' intersects B^*.*

Proof. If Condition 1 holds, then we are done. So let us assume that radius of each of the balls in D' is greater than r^*. B^+ is determined by at most $d + 1$ balls, so there is a ball $b^+ \in D'$ that does not determine B^+.

Let us consider a bisector of b^+ and B^* (See Fig. 3). The bisector subdivides the space into two parts; all the points in one of them are closer to B^*, and all the points in the other part are closet to b^+. Let S^* be the subspace defined by the bisector such that $dist(p, B^*) \leq dist(p, b^+)$ for all points p in S^*. Since $r^* < r(b^+)$, S^* is convex.

All the balls in $D' \setminus \{b^+\}$ intersect B^* and do not intersect interior of b^+. It means that $dist(c(b), B^*) \leq dist(c(b), b^+)$ for all $b \in D' \setminus \{b^+\}$, so $c(b)$ is contained in S^*. B^+ is determined by balls in $D' \setminus \{b^+\}$, so $c(B^+)$ is also contained in S^*.[1]

By the definition of S^*, $dist(c(B^+), B^*) \leq dist(c(B^+), b^+)$, and $dist(c(B^+), b^+) \leq r(B^+)$ by the definition of B^+. Then $dist(c(B^+), B^*) \leq r(B^+)$, which proves the lemma.   □

Our algorithm is as follows. We keep two solutions simultaneously; one is based on the assumption that Condition 1 in Lemma 3 holds, and the other is based on the assumption that Condition 2 holds. We choose the better one among them at the end.

The solution based on Condition 1 can be computed as follows. We keep the first $d + 2$ input balls, and then find the smallest ball b_{min} among them. We set the center of our solution to $c(b_{min})$, and then expand the radius of our solution whenever a new input ball arrives that does not intersect our solution while keeping the center unchanged.

The solution based on Condition 2 can be computed by the almost same way except the way to choose the center of our solution. We keep the first $d + 2$ input balls, and then compute the optimal solution B^+ for them. We set the center of our solution to $c(B^+)$. The remaining parts are the same as our solution based on Condition 1.

Both of our solutions obviously intersects all the input balls. By Lemma 3 and the problem definition, one of b_{min} and B^+ has radius $r \leq r^*$ and intersects B^*, which means that the algorithm guarantees a 3-approximate solution. Our algorithm spends $O(1)$ time [3] to compute B^+ and $O(1)$ update time by using $O(1)$ space.

---

[1] $c(B^+)$ is contained in the convex hull of centers of balls in $D' \setminus \{b^+\}$, and the convex hull is contained in S^*.

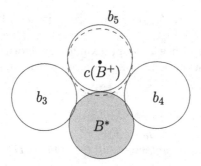

Fig. 4. A tight example of Theorem 3: three input balls b_3, b_4 and b_5, the optimal solution B^*, and the optimal solution B^+ (dashed) for the first five input balls.

Theorem 3. *For streaming balls in fixed constant dimensions d, there is an algorithm that guarantees a 3-approximation to the smallest intersecting ball of disjoint balls problem using $O(1)$ space and $O(1)$ update time.*

Note that the space of our algorithm does not depend on n, and it has polynomial dependency on d.

Now we show that our approximation factor analysis for the algorithm is tight by showing an example in $\mathbb{R}^3$ as follow. See Fig. 4. Let us consider the first five input balls. If they satisfy Condition 1 in Lemma 3, it is trivial to show tightness, so let us assume that they only satisfy Condition 2.

The centers of the first two input balls b_1 and b_2 are at $(0, 0, a)$ and $(0, 0, -a)$, respectively, where a is an arbitrarily large positive constant. We assume that $dist(b_1, b_2) = 2r^* - \varepsilon^*$ where ε^* is an arbitrarily small positive constant, and they determine B^+; $c(B^+) = (0, 0, 0)$. The remaining three input balls b_3, b_4 and b_5 are slightly greater than B^* and their centers are on the xy-plane. Let us assume that c^* is also on the xy-plane, so they look like Fig. 4 on the xy-plane. The figure shows that the five input balls are interior-disjoint, and they do not satisfy Condition 1. As you see, we can make the intersecting area of B^+ and B^* as small as possible, which proves the tightness of the analysis.

5 $(1 + \varepsilon)$-Approximation Algorithm in Fixed Dimensions d

We present a $(1 + \varepsilon)$-approximation algorithm in fixed constant dimensions d in this section. Before we propose our algorithm, we describe a $(1 + \varepsilon)$-approximation algorithm for the smallest enclosing ball of points problem.

Definition 1. *Chan [17] Given a double-argument measure $w(P, x) = \max_{p,q \in P} (p - q) \cdot x$ that is monotone in its first argument, a subset $R \subset P$ is called an ε-core-set of P over all vectors $x \in \mathbb{R}^d$ if $w(R, x) \geq (1 - \varepsilon)w(P, x)$ for all x.*

For streaming points in any fixed dimensions d, one can devise an ε-core-set by maintaining extreme points along a number of different directions using

$O(1/\varepsilon^{(d-1)/2})$ space and $O(1/\varepsilon^{(d-1)/2})$ update time [17]. A $(1+\varepsilon)$-approximate solution can be obtained by increasing the radius of the optimal solution for points in an ε-core-set by $2\varepsilon r_p^*$ where r_p^* is the optimal radius of the smallest enclosing ball of points problem.

The basic idea of our algorithm is as follows. Let $D_<$ be a subset of D such that $r(b) \leq \varepsilon r^*$ for $b \in D_<$, and $C_< = \{c(b) \mid b \in D_<\}$. We also denote by $\varepsilon C_<$ an ε-core-set of $C_<$, and $\varepsilon D_< = \{b \mid c(b) \in \varepsilon C_<\}$. We maintain $\varepsilon D_<$ and all balls in $D \setminus D_<$. For $\varepsilon D_<$, the following lemma hold.

Lemma 4. *Let B be a ball that intersects all the balls in $\varepsilon D_<$, then $dist(B, b) \leq 5\varepsilon r^*$ for any $b \in D_<$ if $0 < \varepsilon < 1$.*

Proof. Let $b \in D_<$ be a ball that does not intersect B. It means that $b \notin \varepsilon D_<$. By Definition 1, there is a ball $b' \in \varepsilon D_<$ such that $(c(b) - c(b')) \cdot x \leq \varepsilon \cdot w(C_<, x)$ where $x = \frac{(c(b)-c(B))}{|c(b)c(B)|}$. Let p be a point in $B \cap b'$. The following equation proves the lemma.

$$dist(B, b) \leq dist(B, c(b)) \leq (c(b) - p) \cdot x = (c(b) - c(b')) \cdot x + (c(b') - p) \cdot x$$
$$\leq \varepsilon \cdot w(C_<, x) + \varepsilon r^* \leq \varepsilon(2r^* + 2\varepsilon r^*) + \varepsilon r^* \leq 5\varepsilon r^*$$

$\square$

So, we can get a $(1 + 5\varepsilon)$-approximate solution by increasing the radius of the optimal solution for balls in $\varepsilon D_<$ and $D \setminus D_<$ by $5\varepsilon r^*$ if $0 < \varepsilon < 1$. We present a 3-approximation algorithm in Sect. 4, so $0 < 5\varepsilon < 2$ is enough.

Now, let us consider our algorithm in detail. We partition D into $O(\varepsilon^d)$ subsets. Let D_i be the subset that contains $\lceil 1/\varepsilon^d \rceil$ input balls in order from the $((i-1) \cdot \lceil 1/\varepsilon^d \rceil + 1)$-th input ball. We process D_i one by one in order.

For D_1, we compute the optimal solution B for balls in it. We find balls in D_1 each ball b of them satisfies $r(b) \leq \varepsilon r(B)$, and then compute $\varepsilon D_<$ by considering their centers. We insert all the other balls in a set $D_>$. We only maintain $\varepsilon D_<$ and $D_>$ for the next step.

For D_2, we compute the optimal solution B for the balls in $D_2 \cup D_> \cup \varepsilon D_<$. Similarly, we find balls each of radius smaller than or equal to $\varepsilon r(B)$ from $D_2 \cup D_>$, and then update $\varepsilon D_<$ by considering them. We delete such balls from D_2 and $D_>$, and then update $D_>$ by inserting all remaining balls in D_2. We repeat this process for all D_i where i is an integer between 1 and $O(\varepsilon^d)$.

At the end of the algorithm, we compute the optimal solution B for balls in $D_> \cup \varepsilon D_<$, and then increase the radius of B by $5\varepsilon r^+$ where r^+ is the radius from the algorithm in Sect. 4. We use the algorithm in Sect. 4 simultaneously to compute r^+. Finally, we report B as our solution.

The correctness of our algorithm immediately follows from Lemma 4. In each step, a ball b that satisfies $r(b) \leq \varepsilon r(B)$ also guarantees that $r(b) \leq \varepsilon r^*$. At the end of the algorithm, we maintain $\varepsilon D_<$ and all the balls in $D \setminus D_<$, so the optimal solution for them guarantees that $r(B) \leq r^*$ and $dist(B, b) \leq 5\varepsilon r^*$ by Lemma 4. The radius r^+ guarantees that $r^* \leq r^+ \leq 3r^*$ by Theorem 3, so $5\varepsilon r^* \leq 5\varepsilon r^+ \leq 15\varepsilon r^*$. Therefore after increasing the radius of B by $5\varepsilon r^+$, B

intersects all the input balls. The approximation factor is $(1 + 15\varepsilon)$, and we can get a $(1 + \varepsilon')$ approximation algorithm by adjusting a parameter $\varepsilon' = 15\varepsilon$.

Let us analyze the complexity of the algorithm. To maintain $\varepsilon D_<$, the algorithm uses $O(1/\varepsilon^{(d-1)/2})$ space and $O(1/\varepsilon^{(d-1)/2})$ update time [7]. The size of $D_>$ can be computed by the following lemma.

Lemma 5. *For given a ball b of radius r, there are at most $O(1/\varepsilon^d)$ interior-disjoint balls that intersect b if radius of each of them is greater than or equal to εr in fixed constant dimensions d.*

Proof. We are going to prove the lemma for the interior-disjoint balls each of radius εr. Obviously, if there are at most $O(1/\varepsilon^d)$ such balls, then the lemma holds. Let us consider a ball $b' = b(c(b), r + 2\varepsilon r)$. A ball that intersects b should be contained in b'. We can compute the maximum number of the interior-disjoint balls that are contained in b' by considering their volumes. The volume of b' is $\Theta((r + 2\varepsilon r)^d)$, and the volume of a ball of radius εr is $\Theta((\varepsilon r)^d)$. The sum of the volumes of all the balls contained in b' can not exceed the volume of b', so the maximum number of the interior-disjoint balls in b' is $O((r + 2\varepsilon r)/\varepsilon r)^d) = O(1/\varepsilon^d)$, which prove the lemma.   □

By Lemma 5, the size of $D_>$ is $O(1/\varepsilon^d)$. In each step the algorithm holds $O(1/\varepsilon^d)$ balls. We compute the optimal solution for them in each step and it takes linear time [18], so we spend $O(1)$ amortized time per update.

Theorem 4. *For streaming balls in fixed constant dimensions d, there is an algorithm that guarantees a $(1 + \varepsilon)$ approximation to the smallest intersecting ball of disjoint balls problem using $O(1/\varepsilon^d)$ space and $O(1/\varepsilon^{(d-1)/2})$ amortized update time.*

6   Conclusion

In this paper, we introduced three approximation algorithms for the smallest intersecting ball of disjoint balls problem. One of them is for the problem in any arbitrarily dimensions, and the others are for the problem in fixed constant dimensions. As the exact problem is very difficult (no polynomial algorithm is known if d is not constant), approximation seems to be the only way forward.

We do not know any better lower bound for the worst-case approximation ratio than $(1 + \sqrt{2})/2 > 1.207$ that is a lower bound for the smallest enclosing ball of points problem in the streaming model if we use space only polynomially bounded in d [6]. We believe that this lower bound is not tight for our problem and it can be improved.

A natural extension of our algorithm is to allow the input balls to overlap. However, this poses a great number of challenges since the size of the optimal answer could be zero (when a point pierces all the balls). But if we allow overlapping with some restrictions it may be possible to solve the problem. Another interesting question is how to improve our results in static setting. Even in the static setting, no approximation algorithms are known for the problem except our results in this paper.

References

1. Mordukhovich, B., Nam, N., Villalobos, C.: The smallest enclosing ball problem and the smallest intersecting ball problem: existence and uniqueness of solutions. Optim. Lett. **7**(5), 839–853 (2013)
2. Löffler, M., van Kreveld, M.: Largest bounding box, smallest diameter, and related problems on imprecise points. Comput. Geom. **43**(4), 419–433 (2010)
3. Matoušek, J., Sharir, M., Welzl, E.: A subexponential bound for linear programming. Algorithmica **16**(4–5), 498–516 (1996)
4. Ahn, H.K., Kim, S.S., Knauer, C., Schlipf, L., Shin, C.S., Vigneron, A.: Covering and piercing disks with two centers. Comput. Geom. **46**(3), 253–262 (2013)
5. Zarrabi-Zadeh, H., Chan, T.: A simple streaming algorithm for minimum enclosing balls. In: Proceedings of the 18th Canadian Conference on Computational Geometry, pp. 139–142 (2006)
6. Agarwal, P.K., Sharathkumar, R.: Streaming algorithms for extent problems in high dimensions. In: Proceedings of the 21st ACM-SIAM Symposium on Discrete Algorithms, SODA 2010, pp. 1481–1489 (2010)
7. Chan, T.M., Pathak, V.: Streaming and dynamic algorithms for minimum enclosing balls in high dimensions. Comput. Geom. **47**(2, Part B), 240–247 (2014)
8. Bâdoiu, M., Clarkson, K.L.: Smaller core-sets for balls. In: Proceedings of the 14th ACM-SIAM Symposium on Discrete Algorithms, SODA 2003, pp. 801–802 (2003)
9. Garey, M., Johnson, D.: Computers and Intractability: A Guide to the Theory of NP-Completeness. W.H. Freeman, New York (1979)
10. Chan, T.: More planar two-center algorithms. Comput. Geom. **13**(3), 189–198 (1999)
11. Agarwal, P., Avraham, R., Sharir, M.: The 2-center problem in three dimensions. In: Proceedings of the 26th ACM Symposium Computational Geometry, pp. 87–96 (2010)
12. Gonzalez, T.: Clustering to minimize the maximum intercluster distance. Theoret. Comput. Sci. **38**, 293–306 (1985)
13. Feder, D., Greene, D.: Optimal algorithms for approximate clustering. In: Proceedings of the 20th ACM Symposium on Theory of Computing, pp. 434–444 (1988)
14. Charikar, M., Chekuri, C., Feder, T., Motwani, R.: Incremental clustering and dynamic information retrieval. SIAM J. Comput. **33**(6), 1417–1440 (2004)
15. Guha, S.: Tight results for clustering and summarizing data streams. In: Proceedings of the 12th International Conference on Database Theory, pp. 268–275 (2009)
16. Matthew McCutchen, R., Khuller, S.: Streaming algorithms for k-center clustering with outliers and with anonymity. In: Goel, A., Jansen, K., Rolim, J.D.P., Rubinfeld, R. (eds.) APPROX and RANDOM 2008. LNCS, vol. 5171, pp. 165–178. Springer, Heidelberg (2008)
17. Chan, T.M.: Faster core-set constructions and data-stream algorithms in fixed dimensions. Comput. Geom. **35**(12), 20–35 (2006)
18. Chazelle, B., Matoušek, J.: On linear-time deterministic algorithms for optimization problems in fixed dimension. J. Algorithms **21**(3), 579–597 (1996)

Multi-player Diffusion Games on Graph Classes

Laurent Bulteau, Vincent Froese, and Nimrod Talmon[(✉)]

Institut Für Softwaretechnik und Theoretische Informatik,
TU Berlin, Berlin, Germany
nimrodtalmon77@gmail.com

Abstract. We study competitive diffusion games on graphs introduced by Alon et al. [1] to model the spread of influence in social networks. Extending results of Roshanbin [7] for two players, we investigate the existence of pure Nash equilibria for at least three players on different classes of graphs including paths, cycles, and grid graphs. As a main result, we answer an open question proving that there is no Nash equilibrium for three players on $m \times n$ grids with $\min\{m, n\} \geq 5$.

1 Introduction

Social networks, and the diffusion of information within them, yields an interesting and well-researched field of study. Among other models, competitive diffusion games have been introduced by Alon et al. [1] as a game-theoretic approach towards modelling the process of diffusion (or propagation) of influence (or information in general) in social networks. Such models have applications in "viral marketing" where several companies (or brands) compete in influencing as many customers (of products) or users (of technologies) as possible by initially selecting only a "small" subset of target users that will "infect" a large number of other users. Herein, the network is modeled as an undirected graph where the vertices correspond to the users, with edges modeling influence relations between them. The companies, being the players of the corresponding diffusion game, choose an initial subset of target vertices which then influence other neighboring vertices via a certain propagation process. More concretely, a vertex adopts a company's product at some specific time during the process if he is influenced by (that is, connected by an edge to) another vertex that already adopted this product. After adopting a product of one company, a vertex will never adopt any other product in the future. However, if a vertex gets influenced by several companies at the same time, then he will not adopt any of them and he is removed from the game (the reason being that the effects of these influencing companies on the customer cancel out each other such that the customer is "too confused" to adopt any of the products). See Sect. 1.3 for the formal definitions of the game.

A full version is available at http://arxiv.org/abs/1412.2544.

L. Bulteau—Supported by the Alexander von Humboldt Foundation, Bonn, Germany.

V. Froese—Supported by the DFG, project DAMM (NI 369/13).

N. Talmon—Supported by DFG Research Training Group MDS (GRK 1408).

R. Jain et al. (Eds.): TAMC 2015, LNCS 9076, pp. 200–211, 2015.
DOI: 10.1007/978-3-319-17142-5_18

In their initial work, Alon et al. [1] studied how the existence of pure Nash equilibria is influenced by the diameter of the underlying graph. Following this line of research, Roshanbin [7] investigated the existence of Nash equilibria for competitive diffusion games with two players on several classes of graphs such as paths, cycles, and grid graphs. Notably, she proved that on sufficiently large grids, there always exists a Nash equilibrium for two players, further conjecturing that there is no Nash equilibrium for three players on grids. We extend the results of Roshanbin [7] for two players to three or more players on paths, cycles, and grid graphs, proving the conjectured non-existence of a pure Nash equilibrium for three players on grids as a main result. An overview of our results is given in Sect. 1.2. After introducing the preliminaries in Sect. 1.3, we discuss our results for paths and cycles in Sect. 2, followed by the proof of our main theorem on grids in Sect. 3. We finish with some statements considering general graphs in Sect. 4.

1.1 Related Work

The study of influence maximization in social networks was initiated by Kempe et al. [5]. Several game-theoretic models have been suggested, including our model of reference, introduced by Alon et al. [1]. Some interesting generalizations of this model are the model by Tzoumas et al. [11], who considered a more complex underlying diffusion process (there, depending on its neighborhood, a general scheme is used to determine whether a vertex adopts a product), and the model studied by Etesami and Basar [3], allowing each player to choose multiple vertices. Dürr and Thang [2] and Mavronicolas et al. [6] studied so-called Voronoi games, which are closely related to our model (but not identical; there, instead of an underlying diffusion process, each vertex is assigned to its closest player and vertices can be shared). Concerning our model, Alon et al. [1] claimed the existence of pure Nash equilibria for any number of players on graphs of diameter at most two, however, Takehara et al. [10] gave a counterexample consisting of a graph with nine vertices and diameter two with no Nash equilibrium for two players.

Our main point of reference is the work of Roshanbin [7], who studied the existence (and non-existence) of pure Nash equilibria mainly for two players on special graph classes (paths, cycles, trees, unicycles, and grids); indeed, our work can be seen as an extension of that work to more than two players. Small [8] already showed that there is a Nash equilibrium for any number of players on any star or clique. Small and Mason [9] proved that there is always a pure Nash equilibrium for two players on a tree, but not always for more than two players. Janssen and Vautour [4] considered safe strategies on trees and spider graphs, where a safe strategy is a strategy which maximizes the minimum pay-off of a certain player, when the minimum is taken over the possible unknown actions of the other players.

1.2 Our Results

We begin by characterizing the existence of Nash equilibria for paths and cycles, showing that, except for three players on paths of length at least six, a Nash equilibrium exists for any number of players playing on any such graph (Theorems 1 and 2). We then prove Conjecture 1 of Roshanbin [7], showing that there is no Nash equilibrium for three players on $G_{m \times n}$, as long as both m and n are at least 5 (Theorem 3). Finally, we investigate the minimum number of vertices such that there is an arbitrary graph with no Nash equilibrium for k players. We prove an upper bound showing that there always exists a tree on $\lfloor \frac{3}{2}k \rfloor + 2$ vertices with no Nash equilibrium for k players (Theorem 4). Due to space constraints, some of the proofs are omitted. Please refer to the full version (available at http://arxiv.org/abs/1412.2544).

1.3   Preliminaries

Notation. For $i, j \in \mathbb{N}$ with $i < j$, we define $[i, j] := \{i, \ldots, j\}$ and $[i] := \{1, \ldots, i\}$. We consider simple, finite, undirected graphs $G = (V, E)$ with vertex set V and edge set $E \subseteq \{\{u, v\} \mid u, v \in V\}$. A path $P_n = (V, E)$ on n vertices is the graph with $V = [n]$ and $E = \{\{i, i+1\} \mid i \in [n-1]\}$. A cycle $C_n = (V, E)$ on n vertices is the graph with $V = [n]$ and $E = \{\{i, i+1\} \mid i \in [n-1]\} \cup \{\{n, 1\}\}$.

For $m, n \in \mathbb{N}$, the $m \times n$ grid $G_{m \times n} = (V, E)$ is a graph with vertices $V = [m] \times [n]$ and edges $E = \{\{(x, y), (x', y')\} \mid |x - x'| + |y - y'| = 1\}$. We use the term *position* for a vertex $x \in V$. We define the *distance* of two positions $x = (x_1, y_1)$, $y = (x_2, y_2) \in V$ as $\|x - y\|_1 := |x_1 - x_2| + |y_1 - y_2|$ (note that this corresponds to the length of a shortest path from x to y in the grid). We denote the number of players by k and enumerate the players as Player 1, ..., Player k.

Diffusion Game on Graphs. A *game* $\Gamma = (G, k)$ is defined by an undirected graph $G = (V, E)$ and a number k of players, each having its distinct color in $[k]$. The *strategy space* of each player is V, such that each Player i selects a single vertex $v_i \in V$ at time 0, which is then colored by her color i. If two players choose the same vertex v, then this vertex is removed from the graph. For Player i, we use the terms strategy and position interchangeably, referring to its chosen vertex. A *strategy profile* is a tuple $(v_1, \ldots, v_k) \in V^k$ containing the initially chosen vertex for each player. The *pay-off* $U_i(v_1, \ldots, v_k)$ of Player i is the number of vertices with color i after the following propagation process. At time $t + 1$, any so far uncolored vertex that has only uncolored neighbors and neighbors colored in i (and no neighbors with other colors $j \in [k] \setminus \{i\}$) is colored in i. Any uncolored vertex with more than two different colors among its neighbors is removed from the graph. The process terminates when the coloring of the vertices does not change between consecutive steps. A strategy profile $(v_1, \ldots, v_k)$ is a (pure) *Nash equilibrium* if, for each Player $i \in [k]$ and each vertex $v' \in V$, it holds that $U_i(v_1, \ldots, v_{i-1}, v', v_{i+1}, \ldots, v_k) \leq U_i(v_1, \ldots, v_k)$.

2   Paths and Cycles

In this section, we fully characterize the existence of Nash equilibria on paths and cycles, for any number k of players.

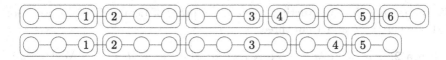

Fig. 1. Illustrations for Theorem 1, showing a Nash equilibrium for 6 players on P_{15} (top) and a Nash equilibrium for 5 players on P_{14} (bottom). The boxes show the colored regions of each player.

Theorem 1. *For any $k \in \mathbb{N}$ and any $n \in \mathbb{N}$, there is a Nash equilibrium for k players on P_n, except for $k = 3$ and $n \geq 6$.*

The general idea of the proof is to pair the players and distribute these pairs evenly. In the rest of this section, we prove three lemmas whose straightforward combination proves Theorem 1.

Lemma 1. *For any even $k \in \mathbb{N}$ and any $n \in \mathbb{N}$, there is a Nash equilibrium for k players on P_n.*

Proof. If $n \leq k$, then any strategy profile where each vertex of the path is chosen by at least one player is clearly a Nash equilibrium.

Otherwise, if $n > k$, then the idea is to build pairs of players, which are then placed such that two paired players are neighboring and the distance of any two consecutive pairs is roughly equal (specifically, differs by at most two). See Fig. 1 for an example. Intuitively, this yields a Nash equilibrium since each player obtains roughly the same pay-off (specifically, differing by at most one), therefore no player can improve. Since we have n vertices, we want each player's pay-off to be at least $z := \lfloor \frac{n}{k} \rfloor$. This leaves $r := n(\mod k)$ other vertices, which we distribute between the first r players such that the pay-off of any player is at most $z + 1$. This can be achieved as follows. Let $p_i \in [n]$ denote the position of Player i, that is, the index of the chosen vertex on the path. We define

$$p_i := \begin{cases} z \cdot i + \min\{i, r\} & \text{if } i \text{ is odd,} \\ p_{i-1} + 1 & \text{if } i \text{ is even.} \end{cases}$$

Note that, by construction, it holds that $p_1 \in \{z, z+1\}$ and $p_k = n - z + 1$. Moreover, for each odd indexed player $i \geq 3$, we have that $2z - 1 \leq p_i - p_{i-1} \leq 2z + 1$. We claim that $u_i := U_i(p_1, \ldots, p_k) \in \{z, z+1\}$ holds for each $i \in [k]$. Clearly, $u_1 = p_1 \in \{z, z+1\}$ and $u_k = n - p_k + 1 = z$. For all odd $i \geq 3$, it is not hard to see that $u_i = u_{i-1} = 1 + \lfloor (p_i - p_{i-1} - 1)/2 \rfloor \in \{z, z+1\}$, proving the claim.

To see that the strategy profile $(p_1, \ldots, p_k)$ is a Nash equilibrium, consider an arbitrary player i and any other strategy $(p_i \neq) p_i' \in [n]$ that she picks. Clearly, we can assume that $p_i' \neq p_j$ holds for all $j \neq i$ since otherwise Player i's pay-off is zero. If $p_i' < p_1$ or $p_i' > p_k$, then Player i gets a pay-off of at most z. If $p_j < p_i' < p_{j+1}$ for some even $j \in [2, k - 2]$, then her pay-off is at most $1 + \lfloor (p_{j+1} - p_j - 2)/2 \rfloor \leq z$. $\square$

We can modify the construction given in the proof of Lemma 1 to also work for odd numbers k greater than three.

Lemma 2. *For any odd $k > 3 \in \mathbb{N}$ and for any $n \in \mathbb{N}$, there is a Nash equilibrium for k players on P_n.*

Proof. We give a strategy profile based on the construction for an even number of players (proof of Lemma 1). The idea is to pair the players, placing the remaining lonely player between two consecutive pairs.

This is best explained using a reduction to the even case. Specifically, given the strategy profile $(p'_1, \ldots, p'_{k+1})$ for an even number $k+1$ of players on P_{n+1} as constructed in the proof of Lemma 1, we define the strategy profile $(p_1, \ldots, p_k) :=$ $(p'_1, \ldots, p'_{k-2}, p'_k - 1, p'_{k+1} - 1)$. To see why this results in a Nash equilibrium, let $z := \lfloor (n+1)/(k+1) \rfloor$ and note that by construction it holds that $p_1 \in \{z, z+1\}$, $p_k = n - z + 1$, and $2z - 1 \leq p_{i+1} - p_i \leq 2z + 1$, for all $i \in [2, k-1]$. Moreover, each player receives a pay-off of at least z, therefore all players (except for Player $(k-2)$) cannot improve by the same arguments as in the proof of Lemma 1. Regarding Player $(k-2)$, note that her pay-off is

$$1 + \lfloor (p_{k-1} - p_{k-2} - 1)/2 \rfloor + \lfloor (p_{k-2} - p_{k-3} - 1)/2 \rfloor \geq 2z - 1.$$

Hence, she clearly cannot improve by choosing any position outside of $[p_{k-3}, p_{k-1}]$. Moreover, she cannot improve by choosing any other position in $[p_{k-3}, p_{k-1}]$. To see this, note that her maximum pay-off from any position in $[p_{k-3}, p_{k-1}]$ is

$$1 + \lfloor (p_{k-1} - p_{k-3} - 2)/2 \rfloor = 1 + \lfloor (p_{k-1} - p_{k-2} - 1 + p_{k-2} - p_{k-3} - 1)/2 \rfloor,$$

which is equal to the above pay-off since $p_{k-1} - p_{k-2}$ and $p_{k-2} - p_{k-3}$ cannot both be even, by construction. $\square$

It remains to discuss the fairly simple (non)-existence of Nash equilibria for three players. Note that Roshanbin [7] already stated without proof that there is no Nash equilibrium for three players on $G_{2\times n}$ and $G_{3\times n}$ and that Small and Mason [9] showed that there is no Nash equilibrium for three players on P_7. For the sake of completeness, we prove the following lemma.

Lemma 3. *For three players, there is a Nash equilibrium on P_n if and only if $n \leq 5$.*

Proof. If $n \leq 3$, then a strategy profile where each vertex of the path is chosen by at least one player is clearly a Nash equilibrium. For $n \in \{4, 5\}$, the strategy profile $(2, 3, 4)$ is a Nash equilibrium.

To see that there is no Nash equilibrium for $n \geq 6$, consider an arbitrary strategy profile (p_1, p_2, p_3). Without loss of generality, we can assume that $p_1 < p_2 < p_3$ and consider the following two cases. First, we assume that $p_2 = p_1 + 1$ and $p_3 = p_2 + 1$. If $p_1 > 2$, then Player 2 increases her pay-off by choosing $p_1 - 1$. Otherwise, it holds that $p_3 < n - 1$ and Player 2 increases her pay-off by moving to $p_3 + 1$. Therefore, this case does not yield a Nash equilibrium.

For the remaining case, it holds that $p_1 < p_2 - 1$ or $p_3 > p_2 + 1$. If $p_1 < p_2 - 1$, then Player 1 increases her pay-off by moving to $p_2 - 1$, while if $p_3 > p_2 + 1$, then Player 3 increases her pay-off by moving to $p_2 + 1$. Thus, this case does not yield a Nash equilibrium as well, and we are done.   □

We close this section with the following result considering cycles. Interestingly, for cycles there exists a Nash equilibrium also for three players.

Theorem 2. *For any $k, n \in \mathbb{N}$, there is a Nash equilibrium for k players on C_n.*

Proof. It is an easy observation that the constructions given in the proofs of Lemmas 1 and 2 also yield Nash equilibria for cycles, that is, when the two endpoints of the path are connected by an edge. Thus, it remains to show a Nash equilibrium for $k = 3$ players for any C_n. We set $p_1 := 1$, $p_2 := n$ and

$$p_3 := \begin{cases} \lfloor n/2 \rfloor & \text{if } n \mod 4 = 1, \\ \lceil n/2 \rceil & \text{otherwise.} \end{cases}$$

It is not hard to check that (p_1, p_2, p_3) is a Nash equilibrium.   □

3   Grid Graphs

In this section we consider three players on the $m \times n$ grid $G_{m \times n}$ and prove the following main theorem.

Theorem 3. *If $n \geq 5$ and $m \geq 5$, then there is no Nash equilibrium for three players on $G_{m \times n}$.*

Before proving the theorem, let us first introduce some general definitions and observations. Throughout this section, we denote the strategy of Player i, that is, the initially chosen vertex of Player i, by $p_i := (x_i, y_i) \in [m] \times [n]$. Note that any strategy profile where more than one player chooses the same position is never a Nash equilibrium since in this case each of these players gets a pay-off of zero, and can improve its pay-off by choosing any free vertex (to obtain a pay-off of at least one). Therefore, we will assume without loss of generality that $p_1 \neq p_2$, $p_2 \neq p_3$, and $p_1 \neq p_3$. Further, note that the game is symmetric with respect to the axes. Specifically, reflecting coordinates along a dimension or rotating the grid by 90 degrees yields the same outcome for the game. Thus, in what follows, we only consider possible cases up to these symmetries.

We define $\Delta_x := \max_{i,j \in [k]} |x_i - x_j|$ and $\Delta_y := \max_{i,j \in [k]} |y_i - y_j|$ to be the maximum coordinate-wise differences among the positions of the players. We say that a player *strictly controls* the other two players, if both reside on the same side of the player, in both dimensions.

Definition 1. *Player i strictly controls the other players, if either*

$$\forall j \neq i : x_i < x_j \wedge y_i < y_j,$$
$$\text{or } \forall j \neq i : x_i < x_j \wedge y_i > y_j,$$
$$\text{or } \forall j \neq i : x_i > x_j \wedge y_i < y_j,$$
$$\text{or } \forall j \neq i : x_i > x_j \wedge y_i > y_j \text{ holds.}$$

The proof of Theorem 3 proceeds as follows.

Proof (Theorem 3). Let $m \geq 5$ and $n \geq 5$. We perform a case distinction based on the relative positions of the three players. As a first case, we consider strategy profiles where the players are playing "far" from each other, that is, there are two players whose positions differ by at least four in some coordinate (formally, $\max\{\Delta_x, \Delta_y\} \geq 3$). For these profiles, we distinguish two subcases, namely, whether there exists a player who strictly controls the others (Lemma 4) or not (Lemma 5). We prove that none of these cases yields a Nash equilibrium by showing that there always exists a player who can improve her pay-off. Notably, the improving player always moves closer to the other two players. We are left with the case where the players are playing "close" to each other, specifically, their positions all lie inside a 3×3 subgrid (that is, $\max\{\Delta_x, \Delta_y\} \leq 2$). For these strategy profiles, we show that there always exists a player who can improve her pay-off (Lemma 6), however the improving position depends not only on the relative positions between the players, but also on the global positioning of this subgrid on the main grid. This leads to a somewhat erratic behaviour, which we overcome by considering all possible close positions (up to symmetries) in the proof of Lemma 6. Altogether, Lemmas 4, 5, and 6, cover all possible strategy profiles (ruling them out as Nash equilibria), thus implying the theorem. □

In order to conclude Theorem 3, it remains to prove the lemmas mentioned in the case distinction discussed above. To this end, we start with two easy preliminary results. First, we observe (as can be easily proven by induction) that a vertex for which the player with the shortest distance to it is unique is colored in that player's color.

Observation 1. *Let $x \in [m] \times [n]$ and $i \in [k]$. If $\|p_i - x\|_1 < \|p_j - x\|_1$ holds for all $j \neq i$, then x will be colored in color i at the end of the propagation process.*

Based on Observation 1, we show that if a player has distance at least three to the other players and both of them are positioned on the same side of that player (with respect to both dimensions), then she can improve her pay-off by moving closer to the others (see Fig. 2 for an illustration).

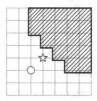

Fig. 2. Example of a strategy profile where Player 1 (white circle) has both other players to her top right with distance at least three (the shaded region denotes the possible positions for Player 2 and 3). Player 1 can increase her pay-off by moving closer to the others (star).

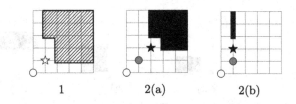

Fig. 3. Possible cases (up to symmetry) for Player 1 (white) strictly controlling Player 2 (gray) and Player 3 (black). Circles denote the player's strategies. The shaded region contains the possible positions of both Player 2 and 3, whereas the black regions denote possible positions for Player 3 only. A star marks the position improving the pay-off of the respective player.

Proposition 1. *If $x_1 \leq x_j$, $y_1 \leq y_j$, and $\|p_1 - p_j\|_1 \geq 3$ holds for $j \in \{2,3\}$, then Player 1 can increase her pay-off by moving to $(x_1 + 1, y_1 + 1)$.*

Proof. Let $p_1' := (x_1 + 1, y_1 + 1)$ and $x \in [x_1] \times [y_1]$. Note that $\|p_1' - x\|_1 = \|p_1 - x\|_1 + 2 < \|p_j - x\|_1 = \|p_1 - p_j\|_1 + \|p_1 - x\|_1 \geq \|p_1 - x\|_1 + 3$ holds for $j \in \{2,3\}$. Hence, Player 1 still has the unique shortest distance to x. By Observation 1, x gets color 1. Moreover, for any other position $x \notin [x_1] \times [y_1]$, there is a shortest path from p_1 to x going through at least one of the positions $(x_1 + 1, y_1)$, $(x_1, y_1 + 1)$, or p_1'. Clearly, there is also a shortest path from p_1' to x of at most the same length going through one of these positions. Thus, if x was colored in color 1 before, then x is still colored in color 1.

To see that Player 1 strictly increases her pay-off, note that $\|p_1' - x\|_1 = \|p_1 - x\|_1 - 2$ holds for all $x \in [x_1 + 1, n] \times [y_1 + 1, m]$. Hence, Player 1 now has the unique shortest distance to all those positions where the distance from p_1 was at most one larger than the shortest distance from any other player (clearly, there exists at least one such position with color $j \neq 1$). By Observation 1, these positions now get color 1, thus Player 1 strictly increases her pay-off. □

We go on to prove the lemmas needed for Theorem 3, starting with the case that the players play far from each other. The following lemma handles the first subcase, that is, where one of the players strictly controls the others.

Lemma 4. *A strategy profile with $\max\{\Delta_x, \Delta_y\} \geq 3$ where one of the players strictly controls the others is not a Nash equilibrium.*

Proof. We assume without loss of generality that Player 1 strictly controls Player 2 and Player 3, specifically, we assume that $x_1 < x_2$ and $y_1 < y_2$ and $x_1 < x_3$ and $y_1 < y_3$ holds. Figure 3 depicts the three possible cases for the positions of Player 2 and Player 3. For each case, we show that a player which can improve her pay-off exists.

Case 1: We assume that $(x_2, y_2) \neq (x_1 + 1, y_1 + 1)$ and $(x_3, y_3) \neq (x_1 + 1, y_1 + 1)$.
By Proposition 1, Player 1 gets a higher pay-off from $(x_1 + 1, y_1 + 1)$.

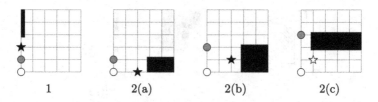

Fig. 4. Possible cases (up to symmetry) when no player strictly controls the others. Circles denote the positions of Player 1 (white) and Player 2 (gray). The black regions contain the possible positions for Player 3. A star marks the position improving the pay-off of the respective player.

Case 2: We assume without loss of generality that $(x_2, y_2) = (x_1 + 1, y_1 + 1)$.
  (a) We assume $x_2 < x_3$ and $y_2 < y_3$. Then, $x_3 > x_2 + 1$ or $y_3 > y_2 + 1$ holds since $\max\{\Delta_x, \Delta_y\} \geq 3$. Note that Player 3 strictly controls Player 1 and Player 2 and that this case is symmetric to Case 1.
  (b) We assume $x_2 \geq x_3$ or $y_2 \geq y_3$. Then, it holds that $x_3 = x_2$ or $y_3 = y_2$. We assume $x_3 = x_2$ (the argument for $y_3 = y_2$ being analogous). Since $\max\{\Delta_x, \Delta_y\} \geq 3$, we have $y_3 > y_2 + 1$, thus Player 3 can improve by moving to $(x_2, y_2 + 1)$ because then all positions in $[m] \times [y_2 + 1, n]$ are colored in color 3, and before only a strict subset of these positions were colored in her color.   □

The other subcase, where no player strictly controls the others, is handled by the following lemma.

Lemma 5. *A strategy profile with* $\max\{\Delta_x, \Delta_y\} \geq 3$ *where no player strictly controls the others is not a Nash equilibrium.*

Proof. If no player strictly controls the others, then it follows that at least two players have the same coordinate in at least one dimension. We perform a case distinction on the cases as depicted in Fig. 4.

Case 1: All three players have the same coordinate in one dimension. We assume that $x_1 = x_2 = x_3$ (the case $y_1 = y_2 = y_3$ is analogous). Without loss of generality also $y_1 < y_2 < y_3$ holds. Since $\max\{\Delta_x, \Delta_y\} \geq 3$, it follows that $y_{i+1} - y_i \geq 2$ holds for some $i \in \{1, 2\}$, say for $i = 2$. Clearly, Player 3 can improve her pay-off by choosing $(x_3, y_2 + 1)$ (analogous to Case 2b in the proof of Lemma 4).

Case 2: There is a dimension where two players have the same coordinate but not all three players have the same coordinate in any dimension. We assume $x_1 = x_2 < x_3$ and $y_1 < y_2$ (all other cases are analogous). We also assume that $y_1 \leq y_3 \leq y_2$, since otherwise Player 3 strictly controls the others, and this case is handled by Lemma 4.
  (a) We assume that $y_2 = y_1 + 1$. Then $x_3 \geq x_1 + 3$ holds since $\max\{\Delta_x, \Delta_y\} \geq 3$. Player 3 increases her pay-off by moving to $(x_1 + 2, y_1)$ (analogous to Case 2b in the proof of Lemma 4).

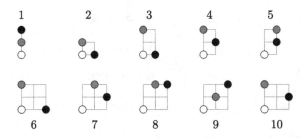

Fig. 5. Possible positions (up to symmetry) of three players playing inside a subgrid of size at most 3×3.

    (b) We assume that $y_2 = y_1 + 2$. Then $x_3 \geq x_1 + 3$ holds since $\max\{\Delta_x, \Delta_y\} \geq$ 3. Player 3 increases her pay-off by moving to $(x_1 + 2, y_1 + 1)$ (analogous to Case 2b in the proof of Lemma 4).

    (c) We assume that $y_2 > y_1 + 2$ and $|y_2 - y_3| \leq |y_1 - y_3|$. That is, without loss of generality, Player 3 is closer to Player 2. Then, by Proposition 1, Player 1 increases her pay-off by moving to $(x_1 + 1, y_1 + 1)$.   □

It remains to consider the cases where the players play close to each other.

Lemma 6. *A strategy profile with* $\max\{\Delta_x, \Delta_y\} \leq 2$ *is not a Nash equilibrium.*

Proof. First, we assume that $\Delta_x + \Delta_y \geq 2$, as otherwise there would be at least two players on the same position (so each one of them can improve by moving to any free vertex). Without loss of generality, we also assume that $\Delta_x \leq \Delta_y$, leaving the cases depicted in Figure 5 for consideration. Due to space constrains, we omit this case analysis. Please refer to the full version.   □

4   General Graphs

In this section, we study the existence of Nash equilibria on arbitrary graphs. Using computer simulations, we found that for two players, a Nash equilibrium exists on any graph with at most $n = 7$ vertices. For $n = 8$, we obtained the graph depicted in Figure 6, for which there is no Nash equilibrium for two players. As it is clear that adding isolated vertices to the graph in Figure 6 does not allow for a Nash equilibrium, we conclude the following.

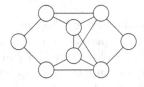

Fig. 6. A graph on 8 vertices with no Nash equilibrium for two players.

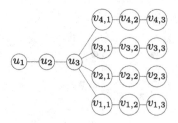

Fig. 7. A tree with no Nash equilibrium for 9 players.

Corollary 1. *For two players, there is a Nash equilibrium on each n-vertex graph if and only if $n \leq 7$.*

For more than two players, we can show the following.

Theorem 4. *For any $k > 2$ and any $n \geq \lfloor \frac{3}{2}k \rfloor + 2$, there exists a tree with n vertices such that there is no Nash equilibrium for k players.*

Proof. We describe a construction only for $n = \lfloor \frac{3}{2}k \rfloor + 2$, as we can add arbitrarily many isolated vertices without introducing a Nash equilibrium.

We first describe the construction for k being odd. We create one P_3, whose vertices we denote by u_1, u_2, and u_3, such that u_2 is the middle vertex of this P_3. For each $i \in [2, \lceil \frac{k}{2} \rceil]$, we create a copy of P_3, denoted by P_i, whose vertices we denote by $v_{i,1}$, $v_{i,2}$, and $v_{i,3}$, such that $v_{i,2}$ is the middle vertex of P_i. For each $i \in [2, \lceil \frac{k}{2} \rceil]$, we connect $v_{i,1}$ to u_3. An example for $k = 9$ is depicted in Fig. 7.

For k being even, we create one P_2, whose vertices we denote by u_1, u_2. For each $i \in [2, \frac{k}{2} + 1]$, we create a copy of P_3, denoted by P_i, whose vertices we denote by $v_{i,1}$, $v_{i,2}$, and $v_{i,3}$, such that $v_{i,2}$ is the middle vertex of P_i. For each $i \in [2, \frac{k}{2} + 1]$, we connect $v_{i,1}$ to u_2. Due to space constraints, the analysis showing that no Nash equilibrium exists for k players playing on these graphs is omitted. Please refer to the full version.   □

5   Conclusion

We studied competitive diffusion games for three or more players on paths, cycles, and grid graphs, answering—as a main contribution—an open question concerning the existence of a Nash equilibrium for three players on grids [7] negatively. Moreover, we provide a first systematic study of this game for more than two players. However, there are several questions left open, of which we mention some here.

An immediate question (generalizing Theorem 3) is whether a Nash equilibrium exists for more than three players on grids. Also, giving a lower bound for the number of vertices n such that there is a graph with n vertices with no Nash equilibrium for k players is an interesting question as it is not clear that the upper bounds given in Theorem 4 are optimal. In other words, is it true that $n \leq \frac{3}{2}k + 1$ implies the existence of a Nash equilibrium for k players?

References

1. Alon, N., Feldman, M., Procaccia, A.D., Tennenholtz, M.: A note on competitive diffusion through social networks. Inf. Process. Lett. **110**(6), 221–225 (2010)
2. Dürr, C., Thang, N.K.: Nash equilibria in voronoi games on graphs. In: Arge, L., Hoffmann, M., Welzl, E. (eds.) ESA 2007. LNCS, vol. 4698, pp. 17–28. Springer, Heidelberg (2007)
3. Etesami, S.R., Basar, T.: Complexity of equilibrium in diffusion games on social networks. In: Proceedings of the 2014 American Control Conference, pp. 2065–2070 (2014)
4. Janssen, J., Vautour, C.: Finding safe strategies for competitive diffusion on trees. Internet Math. (2014) http://arxiv.org/abs/1404.5356
5. Kempe, D., Kleinberg, J., Tardos, É.: Maximizing the spread of influence through a social network. In: Proceedings of the Ninth ACM SIGKDD International Conference on Knowledge Discovery and Data Mining, pp. 137–146. ACM (2003)
6. Mavronicolas, M., Monien, B., Papadopoulou, V.G., Schoppmann, F.: Voronoi games on cycle graphs. In: Ochmański, E., Tyszkiewicz, J. (eds.) MFCS 2008. LNCS, vol. 5162, pp. 503–514. Springer, Heidelberg (2008)
7. Roshanbin, E.: The competitive diffusion game in classes of graphs. In: Gu, Q., Hell, P., Yang, B. (eds.) AAIM 2014. LNCS, vol. 8546, pp. 275–287. Springer, Heidelberg (2014)
8. Small, L.: Information diffusion on social networks. Ph.D. thesis, National University of Ireland Maynooth (2012)
9. Small, L., Mason, O.: Nash equilibria for competitive information diffusion on trees. Inf. Process. Lett. **113**(7), 217–219 (2013)
10. Takehara, R., Hachimori, M., Shigeno, M.: A comment on pure-strategy Nash equilibria in competitive diffusion games. Inf. Process. Lett. **112**(3), 59–60 (2012)
11. Tzoumas, V., Amanatidis, C., Markakis, E.: A game-theoretic analysis of a competitive diffusion process over social networks. In: Goldberg, P.W. (ed.) WINE 2012. LNCS, vol. 7695, pp. 1–14. Springer, Heidelberg (2012)

Reconfiguration of Cliques in a Graph

Takehiro Ito[1], Hirotaka Ono[2], and Yota Otachi[3(✉)]

[1] Graduate School of Information Sciences, Tohoku University,
Aoba-yama 6-6-05, Sendai 980-8579, Japan
takehiro@ecei.tohoku.ac.jp
[2] Faculty of Economics, Kyushu University,
Hakozaki 6-19-1, Higashi-ku, Fukuoka 812-8581, Japan
hirotaka@econ.kyushu-u.ac.jp
[3] School of Information Science, JAIST, Asahidai 1-1,
Nomi, Ishikawa 923-1292, Japan
otachi@jaist.ac.jp

Abstract. We study reconfiguration problems for cliques in a graph, which determine whether there exists a sequence of cliques that transforms a given clique into another one in a step-by-step fashion. As one step of a transformation, we consider three different types of rules, which are defined and studied in reconfiguration problems for independent sets. We first prove that all the three rules are equivalent in cliques. We then show that the problems are PSPACE-complete for perfect graphs, while we give polynomial-time algorithms for several classes of graphs, such as even-hole-free graphs and cographs. In particular, the shortest variant, which computes the shortest length of a desired sequence, can be solved in polynomial time for chordal graphs, bipartite graphs, planar graphs, and bounded treewidth graphs.

1 Introduction

Recently, *reconfiguration problems* attract attention in the field of theoretical computer science. The problem arises when we wish to find a step-by-step transformation between two feasible solutions of a problem such that all intermediate results are also feasible and each step abides by a fixed reconfiguration rule (i.e., an adjacency relation defined on feasible solutions of the original problem). This kind of reconfiguration problem has been studied extensively for several well-known problems, including SATISFIABILITY [10], INDEPENDENT SET [3,11,12,14,17,23], VERTEX COVER [12,13,16,17], CLIQUE, MATCHING [12], VERTEX-COLORING [2], and so on. (See also a recent survey [22].)

It is well known that independent sets, vertex covers and cliques are related with each other. Indeed, the well-known reductions for NP-completeness proofs are essentially the same for the three problems [7]. Despite reconfiguration problems for independent sets and vertex covers are two of the most well studied problems, we have only a few known results for reconfiguration problems for cliques (as we will explain later). In this paper, we thus systematically investigate the complexity status of reconfiguration problems for cliques, and show that

© Springer International Publishing Switzerland 2015
R. Jain et al. (Eds.): TAMC 2015, LNCS 9076, pp. 212–223, 2015.
DOI: 10.1007/978-3-319-17142-5_19

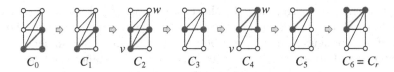

Fig. 1. A sequence $\langle C_0, C_1, \ldots, C_6 \rangle$ of cliques in the same graph, where the vertices in cliques are depicted by large (blue) circles (tokens) (Color figure online).

the problems can be solved in polynomial time for a variety of graph classes, in contrast to independent sets and vertex covers.

1.1   Our Problems and Three Rules

Recall that a *clique* of a graph $G = (V, E)$ is a vertex subset of G in which every two vertices are adjacent. (Figure 1 depicts seven different cliques in the same graph.) Suppose that we are given two cliques C_0 and C_r of G, and imagine that a token is placed on each vertex in C_0. Then, we are asked to transform C_0 into C_r by abiding a prescribed reconfiguration rule on cliques. In this paper, we define three different reconfiguration rules on cliques, which were originally defined as the reconfiguration rules on independents sets [14], as follows:

- *Token Addition and Removal* (TAR rule): We can either add or remove a single token at a time if it results in a clique of size at least a given threshold $k \geq 0$. For example, in the sequence $\langle C_0, C_1, \ldots, C_6 \rangle$ in Fig. 1, every two consecutive cliques follow the TAR rule for the threshold $k = 2$. In order to emphasize the threshold k, we sometimes call this rule the TAR(k) rule.
- *Token Jumping* (TJ rule): A single token in a clique C can "jump" to any vertex in $V \setminus C$ if it results in a clique. For example, consider the sequence $\langle C_0, C_2, C_4, C_6 \rangle$ in Fig. 1, then two consecutive cliques C_{2i} and C_{2i+2} follow the TJ rule for each $i \in \{0, 1, 2\}$.
- *Token Sliding* (TS rule): We can slide a single token on a vertex v in a clique C to another vertex w in $V \setminus C$ if it results in a clique and there is an edge vw in G. For example, consider the sequence $\langle C_2, C_4 \rangle$ in Fig. 1, then two consecutive cliques C_2 and C_4 follow the TS rule, because v and w are adjacent.

A sequence $\langle C_0, C_1, \ldots, C_\ell \rangle$ of cliques of a graph G is called a *reconfiguration sequence* between two cliques C_0 and C_ℓ under TAR(k) (or TJ, TS) if two consecutive cliques C_{i-1} and C_i follow the TAR(k) (resp., TJ, TS) rule for all $i \in \{1, 2, \ldots, \ell\}$. The *length* of a reconfiguration sequence is defined to be the number of cliques in the sequence minus one, that is, the length of $\langle C_0, C_1, \ldots, C_\ell \rangle$ is ℓ.

   Given two cliques C_0 and C_r of a graph G (and an integer $k \geq 0$ for TAR), CLIQUE RECONFIGURATION under TAR (or TJ, TS) is to determine whether there exists a reconfiguration sequence between C_0 and C_r under TAR(k) (resp., TJ, TS). For example, consider the cliques C_0 and $C_r = C_6$ in Fig. 1; let $k = 2$ for TAR. Then, it is a yes-instance under the TAR(2) and TJ rules as illustrated in Fig. 1, but is a no-instance under the TS rule.

In this paper, we also study the shortest variant, called SHORTEST CLIQUE RECONFIGURATION, under each of the three rules which computes the shortest length of a reconfiguration sequence between two given cliques under the rule. We define the shortest length to be infinity for a no-instance, and hence this variant is a generalization of CLIQUE RECONFIGURATION.

1.2 Known and Related Results

Ito et al. [12] introduced CLIQUE RECONFIGURATION under TAR, and proved that it is PSPACE-complete in general. They also considered the optimization problem of computing the maximum threshold k such that there is a reconfiguration sequence between two given cliques C_0 and C_r under TAR(k). This maximization problem cannot be approximated in polynomial time within any constant factor unless P = NP [12].

INDEPENDENT SET RECONFIGURATION is one of the most well-studied reconfiguration problems, defined for independent sets in a graph. Kamiński et al. [14] studied the problem under TAR, TJ and TS. It is well known that a clique in a graph G forms an independent set in the complement $\overline{G}$ of G, and vice versa. Therefore, one may expect that several known results for INDEPENDENT SET RECONFIGURATION can be converted into ones for CLIQUE RECONFIGURATION. However, as far as we checked, only two results can be obtained for CLIQUE RECONFIGURATION by this conversion, because we take the complement of a graph. (These results will be formally discussed in Sect. 3.3.)

In this way, only a few results are known for CLIQUE RECONFIGURATION. In particular, there is almost no algorithmic result, and hence it is desired to develop efficient algorithms for the problem and its shortest variant.

1.3 Our Contribution

In this paper, we embark on a systematic investigation of the computational status of CLIQUE RECONFIGURATION and its shortest variant. Figure 2 summarizes our results, which can be divided into the following four parts.

(1) *Rule equivalence* (Sect. 3): We prove that all rules TAR, TS and TJ are equivalent in CLIQUE RECONFIGURATION. Then, any complexity result under one rule can be converted into the same complexity result under the other two rules. In addition, based on the rule equivalence, we show that CLIQUE RECONFIGURATION under any rule is PSPACE-complete for perfect graphs, and is solvable in linear time for cographs.
(2) *Graphs with bounded clique size* (Sect. 4.1): We show that the shortest variant under any of TAR, TS and TJ can be solved in polynomial time for such graphs, which include bipartite graphs, planar graphs, and bounded treewidth graphs. Interestingly, INDEPENDENT SET RECONFIGURATION under any rule remains PSPACE-complete even for planar graphs [2,11] and bounded treewidth graphs [23]. Therefore, this result shows a nice difference between the reconfiguration problems for cliques and independent sets.

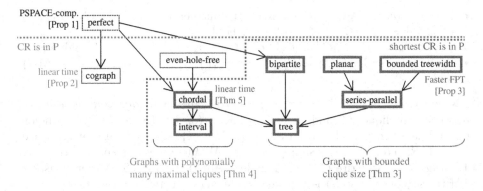

Fig. 2. Our results under all rules TAR, TS and TJ, where CR means CLIQUE RECON-FIGURATION. Each arrow represents the inclusion relationship between graph classes: $A \rightarrow B$ represents that B is properly included in A [4]. Graph classes for which SHORT-EST CLIQUE RECONFIGURATION is solvable in polynomial time are indicated by thick (red) boxes, while the ones for which CLIQUE RECONFIGURATION is solvable in polynomial time are indicated by thin (blue) boxes (Color figure online).

(3) *Graphs with polynomially many maximal cliques* (Sect. 4.2): We show that CLIQUE RECONFIGURATION under any of TAR, TS and TJ can be solved in polynomial time for such graphs, which include even-hole-free graphs, graphs of bounded boxicity, and K_t-subdivision-free graphs.

(4) *Chordal graphs* (Sect. 5): We give a linear-time algorithm to solve the shortest variant under any of TAR, TS and TJ for chordal graphs. Note that the clique size of chordal graphs is not always bounded, and hence this result is independent from Result (2) above.

Due to the page limitation, we omit several proofs from this extended abstract.

2 Preliminaries

In this paper, we assume without loss of generality that graphs are simple and undirected. For a graph G, we sometimes denote by $V(G)$ and $E(G)$ the vertex set and edge set of G, respectively. For a graph G, the *complement* $\overline{G}$ of G is the graph such that $V(\overline{G}) = V(G)$ and $E(\overline{G}) = \{vw \mid v, w \in V(G), vw \notin E(G)\}$. We say that a graph class $\mathcal{G}$ (i.e., a set of graphs) is *closed under taking complements* if $\overline{G} \in \mathcal{G}$ holds for every graph $G \in \mathcal{G}$.

In this paper, we deal with several graph classes systematically, and hence we do not define those graph classes precisely; we simply give the properties used for proving our results, with appropriate references.

2.1 Definitions for CLIQUE RECONFIGURATION

As explained in Introduction, we consider three (symmetric) adjacency relations on cliques in a graph. Let C_i and C_j be two cliques of a graph G. Then, we write

- $C_i \leftrightarrow C_j$ *under* TAR(k) for a nonnegative integer k if $|C_i| \geq k$, $|C_j| \geq k$, and $|C_i \bigtriangleup C_j| = |(C_i \setminus C_j) \cup (C_j \setminus C_i)| = 1$ hold;
- $C_i \leftrightarrow C_j$ *under* TJ if $|C_i| = |C_j|$, $|C_i \setminus C_j| = 1$, and $|C_j \setminus C_i| = 1$ hold; and
- $C_i \leftrightarrow C_j$ *under* TS if $|C_i| = |C_j|$, $C_i \setminus C_j = \{v\}$, $C_j \setminus C_i = \{w\}$, and $vw \in E(G)$ hold.

A sequence $\langle C_1, C_2, \ldots, C_\ell \rangle$ of cliques of G is called a *reconfiguration sequence* between two cliques C_1 and C_ℓ under TAR(k) (or TJ, TS) if $C_{i-1} \leftrightarrow C_i$ holds under TAR(k) (resp., TJ, TS) for all $i \in \{2, 3, \ldots, \ell\}$. A reconfiguration sequence under TAR(k) (or TJ, TS) is simply called a TAR(k)-*sequence* (resp., TJ-*sequence*, TS-*sequence*). We write $C_1 \leftrightsquigarrow C_\ell$ under TAR(k) (or TJ, TS) if there exists a TAR(k)-sequence (resp., TJ-sequence, TS-sequence) between C_1 and C_ℓ.

Let k be a nonnegative integer, and let C and C' be two cliques of a graph G. Then, we define TAR(C, C', k), as follows:

$$\mathsf{TAR}(C, C', k) = \begin{cases} \text{yes} & \text{if } C \leftrightsquigarrow C' \text{ under TAR}(k); \\ \text{no} & \text{otherwise.} \end{cases}$$

Given two cliques C_0 and C_r of a graph G and a nonnegative integer k, CLIQUE RECONFIGURATION under TAR is to compute TAR(C_0, C_r, k). By the definition, TAR(C_0, C_r, k) = no if $|C_0| < k$ or $|C_r| < k$ hold, and hence we may assume without loss of generality that both $|C_0| \geq k$ and $|C_r| \geq k$ hold; we call such an instance simply a TAR-instance, and denote it by (G, C_0, C_r, k).

For two cliques C and C' of a graph G, we similarly define TJ(C, C') and TS(C, C'). Given two cliques C_0 and C_r of G, we similarly define CLIQUE RECONFIGURATION under TJ and TS, and denote their instance by (G, C_0, C_r). Then, we can assume that $|C_0| = |C_r|$ holds in a TJ- or a TS-instance (G, C_0, C_r).

2.2 Definitions for SHORTEST CLIQUE RECONFIGURATION

Given a TAR-instance (G, C_0, C_r, k), let $\mathcal{C} = \langle C_0, C_1, \ldots, C_\ell \rangle$ be a TAR(k)-sequence in G between C_0 and $C_r = C_\ell$. Then, the *length* of $\mathcal{C}$ is defined to be the number of cliques in $\mathcal{C}$ minus one, that is, the length of $\mathcal{C}$ is ℓ. We denote by $\mathsf{dist}_{\mathsf{TAR}}(G, C_0, C_r, k)$ the minimum length of a TAR(k)-sequence in G between C_0 and C_r; we let $\mathsf{dist}_{\mathsf{TAR}}(G, C_0, C_r, k) = +\infty$ if there is no TAR(k)-sequence in G between C_0 and C_r. The shortest variant, SHORTEST CLIQUE RECONFIGURATION, under TAR is to compute $\mathsf{dist}_{\mathsf{TAR}}(G, C_0, C_r, k)$. Similarly, we define $\mathsf{dist}_{\mathsf{TJ}}(G, C_0, C_r)$ and $\mathsf{dist}_{\mathsf{TS}}(G, C_0, C_r)$ for a TJ- and a TS-instance (G, C_0, C_r), respectively. Then, SHORTEST CLIQUE RECONFIGURATION under TJ or TS is defined similarly. We sometimes drop G and simply write $\mathsf{dist}_{\mathsf{TAR}}(C_0, C_r, k)$, $\mathsf{dist}_{\mathsf{TJ}}(C_0, C_r)$ and $\mathsf{dist}_{\mathsf{TS}}(C_0, C_r)$ if it is clear from context.

We note that CLIQUE RECONFIGURATION under any rule is a decision problem asking for the existence of a reconfiguration sequence, and its shortest variant asks for simply computing the shortest length of a reconfiguration sequence. Therefore, the problems do not ask for an actual reconfiguration sequence. However, our algorithms proposed in this paper can be easily modified so that they indeed find a reconfiguration sequence.

3 Rule Equivalence and Complexity

In this section, we first prove that all three rules TAR, TS and TJ are equivalent in CLIQUE RECONFIGURATION. We then discuss some complexity results that can be obtained from known results for INDEPENDENT SET RECONFIGURATION.

3.1 Equivalence of TS and TAR Rules

TS and TAR rules are equivalent, as in the following sense.

Theorem 1. TS *and* TAR *rules are equivalent in* CLIQUE RECONFIGURATION, *as follows*:

(a) *for any* TS-*instance* (G, C_0, C_r), *a* TAR-*instance* (G, C_0', C_r', k') *can be constructed in linear time such that* $\mathsf{TS}(C_0, C_r) = \mathsf{TAR}(C_0', C_r', k')$ *and* $\mathrm{dist}_{\mathsf{TS}}(C_0, C_r) = \mathrm{dist}_{\mathsf{TAR}}(C_0', C_r', k')/2$; *and*

(b) *for any* TAR-*instance* (G, C_0, C_r, k), *a* TS-*instance* (G, C_0', C_r') *can be constructed in linear time such that* $\mathsf{TAR}(C_0, C_r, k) = \mathsf{TS}(C_0', C_r')$.

Proof sketch. We here explain only how to construct the corresponding instances in linear time, and omit the correctness proofs.

(a) Let (G, C_0, C_r) be a TS-instance with $|C_0| = |C_r| = k$. Then, let $C_0' = C_0$, $C_r' = C_r$ and $k' = k$, as the corresponding TAR-instance (G, C_0', C_r', k').

(b) Let (G, C_0, C_r, k) be a TAR-instance; note that $|C_0| \neq |C_r|$ may hold, and both $|C_0| \geq k$ and $|C_r| \geq k$ hold. Then, as the corresponding TS-instance (G, C_0', C_r'), let $C_0' \subseteq C_0$ and $C_r' \subseteq C_r$ be any subsets of size exactly k.   □

By Theorem 1(a), note that the reduction from TS to TAR preserves the shortest length of reconfiguration sequences.

3.2 Equivalence of TJ and TAR Rules

TJ and TAR rules are equivalent, as in the following sense.

Theorem 2. TJ *and* TAR *rules are equivalent in* CLIQUE RECONFIGURATION, *as follows*:

(a) *for any* TJ-*instance* (G, C_0, C_r), *a* TAR-*instance* (G, C_0', C_r', k') *can be constructed in linear time such that* $\mathsf{TJ}(C_0, C_r) = \mathsf{TAR}(C_0', C_r', k')$ *and* $\mathrm{dist}_{\mathsf{TJ}}(C_0, C_r) = \mathrm{dist}_{\mathsf{TAR}}(C_0', C_r', k')/2$; *and*

(b) *for any* TAR-*instance* (G, C_0, C_r, k), *a* TJ-*instance* (G, C_0', C_r') *can be constructed in linear time such that* $\mathsf{TAR}(C_0, C_r, k) = \mathsf{TJ}(C_0', C_r')$.

Proof sketch. We again explain only how to construct the corresponding instances in linear time, and omit the correctness proofs.

(a) Let (G, C_0, C_r) be a TJ-instance with $|C_0| = |C_r| = k$. Then, let $C_0' = C_0$, $C_r' = C_r$ and $k' = k - 1$, as the corresponding TAR-instance (G, C_0', C_r', k').

(b) Let (G, C_0, C_r, k) be a TAR-instance; $|C_0| \neq |C_r|$ may hold, and both $|C_0| \geq k$ and $|C_r| \geq k$ hold. We first claim the following lemma.

Lemma 1. *Let (G, C_0, C_r, k) be a TAR-instance such that $C_0 \neq C_r$. Suppose that there exists an index $j \in \{0, r\}$ such that $|C_j| = k$ and C_j is a maximal clique in G. Then, $\mathsf{TAR}(C_0, C_r, k) = \mathsf{no}$.*

We thus assume without loss of generality that none of C_0 and C_r is a maximal clique in G of size k; note that the maximality of a clique can be determined in linear time. Then, we construct the corresponding TJ-instance (G, C_0', C_r'), as in the following two cases (i) and (ii):

(i) for each $j \in \{0, r\}$ such that $|C_j| \geq k + 1$, let $C_j' \subseteq C_j$ be an arbitrary subset of size exactly $k + 1$; and
(ii) for each $j \in \{0, r\}$ such that $|C_j| = k$, let $C_j' \supset C_j$ be an arbitrary superset of size exactly $k + 1$. □

By Theorem 2(a), note that the reduction from TJ to TAR preserves the shortest length of reconfiguration sequences.

3.3   Results Obtained from INDEPENDENT SET RECONFIGURATION

We here show two complexity results for CLIQUE RECONFIGURATION, which can be obtained from known results for INDEPENDENT SET RECONFIGURATION.

Consider a vertex subset C of a graph G. Then, C forms a clique in G if and only if C forms an independent set in the complement $\overline{G}$ of G. Therefore, the following lemma clearly holds.

Lemma 2. *Let G be a graph, and let C_j be a clique of G for each $j \in \{0, 1, \ldots, \ell\}$. Then, $\langle C_0, C_1, \ldots, C_\ell \rangle$ is a $\mathsf{TAR}(k)$-sequence of cliques in G if and only if $\langle C_0, C_1, \ldots, C_\ell \rangle$ is a $\mathsf{TAR}(k)$-sequence of independent sets in the complement $\overline{G}$ of G.*

By Lemma 2 we can convert a complexity result for INDEPENDENT SET RECONFIGURATION under TAR for a graph class $\mathcal{G}$ into one for CLIQUE RECONFIGURATION under TAR for $\mathcal{G}$ if the graph class $\mathcal{G}$ is closed under taking complements. Note that, by Theorems 1 and 2, any complexity result under one rule can be converted into the same complexity result under the other two rules. Then, we have the following two results.

Proposition 1. CLIQUE RECONFIGURATION *is PSPACE-complete for perfect graphs under all rules* TAR, TS *and* TJ.

Proposition 2. CLIQUE RECONFIGURATION *can be solved in linear time for cographs under all rules* TAR, TS *and* TJ.

4   Polynomial-Time Algorithms

In this section, we show that CLIQUE RECONFIGURATION is solvable in polynomial time for several graph classes. We deal with two types of graph classes, that is, graphs of bounded clique size (in Sect. 4.1) and graphs having polynomially many maximal cliques (in Sect. 4.2).

4.1   Graphs of Bounded Clique Size

In this subsection, we show that SHORTEST CLIQUE RECONFIGURATION can be solved in polynomial time for graphs of bounded clique size. For a graph G, we denote by $\omega(G)$ the size of a maximum clique in G.

Theorem 3. *Let G be a graph with n vertices such that $\omega(G) \leq w$ for a positive integer w. Then,* SHORTEST CLIQUE RECONFIGURATION *under any of* TAR, TS *and* TJ *can be solved in time $O(w^2 n^w)$ for G.*

It is well known that $\omega(G) \leq 4$ for any planar graph G, and $\omega(G') \leq 2$ for any bipartite graph G'. We thus have the following corollary.

Corollary 1. SHORTEST CLIQUE RECONFIGURATION *under* TAR, TS *and* TJ *can be solved in polynomial time for planar graphs and bipartite graphs.*

By the definition of treewidth [1], we have $\omega(G) \leq t+1$ for any graph G whose treewidth can be bounded by a positive integer t. By Theorem 3 this observation gives an $O\big(t^2 n^{t+1}\big)$-time algorithm for SHORTEST CLIQUE RECONFIGURATION. However, for this case, we can obtain a faster fixed-parameter algorithm, where the parameter is the treewidth t, as follows.

Proposition 3. *Let G be a graph with n vertices whose treewidth is bounded by a positive integer t. Then,* SHORTEST CLIQUE RECONFIGURATION *under any of* TAR, TS *and* TJ *can be solved for G in time $O(c^t n)$, where c is some constant.*

Proposition 3 implies that SHORTEST CLIQUE RECONFIGURATION under any of TAR, TS and TJ can be solved in fixed-parameter time $O(c^w n)$ for chordal graphs G when parameterized by the size w of a maximum clique in G, where n is the number of vertices in G and c is some constant; because the treewidth of a chordal graph G can be bounded by the size of a maximum clique in G minus one [18]. However, we give a linear-time algorithm to solve the shortest variant under any rule for chordal graphs in Sect. 5.

4.2   Graphs with Polynomially Many Maximal Cliques

In this subsection, we consider the class of graphs having polynomially many maximal cliques, which properly contains the class of graphs with bounded clique size (in Sect. 4.1). Note that, even if a graph G has a polynomial number of maximal cliques, G may have a super-polynomial number of cliques.

Theorem 4. *Let G be a graph with n vertices and m edges, and let $\mathcal{M}(G)$ be the set of all maximal cliques in G. Then,* CLIQUE RECONFIGURATION *under any of* TAR, TS *and* TJ *can be solved for G in time $O\big(mn|\mathcal{M}(G)| + n|\mathcal{M}(G)|^2\big)$.*

Before proving Theorem 4, we give the following corollary.

Corollary 2. CLIQUE RECONFIGURATION *under* TAR, TS *and* TJ *can be solved in polynomial time for even-hole-free graphs, graphs of bounded boxicity, and K_t-subdivision-free graphs.*

Proof. By Theorem 4 it suffices to show that the claimed graphs have polynomially many maximal cliques. Polynomial bounds on the number of maximal cliques are shown for even-hole-free graphs in [5], for graphs of bounded boxicity in [19], and for K_t-subdivision-free graphs in [15]. □

In this subsection, we prove Theorem 4. However, by Theorems 1(a) and 2(a) it suffices to give such an algorithm only for the TAR rule.

Let (G, C_0, C_r, k) be any TAR-instance. Then, we define the *k-intersection maximal-clique graph* of G, denoted by $MC_k(G)$, as follows:

(i) each node in $MC_k(G)$ corresponds to a (maximal) clique in $\mathcal{M}(G)$; and
(ii) two nodes in $MC_k(G)$ are joined by an edge if and only if $|M \cap M'| \geq k$ holds for the corresponding two maximal cliques M and M' in $\mathcal{M}(G)$.

Note that any maximal clique in $\mathcal{M}(G)$ of size less than k is contained in $MC_k(G)$ as an isolated node. We now give the key lemma to prove Theorem 4.

Lemma 3. *Let G be a graph, and let C and C' be any pair of cliques in G such that $|C| \geq k$ and $|C'| \geq k$. Let $M \supseteq C$ and $M' \supseteq C'$ be arbitrary maximal cliques in $\mathcal{M}(G)$. Then, $C \leftrightsquigarrow C'$ under $\mathsf{TAR}(k)$ if and only if $MC_k(G)$ contains a path between the two nodes corresponding to M and M'.*

Proof of Theorem 4. For any graph G with n vertices and m edges, Tsukiyama et al. [20] proved that the set $\mathcal{M}(G)$ can be computed in time $O(mn|\mathcal{M}(G)|)$. Thus, we can construct $MC_k(G)$ in time $O(mn|\mathcal{M}(G)| + n|\mathcal{M}(G)|^2)$. By the breadth-first search on $MC_k(G)$ which starts from an arbitrary maximal clique (node) $M \supseteq C_0$, we can check in time $O(|\mathcal{M}(G)|^2)$ whether $MC_k(G)$ has a path to a maximal clique $M' \supseteq C_r$. Then, the theorem follows from Lemma 3. □

5 Linear-Time Algorithm for Chordal Graphs

Since any chordal graph is even-hole free, by Corollary 2 CLIQUE RECONFIGURATION is solvable in polynomial time for chordal graphs. Furthermore, we have discussed in Sect. 4.1 that the shortest variant is fixed-parameter tractable for chordal graphs when parameterized by the size of a maximum clique in a graph. However, we give the following theorem in this section.

Theorem 5. SHORTEST CLIQUE RECONFIGURATION *under any of* TAR, TS *and* TJ *can be solved in linear time for chordal graphs.*

In this section, we prove Theorem 5. By Theorems 1(a) and 2(a) it suffices to give a linear-time algorithm for a TAR-instance; recall that the reduction from TS/TJ to TAR preserves the shortest length of reconfiguration sequences.

Our algorithm consists of two phases. The first is a linear-time reduction from a given TAR-instance (G, C_0, C_r, k) for a chordal graph G to a TAR-instance (H, C_0, C_r, k) for an interval graph H such that $\mathsf{dist}_{\mathsf{TAR}}(H, C_0, C_r, k) = \mathsf{dist}_{\mathsf{TAR}}(G, C_0, C_r, k)$. The second is a linear-time algorithm for interval graphs.

5.1   Definitions of Chordal Graphs and Interval Graphs

A graph is a *chordal graph* if every induced cycle is of length three. Recall that $\mathcal{M}(G)$ is the set of all maximal cliques in a graph G, and we denote by $\mathcal{M}(G; v)$ the set of all maximal cliques in G that contain a vertex $v \in V(G)$. A tree T is a *clique tree* of a graph G if it satisfies the following two conditions:

– each node in T corresponds to a maximal clique in $\mathcal{M}(G)$; and
– for each $v \in V(G)$, the subgraph of T induced by $\mathcal{M}(G; v)$ is connected.

It is known that a graph is a chordal graph if and only if it has a clique tree [8]. A clique tree of a chordal graph can be computed in linear time (see [19, Sect. 15.1]).

A graph is an *interval graph* if it can be represented as the intersection graph of intervals on the real line. A *clique path* is a clique tree which is a path. It is known that a graph is an interval graph if and only if it has a clique path [6,9].

5.2   Linear-Time Reduction from Chordal Graphs to Interval Graphs

In this subsection, we describe the first phase of our algorithm.

Let (G, C_0, C_r, k) be any TAR-instance for a chordal graph G, and let T be a clique tree of G. Then, we find an arbitrary pair of maximal cliques M_0 and M_t in G (i.e., two nodes in T) such that $C_0 \subseteq M_0$ and $C_r \subseteq M_t$. Let $(M_0, M_1, \ldots, M_t)$ be the unique path in T from M_0 to M_t. We define a graph H' as the subgraph of G induced by the maximal cliques $M_0, M_1, \ldots, M_t$. Note that H' is an interval graph, because $(M_0, M_1, \ldots, M_t)$ forms a clique path.

The following lemma implies that the interval graph H' has a TAR(k)-sequence $\langle C_0, C_1, \ldots, C_{\ell'} \rangle$ such that $\ell' = \mathsf{dist}_{\mathsf{TAR}}(G, C_0, C_r, k)$, and hence yields that $\mathsf{dist}_{\mathsf{TAR}}(H', C_0, C_r, k) = \mathsf{dist}_{\mathsf{TAR}}(G, C_0, C_r, k)$ holds.

Lemma 4. *Let (G, C_0, C_r, k) be a TAR-instance for a chordal graph G, and let T be a clique tree of G. Suppose that $\langle C_0, C_1, \ldots, C_\ell \rangle$ is a shortest TAR(k)-sequence in G from C_0 to $C_\ell = C_r$. Let $(M_0, M_1, \ldots M_t)$ be the path in T from M_0 to M_t for any pair of maximal cliques $M_0 \supseteq C_0$ and $M_t \supseteq C_r$. Then, there is a monotonically increasing function $f \colon \{0, 1, \ldots, \ell\} \to \{0, 1, \ldots, t\}$ such that $C_i \subseteq M_{f(i)}$ for each $i \in \{0, 1, \ldots, \ell\}$.*

Although Lemma 4 implies that $\mathsf{dist}_{\mathsf{TAR}}(H', C_0, C_r, k) = \mathsf{dist}_{\mathsf{TAR}}(G, C_0, C_r, k)$ holds for the interval graph H', it seems difficult to find two maximal cliques $M_0 \supseteq C_0$ and $M_t \supseteq C_r$ (and hence construct H' from G) in linear time. However, by a small trick, we can construct an interval graph H in linear time such that $\mathsf{dist}_{\mathsf{TAR}}(H, C_0, C_r, k) = \mathsf{dist}_{\mathsf{TAR}}(G, C_0, C_r, k)$, as follows.

Lemma 5. *Given a TAR-instance (G, C_0, C_r, k) for a chordal graph G, one can obtain a subgraph H of G in linear time such that H is an interval graph, $C_0, C_r \subseteq V(H)$ and $\mathsf{dist}_{\mathsf{TAR}}(H, C_0, C_r, k) = \mathsf{dist}_{\mathsf{TAR}}(G, C_0, C_r, k)$.*

5.3   Linear-Time Algorithm for Interval Graphs

In this subsection, we describe the second phase of our algorithm.

Let H be a given interval graph, and we assume that its clique path $\mathcal{P}$ has $V(\mathcal{P}) = \mathcal{M}(H) = \{M_0, M_1, \ldots, M_t\}$ and $E(\mathcal{P}) = \{M_i M_{i+1} \mid 0 \le i < t\}$. Note that we can assume that $t \ge 1$, that is, H has at least two maximal cliques; otherwise we can easily solve the problem in linear time. For a vertex v in H, let $l_v = \min\{i \mid v \in M_i\}$ and $r_v = \max\{i \mid v \in M_i\}$; the indices l_v and r_v are called the *l-value* and *r-value* of v, respectively. Note that $v \in M_i$ if and only if $l_v \le i \le r_v$. For an interval graph H, such a clique path $\mathcal{P}$ and the indices l_v and r_v for all vertices $v \in V(H)$ can be computed in linear time [21].

Let (H, C_0, C_r, k) be a TAR-instance. We assume that $C_0 \subseteq M_0$, $C_0 \not\subseteq M_1$ and $C_r \subseteq M_t$; otherwise we can remove the maximal cliques M_i such that $i < \min\{r_v \mid v \in C_0\}$ or $i > \max\{l_v \mid v \in C_r\}$ in linear time. Our algorithm greedily constructs a shortest TAR(k)-sequence from C_0 to C_r, as follows:

(1) if $C_0 \not\subseteq C_r$ and $|C_0| \ge k+1$, then remove a vertex with the minimum r-value in $C_0 \setminus C_r$ from C_0;
(2) otherwise add a vertex in $(C_r \setminus C_0) \cap M_0$ if any; if no such vertex exists, add a vertex with the maximum r-value in $M_0 \setminus C_0$.

We regard the clique obtained by one of the operations above as C_0; if necessary, we shift the indices of M_i so that $C_0 \subseteq M_0$ and $C_0 \not\subseteq M_1$ hold; and repeat. If $C_0 \ne C_r$ and none of the operations above is possible, we can conclude that (H, C_0, C_r, k) is a no-instance.

We omit the correctness proof of this greedy algorithm and the estimation of its running time. This completes the proof sketch of Theorem 5.   □

6   Conclusion

In this paper, we have systematically shown that CLIQUE RECONFIGURATION and its shortest variant can be solved in polynomial time for several graph classes. As far as we know, this is the first example of a reconfiguration problem such that it is PSPACE-complete in general, but is solvable in polynomial time for such a variety of graph classes.

Acknowledgments. This work is partially supported by MEXT/JSPS KAKENHI 25106504 and 25330003 (T. Ito), 25104521 and 26540005 (H. Ono), and 24106004 and 25730003 (Y. Otachi).

References

1. Bodlaender, H.L., Drange, P.G., Dregi, M.S., Fomin, F.V., Lokshtanov, D., Pilipczuk, M.: An $O(c^k n)$ 5-approximation algorithm for treewidth. In: Proceedings of FOCS 2013, pp. 499–508 (2013)

2. Bonsma, P., Cereceda, L.: Finding paths between graph colourings: PSPACE-completeness and superpolynomial distances. Theoret. Comput. Sci. **410**, 5215–5226 (2009)
3. Bonsma, P.: Independent set reconfiguration in cographs. In: Kratsch, D., Todinca, I. (eds.) WG 2014. LNCS, vol. 8747, pp. 105–116. Springer, Heidelberg (2014)
4. Brandstädt, A., Le, V.B., Spinrad, J.P.: Graph Classes: A Survey, SIAM, Philadelphia (1999)
5. da Silva, M.V.G., Vušković, K.: Triangulated neighborhoods in even-hole-free graphs. Discrete Math. **307**, 1065–1073 (2007)
6. Fulkerson, D.R., Gross, O.A.: Incidence matrices and interval graphs. Pac. J. Math. **15**, 835–855 (1965)
7. Garey, M.R., Johnson, D.S.: Computers and Intractability: A Guide to the Theory of NP-Completeness. Freeman, San Francisco (1979)
8. Gavril, F.: The intersection graphs of subtrees in trees are exactly the chordal graphs. J. Comb. Theory, Ser. B **16**, 47–56 (1974)
9. Gilmore, P.C., Hoffman, A.J.: A characterization of comparability graphs and of interval graphs. Can. J. Math. **16**, 539–548 (1964)
10. Gopalan, P., Kolaitis, P.G., Maneva, E.N., Papadimitriou, C.H.: The connectivity of Boolean satisfiability: computational and structural dichotomies. SIAM J. Comput. **38**, 2330–2355 (2009)
11. Hearn, R.A., Demaine, E.D.: PSPACE-completeness of sliding-block puzzles and other problems through the nondeterministic constraint logic model of computation. Theoret. Comput. Sci. **343**, 72–96 (2005)
12. Ito, T., Demaine, E.D., Harvey, N.J.A., Papadimitriou, C.H., Sideri, M., Uehara, R., Uno, Y.: On the complexity of reconfiguration problems. Theoret. Comput. Sci. **412**, 1054–1065 (2011)
13. Ito, T., Nooka, H., Zhou, X.: Reconfiguration of vertex covers in a graph. In: Proceedings of IWOCA 2014 (2014, To appear)
14. Kamiński, M., Medvedev, P., Milanič, M.: Complexity of independent set reconfigurability problems. Theoret. Comput. Sci. **439**, 9–15 (2012)
15. Lee, C., Oum, S.: Number of cliques in graphs with forbidden subdivision. arXiv:1407.7707 (2014)
16. Mouawad, A.E., Nishimura, N., Raman, V.: Vertex cover reconfiguration and beyond. In: Ahn, H.-K., Shin, C.-S. (eds.) ISAAC 2014. LNCS, vol. 8889, pp. 447–458. Springer, Heidelberg (2014)
17. Mouawad, A.E., Nishimura, N., Raman, V., Simjour, N., Suzuki, A.: On the parameterized complexity of reconfiguration problems. In: Gutin, G., Szeider, S. (eds.) IPEC 2013. LNCS, vol. 8246, pp. 281–294. Springer, Heidelberg (2013)
18. Robertson, N., Seymour, P.D.: Graph minors. II. Algorithmic aspects of tree-width. J. Algorithms **7**, 309–322 (1986)
19. Spinrad, J.P.: Efficient Graph Representations. American Mathematical Society, Providence (2003)
20. Tsukiyama, S., Ide, M., Ariyoshi, H., Shirakawa, I.: A new algorithm for generating all the maximal independent sets. SIAM J. Comput. **6**, 505–517 (1977)
21. Uehara, R., Uno, Y.: On computing longest paths in small graph classes. Int. J. Found. Comput. Sci. **18**, 911–930 (2007)
22. van den Heuvel, J.: The complexity of change. Surveys in Combinatorics 2013, London Mathematical Society Lecture Notes Series 409 (2013)
23. Wrochna, M.: Reconfiguration in bounded bandwidth and treedepth. arXiv:1405.0847 (2014)

The Complexity of Finding Effectors

Laurent Bulteau, Stefan Fafianie, Vincent Froese, Rolf Niedermeier,
and Nimrod Talmon[✉]

Institut Für Softwaretechnik Und Theoretische Informatik,
TU Berlin, Berlin, Germany
{l.bulteau,nimrodtalmon77}@gmail.com,
{stefan.fafianie,vincent.froese,rolf.niedermeier}@tu-berlin.de

Abstract. The NP-hard EFFECTORS problem on directed graphs is
motivated by applications in network mining, particularly concerning
the analysis of (random) information-propagation processes. In the corresponding model the arcs carry probabilities and there is a probabilistic
diffusion process activating nodes by neighboring activated nodes with
probabilities as specified by the arcs. The point is to explain a given network activation state best possible using a minimum number of "effector
nodes"; these are selected before the activation process starts.

We complement and extend previous work from the data mining community by a more thorough computational complexity analysis of EFFEC-
TORS, identifying both tractable and intractable cases. To this end, we
also exploit a parameterization measuring the "degree of randomness"
(the number of 'really' probabilistic arcs) which might prove useful for
analyzing other probabilistic network diffusion problems.

1 Introduction

To understand and master the dynamics of information propagation in networks
(biological, chemical, computer, information, social) is a core research topic in
data mining and related fields. A prominent problem in this context is the NP-
hard problem EFFECTORS [10]: The input is a directed (influence) graph with a
subset of nodes marked as active (the target nodes) and each arc of the graph
carries an influence probability between 0 and 1. The task is to find few "effector nodes" that can "best explain" the set of given active nodes, that is, the
activation state of the graph; herein, in one round (this is known as the independent cascade model [9]) an activated node (initially consisting only the chosen
effectors) can activate every out-neighbor with the corresponding arc probability; see Sect. 2 for a formal model and problem definition. It is important to

A full version is available at http://arxiv.org/abs/1411.7838.

Laurent Bulteau—Supported by the Alexander von Humboldt Foundation, Bonn,
Germany.

Stefan Fafianie—Supported by the DFG Emmy Noether-program (KR 4286/1).

Vincent Froese—Supported by the DFG, project DAMM (NI 369/13).

Nimrod Talmon—Supported by DFG Research Training Group MDS (GRK 1408).

R. Jain et al. (Eds.): TAMC 2015, LNCS 9076, pp. 224–235, 2015.
DOI: 10.1007/978-3-319-17142-5_20

note that we allow effectors to be chosen from the *whole* set of graph nodes and not only from the set of target nodes. This makes our model, in a sense, more general than the original one by Lappas et al. [10].[1] Our main contribution is to extend and clarify research on the computational complexity status of EFFECTORS, which has been initiated by Lappas et al. [10]. In particular, as *probabilistic* information propagation is central in this model as well as in other network diffusion models, we put particular emphasis on studying how the "degree of randomness" in the network governs the computational complexity. Moreover, compared to previous work, we make an effort to present the results in a more formal setting conducting a rigorous mathematical analysis.

Informally speaking (concrete statements of our results appear in Sect. 2 after having provided formal definitions), we gained the following main insights.

- With unlimited degree of randomness, finding effectors is computationally very hard. In fact, even computing the "cost" (how well does a set of effectors explain a given activation state) of a *given* set of effectors is intractable. This significantly differs from deterministic models.
- Even if the directed input graph is acyclic, then this does *not* lead to a significant decrease of the computational complexity.
- Bounding the degree of randomness (in other words, bounding the number of arcs with probability different from 0 or 1), that is, *parameterizing on the degree of randomness*, yields some encouraging (fixed-parameter) tractability results for otherwise intractable cases.
- We identify some flaws in the work of Lappas et al. [10] (see the last part of Sect. 4.3 for details), who claim one case to be intractable which in fact is tractable and one case the other way around.

Admittedly, in real-world applications (where influence probabilities are determined through observation and simulation, often involving noise) the number of probabilistic arcs may be high, thus rendering the parameter "number of probabilistic arcs" doubtful. However, note that finding effectors is computationally very hard (also in terms of polynomial-time approximability) and so in order to make the computation of a solution more feasible one might round up (to 1) arc probabilities which are close to 1 and round down (to 0) arc probabilities which are close to 0. Thus, we can achieve a tradeoff between running time and accuracy of the result. Depending on the degree of rounding (as much as a subsequent fixed-parameter algorithm exploiting the mentioned parameter would "allow") in this way one might at least find a good approximation of an optimal set of effectors in reasonable time.

Related Work. Our main point of reference is the work of Lappas et al. [10]. Indeed, we use a slightly different (and more general) problem definition: they define the effectors to be necessarily a subset of the target nodes, whereas we

---

[1] We conjecture that both models coincide if we have unlimited budget, that is, if the number of chosen effectors does not matter. On the contrary, they do not coincide if we have limited budget, see Sect. 2.

allow the effectors to form an arbitrary subset of the nodes. It turns out that these two definitions really yield different problems (see Sect. 2 for an extensive discussion of the two models). The also NP-hard special case where all nodes are target nodes (and hence where the two models above clearly coincide) is called INFLUENCE MAXIMIZATION and is also well studied in the literature [4,6,9]. Finally, a closely related deterministic version (called TARGET SET SELECTION) with the additional difference of having node-individual thresholds specifying how many neighboring nodes need to be active to make a node active has also been extensively studied, in particular from a parameterized complexity point of view [3,5,11].

2   Preliminaries and Model Discussion

In this section, we first provide the formal framework, overview our results, and explain our modeling, particularly discussing the difference between our model and the one by Lappas et al. [10].

Preliminaries. We basically use the same definitions as Lappas et al. [10] except for few differences in notation.

Influence Graphs. An *influence graph* $G = (V, E, w)$ is a simple directed graph equipped with a function $w : E \to (0, 1] \cap \mathbb{Q}$ assigning an *influence weight* to each arc $(u \to v) \in E$ which represents the *influence of node u on node v*. We denote the number of nodes in G by $n := |V|$ and the number of arcs in G by $m := |E|$.

Information Propagation. We consider the following information-propagation process, called the *Independent Cascade (IC)* model [9]. Within this model, each node is in one of two states: *active* or *inactive*. When a node u becomes active for the first time, at time step t, it gets a single chance to activate its out-neighbors. Specifically, u succeeds in activating a neighbor v with probability $w(u \to v)$. If u succeeds, then v will become active at step $t+1$. Otherwise, u cannot make any more attempts to activate v in any subsequent round. As usual, we assume that the precision of the probabilities determined by the function w is polynomially bounded in the number n of nodes in the input graph.

Cost Function. For a given influence graph $G = (V, E, w)$, subset $X \subseteq V$ of effectors, and subset $A \subseteq V$ of active nodes, we define a cost function

$$C_A(G, X) := \sum_{v \in A} (1 - p(v|X)) + \sum_{v \in V \setminus A} p(v|X),$$

where for each $v \in V$, we define $p(v|X)$ to be the probability of v being active after the termination of the information-propagation process starting with X as the active nodes. An alternative definition is that $C_A(G, X) := \sum_{v \in V} C_A(v, X)$, where $C_A(v, X) := 1 - p(v|X)$ if $v \in A$ and $C_A(v, X) := p(v|X)$ if $v \notin A$.

Main Problem Definition. Our central problem EFFECTORS is formulated as a decision problem—it relates to finding few nodes which best explain (lowest cost) the given network activation state specified by a subset $A \subseteq V$ of nodes.

EFFECTORS

Input: An influence graph $G = (V, E, w)$, a set of target nodes $A \subseteq V$, a budget $b \in \mathbb{N}$, and a cost $c \in \mathbb{Q}$.

Question: Is there a subset $X \subseteq V$ of effectors such that $|X| \le b$ and $C_A(G, X) \le c$?

We will additionally consider the related problem EFFECTORS-COST (see Sect. 3) where the set X of effectors is already given and one has to determine its cost.

Parameters. The most natural parameters to consider for a parameterized complexity analysis are the maximum number b of effectors, the cost value c, and the number $a := |A|$ of target nodes. Moreover, we will be especially interested in quantifying the amount of randomness in the influence graph. To this end, consider an arc $(u \to v) \in E$: if $w(u \to v) = 1$, then this arc is not probabilistic. We define the parameter number r of probabilistic arcs, that is, $r := |\{(u \to v) \in E : 0 < w(u \to v) < 1\}|$. We will also briefly discuss the parameterization by the treewidth of the underlying undirected graph.

Graph Theory. We use the acronym DAG for directed acyclic graphs. The DAG of strongly connected components of a directed graph is called its *condensation*. A *directed tree* is an arbitrary orientation of an undirected tree.

Computational Complexity. We assume familiarity with the basic notions of algorithms and complexity. Several of our results will be cast using the framework of parameterized complexity analysis. An instance (I, k) of a parameterized problem consists of the actual instance I and an integer k being the *parameter* [7,8,12]. A parameterized problem is called *fixed-parameter tractable* (FPT) if there is an algorithm solving it in $f(k) \cdot |I|^{O(1)}$ time, whereas an algorithm with running time $O(|I|^{f(k)})$ only shows membership in the class XP (clearly, FPT $\subseteq$ XP). Thus, achieving fixed-parameter tractability is computationally much more attractive. One can show that a parameterized problem L is (presumably) not fixed-parameter tractable by devising a *parameterized reduction* from a W[1]-hard or W[2]-hard problem (such as CLIQUE or SET COVER, respectively, each parameterized by the solution size) to L. A parameterized reduction from a parameterized problem L to another parameterized problem L' is a function that, given an instance (I, k), computes in $f(k) \cdot |I|^{O(1)}$ time an instance (I', k') (with $k' \le g(k)$) such that $(I, k) \in L \Leftrightarrow (I', k') \in L'$. We will also consider counting problems of the form "compute func(x)". Informally speaking, we can associate a decision problem in NP to a counting problem in #P. Then, analogously to NP-hardness, showing that a counting problem is #P-hard gives strong evidence for the intractability of the counting problem.

Our Results. Before we discuss our model and the one by Lappas et al. [10], we overview our main results. We will treat the subproblem EFFECTORS-COST in Sect. 3, and EFFECTORS in Sect. 4. Our results are summarized in Table 1. Note that most of our results transfer to the model of Lappas et al. [10]. In particular, this implies that their claims that the "zero-cost" special case is NP-hard [10, Lemma 1] and that the deterministic version is polynomial-time solvable are

Table 1. Computational complexity of the different variants of EFFECTORS. Note that all hardness results hold also for DAGs. The parameter a stands for the number of active nodes, b for the budget, c for the cost value, and r for the number of probabilistic arcs.

	Deterministic ($r = 0$)	Parameterized (by r)	Probabilistic (arbitrary r)
EFFECTORS-COST	FPT [wrt. r], Theorem 2		#P-hard, Corollary 1
EFFECTORS (general case)	XP [wrt. $\min(a, b, c)$], Proposition 1	W[2]-hard [wrt. $b + c$], Theorem 3 W[1]-hard [wrt. $a + b + c$], Theorem 3	
Infinite budget ($b = \infty$)	FPT [wrt. r], Theorem 5		NP-hard, Theorem 4
Influence Maximization ($A = V$)	FPT [wrt. $b + c$], Theorem 6 FPT [wrt. treewidth], [3]	W[1]-hard [wrt. $\min(b, c)$], Theorem 6	

both flawed, because from our results exactly the opposite follows (see the last part of Sect. 4.3 for details). Due to lack of space, most of the proofs are omitted. For full formal proofs, refer to the full version (available at http://arxiv.org/abs/1411.7838).

Model Discussion. Our definition of EFFECTORS differs from the problem definition of Lappas et al. [10] in that we do not require the effectors to be chosen among the target nodes. Before pointing out possible advantages and motivating our problem definition, we give a simple example illustrating the difference between these two problems.

Consider the influence graph in Fig. 1, consisting of one non-target node (white) having three outgoing arcs of weight 1 to three target nodes (black). Clearly, for $b = c = 1$, this is a "no"-instance if we are only allowed to pick target nodes as effectors since the probability of being active will be 0 for two of the three target nodes in any case, which yields a cost of at least 2. According to our problem definition, however, we are allowed to select the non-target node, which only incurs a cost of 1, showing that this is a "yes"-instance. We think that our model captures the natural assumption that an effector node does not have to remain active forever. Indeed, the modeling of Lappas et al. [10] might be interpreted as a "monotone version" as for example discussed by Askalidis et al. [2], while in this sense our model allows for "non-monotone explanations".

Clearly, if all nodes are target nodes (this particular setting is called *Influence Maximization*), then the two models coincide. Furthermore, we strongly conjecture that if we have an unlimited budget, then it suffices to search for a solution among the target nodes, that is, also for $b = \infty$, we believe that the two problem definitions are equivalent.

Fig. 1. A small example where it is optimal to choose a non-target node as effector (Color figure online).

Conjecture 1. For $b = \infty$, it holds that for every "yes"-instance (G, A, b, c) of EFFECTORS there exists a solution $X \subseteq A$.

At least for directed trees (that is, the underlying undirected graph is a tree) we can prove Conjecture 1. The idea of proof is that if an optimal solution contains a non-target node, then this node only influences nodes reachable from it via paths that do not visit other nodes in the solution. Within this smaller tree of influenced nodes there must be some subtrees rooted at target nodes such that the expected cost for such a subtree is smaller if its target root node is activated during the propagation process compared to the case when it is not. Choosing these target nodes directly as effectors, replacing the non-target node, yields another optimal solution with fewer non-target nodes.

Theorem 1. *Conjecture 1 holds for directed trees.*

3   Computing the Cost Function

We consider the problem of computing the cost for a given set of effectors.

> EFFECTORS-COST
> **Input:** An influence graph $G = (V, E, w)$, a set of target nodes $A \subseteq V$, and a set of effectors $X \subseteq V$.
> **Compute:** The cost $C_A(G, X)$.

EFFECTORS-COST is polynomial-time solvable on directed trees [10]. On the contrary, EFFECTORS-COST is unlikely to be polynomial-time solvable already on DAGs. This follows from a result by Wang et al. [13, Theorem 1]. They show that computing the expected number of activated nodes for a single given effector is #P-hard on DAGs. Note that for the case $A = \emptyset$, the cost equals the expected number of activated nodes at the end of the propagation process.

Corollary 1. EFFECTORS-COST on DAGs is #P-hard for $|A| = 0$ and $|X| = 1$.

On the positive side, EFFECTORS-COST is fixed-parameter tractable with respect to the number r of probabilistic arcs. The general idea is to recursively simulate the propagation process, branching over the probabilistic arcs, and to compute a weighted average of the final activation state of the graph.

Theorem 2. EFFECTORS-COST *can be solved in* $O(2^r \cdot n(n + m))$ *time, where* r *is the number of probabilistic arcs.*

4   Finding Effectors

We treat the general variant of EFFECTORS in Sect. 4.1, the special case of unlimited budget in Sect. 4.2, and the special case of influence maximization in Sect. 4.3.

4.1   General Model

We consider the parameters number a of target nodes, the budget b, and the cost c. We first notice that if at least one of them equals zero, then EFFECTORS is polynomial-time solvable. This holds trivially for parameters a and b; simply choose the empty set as a solution. This is optimal for $a = 0$, and the only feasible solution for $b = 0$. For parameter c, the following holds, using a simple decomposition into strongly connected components.

Lemma 1. For $c = 0$, EFFECTORS can be solved in $O(n + m)$ time.

Based on Lemma 1, by basically checking all possibilities in a brute-force manner, we obtain simple polynomial-time algorithms for EFFECTORS in the cases of a constant number a of target nodes, budget b, or cost c.

Proposition 1. For $r = 0$, EFFECTORS is in XP with respect to each of the parameters a, b, and c.

In the following, we show that, even for $r = 0$, EFFECTORS is W[1]-hard with respect to the *combined* parameter (a, b, c), and even W[2]-hard with respect to the *combined* parameter (b, c). We briefly sketch the proof of the first statement, and mention that the second statement is proven by a reduction from the W[2]-complete DOMINATING SET problem.

Theorem 3. *The following statements hold.*

1. EFFECTORS, *parameterized by the combined parameter* (a, b, c), *is* W[1]-*hard, even if* $r = 0$ *and the influence graph is a DAG.*
2. EFFECTORS, *parameterized by the combined parameter* (b, c), *is* W[2]-*hard, even if* $r = 0$ *and the influence graph is a DAG.*

Proof (Sketch for the first statement). We describe a parameterized reduction from the W[1]-hard problem MULTI-COLORED CLIQUE, which asks for the existence of a colorful clique of size k in a simple and undirected graph whose vertices are colored with k colors. Given an instance of MULTI-COLORED CLIQUE($G = (V, E), k$), we construct an instance of EFFECTORS with $b = \binom{k}{2}$, $c = \binom{k}{2} + k$, and an influence graph defined as follows. Add $\binom{k}{2} + k + 1$ nodes for each pair of distinct colors. Let us call these nodes *color-pair nodes*. Now, add a *vertex node* n_v for each $v \in V$, add an *edge node* $e_{u,v}$ for each $e = \{u, v\} \in E$, and add arcs $\{e_{u,v} \to n_u, e_{u,v} \to n_v\}$. For each edge, let L be the color-pair nodes corresponding to the colors of u and v, and add arcs $\{e_{u,v} \to l \mid l \in L\}$. Finally, let the set of target nodes A contain all color-pair nodes and set the influence weights of all arcs to 1. □

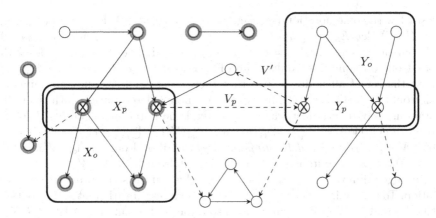

Fig. 2. Illustration for Theorem 5. Effectors of a solution are marked with an aura. Probabilistic arcs are dashed, and nodes of V_p (with an outgoing probabilistic arc) are marked with a cross. For readability, target nodes are not represented. Intuitively, the algorithm guesses the partition of V_p into X_p (effectors) and Y_p (non-effectors). Node set X_p (respectively, Y_p) is then extended to its closure X_o (respectively, its closure Y_o in the reverse graph). The remaining nodes form a deterministic subgraph $G[V']$, in which effectors, forming the set X', are selected by solving an instance of MAXIMUM WEIGHT CLOSURE.

4.2 Special Case: Unlimited Budget

Here, we concentrate on a model variant where we are allowed to choose any number of effectors, that is, the goal is to minimize the overall cost with an unlimited budget of effectors. In general, EFFECTORS with unlimited budget remains intractable, though (via reduction from a #P-hard counting problem).

Theorem 4. *If* P $\neq$ NP, *then* EFFECTORS, *even with unlimited budget, is not polynomial-time solvable on DAGs.*

However, with unlimited budget, EFFECTORS is fixed-parameter tractable with respect to the number r of probabilistic arcs.

Theorem 5. *If* $b = \infty$, *then* EFFECTORS *is solvable in* $O(4^r \cdot n^4)$ *time, where* r *is the number of probabilistic arcs.*

Proof. The general idea is to fully determine the probabilistic aspects of the graph, and then to remove all of the corresponding nodes and arcs. We can show that this leaves an equivalent "deterministic graph" that we can solve using a reduction to the problem MAXIMUM WEIGHT CLOSURE, which is itself polynomial-time solvable by a polynomial-time reduction to a flow maximization problem [1, Chapter 19].

MAXIMUM WEIGHT CLOSURE
Input: A directed graph $G = (V, E)$ with weights on the vertices.
Compute: A maximum-weight set of vertices $X \subseteq V$ with no arcs going out of the set.

We start with some notation (see Fig. 2 for an illustration). For an input graph $G = (V, E)$, let $E_p := \{(u \to v) \in E \mid w(u \to v) < 1\}$ denote the set of probabilistic arcs and let $V_p := \{u \mid (u \to v) \in E_p\}$ denote the set of nodes with at least one outgoing probabilistic arc. For a node $v \in V$, let $\mathrm{cl}_{\mathrm{det}}(v)$ ($\mathrm{cl}_{\mathrm{det}}^{-1}(v)$) denote the set of all nodes u such that there exists at least one deterministic path from v to u (respectively, from u to v), where a deterministic path is a path containing only deterministic arcs. We extend the notation to subsets V' of V and write $\mathrm{cl}_{\mathrm{det}}(V') = \bigcup_{v \in V'} \mathrm{cl}_{\mathrm{det}}(v)$ and $\mathrm{cl}_{\mathrm{det}}^{-1}(V') = \bigcup_{v \in V'} \mathrm{cl}_{\mathrm{det}}^{-1}(v)$. We call a subset $V' \subseteq V$ of nodes *deterministically closed* if and only if $\mathrm{cl}_{\mathrm{det}}(V') = V'$, that is, there are no outgoing deterministic arcs from V' to $V \setminus V'$.

Our algorithm will be based on a closer analysis of the structure of an optimal solution. To this end, let $G = (V, E, w)$ be an input graph with a set $A \subseteq V$ of target nodes and let $X \subseteq V$ be an optimal solution with minimum cost $C_A(G, X)$. Clearly, we can assume that X is deterministically closed, that is, $\mathrm{cl}_{\mathrm{det}}(X) = X$, since we have an infinite budget $b = \infty$.

We write V_p as a disjoint union of $X_p := V_p \cap X$ and $Y_p := V_p \setminus X$. We also use $X_o := \mathrm{cl}_{\mathrm{det}}(X_p)$, $Y_o := \mathrm{cl}_{\mathrm{det}}^{-1}(Y_p)$ and $V_o = X_o \cup Y_o$. Since X is deterministically closed, we have that $X_o \subseteq X$ and $Y_o \cap X = \emptyset$. We write $V' := V \setminus V_o$ and $X' := X \setminus X_o = X \cap V'$. Note that X' is deterministically closed in $G[V']$ and that $G[V']$ contains only deterministic arcs. Moreover, note that the sets X_o, Y_p, Y_o, V_o, and V', are directly deduced from the choice of X_p, and that for a given X_p, the set X' can be any closed subset of V'.

We first show that the nodes in V_o are only influenced by effectors in X_o, that is, for any node $v \in V_o$, it holds that $p(v|X) = p(v|X_o)$. This is clear for $v \in X_o$, since in this case $p(v|X) = p(v|X_o) = 1$. Assume now that there is a node $x \in X'$ with a directed path to $v \in Y_o$ that does not contain any node from X_o. Two cases are possible, depending on whether this path is deterministic. If it is, then since $v \in \mathrm{cl}_{\mathrm{det}}^{-1}(Y_p)$, then there exists a deterministic path from x to some $u \in Y_p$, via v. Hence, $x \in \mathrm{cl}_{\mathrm{det}}^{-1}(Y_p) = Y_o$, yielding a contradiction. Assume now that the path from x to v has a probabilistic arc and write $u \to u'$ for the first such arc. Hence, $x \in \mathrm{cl}_{\mathrm{det}}^{-1}(u)$ and $u \in V_p$. Since we assumed that the path does not contain any node from X_o, we have $u \notin X_p$, and therefore $u \in Y_p$. Again, we have $x \in \mathrm{cl}_{\mathrm{det}}^{-1}(Y_p)$, yielding a contradiction.

Hence, the nodes in V_o are not influenced by the nodes in X'. Now consider nodes in V'. Note that we have $p(v|X) = 1$ for $v \in X'$ and $p(v|X) = p(v|X_o)$ for $v \in V' \setminus X'$, since $G[V']$ is deterministic and X' is deterministically closed.

Overall, $C_A(v, X) = C_A(v, X_o)$ for all $v \in V \setminus X'$. The total cost of solution X can now be written as

$$C_A(G, X) = \sum_{v \in V \setminus X'} C_A(v, X_o) + \sum_{v \in X'} C_A(v, X)$$

$$= \sum_{v \in V} C_A(v, X_o) - \sum_{v \in X'} (C_A(v, X_o) - C_A(v, X))$$

$$= \alpha(X_o) - \beta(X_o, X'),$$

where

$$\alpha(X_o) := \sum_{v \in V} C_A(v, X_o) \quad \text{and} \quad \beta(X_o, X') := \sum_{v \in X'} (C_A(v, X_o) - C_A(v, X)).$$

We further define, for all $v \in V'$, $\gamma(v, X_o) := 1 - p(v|X_o)$ if $v \in A$, and $\gamma(v, X_o) = p(v|X_o) - 1$ if $v \notin A$. Note that, for $v \in X'$, the difference $C_A(v, X_o) - C_A(v, X)$ is exactly $\gamma(v, X_o)$, hence $\beta(X_o, X') := \sum_{v \in X'} \gamma(v, X_o)$.

The algorithm can now be described directly based on the above formulas. Specifically, we branch over all subsets $X_p \subseteq V_p$ (note that the number of these subsets is upper-bounded by 2^r). For each such subset $X_p \subseteq V_p$, we can compute X_o and Y_o in linear time because this involves propagation only through deterministic arcs (outgoing for X_o and ingoing for Y_o). Then, for each node $v \in V$, we compute $p(v|X_o)$ using Theorem 2 in $O(2^r \cdot n(n + m))$ time. This yields the values $\alpha(X_o)$ and $\gamma(v, X_o)$ for each $v \in V'$. The closed subset X' of V' maximizing $\beta(X_o, X')$ is then computed as the solution of MAXIMUM WEIGHT CLOSURE on $G[V']$ (which is solved by a maximum flow computation in $O(n^3)$ time), where the weight of any $v \in V'$ is $\gamma(v, X_o)$. Finally, we return the set $X_o \cup X'$ that yields the minimum value for $\alpha(X_o) - \beta(X_o, X')$. $\qquad \square$

4.3 Special Case: Influence Maximization

In this section, we consider the special case of EFFECTORS, called INFLUENCE MAXIMIZATION, where all nodes are targets $(A = V)$. Note that in this case the variant with unlimited budget and the parameterization by the number of target nodes are irrelevant.

In the influence maximization case, on deterministic instances, one should intuitively choose effectors among the "sources" of the influence graph, that is, nodes without incoming arcs (or among strongly connected components without incoming arcs). Moreover, the budget b bounds the number of sources that can be selected, and the cost c bounds the number of sources that can be left out. In the following theorem, we prove that deterministic EFFECTORS remains intractable even if either one of these parameters is small, but, on the contrary, having $b + c$ as a parameter yields fixed-parameter tractability in the deterministic case. We mention that the first statement is proven by a reduction from the W[2]-hard SET COVER problem, while the second statement is proven by a reduction from the W[1]-hard INDEPENDENT SET problem.

Theorem 6. *The following holds.*

1. EFFECTORS, *parameterized by the maximum number b of effectors, is* W[2]-*hard, even if G is a deterministic $(r = 0)$ DAG and all nodes are target nodes $(A = V)$.*
2. EFFECTORS, *parameterized by the cost c, is* W[1]-*hard, even if G is a deterministic $(r = 0)$ DAG and all nodes are target nodes $(A = V)$.*
3. *If $r = 0$ and $A = V$, then* EFFECTORS *can be solved in $O(\binom{b+c}{b} \cdot (n+m))$ time.*

Treewidth as a Parameter. As EFFECTORS is in general not polynomial-time solvable (unless P = NP (Theorem 4)), but polynomial-time solvable on trees, it is natural to consider the treewidth of the underlying undirected graph as a parameter. Indeed, treewidth is a well-known concept in algorithmic graph theory. Informally, treewidth measures how "tree-like" a graph is—trees have treewidth one. We note that for deterministic influence graphs ($r = 0$) under the influence maximization model ($A = V$), EFFECTORS corresponds to a special case of a related problem, namely TARGET SET SELECTION (with constant thresholds), for which fixed-parameter tractability for the parameter treewidth is already known [3]. It is basically straightforward—but tedious and technical—to extend this algorithm to the case where some nodes are non-targets ($A \subsetneq V$). We conjecture that, for influence graphs with $r > 0$ probabilistic arcs, the problem is still fixed-parameter tractable for the combined parameter treewidth and r. The most challenging open question is whether EFFECTORS is fixed-parameter tractable when parameterized by the treewidth, even with an unbounded number of probabilistic arcs.

Results in Contradiction with Lappas et al. [10]. The following two claims from the literature are contradicted by the results presented in this paper.

According to Lappas et al. [10, Lemma 1], in the INFLUENCE MAXIMIZATION case with $c = 0$, EFFECTORS is NP-complete. The reduction is incorrect: it uses a target node ℓ which influences all other vertices with probability 1 (in at most two steps). It suffices to select ℓ as an effector in order to activate all vertices, so such instances always have a trivial solution ($X = \{\ell\}$), and the reduction collapses. On the contrary, we prove in Lemma 1 that all instances with $c = 0$ can be solved in linear time.

According to the discussion of Lappas et al. [10] following their Corollary 1, there exists a polynomial-time algorithm for EFFECTORS with deterministic instances (with $r = 0$). Note that the selection model corresponds to our own model in the case of INFLUENCE MAXIMIZATION. However, the given algorithm is flawed: it does not consider the influence *between* different strongly connected components. Indeed, as we prove in Theorem 6, finding effectors under the deterministic model is NP-hard, even in the case of INFLUENCE MAXIMIZATION.

5   Conclusion

We leave several challenges for future research. First, it remains to (dis)prove that Conjecture 1 also holds for arbitrary directed graphs. Further, we have made some unproven claims about (fixed-parameter) tractability when restricting EFFECTORS to directed graphs whose underlying graphs have bounded treewidth. Two more general directions could be to extend our results concerning the parameter "degree of randomness" to other probabilistic diffusion models or to make the considered probabilistic information-propagation problems more tractable by developing simpler (and better to analyze) "linearized models"—the non-linearity in computing the activation probabilities of nodes appears to be an important cause for computational hardness.

References

1. Ahuja, R.K., Magnanti, T.L., Orlin, J.B.: Network Flows: Theory, Algorithms, and Applications. Prentice Hall, Upper Saddle River (1993)
2. Askalidis, G., Berry, R.A., Subramanian, V.G.: Explaining snapshots of network diffusions: structural and hardness results. In: Cai, Z., Zelikovsky, A., Bourgeois, A. (eds.) COCOON 2014. LNCS, vol. 8591, pp. 616–625. Springer, Heidelberg (2014)
3. Ben-Zwi, O., Hermelin, D., Lokshtanov, D., Newman, I.: Treewidth governs the complexity of target set selection. Discrete Optim. **8**(1), 87–96 (2011)
4. Bharathi, S., Kempe, D., Salek, M.: Competitive influence maximization in social networks. In: Deng, X., Graham, F.C. (eds.) WINE 2007. LNCS, vol. 4858, pp. 306–311. Springer, Heidelberg (2007)
5. Chopin, M., Nichterlein, A., Niedermeier, R., Weller, M.: Constant thresholds can make target set selection tractable. Theory Comput. Syst. **55**(1), 61–83 (2014)
6. Domingos, P., Richardson, M.: Mining the network value of customers. In: Proceedings of the Seventh ACM SIGKDD International Conference on Knowledge Discovery and Data Mining, pp. 57–66. ACM (2001)
7. Downey, R.G., Fellows, M.R.: Fundamentals of Parameterized Complexity. Springer, London (2013)
8. Flum, J., Grohe, M.: Parameterized Complexity Theory. Springer, Heidelberg (2006)
9. Kempe, D., Kleinberg, J., Tardos, É.: Maximizing the spread of influence through a social network. In: Proceedings of the Ninth ACM SIGKDD International Conference on Knowledge Discovery and Data Mining, pp. 137–146. ACM (2003)
10. Lappas, T., Terzi, E., Gunopulos, D., Mannila, H.: Finding effectors in social networks. In: Proceedings of the 16th ACM SIGKDD International Conference on Knowledge Discovery and Data Mining, pp. 1059–1068. ACM (2010)
11. Nichterlein, A., Niedermeier, R., Uhlmann, J., Weller, M.: On tractable cases of target set selection. Soc. Netw. Anal. Min. **3**(2), 233–256 (2013)
12. Niedermeier, R.: Invitation to Fixed-Parameter Algorithms. Oxford University Press, New York (2006)
13. Wang, C., Chen, W., Wang, Y.: Scalable influence maximization for independent cascade model in large-scale social networks. Data Min. Knowl. Disc. **25**(3), 545–576 (2012)

Common Developments of Three Incongruent Boxes of Area 30

Dawei Xu[1], Takashi Horiyama[2]([✉]), Toshihiro Shirakawa[1,2],
and Ryuhei Uehara[1]

[1] School of Information Science,
Japan Advanced Institute of Science and Technology, Nomi, Japan
{xudawei,uehara}@jaist.ac.jp
[2] Information Technology Center, Saitama University, Saitama, Japan
horiyama@al.ics.saitama-u.ac.jp

Abstract. We investigate common developments that can fold into plural incongruent orthogonal boxes. Recently, it was shown that there are infinitely many orthogonal polygons that folds into three boxes of different size. However, the smallest one that folds into three boxes consists of 532 unit squares. From the necessary condition, the smallest possible surface area that can fold into two boxes is 22, which admits to fold into two boxes of size $1 \times 1 \times 5$ and $1 \times 2 \times 3$. On the other hand, the smallest possible surface area for three different boxes is 46, which may admit to fold into three boxes of size $1 \times 1 \times 11$, $1 \times 2 \times 7$, and $1 \times 3 \times 5$. For the area 22, it has been shown that there are 2,263 common developments of two boxes by exhaustive search. However, the area 46 is too huge for search. In this paper, we focus on the polygons of area 30, which is the second smallest area of two boxes that admits to fold into two boxes of size $1 \times 1 \times 7$ and $1 \times 3 \times 3$. Moreover, when we admit to fold along diagonal lines of rectangles of size 1×2, the area may admit to fold into a box of size $\sqrt{5} \times \sqrt{5} \times \sqrt{5}$. That is, the area 30 is the smallest candidate area for folding three different boxes in this manner. We perform two algorithms. The first algorithm is based on ZDDs, zero-suppressed binary decision diagrams, and it computes in 10.2 days on a usual desktop computer. The second algorithm performs exhaustive search, however, straightforward implementation cannot be run even on a supercomputer since it causes memory overflow. Using a hybrid search of DFS and BFS, it completes its computation in 3 months on a supercomputer. As results, we obtain (1) 1,080 common developments of two boxes of size $1 \times 1 \times 7$ and $1 \times 3 \times 3$, and (2) 9 common developments of three boxes of size $1 \times 1 \times 7$, $1 \times 3 \times 3$, and $\sqrt{5} \times \sqrt{5} \times \sqrt{5}$.

1 Introduction

Since Lubiw and O'Rourke posed the problem in 1996 [10], polygons that can fold into a (convex) polyhedron have been investigated in the area of computational geometry. In general, we can state the development/folding problem as follows:

© Springer International Publishing Switzerland 2015
R. Jain et al. (Eds.): TAMC 2015, LNCS 9076, pp. 236–247, 2015.
DOI: 10.1007/978-3-319-17142-5_21

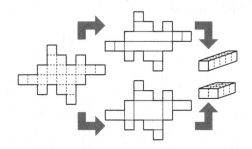

Fig. 1. Cubigami.

Fig. 2. A polygon folding into two boxes of size $1 \times 1 \times 5$ and $1 \times 2 \times 3$ in [12].

Input : A polygon P and a polyhedra Q
Output: Determine whether P can fold into Q or not

When Q is a tetramonohedron (a tetrahedron with four congruent triangular faces), Akiyama and Nara gave a complete characterization of P by using the notion of tiling [2,3]. Except that, we have quite a few results from the mathematical viewpoint. Hence we can tackle this problem from the viewpoint of computational geometry and algorithms.

From the viewpoint of computation, one natural restriction is that considering the orthogonal polygons and polyhedra which consist of unit squares and unit cubes, respectively. Such polygons have wide applications including packaging and puzzles, and some related results can be found in the books on geometric folding algorithms by Demaine and O'Rourke [6,14]. However, this problem is counterintuitive. For example, the puzzle "cubigami" (Fig. 1) is a common development of all tetracubes except one (since the last one has surface area 16, while the others have surface area 18), which is developed by Miller and Knuth. One of the many interesting problems in this area asks whether there exists a polygon that folds into plural incongruent orthogonal boxes. This folding problem is very natural but still counterintuitive; for a given polygon that consists of unit squares, and the problem asks are there two or more ways to fold it into simple convex orthogonal polyhedra (Fig. 2). Biedl et al. first gave two polygons that fold into two incongruent orthogonal boxes [5] (see also Fig. 25.53 in the book by Demaine and O'Rourke [6]). Later, Mitani and Uehara constructed infinite families of orthogonal polygons that fold into two incongruent orthogonal boxes [12]. Recently, Shirakawa and Uehara extended the result to three boxes in a nontrivial way; that is, they showed infinite families of orthogonal polygons that fold into three incongruent orthogonal boxes [16]. However, the smallest polygon by their method contains 532 unit squares, and it is open if there exists much smaller polygon of several dozens of squares that folds into three (or more) different boxes.

Table 1. A part of possible size $a \times b \times c$ of boxes and its common surface area $2(ab + bc + ca)$.

$2(ab + bc + ca)$	$a \times b \times c$
22	$1 \times 1 \times 5$, $1 \times 2 \times 3$
30	$1 \times 1 \times 7$, $1 \times 3 \times 3$
34	$1 \times 1 \times 8$, $1 \times 2 \times 5$
38	$1 \times 1 \times 9$, $1 \times 3 \times 4$
46	$1 \times 1 \times 11$, $1 \times 2 \times 7$, $1 \times 3 \times 5$
54	$1 \times 1 \times 13$, $1 \times 3 \times 6$, $3 \times 3 \times 3$
58	$1 \times 1 \times 14$, $1 \times 2 \times 9$, $1 \times 4 \times 5$
62	$1 \times 1 \times 15$, $1 \times 3 \times 7$, $2 \times 3 \times 5$
64	$1 \times 2 \times 10$, $2 \times 2 \times 7$, $2 \times 4 \times 4$
70	$1 \times 1 \times 17$, $1 \times 2 \times 11$, $1 \times 3 \times 8$, $1 \times 5 \times 5$
88	$1 \times 2 \times 14$, $1 \times 4 \times 8$, $2 \times 2 \times 10$, $2 \times 4 \times 6$

It is easy to see that two boxes of size $a \times b \times c$ and $a' \times b' \times c'$ can have a common development only if they have the same surface area, i.e., when $2(ab + bc + ca) = 2(a'b' + b'c' + c'a')$ holds. We can compute small surface areas that admit to fold into two or more boxes by a simple exhaustive search. We show a part of the table for $1 \leq a \leq b \leq c \leq 50$ in Table 1. From the table, we can say that the smallest surface area is at least 22 to have a common development of two boxes, and their sizes are $1 \times 1 \times 5$ and $1 \times 2 \times 3$. In fact, Abel et al. have confirmed that there exist 2,263 common developments of two boxes of size $1 \times 1 \times 5$ and $1 \times 2 \times 3$ [1]. On the other hand, the smallest surface area that may admit to fold into three boxes is 46, which may fold into three boxes of size $1 \times 1 \times 11$, $1 \times 2 \times 7$, and $1 \times 3 \times 5$. However, the number of polygons of area 46 seems to be too huge to search. This number is strongly related to the enumeration and counting of polyominoes, namely, orthogonal polygons that consist of unit squares [7]. The number of polyominoes of area n is well investigated in the puzzle society, but it is known up to $n = 45$, which is given by the third author (see the OEIS (https://oeis.org/A000105) for the references). Since their common area consists of 46 unit squares, it seems to be hard to enumerate all common developments of three boxes of size $1 \times 1 \times 11$, $1 \times 2 \times 7$, and $1 \times 3 \times 5$.

One natural step is the next one of the surface area 22 in Table 1. The next area of 22 in the table is 30, which admits to fold into two boxes of size $1 \times 1 \times 7$ and $1 \times 3 \times 3$. When Abel et al. had confirmed the area 22 in 2011, it takes around 10 h. Thus we cannot use the straightforward way in [1] for the area 30. We first employ a nontrivial extention of the method based on a zero-suppressed binary decision diagram (ZDD) used in [4], which is so-called frontier-based search algorithm for enumeration [9]. Our first algorithm based on ZDD runs in around 10 days on an ordinary PC. To perform double-check, we also use supercomputer (CRAY XC30). We note that we cannot use the same way as one for area 22 shown in [1] since it takes too huge memory even on a supercomputer. Therefore, we use

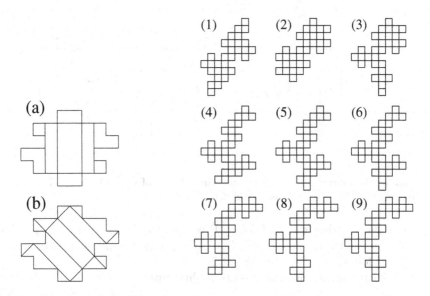

Fig. 3. The common development shown in [5]. (a) It folds into a box of size $1 \times 2 \times 4$ and (b) it also folds into a box of size $\sqrt{2} \times \sqrt{2} \times 3\sqrt{2}$.

Fig. 4. Nine polygons that fold into three boxes of size $1 \times 1 \times 7$, $1 \times 3 \times 3$, and $\sqrt{5} \times \sqrt{5} \times \sqrt{5}$. The last one can fold into the third box in two different ways (Fig. 5).

a hybrid search of the breadth first search and the depth first search. Our first result is the number of common developments of two boxes of size $1 \times 1 \times 7$ and $1 \times 3 \times 3$, which is 1,080.

Based on the obtained common developments, we next change the scheme. In [5], they also considered folding along 45° lines, and showed that there was a polygon that folded into two boxes of size $1 \times 2 \times 4$ and $\sqrt{2} \times \sqrt{2} \times 3\sqrt{2}$ (Fig. 3). In this context, we can observe that the area 30 may admit to fold into another box of size $\sqrt{5} \times \sqrt{5} \times \sqrt{5}$ by folding along the diagonal lines of rectangles of size 1×2. This idea leads us to the problem that asks if there exist common developments of three boxes of size $1 \times 1 \times 7$, $1 \times 3 \times 3$, and $\sqrt{5} \times \sqrt{5} \times \sqrt{5}$ among the common developments of two boxes of size $1 \times 1 \times 7$ and $1 \times 3 \times 3$.

We remark that this is a special case of the development/folding problem above. In our case, P is one of the 1,080 polygons that consist of 30 unit squares, and Q is the cube of size $\sqrt{5} \times \sqrt{5} \times \sqrt{5}$. We note that we can use a pseudopolynomial time algorithm for Alexandrov's Theorem proposed in [8], however, it runs in $O(n^{456.5})$ time, and it is not practical. Therefore, we develop the other efficient algorithm specialized in our case that checks if a polyomino P of area 30 can fold into a cube Q of size $\sqrt{5} \times \sqrt{5} \times \sqrt{5}$. Using the algorithm, we check if these common developments of two boxes of size $1 \times 1 \times 7$ and $1 \times 3 \times 3$ can also fold into the third box of size $\sqrt{5} \times \sqrt{5} \times \sqrt{5}$, and give an affirmative answer. We find that nine of 1,080 common developments of two boxes can fold into the third box (Fig. 4). Moreover, one of the nine common developments of three

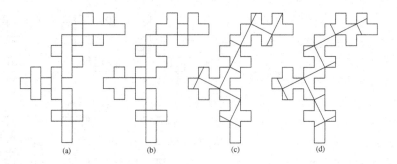

Fig. 5. The unique polygon folds into three boxes of size (a) $1 \times 1 \times 7$, (b) $1 \times 3 \times 3$, and (c)(d) $\sqrt{5} \times \sqrt{5} \times \sqrt{5}$ in four different ways.

boxes has another way of folding. Precisely, the last one (Fig. 4(9)) admits to fold into the third box of size $\sqrt{5} \times \sqrt{5} \times \sqrt{5}$ in two different ways. These four ways of folding are depicted in Fig. 5.

We summarize the main results in this paper:

Theorem 1. *(1) There are 1,080 polyominoes of area 30 that admit to fold (along the edges of unit squares) into two boxes of size $1 \times 1 \times 7$ and $1 \times 3 \times 3$. (2) Among the above 1,080, nine polyominoes can fold into the third box of size $\sqrt{5} \times \sqrt{5} \times \sqrt{5}$ if we admit to fold along diagonal lines (Fig. 4). (3) Among these nine polyominoes, one can fold into the third box in two different ways (Fig. 5).*

2   Preliminaries

2.1   Problem Definitions

Demaine and O'Rourke [6, Chap. 21] give a formal definition of the *development* of a polyhedron as the *net*[1]. Briefly, the development is the unfolding obtained by slicing the surface of the polyhedron, and it forms a single connected simple polygon without self-overlap. The *common development* of two (or more) polyhedra is the development that can fold into both (or all) of them. We only consider connected orthogonal polygons that consist of unit squares, which are called *polyominoes* [7], as developments. Polyominoes obtained from a development by removing some unit squares are called *partial developments* of it. We call a convex orthogonal polyhedron (folded from a polyomino) a *box*.

The cut edges of an edge development of a convex polyhedron form a spanning tree of the 1-skeleton (i.e., the graph formed by the vertices and the edges) of the polyhedron (See e.g., [6, Lemma 22.1.1]). Figure 6(a) and (b) are the 1-skeleton of a cube and its spanning tree, respectively. In our problem, given a box of size $a \times b \times c$, we divide the faces into unit squares, and cut the surface along edges of the unit squares. We call such a development a *unit square development*.

---

[1] Since the word "net" has several meaning, we use "development" instead of it to make clear.

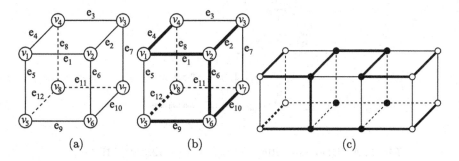

Fig. 6. 1-skeletons and spanning trees of a cube and a box of size $1 \times 1 \times 3$.

In Fig. 6(c), we regard the eight vertices (colored in white) as special, where the angle sum at each corner is 270°. We call them *corners*. The 1-skeleton of a box is given as $G = (V_c \cup V_o, E)$, where V_c and V_o denote the sets of eight corners and others, respectively, and E denote the set of edges of unit length. The cut edges of a unit square development form a tree spanning to the eight corners.

Now, we go back to the common development. It is easy to see that two boxes of size $a \times b \times c$ and size $a' \times b' \times c'$ have a common unit square development only if they have the same surface area, i.e., $2(ab+bc+ca) = 2(a'b'+b'c'+c'a')$. Such 3-tuples (a, b, c) can be computed by a simple enumeration for small areas (Table 1), but it seems that we have many corresponding 3-tuples for large area. In fact, this intuition can be proved as follows:

Theorem 2 [13]. *We say two 3-tuples (a, b, c) and (a', b', c') are* distinct *if and only if $a \neq a'$, $b \neq b'$, or $c \neq c'$. For any positive integer p, there are p distinct 3-tuples (a_i, b_i, c_i) for $i = 1, 2, \ldots, p$ such that $a_i b_i + b_i c_i + c_i a_i = a_j b_j + b_j c_j + c_j a_j$ for any $1 \leq i, j \leq p$.*

Proof. For a given p, we let $a_i = 2^i - 1$, $b_i = 2^{2p-i} - 1$, $c_i = 1$ for $i = 1, 2, \ldots, p$. Then we have $a_i b_i + b_i c_i + c_i a_i = (2^{2p} - 2^i - 2^{2p-i} + 1) + (2^{2p-i} - 1) + (2^i - 1) = 2^{2p} - 1$ for any i. It is easy to see that all 3-tuples (a_i, b_i, c_i) are distinct. Thus we have the theorem.   □

By Theorem 2, we can consider any number of boxes that may share the common developments.

2.2   Enumeration by Zero-Suppressed Binary Decision Diagrams

A *zero-suppressed binary decision diagram (ZDD)* [11] is directed acyclic graph that represents a family of sets. As illustrated in Fig. 7, it has the unique source node[2], called *the root node*, and has two sink nodes 0 and 1, called *the 0-node* and *the 1-node*, respectively (which are together called the constant nodes). Each of the other nodes is labeled by one of the variables $x_1, x_2, \ldots, x_n$, and has exactly two outgoing edges, called *0-edge* and *1-edge*, respectively. On every path from the root node to a constant node in a ZDD, each variable appears at most once in the same order. The size of a ZDD is the number of nodes in it.

---

[2] We distinguish *nodes* of a ZDD from *vertices* of a graph (or a 1-skeleton).

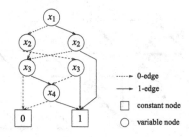

Fig. 7. A ZDD representing $\{\{1,2\},\{1,3,4\},\{2,3,4\},\{3\},\{4\}\}$.

Every node v of a ZDD represents a family of sets $\mathcal{F}_v$, defined by the subgraph consisting of those edges and nodes reachable from v. If node v is the 1-node (respectively, 0-node), $\mathcal{F}_v$ equals to $\{\{\}\}$ (respectively, $\{\}$). Otherwise, $\mathcal{F}_v$ is defined as $\mathcal{F}_{0\text{-}succ(v)} \cup \{S \mid S = \{var(v)\} \cup S', S' \in \mathcal{F}_{1\text{-}succ(v)}\}$, where $0\text{-}succ(v)$ and $1\text{-}succ(v)$, respectively, denote the nodes pointed by the 0-edge and the 1-edge from node v, and $var(v)$ denotes the label of node v. The family $\mathcal{F}$ of sets represented by a ZDD is the one represented by the root node. Figure 7 is a ZDD representing $\mathcal{F} = \{\{1,2\},\{1,3,4\},\{2,3,4\},\{3\},\{4\}\}$. Each path from the root node to the 1-node, called *1-path*, corresponds to one of the sets in $\mathcal{F}$.

Now, we focus on the enumeration of developments by ZDDs. As denoted in Sect. 2.1, the cut edges of an edge development form a spanning tree of the 1-skeleton (e.g., edges $\{e_1, e_2, e_4, e_7, e_6, e_9, e_{10}\}$ in Fig. 6(b)). This conditions can be interpreted as follows:

Property 1. Given the 1-skeleton $G = (V, E)$ of a polyhedron, the cut edges of its edge development is the set of edges E_d $(\subseteq E)$ satisfying: (1) E_d has no cycle. (2) Subgraph of G induced by E_d has only one connected component. (3) Each vertex in V is adjacent to at least one edge in E_d.

Algorithm 1 [4] gives the frontier-based search [9] to construct a ZDD representing a family of spanning trees. It can be considered as one of DP-like algorithms. Each search node in the algorithm corresponds to a subgraphs of the given graph G. The search begins with node$_{\text{root}}$ (i.e., the root node of the resulting ZDD) corresponding to $(V, \{\})$. In the search, we check whether we can adopt edge e_i or not, in the order of $i = 1, 2, \ldots, m$, where m is the number of edges in G. In Line 4 of Algorithm 1, current search node is $\hat{n}$, and in case $x = 1$ (respectively, $x = 0$), we adopt (respectively, do not adopt) e_i. Search node n' corresponds to the resulting graph, and is pointed by the x-edge of $\hat{n}$ in Line 13.

The key is to share nodes of the constructing ZDD (in Lines 9 and 10) by simple "knowledge" of subgraphs, and not to traverse the same subproblems more than once. Each search node $\hat{n}$ in the algorithm has an array $\hat{n}.comp[]$ as an knowledge, where $\hat{n}.comp[v_j]$ indicates the ID of the connected component v_j belonging to. We can reduce the size of knowledge by maintaining the values of $\hat{n}.comp[]$ just for vertices incident to both a processed and an unprocessed edges. Such set of vertices are called the i-th *frontier* F_i $(\in V)$, which is formally

Algorithm 1. Construct ZDD

Input : Graph $G = (V, E)$ with n vertices and m edges
Output: ZDD representing a family of spanning trees in G

1 $N_1 := \{node_{root}\}$. $N_i := \{\}$ for $i = 2, 3, \ldots, m+1$
2 **for** $i := 1, 2, \ldots m$ **do**
3   **foreach** $\hat{n} \in N_i$ **do**
4   **foreach** $x \in \{0, 1\}$ **do**   // 0-edge and 1-edge
5           $n' := \text{CheckTerminal}(\hat{n}, i, x)$ // returns 0, 1, or nil
6   **if** $n' = nil$ **then**   // n' is neither 0 nor 1
7   Copy $\hat{n}$ to n'
8              $\text{UpdateInfo}(n', i, x)$
9   **if** *there exists* $n'' \in N_{i+1}$ *that is identical to* n' **then**
10                $n' := n''$
11   **else**
12                $N_{i+1} := N_{i+1} \cup \{n'\}$
13   Create the x-edge of $\hat{n}$ and make it point at n'

---

defined as $F_i = (\cup_{j=1,\ldots,i}\, e_j) \cap (\cup_{j=i+1,\ldots,m}\, e_j)$, $F_0 = F_m = \{\}$. We check whether the subgraph corresponding to the search node $\hat{n}$ consists a spanning tree in Procedure CheckTerminal. For more detail, see [9].

3 Algorithms for the First Two Boxes of Size $1 \times 1 \times 7$ and $1 \times 3 \times 3$

3.1 Algorithm Based on ZDDs

We first describe how to obtain all common unit cube developments of two incongruent boxes of sizes $1 \times 1 \times 7$ and $1 \times 3 \times 3$ by ZDDs. The strategy is simple: For each box, we enumerate sets of cut edges corresponding to unit cube developments, and convert them to the shapes of the developments, each of which is represented by a sequence of interior angles of a polyomino. Then, we obtain common developments that appear in both of the two boxes. The important thing is to enumerate the sets of cut edges efficiently. For obtaining unit cube developments, we generalize the algorithm given in Sect. 2.2. Once a ZDD is obtained, each of its 1-paths represents a set of cut edges. By traversing the ZDD, we can obtain 1-paths, and thus obtain the shapes of developments. The difference between the problem in Sect. 2.2 and ours can be seen in Fig. 6(b) and (c). In our problem, faces of our boxes are divided into unit squares, and we need to make a tree spanning to the eight corners, not spanning to all vertices. The cut edges of a unit square development of our box has the following property:

Property 2. Given the 1-skeleton $G = (V_c \cup V_o, E)$ of a box, the cut edges of its unit square development is the set of edges E_d $(\subseteq E)$ satisfying: (1) E_d has no cycle. (2) Subgraph of G induced by E_d has only one connected component of size greater than 1. (3) Each vertex in V_c is adjacent to at least one edge in E_d. (4) No vertex in V_o is adjacent to exactly one edge in E_d.

Conditions (1) and (2) are essentially equivalent to those in Property 1. Condition (3) is to flatten the corners of the box into a plane. Conditions (2) and (3) guarantees that all vertices in V_c are connected. Condition (4) is to avoid a vertex in V_o adjacent to exactly one edge in E_d. (If there exists such an edge, we can eliminate it from E_d.) Conditions (2) and (4) guarantees that all vertices in V_o adjacent to two or more edges are connected to the vertices in V_c. Thus, we have a tree spanning the vertices in V_c.

To check the above conditions, we modify Procedures UpdateInfo and Check-Terminal. For counting the number of adopted edges adjacent to v_j and the size of connected component v_j belonging to, we prepare two arrays $\hat{n}.\mathrm{deg}[\ ]$ and $\hat{n}.\mathrm{size}[\ ]$. In Procedure 3, we initialize $\hat{n}.\mathrm{deg}[v_j] := 0$ (i.e., the number of adopted edges in E_d adjacent to v_j is 0) and $\hat{n}.\mathrm{size}[v_j] := 1$ (i.e., vertex v_j is a singleton) in Line 3. If edge $e_i = (v_{i_1}, v_{i_2})$ is adopted to E_d (i.e., $x = 1$), we update the degrees of v_{i_1} and v_{i_2}, and the size of their connected components in Lines 8, 9 and 12.

In Procedure 2, Condition (1) is checked in Lines 2–4. If vertex v_j leaves from the frontier, we have no chance to adopt its adjacent edges, which means the degree of v_j does not change. Thus, we check Conditions (3) and (4) in Lines 8 and 9, respectively. At the same time, we have no chance to grow the size of v_j's connected components. Thus, we check whether we have two or more connected components in Lines from 14 to 16, and terminate the search if it holds. Otherwise, we have only one connected component, and hence we cannot adopt any edges in the remaining search. Thus, we check Conditions (3) and (4) in Lines from 17 to 22, and returns the result.

3.2   Algorithm Based on Exhaustive Search

Here we describe the exhaustive algorithm for generating all common developments of two boxes of size $1 \times 1 \times 7$ and $1 \times 3 \times 3$. The basic idea is similar to one in [1]: Let L_i be the set of all common partial developments of area i of two boxes. Then L_1 consists of a unit square, and each L_i with $i > 1$ is a subset of polyominoes of size i that can be computed from L_{i-1} by the breadth first search. Each L_i is maintained by a huge hash table, which means that we use $O(\max_i\{i|L_i| + (i-1)|L_{i-1}|\})$ space for the computation of step i.

This simple idea works up to 22 for two boxes of size $1 \times 1 \times 5$ and $1 \times 2 \times 3$ in [1] since the maximum number of $|L_i \cup L_{i-1}|$ takes 1.01×10^7 when $i = 18$. However, for the surface area 30, it does not work even on a supercomputer (CRAY XC30) due to memory overflow when $i = 22$.

Thus we divide the computation into two phases. In the first phase, we compute L_i for each $i = 2, \ldots, 16$. As a result, we have L_{16} that consists of 7,486,799 common partial developments of two boxes of size $1 \times 1 \times 7$ and $1 \times 3 \times 3$. In the second phase, we partition L_{16} into 75 disjoint subsets L_{16}^j with $1 \leq j \leq 75$. For each L_{16}^j, we independently compute up to L_{30}^j in parallel by the BFS algorithm again. In the final step, we merge L_{30}^j with $1 \leq j \leq 75$, remove duplicates, and obtain L_{30}.

Procedure 2. CheckTerminalRevised($\hat{n}, i, x$)

1 Let (v_{i_1}, v_{i_2}) denote $e_i \in E$
2 if $x = 1$ then
3 $\quad$ if $\hat{n}.\mathrm{comp}[v_{i_1}] = \hat{n}.\mathrm{comp}[v_{i_2}]$ then $\quad$ // v_{i_1}, v_{i_2} are in the same component
4 $\quad\quad$ return 0 $\quad\quad\quad\quad\quad\quad\quad\quad\quad\quad$ // we have a cycle by adding e_i

5 Copy $\hat{n}$ to n'
6 UpdateInfo(n', i, x)
7 foreach $v_j \in \{v_{i_1}, v_{i_2}\}$ satisfying $v_j \notin F_i$ do // v_j is leaving from the frontier
8 $\quad$ // Check the degree constraints for v_j
9 $\quad$ if (v_j is in V_c) and ($\hat{n}.\mathrm{deg}[v_j] = 0$) then return 0
10 $\quad$ if (v_j is in V_o) and ($\hat{n}.\mathrm{deg}[v_j] = 1$) then return 0
11 $\quad$ if ($\forall v_k \in F_i$ $\hat{n}.\mathrm{comp}[v_j] \neq \hat{n}.\mathrm{comp}[v_k]$) then
12 $\quad\quad$ // v_j's connected component cannot connect to any other components
13 $\quad\quad$ if ($\hat{n}.\mathrm{size}[v_j] > 1$) then
14 $\quad\quad\quad$ if ($\exists v_\ell \in F_i$ ($\hat{n}.\mathrm{size}[v_\ell] > 1$)) then
15 $\quad\quad\quad\quad$ // we have two or more connected components of size > 1
16 $\quad\quad\quad\quad$ return 0

17 $\quad\quad$ else // We cannot adopt any edges
18 $\quad\quad\quad$ foreach $v_{j'} \in \cup_{i'=i+1,...,m} e_{i'}$ do
19 $\quad\quad\quad\quad$ // Check the degree constraints for remaining vertices
20 $\quad\quad\quad\quad$ if ($v_{j'}$ is in V_c) and ($\hat{n}.\mathrm{deg}[v_{j'}] = 0$) then return 0
21 $\quad\quad\quad\quad$ if ($v_{j'}$ is in V_o) and ($\hat{n}.\mathrm{deg}[v_{j'}] = 1$) then return 0

22 $\quad\quad\quad$ return 1
23

24 $\quad$ $F_i := F_i \setminus \{v_j\}$
25 return *nil*

Procedure 3. UpdateInfoRevised($\hat{n}, i, x$)

1 Let (v_{i_1}, v_{i_2}) denote $e_i \in E$
2 foreach $v_j \in \{v_{i_1}, v_{i_2}\}$ such that $v_j \notin F_{i-1}$ do // v_j is entering the frontier
3 $\quad$ $\hat{n}.\mathrm{comp}[v_j] := j$ $\quad\quad\quad\quad\quad$ // The initial component ID is the index of v_j
4 $\quad$ $\hat{n}.\mathrm{deg}[v_j] := 0$, $\hat{n}.\mathrm{size}[v_j] := 1$

5 if $x = 1$ then // Merge two components of v_{i_1}, v_{i_2}
6 $\quad$ $c_{\min} := \min\{\hat{n}.\mathrm{comp}[v_{i_1}], \hat{n}.\mathrm{comp}[v_{i_2}]\}$
7 $\quad$ $c_{\max} := \max\{\hat{n}.\mathrm{comp}[v_{i_1}], \hat{n}.\mathrm{comp}[v_{i_2}]\}$
8 $\quad$ $\hat{n}.\mathrm{deg}[v_{i_1}] := \hat{n}.\mathrm{deg}[v_{i_1}] + 1$, $\hat{n}.\mathrm{deg}[v_{i_2}] := \hat{n}.\mathrm{deg}[v_{i_2}] + 1$
9 $\quad$ $s = \hat{n}.\mathrm{size}[v_{i_1}] + \hat{n}.\mathrm{size}[v_{i_2}]$
10 $\quad$ foreach $v_j \in F_i$ do
11 $\quad\quad$ if $\hat{n}.\mathrm{comp}[v_j] = c_{\max}$ then $\hat{n}.\mathrm{comp}[v_j] := c_{\min}$
12 $\quad\quad$ if ($\hat{n}.\mathrm{comp}[v_j] = c_{\min}$) or ($\hat{n}.\mathrm{comp}[v_j] = c_{\max}$) then $\hat{n}.\mathrm{size}[v_j] := s$

13 foreach $v_j \in \{v_{i_1}, v_{i_2}\}$ such that $v_j \notin F_i$ do // v_j is leaving the frontier
14 $\quad$ Forget $\hat{n}.\mathrm{comp}[v_j]$, $\hat{n}.\mathrm{deg}[v_j]$ and $\hat{n}.\mathrm{size}[v_{i_2}]$

4 Algorithm for the Third Box

Let L_{30} be the set of all common developments of two boxes of size $1 \times 1 \times 7$ and $1 \times 3 \times 3$. We here note that if we can compute L_{30} efficiently, we can check in the same manner; that is, we generate all developments of the cube of size $\sqrt{5} \times \sqrt{5} \times \sqrt{5}$ by cutting along the line of unit squares, and check if each one appears in L_{30} or not. Thus, in the first method based on ZDDs, we can use the same way again; we construct all developments of the cube of size $\sqrt{5} \times \sqrt{5} \times \sqrt{5}$ based on the connection network on unit squares, and check if each one appears in L_{30} or not. In the second method based on the exhaustive search for two boxes, we check if each development in L_{30} can be folded into a cube of size $\sqrt{5} \times \sqrt{5} \times \sqrt{5}$.

The program of the first method based on ZDDs runs on a usual desktop computer with Intel Xeon E5-2643 and 128 GB memory. It takes 0.10 and 71.53 s for obtaining the sets of cut edges of two boxes of size $1 \times 1 \times 7$ and $1 \times 3 \times 3$, respectively, and 7.7 days for converting the cut edges into the shapes of developments and for obtaining the common developments. For the third box of size $\sqrt{5} \times \sqrt{5} \times \sqrt{5}$, It takes 354.64 s for obtaining cut edges, and 2.5 days for obtaining the common developments of the three boxes. It takes 10.2 days in total. The program of the second method runs, in total, in 3 months on the supercomputer (CRAY XC30), and we obtain 1,080 common developments in L_{30} of two boxes of size $1 \times 1 \times 7$ and $1 \times 3 \times 3^3$ and 9 common developments of three boxes of size $1 \times 1 \times 7$, $1 \times 3 \times 3$ and $\sqrt{5} \times \sqrt{5} \times \sqrt{5}$.

5 Concluding Remarks

Recently, Shirakawa and Uehara showed infinite families of orthogonal polygons that fold into three incongruent orthogonal boxes [16]. However, the smallest polygon contains 532 unit squares. In this paper, we show that there exist orthogonal polygons of 30 unit squares that fold into three incongruent orthogonal boxes if we allow us to fold along slanted lines. In the original framework in [16], the smallest possible surface area that may fold into three different boxes is 46, which may produce three boxes of size $1 \times 1 \times 11$, $1 \times 2 \times 7$, and $1 \times 3 \times 5$. We conjecture that there exists an orthogonal polygon of 46 unit squares that admits to fold these three boxes. Some nontrivial properties in Figs. 4 and 5 may help to find it.

There are many future work in this area. For example, does there exist a polyomino that folds into four or more boxes? Is there some upper bound of the number of boxes which can be folded from one polyomino? We remind that Theorem 2 says that we have no upper bound by the constraint of the surface areas. But it is hard to imagine that one polyomino can fold into, say, 10,000 different boxes. General development/folding problems are also remained open. For example, Shirakawa et al. found a common development of a unit cube and an almost regular tetrahedron (with relative error $< 2.89200 \times 10^{-1796}$) [15], however, a common development of two Platonic solids are still open.

---

[3] We note that the maximum number of partial developments is given when $j = 24$.

References

1. Abel, Z., Demaine, E., Demaine, M., Matsui, H., Rote, G., Uehara, R.: Common development of several different orthogonal boxes. In: 23rd Canadian Conference on Computational Geometry (CCCG 2011), pp. 77–82 (2011)
2. Akiyama, J.: Tile-makers and semi-tile-makers. Math. Assoc. Amerika **114**, 602–609 (2007)
3. Akiyama, J., Nara, C.: Developments of polyhedra using oblique coordinates. J. Indonesia. Math. Soc. **13**(1), 99–114 (2007)
4. Araki, Y., Horiyama, T., Uehara, R.: Common unfolding of regular Tetrahedron and Johnson-Zalgaller solid. In: Rahman, M.S., Tomita, E. (eds.) WALCOM 2015. LNCS, vol. 8973, pp. 294–305. Springer, Heidelberg (2015)
5. Biedl, T., Chan, T., Demaine, E., Demaine, M., Lubiw, A., Munro, J.I., Shallit, J.: Notes from the University of Waterloo Algorithmic Problem Session, 8 September 1999
6. Demaine, E.D., O'Rourke, J.: Geometric Folding Algorithms: Linkages, Origami, Polyhedra. Cambridge University Press, Cambridge (2007)
7. Golomb, S.W.: Polyominoes. Princeton University Press, Princeton (1994)
8. Kane, D., Price, G.N., Demaine, E.D.: A pseudopolynomial algorithm for Alexandrov's theorem. In: Dehne, F., Gavrilova, M., Sack, J.-R., Tóth, C.D. (eds.) WADS 2009. LNCS, vol. 5664, pp. 435–446. Springer, Heidelberg (2009)
9. Kawahara, J., Inoue, T., Iwashita, H., Minato, S.: Frontier-based search for enumerating all constrained subgraphs with compressed representation. Technical report TCS-TR-A-14-76, Division of Computer Science, Hokkaido University (2014)
10. Lubiw, A., O'Rourke, J.: When can a polygon fold to a polytope? Technical report 048. Department of Computer Science, Smith College (1996)
11. Minato, S.: Zero-suppressed bdds for set manipulation in combinatorial problems. In: 30th ACM/IEEE Design Automation Conference (DAC 1993), pp. 272–277 (1993)
12. Mitani, J., Uehara, R.: Polygons folding to plural incongruent orthogonal boxes. In: Canadian Conference on Computational Geometry (CCCG 2008), pp. 39–42 (2008)
13. Okumura, T.: Personal communication, August 2014
14. O'Rourke, J.: How to Fold It: The Mathematics of Linkage, Origami and Polyhedra. Cambridge University Press, Cambridge (2011)
15. Shirakawa, T., Horiyama, T., Uehara, R.: Construct of common development of regular tetrahedron and cube. In: 27th European Workshop on Computational Geometry (EuroCG 2011), pp. 47–50, 28–30 March 2011
16. Shirakawa, T., Uehara, R.: Common developments of three incongruent orthogonal boxes. Int. J. Comput. Geom. Appl. **23**(1), 65–71 (2013)

Finding Connected Dense k-Subgraphs

Xujin Chen$^{(\boxtimes)}$, Xiaodong Hu, and Changjun Wang

Institute of Applied Mathematics, AMSS,
Chinese Academy of Sciences, Beijing 100190, China
{xchen,xdhu,wcj}@amss.ac.cn

Abstract. Given a connected graph G on n vertices and a positive integer $k \leq n$, a subgraph of G on k vertices is called a k-subgraph in G. We design combinatorial approximation algorithms for finding a connected k-subgraph in G such that its density is at least a factor $\Omega(\max\{n^{-2/5}, k^2/n^2\})$ of the density of the densest k-subgraph in G (which is not necessarily connected). These particularly provide the first non-trivial approximations for the densest connected k-subgraph problem on general graphs.

Keywords: Densest k-subgraphs · Connectivity · Combinatorial approximation algorithms

1 Introduction

Let $G = (V, E)$ be a connected simple undirected graph with n vertices, m edges, and nonnegative edge weights. The (*weighted*) *density* of G is defined as its average (weighted) degree. Let $k \leq n$ be a positive integer. A subgraph of G is called a k-*subgraph* if it has exactly k vertices. The *densest k-subgraph problem* (DkSP) is to find a k-subgraph of G that has the maximum density, equivalently, a maximum number of edges. If the k-subgraph requires to be connected, then the problem is referred to as the *densest connected k-subgraph problem* (DCkSP). Both DkSP and DCkSP have their weighted generalizations, denoted respectively as HkSP and HCkSP, which ask for a heaviest (connected) k-subgraph, i.e., a (connected) k-subgraph with a maximum total edge weight. Identifying k-subgraphs with high densities is a useful primitive, which arises in diverse applications – from social networks, to protein interaction graphs, to the world wide web, etc. While dense subgraphs can give valuable information about interactions in these networks, the additional connectivity requirement turns out to be natural in various scenarios.

Related Work. An easy reduction from the maximum clique problem shows that DkSP, DCkSP and their weighted generalizations are all NP-hard in general. The NP-hardness remains even for some very restricted graph classes such as

Research supported in part by by NNSF of China under Grant No. 11222109, 11021161 and 10928102, by 973 Project of China under Grant No. 2011CB80800, and by CAS Program for Cross & Cooperative Team of Science & Technology Innovation.

© Springer International Publishing Switzerland 2015
R. Jain et al. (Eds.): TAMC 2015, LNCS 9076, pp. 248–259, 2015.
DOI: 10.1007/978-3-319-17142-5_22

chordal graphs, triangle-free graphs, comparability graphs and bipartite graphs of maximum degree three.

Most literature on finding dense subgraphs focus on the versions without requiring the subgraphs to be connected. For DkSP and its generalization HkSP, narrowing the large gap between the lower and upper bounds on the approachability is an important open problem.On the negative side, Feige [9] showed that computing a $(1 + \varepsilon)$ approximation for DkSP is at least as hard as refuting random 3-SAT clauses for some $\varepsilon > 0$. Khot [15] showed that there does not exist any polynomial time approximation scheme (PTAS) for DkSP assuming NP does not have randomized algorithms that run in sub-exponential time. Recently, constant factor approximations in polynomial time for DkSP have been ruled out by Raghavendra and Steurel [20] under Unique Games with Small Set Expansion conjecture. On the positive side, considerable efforts have been devoted to finding good quality approximations for HkSP. Improving the $O(n^{0.3885})$-approximation of Kortsarz and Peleg [17], Feige et al. [11] proposed a combinatorial algorithm with approximation ratio $O(n^\delta)$ for some $\delta < 1/3$. The latest algorithm of Bhaskara et al. [4] provides an $O(n^{1/4+\varepsilon})$-approximation in $n^{O(1/\varepsilon)}$ time. If allowed to run for $n^{O(\log n)}$ time, their algorithm guarantees an approximation ratio of $O(n^{1/4})$. The $O(n/k)$-approximation algorithm by Asahiro et al. [3] is remarkable for its simple greedy removal method. Linear and semidefinite programming relaxation approaches have been adopted in [10,13,21] to design randomized rounding algorithms.

For some special graph classes, better approximations have been obtained for DkSP and HkSP. Arora et al. [2] gave a PTAS for the restricted DkSP where $m = \Omega(n^2)$ and $k = \Omega(n)$, or each vertex of G has degree $\Omega(n)$. Demaine et al. [8] developed a 2-approximation algorithm for DkSP on H-minor-free graphs, where H is any given fixed graph. Chen et al. [5] showed that DkSP on a large family of intersection graphs admits constant factor approximations.

The work on approximating densest/heaviest connected k-subgraphs are relatively very limited. To the best of our knowledge, the existing polynomial time algorithms deal only with special graphical topologies, including: (a) 2-approximation for the metric HkSP (HCkSP) [14], where the underlying graph G is complete, and the connectivity is trivial; (b) exact algorithms for HkSP and HCkSP on trees [7], for DkSP and DCkSP on h-trees, cographs and split graphs [7], and for DCkSP on interval graphs whose clique graphs are simple paths [19].

Among the well-known relaxations of DkSP and HkSP is the problem of finding a (connected) subgraph of maximum weighted density that does not have any cardinality constraint. It is strongly polynomial time solvable using max-flow based techniques [12,18]. Andersen and Chellapilla [1] and Khuller and Saha [16] studied two relaxed variants of HkSP for finding a weighted densest subgraph with at least or at most k vertices. The former variant was shown to be NP-hard even in the unweighted case, and admit 2-approximation in the weighted setting. The approximation of the latter variant was proved to be as hard as that of DkSP/HkSP up to a constant factor.

Our Results. Given the interest in finding densest/heaviest connected k-subgraphs from both the theoretical and practical point of view, a better understanding of the problems is an important challenge for the field. In this paper, we design $O(mn \log n)$ time combinatorial approximation algorithms for finding a connected k-subgraph of G whose density (weighted density) is at least a factor $\Omega(\max\{n^{-2/5}, k^2/n^2\})$ $(\Omega(\max\{1/k, k^2/n^2\}))$ of the density (weighted density) of the densest (heaviest) k-subgraph of G which is not necessarily connected. These particularly provide the first non-trivial approximation ratios for DCkSP and HCkSP on general graphs: $O(\min\{n^{2/5}, n^2/k^2\})$ for DCkSP and $O(\min\{k, n^2/k^2\})$ for HCkSP. Note that $\min\{k, n^2/k^2\} \le n^{2/3}$.

To evaluate the quality of our algorithms' performance guarantees $O(n^{2/5})$ and $O(n^{2/3})$, which are compared with the optimums of DkSP and HkSP, we investigate the maximum ratio Λ (resp. Λ_w), over all graphs G (resp. over all graphs G and all nonnegative edge weights), between the maximum density (resp. weighted density) of *all* k-subgraphs and that of *all connected* k-subgraphs in G. The following examples show $\Lambda \ge \frac{1}{3}n^{1/3}$ and $\Lambda_w \ge \frac{1}{2}n^{1/2}$.

Example 1. (a) The graph G is formed from ℓ vertex-disjoint ℓ-cliques $L_1, \ldots, L_\ell$ by adding, for each $i = 1, \ldots, \ell - 1$, a path P_i of length $\ell^2 + 1$ to connect L_i and L_{i+1}, where P_i intersects all the ℓ cliques only at a vertex in L_i and a vertex in L_{i+1}. Let $k = \ell^2$. Note that G has $n = \ell^2 + \ell^2(\ell - 1) = \ell^3$ vertices. The unique densest k-subgraph of G is the disjoint union of $L_1, \ldots, L_\ell$ and has density $\ell - 1$. One of densest connected k-subgraphs of G is induced by the ℓ vertices in L_1 and certain $\ell^2 - \ell$ vertices in P_1, and has density $(\ell(\ell - 1) + 2(\ell^2 - \ell))/\ell^2$. Hence $\Lambda \ge \ell^2/(\ell + 2\ell) = \frac{1}{3}n^{1/3}$.

(b) The graph G is a tree formed from a star on $\ell + 1$ vertices by dividing each edge into a path of length $\ell + 1$. All pendant edges have weight 1 and other edges have weight 0. Let $k = 2\ell$. Note that G has $n = \ell^2 + 1$ vertices. The unique heaviest k-subgraph of G is induced by the ℓ pendant edges of G, and has weighted density 1. Every heaviest connected k-subgraph of G is a path containing exactly one pendant edge of G, and has weighted density $1/\ell$. Hence $\Lambda_w \ge \ell \ge \frac{1}{2}n^{1/2}$.

The remainder of this paper is organized as follows. Section 2 gives notations, definitions and basic properties necessary for our discussion. Section 3 is devoted to designing approximation algorithms for finding connected dense k-subgraphs. Section 4 discusses extension to the weighted case, and future research directions. The omitted details can be found in [6].

2   Preliminaries

Graphs studied in this paper are simple and undirected. For any graph $G' = (V', E')$ and any vertex $v \in V'$, we use $d_{G'}(v)$ to denote v's degree in G'. The *density* $\sigma(G')$ of G' refers to its average degree, i.e., $\sigma(G') = \sum_{v \in V'} d_{G'}(v)/|V'| = 2|E'|/|V'|$. Following convention, we define $|G'| = |V'|$. By a *component* of G' we mean a maximal connected subgraph of G'.

Throughout let $G = (V, E)$ be a connected graph on n vertices and m edges, and let $k \in [3, n]$ be an integer. Our goal is to find a connected k-subgraph C of G such that its density $\sigma(C)$ is as large as possible. Let $\sigma^*(G)$ and $\sigma_k^*(G)$ denote the maximum densities of a subgraph and a k-subgraph of G, respectively, where the subgraphs are not necessarily connected. It is clear that

$$\sigma^*(G) \geq \sigma_k^*(G) \text{ and } n - 1 \geq \sigma(G) \geq k \cdot \sigma_k^*(G)/n. \tag{2.1}$$

Let S be a subset of V or a subgraph of G. We use $G[S]$ to denote the subgraph of G induced by the vertices in S, and use $G \setminus S$ to denote the graph obtained from G by removing all vertices in S and their incident edges. If S consists of a single vertex v, we write $G \setminus v$ instead of $G \setminus \{v\}$.

The vertices whose removals increase the density of the graph play an important role in our algorithm design.

Definition 1. A vertex $v \in V$ is called *removable* in G if $\sigma(G \setminus v) > \sigma(G)$.

Since $\sigma(G \setminus v) = 2(|E| - d_G(v))/(|V| - 1)$, the following lemma is straightforward. It also provides an efficient way for identifying removable vertices.

Lemma 1. *A vertex $v \in V$ is removable in G if and only if $d_G(v) < \sigma(G)/2$.* □

Lemma 2. *Let G_1 be a connected k-subgraph of G. For any connected subgraph G_2 of G_1, it holds that $\sigma(G_1) \geq \sigma(G_2)/\sqrt{k}$.*

Proof. Suppose that G_2 is a k_2-subgraph of G with m_2 edges. By the definition of density, $\sigma(G_2) \leq k_2 - 1$. The connectivity of G_1 implies $|E(G_1)| \geq |E(G_2)| + |V(G_1 \setminus G_2)|$, and

$$\sigma(G_1) \geq \frac{2(m_2 + k - k_2)}{k} = \frac{k_2 \cdot \sigma(G_2) + 2(k - k_2)}{k}.$$

In case of $k_2 \geq \sqrt{k}$, we have $\sigma(G_1) \geq k_2 \cdot \sigma(G_2)/k \geq \sigma(G_2)/\sqrt{k}$. In case of $k_2 < \sqrt{k}$, since $k \geq 3$, it follows that G_1 has no isolated vertices, and $\sigma(G_1) \geq 1 > k_2/\sqrt{k} > \sigma(G_2)/\sqrt{k}$. □

For a cut-vertex v of G, we use G_v to denote a densest component of $G \setminus v$, and use G_{v+} to denote the connected subgraph of G induced by $V(G_v) \cup \{v\}$. Note that $G \setminus G_v$ is a connected subgraph of G.

3 Algorithms

We design an $O(n^2/k^2)$-approximation algorithm (in Sect. 3.1) and further an $O(n^{2/5})$-approximation algorithm (in Sect. 3.2) for DkSP that always finds a connected k-subgraph of G. For ease of description we assume k is even. The case of odd k can be treated similarly. Alternatively, if k is odd, we can first find a connected $(k - 1)$-subgraph G_1 satisfying $\sigma_{k-1}^*(G)/\sigma(G_1) \leq O(\alpha)$, where $\alpha \in \{n^2/k^2, n^{2/5}\}$. Notice that $\sigma_k^*(G) \leq 3 \cdot \sigma_{k-1}^*(G)$ [6]. It follows that $\sigma_k^*(G)/\sigma(G_1) \leq O(\alpha)$. Then we attach an appropriate vertex to G_1, making a connected k-subgraph G_2 with density $\sigma(G_2) \geq \frac{k-1}{k}\sigma(G_1) \geq \frac{2}{3}\sigma(G_1)$. This guarantees that the approximation ratio is still $\sigma_k^*(G)/\sigma(G_2) \leq O(\alpha)$.

252	X. Chen et al.

3.1 $O(n^2/k^2)$-Approximation

We first give an outline of our algorithm (see Algorithm 1) for finding a connected k-subgraph C of G with density $\sigma(C) \geq \Omega(k^2/n^2) \cdot \sigma_k^*(G)$ (see Theorem 1).

Outline. We start with a connected graph $G' \leftarrow G$ and repeatedly delete removable vertices from G' to increase its density without destroying its connectivity.

– If we can reach G' with $|G'| = k$ in this way, we output C as the resulting G'.
– If we can find a removable cut-vertex r in G' such that $|G'_r| \geq k$, then we recurse with $G' \leftarrow G'_r$.
– If we stop at a G' without any removable vertices, then we construct C from an arbitrary connected $(k/2)$-subgraph by greedily attaching $k/2$ more vertices (see Procedure 1).
– If we are in none of the above three cases, we find a connected subgraph of G' induced by a set S of at most $k/2$ vertices, and then expand the subgraph in two ways: (1) attaching G'_r for all removable vertices r of G' which are contained in S, and (2) greedily attaching no more than $k/2$ vertices. From the resulting connected subgraphs, we choose the one that has more edges (breaking ties arbitrarily), and further expand it to be a connected k-subgraph (see Procedure 2), which is returned as the output C.

Greedy Attachment. We describe how the greedy attaching mentioned in the above outline proceeds. Let S and T be disjoint nonempty vertex subsets (or subgraphs) of G. Note that $1 \leq |S| < n$. The set of edges of G with one end in S and the other in T is written as $[S, T]$. For any positive integer $j \leq n - |S|$, a set S^* of j vertices in $G \setminus S$ with *maximum* $|[S, S^*]|$ can be found greedily by sorting the vertices in $G \setminus S$ as $v_1, v_2, \ldots, v_j, \ldots$ in a non-increasing order of the number of neighbors they have in S. For each $i = 1, 2 \ldots, j$, it can be guaranteed that v_i has either a neighbor in S or a neighbor in $\{v_1, v_2 \ldots, v_{i-1}\}$; in the latter case $i \geq 2$. Setting $S^* = \{v_1, v_2, \ldots, v_j\}$. It is easy to see that

$$|[S, S^*]| \geq \tfrac{j}{n} \cdot |[S, G \setminus S]|. \tag{3.1}$$

Moreover, if $G[S]$ is connected, the choices of v_i's guarantee that $G[S \cup S^*]$ is connected. We refer to this S^* as a *j-attachment* of S in G. Given S, finding a j-attachment of S takes $O(m+n \log n)$ time, which implies the following procedure runs in $O(|E(G')| + |G'| \cdot \log |G'|)$ time.

Procedure 1. Input: a connected graph G' without removable vertices, where $|G'| > k$. Output: a connected k-subgraph of G', written as $\text{PRC1}(G')$.

---
1. $G_1 = (V_1, E_1) \leftarrow$ an arbitrary connected $(k/2)$-subgraph of G'
2. $V_1^* \leftarrow$ a $(k/2)$-attachment of V_1 in G'
3. Output $\text{PRC1}(G') \leftarrow G[V_1 \cup V_1^*]$

Note that the definition of attachment guarantees that $V_1 \cap V_1^* = \emptyset$, $|[V_1, V_1^*]|$ is maximum, and $G[V_1 \cup V_1^*]$ is connected.

Lemma 3. $\sigma(\text{PRC1}(G')) \geq \frac{k}{4|G'|} \cdot \sigma(G')$.

Proof. Since G' has no removable vertices, we deduce from Lemma 1 that every vertex of G' has degree at least $\sigma(G')/2$. Therefore $|[G_1, G' \setminus G_1]| \geq \frac{k}{2} \cdot \frac{\sigma(G')}{2} - 2|E_1|$. Recalling (3.1), we see that the number of edges in $\text{PRC1}(G')$ is at least $|[V_1, V_1^\star]| \geq (\frac{k \cdot \sigma(G')}{4} - 2|E_1|) \cdot \frac{k/2}{|G'|} + |E_1| \geq \frac{k^2}{8|G'|} \cdot \sigma(G')$, proving the lemma. $\square$

Procedure 2. Input: a connected graph G' with $|G'| > k$, where every removable vertex r is a cut-vertex and satisfies $|G'_r| < k$.   Output: a connected k-subgraph of G', written as $\text{PRC2}(G')$.

---

1. $H \leftarrow G'$, $R' \leftarrow R = $ the set of removable vertices of G'
2. **While** $R' \neq \emptyset$ **do**
3.   Take $r \in R'$
4. $H \leftarrow H \setminus V(G'_r)$, $R' \leftarrow R' \setminus V(G'_{r+})$
5. **End-While**
6. For each $v \in V(H)$, define $\theta(v) = |G'_{v+}|$ if $v \in R$, and $\theta(v) = 1$ otherwise
7. Let S be a *minimal* subset of $V(H)$ s.t. $H[S]$ is connected & $\sum_{v \in S} \theta(v) \geq \frac{k}{2}$
8. Let S^* be a $\min\{k/2, |H \setminus S|\}$-attachment of S in H
9. $V_1 \leftarrow S \cup (\cup_{r \in R \cap S} V(G'_r))$, $V_2 \leftarrow S \cup S^\star$
10. Let H' be one of $G'[V_1]$ and $G'[V_2]$ whichever has more edges (break ties arbitrarily)
11. Expand H' to be a connected k-subgraph of G'
12. Output $\text{PRC2}(G') \leftarrow H'$

Under the condition that the resulting graph is connected, the expansion in Step 11 can be done in an arbitrary way. It is easy to see that Procedure 2 runs in $O(|G'| \cdot |E(G')|)$ time.

Lemma 4. *At the end of the while-loop (Step 5) in Procedure 2, we have*

(i) *H is a connected subgraph of G'.*

(ii) *If H contains two distinct vertices r and s that are removable in G', then (by the condition of the procedure both r and s are cut-vertices of G', and moreover) G'_r and G'_s are vertex-disjoint.*

Proof. Note that in every execution of the while-loop, $r \in R'$ is a cut-vertex of H, and $V(H) \cap V(G'_r)$ induces a component of $H \setminus r$. Thus H is connected throughout the procedure. For any two removable vertices r, s of G' with $|G'_r| \leq |G'_s|$ and $r, s \in V(H)$, if G'_r and G'_s are not vertex-disjoint, then $V(G'_r) \cup \{r\} \subseteq V(G'_s)$. It follows that all vertices of $V(G'_r) \cup \{r\}$ have been removed by Step 4 delete when considering $s \in R'$, a contradiction. $\square$

Observe that for any two distinct $r, s \in R$, either G'_{r+} and G'_{s+} are vertex-disjoint, or G'_{r+} contains G'_{s+}, or G'_{s+} contains G'_{r+}. This fact, along with an inductive argument, shows that, throughout Procedure 2, for any $s \in R \setminus V(H)$, there exists at least a vertex $r \in V(H) \cap R$ such that G'_{r+} contains G'_{s+}, implying that $(U_{r \in R \cap V(H)} V(G_{r+})) \cup (V(H) \backslash R) = V(G')$ holds always. By Lemma 4(ii), in Step 7, we see that $V(G')$ is the disjoint union of $V(G_{r+})$, $r \in R \cap V(H)$ and $V(H) \backslash R$, giving $\sum_{v \in V(H)} \theta(v) = |G'| > k$. Hence, the connectivity of H (Lemma 4 (i)) implies that the set S in Step 7 does exist.

Take $u \in S$ such that u is not a cut-vertex of H. If $|S| \geq (k/2) + 1$, then we have $\sum_{v \in S \backslash \{u\}} \theta(v) \geq |S \backslash \{u\}| \geq k/2$, a contradiction to the minimality of S. Hence $|S| \leq k/2$.

Since Step 4 has removed from H all vertices in $V(G'_r)$ for all $r \in R$, we see that V_1 is the disjoint union of S and $\cup_{r \in R \cap S} V(G'_r)$ Recall that $|G'_r| < k$ for all $r \in R \cap S$. If $|V_1| > k$, then $|S| \geq 2$, and either $\theta_u \geq k/2$ or $\sum_{v \in S \backslash \{u\}} \theta(v) \geq k/2$, contradicting to the minimality of S. Noting that $|V_1| = \sum_{v \in S} \theta(v)$, we have

$$k/2 \leq |V_1| \leq k. \tag{3.2}$$

We deduce that the output of Procedure 2 is indeed a connected k-subgraph of G'.

Algorithm 1. Input: connected graph $G = (V, E)$ with $|V| \geq k$.
Output: a connected k-subgraph of G, written as $\text{ALG1}(G)$.

---

1. $G' \leftarrow G$
2. **While** $|G'| > k$ and G' has a removable vertex r that is not a cut-vertex **do**
3. $G' \leftarrow G' \backslash r$
4. **End-While**   // either $|G'| = k$ or any removable vertex of G' is a cut-vertex
5. **If** $|G'| = k$ **then** output $\text{ALG1}(G) \leftarrow G'$
6. **If** $|G'| > k$ and G' has no removable vertices
    then output $\text{ALG1}(G) \leftarrow \text{PRC1}(G')$
7. **If** $|G'| > k$ and $|G'_r| < k$ for each removable vertex r of G'
    then output $\text{ALG1}(G) \leftarrow \text{PRC2}(G')$
8. **If** $|G'| > k$ and $|G'_r| \geq k$ for some removable vertex r of G'
    then output $\text{ALG1}(G) \leftarrow \text{ALG1}(G'_r)$

In the while-loop, we repeatedly delete removable non-cut vertices from G' until $|G'| = k$ or G' has no removable non-cut vertex anymore. The deletion process keeps G' connected, and its density $\sigma(G')$ increasing (cf. Definition 1). When the deletion process finishes, there are four possible cases, which are handled by Steps 5, 6, 7 and 8, respectively.

- In case of Step 5, the output G' is clearly a connected k-subgraph of G.
- In case of Step 6, G' qualifies to be an input of Procedure 1. With this input, Procedure 1 returns the connected k-subgraph $\text{PRC1}(G')$ of G' as the algorithm's output.
- In case of Step 7, G' qualifies to be an input of Procedure 2. With this input, Procedure 2 returns the connected k-subgraph $\text{PRC2}(G')$ of G' as the algorithm's output.
- In case of Step 8, the algorithm recurses with smaller input G'_r, which satisfies $\sigma(G'_r) \geq \sigma(G') \geq \sigma(G)$ and $k \leq |G'_r| < |G'| \leq |G|$.

Hence after $O(n)$ recursions, the algorithm terminates at one of Steps 5 – 7 and outputs a connected k-subgraph of G.

Theorem 1. *Algorithm 1 finds in $O(mn)$ time a connected k-subgraph C of G such that $\sigma_k^*(G)/\sigma(C) \le 12n^2/k^2$.*

Proof. Let $C = \text{ALG1}(G)$ be the output connected k-subgraph of G. If C is output at Step 5, then its density is $\sigma(C) \ge \sigma(G) \ge (k/n) \cdot \sigma_k^*(G)$, where the last inequality is by (2.1). If C is output by Procedure 1 at Step 6, then from Lemma 3 we know its density is at least $\frac{k}{4|G'|} \cdot \sigma(G') \ge \frac{k}{4n} \cdot \sigma(G) \ge \frac{k^2}{4n^2} \cdot \sigma_k^*(G)$.

Now we are only left with the case that $C = \text{PRC2}(G')$ is output by Procedure 2 at Step 7 of Algorithm 1. Let R denote the set of removable vertices of G'. For every $r \in R$, we see that r is a cut-vertex of G' (cf. the note at Step 4 of the algorithm), and $\sigma(G_r') \ge \sigma(G' \setminus r) > \sigma(G')$, where the first inequality is from the definition of G_r' (it is the densest component of $G' \setminus r$), and the second inequality is due to the removability of r. Thus

$$\sigma(G_{r+}') > \sigma(G_r') \cdot |G_r'|/(|G_r'| + 1) \ge \sigma(G')/2 \quad \text{for every } r \in R.$$

Using the notations in Procedure 2, we note that each vertex of $S \setminus R$ is non-removable in G', and therefore has degree at least $\sigma(G')/2$ in G' by Lemma 1. Since $V_1 = S \cup (\cup_{r \in R \cap S} V(G_r')) = (S \setminus R) \cup (\cup_{r \in S \cap R} V(G_{r+}'))$ contains at least $k/2$ vertices (recall (3.2)), it follows that G' contains at least $(\frac{k}{2} \cdot \frac{\sigma(G')}{2})/2 \ge \frac{k}{8} \cdot \sigma(G) \ge \frac{k^2}{8n} \cdot \sigma_k^*(G)$ edges each with at least one end in V_1.

If there are at least $\frac{k^2}{24n} \cdot \sigma_k^*(G)$ edges with both ends in V_1, then by Step 10 of Procedure 2 we have $|E(C)| \ge \frac{k^2}{24n} \cdot \sigma_k^*(G)$ and $\sigma(C) = 2|E(C)|/k \ge \frac{k}{12n} \cdot \sigma_k^*(G) \ge \frac{k^2}{12n^2} \cdot \sigma_k^*(G)$. It remains to consider the case where G' contains at least $\frac{k^2}{12n} \cdot \sigma_k^*(G)$ edges between V_1 and $G' \setminus V_1$. All these edges are between S and $G' \setminus V_1 = H \setminus S$, since each edge incident with any vertex in G_r' ($r \in R$) must have both ends in V_1. So, by the definition of S^* at Step 8 of Procedure 2, we deduce from (3.1) that there are at least a number $|[S, S^*]| \ge \frac{k/2}{n} \cdot |[S, H \setminus S]| \ge \frac{k^3}{24n^2} \cdot \sigma_k^*(G)$ of edges in the subgraph of G' induced by $V_2 = S \cup S^*$. Hence $\sigma(C) \ge 2|[S, S^*]|/k \ge \frac{k^2}{12n^2} \cdot \sigma_k^*(G)$, justifying the performance of the algorithm. See [6] for the runtime analysis. $\square$

3.2 $O(n^{2/5})$-Approximation

In this subsection we design algorithms for finding connected k-subgraphs of G that jointly provide an $O(n^{2/5})$-approximation to DkSP. Among the outputs of all these algorithms (with input G), we select the densest one, denoted as C. Then it can be guaranteed that $\sigma_k^*(G)/\sigma(C) \le O(n^{2/5})$. In view of the $O(n^2/k^2)$-approximation of Algorithm 1, we may focus on the case of $k < n^{4/5}$. (Note that $n^2/k^2 \le n^{2/5}$ if $k \ge n^{4/5}$.)

Let D be a densest connected subgraph of G, which is computable in time $O(mn \log(n^2/m))$ [12,18], because every component of a densest subgraph of G is also a densest subgraph of G. Thus

$$\sigma(D) = \sigma^*(G) \ge \sigma_k^*(G).$$

Moreover, the maximality of $\sigma(D)$ implies that D has no removable vertices.

Algorithm 2. Input: connected graph G along with its densest connected subgraph D. Output: a connected k-subgraph of G, denoted as $\text{ALG2}(G)$.

---

1. **If** $|D| \leq k$ **then** Expand D to be a connected k-subgraph H of G
   $$\text{Output } \text{ALG2}(G) \leftarrow H$$
2. **Else** Output $\text{ALG2}(G) \leftarrow \text{PRC1}(D)$

Lemma 5. *If $k < n^{4/5}$, then $\sigma(\text{ALG2}(G)) \geq \min\{k/(4n), n^{-2/5}\} \cdot \sigma^*(G)$.*

Proof. In case of $|D| \leq k$, by Lemma 2, it follows from $\sigma^*(G) \geq \sigma_k^*(G)$ that the density of the output subgraph $\sigma(H) \geq \sigma(D)/\sqrt{k} = \sigma^*(G)/\sqrt{k}$. Since $k \leq n^{4/5}$, we see that $\sigma(H) \geq n^{-2/5} \cdot \sigma^*(G)$.

In case of $|D| > k$, we deduce from Lemma 3 that the connected k-subgraph $\text{ALG2}(G) = \text{PRC1}(D)$ of D has density at least $\frac{k}{4|D|} \cdot \sigma(D) \geq \frac{k}{4n} \cdot \sigma^*(G)$. $\square$

Our next algorithm is an expansion of Procedure 2 by Feige et al. [11]. Let V_h be a set of $k/2$ vertices of highest degrees in G, and let $d_h = \frac{2}{k}\sum_{v \in V_h} d_G(v)$ denote the average degree of the vertices in V_h.

Algorithm 3. Input: connected graph G with $|G| \geq k$.
Output: a connected k-subgraph of G, denoted as $\text{ALG3}(G)$.

---

1. $V_h^\star \leftarrow$ a $(k/2)$-attachment of V_h in G
2. $H \leftarrow$ a densest component of $G[V_h \cup V_h^\star]$
3. Output $\text{ALG3}(G) \leftarrow$ a k-connected subgraph of G that is expanded from H

In the above algorithm, the subgraph $G[V_h \cup V_h^\star]$ is exactly the output of Procedure 2 in [11], for which it has been shown (cf, Lemma 3.2 of [11]) that

$$\bar{\sigma} := \sigma(G[V_h \cup V_h^\star]) \geq kd_h/(2n).$$

Recalling Lemma 2, we have $\sigma(\text{ALG3}(G)) \geq \sigma(H)/\sqrt{k} \geq \bar{\sigma}/\sqrt{k}$, which implies the following result.

Lemma 6. $\sigma(\text{ALG3}(G)) \geq \frac{\bar{\sigma}}{\sqrt{k}} \geq \frac{\sqrt{k}}{2n} \cdot d_h.$ $\square$

Our last algorithm is a slight modification of Procedure 3 in [11], where we link things up via a "hub" vertex. For vertices u, v of G, let $W(u,v)$ denote the number of walks of length 2 from u to v in G.

Algorithm 4. Input: connected graph $G = (V, E)$ with $|G| \geq k$.
Output: a connected k-subgraph of G, denoted as $\text{ALG4}(G)$.

---

1. $G_\ell \leftarrow G[V \setminus V_h]$.
2. Compute $W(u,v)$ for all pairs of vertices u, v in G_ℓ.
3. For every $v \in V \setminus V_h$, construct a connected k-subgraph C^v of G as follows:

- Sort the vertices $u \in V \setminus V_h \setminus \{v\}$ with positive $W(v,u)$ as $v_1, v_2, \ldots, v_t$ such that $W(v,v_1) \geq W(v,v_2) \geq \cdots \geq W(v,v_t) > 0$.
- $P^v \leftarrow \{v_1, \ldots, v_{\min\{t, k/2-1\}}\}$
- $B^v \leftarrow$ a set of $\min\{d_{G_\ell}(v), k/2\}$ neighbors of v in G_ℓ such that the number of edges between B^v and P^v is maximized.
- $C^v \leftarrow$ the component of $G_\ell[\{v\} \cup B^v \cup P^v]$ that contains v
- Expand C^v to be a connected k-subgraph of G

4. Output $\text{ALG4}(G) \leftarrow$ the densest C^v for $v \in V \setminus V_h$

In the above algorithm, B^v can be found in $O(m + n\log n)$ time, and v is the "hub" vertex ensuring that C^v is connected. Hence the algorithm is correct, and runs in $O(mn + n^2 \log n)$ time, where Step 2 finishes in $O(n^2 \log n)$ time. The key point here is that C^v contains all edges between B^v and P^v, where B^v and P^v are not necessarily disjoint. Using a similar analysis to that in [11] (see [6]), we obtain the following.

Lemma 7. *If* $k \leq \frac{2}{3}n$, *then* $\sigma(\text{ALG4}(G)) \geq \frac{(\sigma_k^*(G)-2\bar{\sigma})^2}{2\max\{k, 2d_h\}} \cdot \frac{k-2}{k} \geq \frac{(\sigma_k^*(G)-2\bar{\sigma})^2}{6\max\{k, 2d_h\}}$. $\square$

We are now ready to prove that the four algorithms given above jointly guarantee an $O(n^{2/5})$-approximation.

Theorem 2. *A connected k-subgraph C of G can be found in $O(mn\log n)$ time such that $\sigma_k^*(G)/\sigma(C) \leq O(n^{2/5})$.*

Proof. Let C be the densest connected k-subgraph of G among the outputs of Algorithms 1 – 4. As mentioned at the beginning of Sect. 3.2, it suffices to consider the case of $k < n^{4/5}$. The connectivity of C gives $\sigma(C) \geq 1$. Clearly, we may assume $n \geq 8$, which along with $k < n^{4/5}$ implies $k \leq 2n/3$. By Lemmas 5–7, we may assume that

$$\sigma(C) \geq \max\left\{ 1, \frac{k\sigma^*(G)}{4n}, \frac{\bar{\sigma}}{\sqrt{k}}, \frac{\sqrt{k}d_h}{2n}, \frac{(\sigma_k(G)-2\bar{\sigma})^2}{6\max\{k, 2d_h\}} \right\}.$$

If $k \geq n^{3/5}$, then $\sigma(C) \geq k \cdot \sigma^*(G)/(4n) \geq \sigma^*(G)/(4n^{2/5}) \geq \sigma_k^*(G)/(4n^{2/5})$. If $k \leq n^{2/5}$, then $\sigma(C) \geq 1 \geq \sigma_k^*(G)/k \geq \sigma_k^*(G)/n^{2/5}$. So we are only left with the case of $n^{2/5} \leq k \leq n^{3/5}$.

Since $\sigma(C) \geq \bar{\sigma}/\sqrt{k} \geq \bar{\sigma}/n^{3/10} \geq \bar{\sigma}/n^{2/5}$, we may assume $\bar{\sigma} < \sigma_k^*(G)/4$, and hence $\sigma_k^*(G) - 2\bar{\sigma} \geq \sigma_k^*(G)/2$. Next we use the geometric mean to prove the performance guarantee as claimed.

In case of $k \geq 2d_h$, since $\sigma^*(G) \geq \sigma_k^*(G)$, we have

$$\sigma(C) \geq \left(1 \cdot \frac{k\sigma^*(G)}{4n} \cdot \frac{(\sigma_k^*(G)/2)^2}{6k} \right)^{1/3} \geq \frac{\sigma_k^*(G)}{5n^{2/5}},$$

In case of $k < 2d_h$, we have

$$\sigma(C) \geq \left(1 \cdot \frac{\sqrt{k}d_h}{2n} \cdot \frac{(\sigma_k^*(G)/2)^2}{12d_h} \cdot \frac{\sqrt{k}d_h}{2n} \cdot \frac{(\sigma_k^*(G)/2)^2}{12d_h} \right)^{1/5} \geq \frac{\sigma_k^*(G)}{7n^{2/5}},$$

where the last inequality follows from the fact that $k \geq \sigma_k^*(G)$. $\square$

4   Conclusion

In Sect. 3, we have given four strongly polynomial time algorithms that jointly guarantee an $O(\min\{n^{2/5}, n^2/k^2\})$-approximation for the unweighted problem – DCkSP. The approximation ratio is compared with the maximum density of *all* k-subgraphs, and in this case no $O(n^{1/3-\varepsilon})$-approximation for any $\varepsilon > 0$ can be expected (recall $\Lambda \geq \frac{1}{3}n^{1/3}$ in Example 1(a)). When studying the weighted generalization – HCkSP, we can extend the techniques developed in Sect. 3.1, and obtain an $O(n^2/k^2)$-approximation for the weighted case. Besides, a simple greedy approach can achieve a $(k/2)$-approximation [6]. As $\min\{n^2/k^2, k\} \leq n^{2/3}$, the following result implies an $O(n^{2/3})$-approximation for HCkSP.

Theorem 3. *For any connected graph $G = (V, E)$ with weight $w \in \mathbb{Z}_+^E$, a connected k-subgraph H of G can be found in $O(nm)$ time such that $\sigma_k^*(G, w)/\sigma(H, w) \leq O(\min\{n^2/k^2, k\})$, where $\sigma(H, w)$ is the weighted density of H, and $\sigma_k^*(G, w)$ is the weighted density of a heaviest k-subgraph of G (which is not necessarily connected).*   □

Since the weighted density of a graph is not necessarily related to its number of edges or vertices, a couple of the results in the previous sections (such as Lemmas 2, 6 and 7) do not hold for the general weighted case. Neither the techniques of extending unweighted case approximations to weighted cases in [11, 17] apply to our setting due to the connectivity constraint. An immediate question is whether an $O(n^{2/5})$-approximation algorithm exists for HCkSP. Note from $\Lambda_w \geq \frac{1}{2}n^{1/2}$ in Example 1(b) that no one can achieve an $O(n^{1/2-\varepsilon})$-approximation for any $\varepsilon > 0$ if the solution value is compared with the maximum weighted density of *all* k-subgraphs. Among other algorithmic approaches, analyzing the properties of densest/heaviest *connected* k-subgraphs is an important and challenging task in obtaining improved approximation ratios for DCkSP and HCkSP.

References

1. Andersen, R., Chellapilla, K.: Finding dense subgraphs with size bounds. In: Avrachenkov, K., Donato, D., Litvak, N. (eds.) WAW 2009. LNCS, vol. 5427, pp. 25–37. Springer, Heidelberg (2009)
2. Arora, S., Karger, D., Karpinski, M.: Polynomial time approximation schemes for dense instances of NP-hard problems. In: Proceedings of the 27th Annual ACM Symposium on Theory of Computing, pp. 284–293 (1995)
3. Asahiro, Y., Iwama, K., Tamaki, H., Tokuyama, T.: Greedily finding a dense subgraph. J. Algorithms **34**(2), 203–221 (2000)
4. Bhaskara, A., Charikar, M., Chlamtac, E., Feige, U., Vijayaraghavan, A.: Detecting high log-densities: an $O(n^{1/4})$ approximation for densest k-subgraph. In: Proceedings of the 42nd Annual ACM Symposium on Theory of Computing, pp. 201–210 (2010)
5. Chen, Danny Z., Fleischer, Rudolf, Li, Jian: Densest k-subgraph approximation on intersection graphs. In: Jansen, Klaus, Solis-Oba, Roberto (eds.) WAOA 2010. LNCS, vol. 6534, pp. 83–93. Springer, Heidelberg (2011)

6. Chen, X., Hu, X., Wang, C.: Finding connected dense k-subgraphs. CoRR abs/ 1501.07348 (2015)
7. Corneil, D.G., Perl, Y.: Clustering and domination in perfect graphs. Discrete Appl. Math. **9**(1), 27–39 (1984)
8. Demaine, E.D., Hajiaghayi, M., Kawarabayashi, K.i.: Algorithmic graph minor theory: decomposition, approximation, and coloring. In: Proceedings of the 46th Annual IEEE Symposium on Foundations of Computer Science, pp. 637–646 (2005)
9. Feige, U.: Relations between average case complexity and approximation complexity. In: Proceedings of the 34th Annual ACM Symposium on Theory of Computing, pp. 534–543 (2002)
10. Feige, U., Langberg, M.: Approximation algorithms for maximization problems arising in graph partitioning. J. Algorithms **41**(2), 174–211 (2001)
11. Feige, U., Peleg, D., Kortsarz, G.: The dense k-subgraph problem. Algorithmica **29**(3), 410–421 (2001)
12. Goldberg, A.V.: Finding a Maximum Density Subgraph. University of California Berkeley, CA (1984)
13. Han, Q., Ye, Y., Zhang, J.: An improved rounding method and semidefinite programming relaxation for graph partition. Math. Program. **92**(3), 509–535 (2002)
14. Hassin, R., Rubinstein, S., Tamir, A.: Approximation algorithms for maximum dispersion. Oper. Res. Lett. **21**(3), 133–137 (1997)
15. Khot, S.: Ruling out ptas for graph min-bisection, dense k-subgraph, and bipartite clique. SIAM J. Comput. **36**(4), 1025–1071 (2006)
16. Khuller, S., Saha, B.: On finding dense subgraphs. In: Albers, S., Marchetti-Spaccamela, A., Matias, Y., Nikoletseas, S., Thomas, W. (eds.) ICALP 2009, Part I. LNCS, vol. 5555, pp. 597–608. Springer, Heidelberg (2009)
17. Kortsarz, G., Peleg, D.: On choosing a dense subgraph. In: Proceedings of the 34th Annual IEEE Symposium on Foundations of Computer Science, pp. 692–701 (1993)
18. Lawler, E.L.: Combinatorial Optimization: Networks and Matroids. Courier Dover Publications, New York (1976)
19. Liazi, M., Milis, I., Zissimopoulos, V.: Polynomial variants of the densest/heaviest k-subgraph problem. In: Proceedings of the 20th British Combinatorial Conference, Durham (2005)
20. Raghavendra, P., Steurer, D.: Graph expansion and the unique games conjecture. In: Proceedings of the 42nd Annual ACM Symposium on Theory of Computing, pp. 755–764 (2010)
21. Srivastav, A., Wolf, K.: Finding dense subgraphs with semidefinite programming. In: Jansen, K., Rolim, J.D.P. (eds.) APPROX 1998. LNCS, vol. 1444, pp. 181–191. Springer, Heidelberg (1998)

The Complexity of Degree Anonymization by Graph Contractions

Sepp Hartung and Nimrod Talmon$^{(\boxtimes)}$

Institut Für Softwaretechnik und Theoretische Informatik, TU Berlin,
Berlin, Germany
nimrodtalmon77@gmail.com

Abstract. We study the computational complexity of k-anonymizing a given graph by as few graph contractions as possible. A graph is said to be k-anonymous if for every vertex in it, there are at least $k-1$ other vertices with exactly the same degree. The general degree anonymization problem is motivated by applications in privacy-preserving data publishing, and was studied to some extent for various graph operations (most notable operations being edge addition, edge deletion, vertex addition, and vertex deletion). We complement this line of research by studying several variants of graph contractions, which are operations of interest, for example, in the contexts of social networks and clustering algorithms. We show that the problem of degree anonymization by graph contractions is NP-hard even for some very restricted inputs, and identify some fixed-parameter tractable cases.

1 Introduction

Motivated by concerns of data privacy in social networks, Clarkson et al. [10] introduced the general degree anonymization problem, defined as follows. Given an input graph G and an allowed operation O, the task is to transform G into a k-anonymous graph by performing as few O operations as possible; a graph is said to be k-anonymous if for every vertex in it, there are at least $k-1$ other vertices with exactly the same degree. This problem has been studied, both theoretically and practically, for several graph modification operations such as edge addition [10,17,20], edge deletion [8], vertex addition [5,9], and vertex deletion [4]. This paper can be seen as complementing this line of research by considering graph contractions, as a natural graph modification operation, specifically studying the (parameterized) complexity of this degree anonymization problem.

This paper also complements research done on the following problem: given an input graph $G = (V, E)$ and a family $\mathcal{F}$ of graphs, find a minimum-size subset of edges $E' \subseteq E$, such that after contracting the edges in E', G would be in the family $\mathcal{F}$ (indeed, in our case, $\mathcal{F}$ is the family of all k-anonymous graphs). Asano and Hirata [1] defined a set of conditions on $\mathcal{F}$, which is sufficient for the NP-hardness of this problem. Others studied specific graph classes

---

A full version is available at http://fpt.akt.tu-berlin.de/talmon/abcfv.pdf.

N. Talmon—Supported by DFG Research Training Group MDS (GRK 1408).

R. Jain et al. (Eds.): TAMC 2015, LNCS 9076, pp. 260–271, 2015.
DOI: 10.1007/978-3-319-17142-5_23

(as $\mathcal{F}$), such as planar graphs [15], bipartite graphs [18], paths [18], trees [16], and d-regular graphs [3]. This last work is of particular interest, as the concept of k-anonymity is a generalization of the notion of regularity (in particular, a graph is n-anonymous if and only if it is regular).

Studying graph contractions in the context of degree anonymization is interesting for several reasons. First, some variants of contractions can preserve original properties of the input graph (for example, connectivity). Second, vertex contraction, where also non-adjacent vertices can be contracted, is the inverse operation of vertex cleaving (as defined by Oxley [22, Chapter 3]), which was studied in the context of degree anonymization by Bredereck et al. [5] (there, called *vertex cloning*). We mention also the relation of graph contractions to communities detection in social networks and to clustering (see, for example, Delling et al. [11]).

2   Preliminaries

We assume familiarity with standard notions regarding algorithms, computational complexity, and graph theory. For a non-negative integer z, we denote $\{1, \ldots, z\}$ by $[z]$.

2.1   Parameterized Complexity

An instance (I, k) of a parameterized problem consists of the "classical" problem instance I and an integer k being the *parameter* [13,21]. A parameterized problem is called *fixed-parameter tractable* (FPT) if there is an algorithm solving it in $f(k) \cdot |I|^{O(1)}$ time, for an arbitrary computable function f only depending on the parameter k. In difference to that, algorithms running in $|I|^{f(k)}$ time prove membership in the class XP (clearly, FPT $\subseteq$ XP). One can show that a parameterized problem L is (presumably) not fixed-parameter tractable by devising a *parameterized reduction* from a W[1]-hard or a W[2]-hard problem to L. A parameterized reduction from a parameterized problem L to another parameterized problem L' is a function that, given an instance (I, k), computes in $f(k) \cdot |I|^{O(1)}$ time an instance (I', k') such that $k' \leq g(k)$ and $(I, k) \in L \Leftrightarrow (I', k') \in L'$. A parameterized problem which is NP-hard even for instances for which the parameter is a constant is said to be Para-NP-hard.

2.2   Graph Theory and Contractions

Given a graph $G = (V, E)$, which may have self-loops and parallel edges, we denote the degree of a vertex $v \in V$ by $\deg(v)$, and define $B_d = \{v \in V : \deg(v) = d\}$ as the set of vertices of degree d (called the *block* of degree d). As usual, we define the degree of a vertex v with x neighbors and y self-loops to be $x + 2y$ (in particular, we count a self-loop twice). We define a *path-star* of degree d and length l to be the graph consisting of one center vertex, connected to d disjoint paths of length l each (indeed, this is a spider graph with equal-length legs). A *caterpillar-tree* is a tree for which removing the leaves and their incident edges leaves a path graph (formally, a path graph is a tree with no vertices of degree larger than 2).

Given an undirected graph $G = (V, E)$ and two adjacent vertices, u and v, contracting the vertices u and v (usually referred to as contracting the edge $e = \{u, v\}$), means removing u and v from V, replacing them by one new vertex (denoted by $u \oplus v$), adjacent to exactly those vertices that were adjacent to u, to v, or to both. The resulting graph is denoted by G/e. In general, given a set of edges $E_1 \subseteq E$, we denote by G/E_1 the graph obtained from G after contracting all the edges in E_1. A graph $G = (V, E)$ is said to be l-contractible to a graph $G' = (V', E')$ if there is a set of edges $E_1 \subseteq E$ of size at most l, such that $G/E_1 = G'$. It follows that $G = (V, E)$ is contractible to $G' = (V', E')$ if and only if there exists a *witness structure* $V = V_1 \cup \ldots \cup V_{|V'|}$, where each V_i is called a *witness set*, such that for each V_i (for $1 \le i \le |V'|$) the subgraph of G induced by V_i is connected and for each pair of witness sets, V_i and V_j $(1 \le i \ne j \le |V'|)$ we have that $\{V_i, V_j\} \in E' \iff \exists v_i \in V_i, v_j \in V_j : \{v_i, v_j\} \in E$ (indeed, the vertices in each part V_i are contracted to form a single vertex). We denote by $\deg(V_i)$ the resulting degree of the vertex corresponding to the contraction of the witness set and we call graph G' the witness graph.

We also define the closely related operation of *vertex contraction*, which is defined similarly to edge contraction, with the only difference that it is allowed to contract non-adjacent vertices as well (indeed, the vertices of a witness set from a vertex-contracted graph are not assumed to be connected). It is clear that a graph contraction operation can sometimes introduce self-loops and parallel edges. We define three variants of edge contractions and vertex contractions, differing by how self-loops and parallel edges are treated:

- *Simple Contraction:* Both self-loops and parallel edges are removed.
- *Hybrid Contraction:* Only self-loops are removed.
- *Non-Simple Contraction:* Nothing is removed.

For the Hybrid and Non-Simple variants, we allow the input graph to be non-simple. See Fig. 1 for some examples.

2.3 Main Problem

Given an undirected input graph G, we are interested in k-anonymizing it by performing at most c edge contractions (where a graph is said to be k-anonymous if every vertex degree in it occurs at least k times; equivalently, if $\forall i \in [n] : |B_i| = 0 \lor |B_i| \ge k$).

> DEGREE ANONYMIZATION BY GRAPH CONTRACTIONS
> **Input:** An undirected graph $G = (V, E)$, a budget $c \in \mathbb{N}$, and an anonymization level $k \in \mathbb{N}$.
> **Question:** Can G be made k-anonymous by performing at most c contractions?

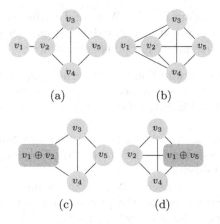

(a)   (b)

(c)   (d)

Fig. 1. Example of 2-anonymizing an input graph. The input graph is depicted in (a), an optimal 2-anonymized graph with respect to edge addition is depicted in (b), an optimal 2-anonymized graph with respect to simple edge contraction or hybrid edge contraction is depicted in (c) (by contracting v_1 and v_2), and an optimal 2-anonymized graph with respect to non-simple vertex contraction is depicted in (d) (by contracting v_1 and v_5). Notice that there is no solution with respect to non-simple edge contraction, and the solution with respect to edge addition is less efficient than the solutions by edge contractions.

When the contraction operation is a simple (hybrid, non-simple) edge contraction operation, we denote the corresponding degree anonymization problem by SEC-A (respectively: HEC-A, NEC-A). Similarly, when the contraction operation is a simple (hybrid, non-simple) vertex contraction operation, we denote the corresponding degree anonymization problem by SVC-A (respectively: HVC-A, NVC-A).

Interestingly, it is not always possible to anonymize a graph by performing only graph contractions. As an example, consider n-anonymizing a complete graph which has one missing edge: as the input graph is not n-anonymized, at least one edge needs to be contracted, but then the number of remaining vertices will be strictly less than n, thus the graph cannot be further made n-anonymous. This phenomenon stands in contrast to anonymization by edge additions, as completing any graph, by adding all missing edges to it, makes it n-anonymous. However, some graphs can be anonymized more efficiently by using edge contractions rather than edge additions (see Fig. 1 for an example).

2.4   Overview

We study the parameterized complexity of degree anonymization by graph contractions, considering the solution size c, the anonymity level k, and the maximum degree Δ, as the most natural parameters. From the variants defined in Sect. 2.3, we consider SEC-A and HEC-A as these are the most common (see, for example, Diestel [12, Chapter 1.7] and Wolle and Bodlaender [23]), and we consider NVC-A as it is equivalent to the underlying number problem (as defined in Sect. 3). We state some important points of our work:

- Contrary to degree anonymization by some other graph operations (for example, by edge addition), here even the underlying number problem (NVC-A) is NP-hard. Moreover, SEC-A, HEC-A, and NVC-A, are all NP-hard even on caterpillar trees.
- Parameterizing by either the solution size c, the maximum degree Δ, or the anonymity level k, does not help for tractability. However, combining Δ with c does help for tractability.
- Combining the maximum degree Δ with the anonymity level k helps for tractability for some variants of the problem, and we could show evidence suggesting intractability for some other variants.

Table 1 gives an overview of our results. Due to space constraints, some of the proofs are omitted. Please refer to the full version (available at http://fpt.akt. tu-berlin.de/talmon/abcfv.pdf).

Table 1. Parameterized complexity landscape of DEGREE ANONYMIZATION BY GRAPH CONTRACTIONS. Rows and columns correspond to parameters, such that each cell corresponds to the combination of the corresponding parameters.

	solution size c	anonymization level k	maximum degree Δ
c	W-h^a (Th. 3) XP (Obs. 1)	W-h^a (Th. 3) XP (Obs. 1)	FPT (Th. 5)
k		Para-NP-h^a (Th. 3)	FPT^b (Cor. 1)
Δ			Para-NP-h^a (Th. 4)

^{a} Only for SEC-A and HEC-A.
^{b} Only for NVC-A.

3 NP-hardness

We begin by considering NVC-A which, surprisingly, reduces to a number problem formed by the degrees in the input graph. This holds because (1) any two vertices can be contracted, and (2) the degree sequence of the resulting graph after performing a contraction only depends on the original degrees of the contracted vertices (indeed, as self-loops and parallel edges are not removed). It follows that NVC-A is equivalent to the following number problem. Therein, a multiset of integers is k-anonymous if each integer in it occurs at least k times.

AN EQUIVALENT FORMULATION OF NVC-A
Input: A set $V = \{d_1, \ldots, d_n\}$ of n integers such that $\forall i : 0 \leq d_i \leq \Delta$ and two integers $k, c \in \mathbb{N}$.
Question: Is there a partition $V = \bigcup_{j \in [z]} V_j$ (where $V_{j_1} \cap V_{j_2} = \emptyset$ for $1 \leq j_1 \neq j_2 \leq z$) such that the multiset $S = \{\sum_{d_i \in V_j} d_i : j \in [z]\}$ is k-anonymous and $\sum_{j \in [z]} (|V_j| - 1) \leq c$?

Informally, the above number problem is in the heart of the graph anonymization problem (for this reason we call it the underlying number problem). Interestingly, contrary to the situation for other operations (such as edge addition), here the underlying number problem is intractable (for formal correctness, we define the input of this number problem to be in unary; this does not cause problems, as we next prove a reduction from a *strongly* NP-hard problem).

Theorem 1. NVC-A *is* NP-*hard even on caterpillar trees.*

Proof. We provide a reduction from the following strongly NP-hard problem [14]:

STRICTLY THREE PARTITION
Input: A set of numbers $S = \{a_1, \ldots, a_{3m}\}$ such that $\sum_{a_i \in S} a_i = mB$ and $\forall i \in [3m] : B/4 < a_i < B/2$.
Question: Are there m disjoint sets $S_1, \ldots, S_m$, each of size 3, such that $\forall j \in [m] : \sum_{a_i \in S_j} a_i = B$?

Given an instance for STRICTLY THREE PARTITION, we create an instance for NVC-A. Intuitively, the idea is to create a set of $3m$ vertices, such that each number a_i would have a corresponding vertex whose degree is proportional to a_i. Then, we will add a distinguished vertex with degree proportional to B, making sure that the only way of anonymizing the block containing this distinguished vertex is by contracting m triplets of vertices corresponding to triplets of numbers, each of sum exactly m. Details follow.

We first scale the input numbers. Specifically, we define $a'_i = a_i \cdot mB$ and $B' = B \cdot mB$. We set $k := m+1$ and $c := 2m$. Then, for each number a'_i, we create a node $v_{a'_i}$ and connect it to a'_i paths of length $c+1$ (consisting of new vertices), such that $\deg(v_{a'_i}) = a'_i$ holds for each i. We add a path-star of degree B' and length $c+1$ (indeed, G is a forest; we can easily transform it into a caterpillar tree by placing all $v_{a'_i}$'s on a path together with the path-star, adjusting the number of additional new vertices connected to each $v_{a'_i}$ accordingly). The correctness proof is omitted, due to space constraints. Please refer to the full version.   □

We can show NP-hardness on caterpillar trees also for SEC-A and HEC-A.

Theorem 2. SEC-A *and* HEC-A *are both* NP-*hard even on caterpillar trees.*

Proof. We provide a reduction from the following strongly NP-hard problem [14]:

NUMERICAL MATCHING WITH TARGET SUMS
Input: Three sets of integers: $A = \{a_1, \ldots, a_n\}$, $B = \{b_1, \ldots, b_n\}$, and $C = \{c_1, \ldots, c_n\}$.
Question: Can the elements of A and B be paired such that, for each $i \in [n]$, c_i is the sum of the ith pair?

The variant where all $3n$ input integers are distinct is also known to be NP-hard in the strong-sense [19]. Without loss of generality, we assume all input integers to be greater than 3. Given an instance for NUMERICAL MATCHING WITH TARGET SUMS, we create an instance for SEC-A and HEC-A. Intuitively, the idea is to create a set of $k - 1$ vertices for each c_i and a pair of vertices for each pair of a_i and b_j, such that the only possibility of anonymizing the vertices corresponding to the c_i's is to contract the correct pairs of a_i's and b_j's. Details follow.

We set $k := n - 1$ and $c := n$. We construct some c-gadgets: for each c_i, we create $k - 1$ path-stars of degree $c_i - 2$ and length $c + 1$. We construct some ab-gadgets: for each pair of integers, $i \in [n]$ and $j \in [n]$, we create two path-stars, one of degree a_i and another of degree b_j, both of length $c + 1$, and connect them by an edge (indeed, the construction as such is a forest; we can transform it into a caterpillar tree by carefully connecting each pair of disconnected components by a path of length $c + 1$). The correctness proof is omitted due to space constraints. Please refer to the full version.   □

4   Non-structural Parameters

Following the hardness results from the last section, we continue our quest for tractability by considering non-structural parameters. We observe first that for constant solution size c we can simply enumerate all possible solutions, therefore concluding the following.

Observation 1. DEGREE ANONYMIZATION BY GRAPH CONTRACTIONS *is* XP *with respect to* c.

However, there is no hope for fixed-parameter tractability with respect to this parameter, as even combining it with the anonymity level k does not help for tractability.

Theorem 3. *Both* SEC-A *and* HEC-A *are* NP-*hard and* W-*hard with respect to* c, *even if* $k = 2$.

Proof. For SEC-A, we provide a reduction from the following W[2]-hard problem, parameterized by the solution size [13]:

SET COVER
Input: Sets $S_1, \ldots, S_m$ containing elements from $x_1, \ldots, x_n$ and $h \in \mathbb{N}$.
Question: Is there a set of at most h sets covering all elements?

Given an instance for SET COVER, we create an instance for SEC-A. We set $k := 2$ and $c := h$. For each x_i we create a new vertex x_i'. For each S_j we create two new vertices, S_j' and S_j'', and connect them by an edge. Each S_j' and S_j'' (corresponding to a set S_j) are connected to all x_i''s which correspond to elements $x_i \in S_j$. We add several paths of length $c + 1$ to each x_i' such that the degree of each x_i' will be $f(i) = i(c + 1) + 2$. Similarly, we add several paths of length $c + 1$ to each S_j' and S_j'', such that the degree of each S_j' and S_j'' will be

$f(n+1)$. For every $i \in [n]$ and $z \in [h]$, we add a path-star of degree $f(i) - z$ and length $c + 1$. We add k path-stars of degree $f(n + 1)$ and length $c + 1$ in order to anonymize the vertices corresponding to the sets.

Given a set cover, it is possible to anonymize the input graph, by contracting together each pair of S'_j and S''_j which correspond to a set S_j in the cover: the degrees of each x'_i will decrease by the number of sets covering it, therefore the graph would be anonymized. For the other direction, notice that each x_i needs to be anonymized, therefore, by a simple exchange argument, a solution must correspond to a set cover.

For HEC-A, we provide a reduction from the following W[1]-hard problem, parameterized by the solution size h (an h-coloring is a function $color : V \to [h]$, assigning to each vertex v a color $color(v) \in [h])$ [13]:

MULTI-COLORED CLIQUE
Input: An undirected graph $G = (V, E)$ and an h-coloring of its vertices.
Question: Is there a size-h clique which includes vertices of all h colors?

Cai [6] showed that MULTI-COLORED CLIQUE remains hard even on regular graphs. We assume, without loss of generality, that there are no monochromatic edges. Given an instance for MULTI-COLORED CLIQUE, we create an instance for HEC-A. We define the following function, $f(i) = 2^i \cdot 2\binom{h}{2}$, whose domain is the set of colors (that is, $i \in [h]$).

We set $k := 2$ and $c := h - 1$. For every vertex v, we add $(f(color(v)) - \deg(v))$ paths of length $c + 1$ such that the degree of each vertex colored in color $i \in [h]$ is $f(i)$. We construct $k + 1$ copies of this modified graph. We add $k - 1$ path-stars of degree $((\sum_{i \in [h]} f(i)) - 2\binom{h}{2})$ and length $c + 1$.

Given a multi-colored clique of size h, it is possible to contract the vertices of the clique into one vertex: the degree of the new vertex will be equal to the degree of the $k - 1$ path-stars, resulting in an anonymized graph, due to the $k + 1$ copies.

For the other direction, notice that contracting edges of a path-star does not change its degree. Moreover, as there are no monochromatic edges, we can only contract edges of different colors. Due to the way we defined $f(i)$, the only possible way of reaching the degree of the path-star (that is, $\sum_{i \in [h]} f(i) - 2\binom{h}{2}$) is by contracting a multi-colored clique, because all colors are needed for the first part (that is, $\sum_{i \in [h]} f(i)$) and all edges between the colors are needed for the second part (that is, $2\binom{h}{2}$). □

5 Structural Parameters

We go on to consider the maximum degree Δ of the input graph, as a natural structural parameter. For example, the edge addition variant admits an FPT-algorithm and even a polynomial kernel with respect to Δ [17]. In contrast, we next show that for our case (that is, for edge contractions), parameter Δ alone does not help for tractability. The reductions, from VERTEX COVER ON

CUBIC GRAPHS (for SEC-A) and PARTITION INTO TRIANGLES (for HEC-A), are omitted due to space constraints.

Theorem 4. *Both* SEC-A *and* HEC-A *are* Para-NP-*hard with respect to* Δ.

Contrary to the above hardness results, combining the maximum degree with the solution size does help for tractability, for all variants of DEGREE ANONYMIZATION BY GRAPH CONTRACTIONS.

Theorem 5. DEGREE ANONYMIZATION BY GRAPH CONTRACTIONS *is* FPT *with respect to* (Δ, c).

Proof. Consider a yes-instance for DEGREE ANONYMIZATION BY GRAPH CONTRACTIONS. There exists a set E' of at most c edges such that contracting them would result in a k-anonymous graph. Consider the set V' of vertices, containing all the endpoints of the edges in E', including also all of their neighbors (formally, $V' := N[\{u, v | \{u, v\} \in E'\}]$, where $N[U]$ denotes the closed neighborhood of $U \subseteq V$). As each edge has two endpoints and each vertex has at most Δ neighbors, it follows that $|V'| \leq 2c(\Delta + 1)$. Consider the set V'' containing all vertices whose degree will be changed as a result of contracting the edges in E'. Roughly speaking, as it holds that $V'' \subseteq V'$, it is enough to find the subgraph induced by V'.

To this end, we consider all possible graphs H containing at most $2c(\Delta + 1)$ vertices. For each such graph H, we consider all possible sets C of at most c edges to be contracted. For each such pair of a graph H and a set C, we compute the degree changes in H incurred by contracting the edges in C. If these degree changes make the graph k-anonymous, then we try to find this graph H as a subgraph in G. This step can be performed by using, for example, the result of Cai et al. [7]. □

We consider now the combined parameter Δ and k. The situation here is more involved. First, for NVC-A, we can bound c in these parameters by a function dependent only on Δ and k.

Lemma 1. *For any yes-instance* (V, k, c) *of* NVC-A *it holds that* (V, k, c'), *with* $c' = k \cdot (\Delta \cdot \Delta!)^{\Delta}$, *is also a yes-instance.*

Proof. Let (V, k, c) be a yes-instance of NVC-A and denote by $c_{\mathrm{opt}} \leq c$ the smallest number such that (V, k, c_{opt}) is still a yes-instance. Moreover, let the partition $P = \{V_1, \ldots, V_i\}$ of V be a solution which corresponds to c_{opt} (that is, P is the witness structure corresponding to a solution of (V, k, c_{opt})). In the following we define two operations on P with the property that applying each of them, when it is applicable, results in another solution with less than c_{opt} contractions. Since we show that at least one of them is applicable in case $c_{\mathrm{opt}} > k \cdot (\Delta \cdot \Delta!)^{\Delta}$, this proves Lemma 1.

To formally describe our operations, we associate with each witness set V_i a witness vector $\vec{v_i} \in \mathbb{N}^{\Delta}$ with $\vec{v_i}[j]$ being equal to the number of vertices of degree j in the witness set V_i. The *degree* of a witness set is defined to be the sum of the degrees of the vertices in the witness set (that is, the degree of the vertex corresponding to contracting all of the vertices in the witness set).

Operation 1: This operation is applicable to P if there are at least k witness sets in P, all of equal degree, such that in each of them, say V_i, there is at least one j with $\vec{v}_i[j] \geq \Delta!$. If there exists such a collection of witness sets, then consider such a collection P which is maximal with respect to containment, and do the following. For each witness set V_i in this collection, let j be an integer with $\vec{v}_i[j] \geq \Delta!$. remove $(\Delta!/j)$-many vertices of degree j from V_i (notice that $\Delta!/j$ is always an integer), and form a new witness set containing these vertices.

We introduced at least k new witness sets, all of them of degree exactly $\Delta!$, and we decreased the degree of each of the initial witness sets by the same number $\Delta!$. Since there are at least k of such witness sets, it follows that performing this operation results in a partition of V that is still a solution for (V, k, c_{opt}), while requiring less edge contractions than P requires.

Operation 2: This operation is applicable to P if there is a collection of at least k witness sets in P, such that the witness sets in the collection all have the same witness vector, and this witness vector is of hamming weight of at least 2 (that is, these are not singletons). If such a collection exists, then choose an arbitrary integer j occurring in this witness vector. Then, for each witness set V_i in this collection, remove one vertex of degree j from V_i and form a new witness set containing only this vertex of degree j (that is, form a new singleton witness set).

Since there are at least k witness sets where a vertex of the same degree j is cut out from them, the resulting partition is a solution for V which requires less edge contractions than P requires.

Applicability: It remains to argue that in case of $c_{\mathrm{opt}} > k \cdot (\Delta \cdot \Delta!)^{\Delta}$, at least one of the two operations described above is applicable. First, assume that P contains a witness set V_i of degree at least $(\Delta \cdot \Delta!)$. Then, since P is k-anonymous, it holds that there are at least k witness sets of the same degree, which is at least $(\Delta \cdot \Delta!)$. It follows that each of these witness sets must contain at least one integer j which occurs at least $\Delta!$ times in it. Thus, Operation 1 is applicable.

So, let us assume now that the degree of each witness set in P is at most $(\Delta \cdot \Delta!)$. Then, we have that there are at most $(\Delta \cdot \Delta!)^{\Delta}$ different witness vectors, none of them with degree greater or equal to $(\Delta \cdot \Delta!)$. Hence, if P contains at least $k \cdot (\Delta \cdot \Delta!)^{\Delta}$ witness sets of size at least two, then Operation 2 is applicable.

Finally, a solution for which $c_{\mathrm{opt}} > k \cdot (\Delta \cdot \Delta!)^{\Delta}$ edge contractions have been performed either contains a set of size at least $(\Delta \cdot \Delta!)$ or it contains at least

$$\frac{k \cdot (\Delta \cdot \Delta!)^{\Delta}}{(\Delta \cdot \Delta!)} = k \cdot (\Delta \cdot \Delta!)^{\Delta}$$

witness sets of size at least two. □

Using the above Lemma, we can show the following.

Corollary 1. NVC-A *is* FPT *with respect to* (Δ, k).

Proof. For a given instance (V, k, c) of NVC-A we decide the instance $(V, k,$ $\min\{c, k \cdot (\Delta \cdot \Delta!)^\Delta\})$ using the FPT-algorithm with respect to (Δ, c) (Theorem 5). By Lemma 1, these two instances are equivalent and the corresponding running time proves fixed-parameter tractability with respect to (Δ, k). $\square$

6   Conclusion

We investigated the (parameterized) complexity of degree anonymization by several variants of graph contractions. We showed that most of the variants are intractable even on very restricted graph classes (indeed, even the underlying number problem) and we could identify some fixed-parameter tractable cases.

For further research, one could consider related graph operations, such as *structure contraction* (contracting a whole subgraph at unit cost), *edge twisting* (see [22, Chapter 3]), and *vertex dissolution* (see [22, Chapter 3]).

Bazgan and Nichterlein [2] studied graph anonymization with edge/vertex deletions from the viewpoint of approximation algorithms, while mainly obtaining inapproximability result for the variant of minimizing the number of edit operations. One way of extending this line of research would be to study whether their results transfer to edge contractions and to look at different notions of approximations. For example, partially anonymizing an input graph (only *some* of the vertices are anonymized) or almost anonymizing an input graph (for each vertex, there are at least $k - 1$ other vertices of *roughly* the same degree).

References

1. Asano, T., Hirata, T.: Edge-contraction problems. J. Comput. Syst. Sci. **26**(2), 197–208 (1983)
2. Bazgan, C., Nichterlein, A.: Parameterized inapproximability of degree anonymization. In: Cygan, M., Heggernes, P. (eds.) IPEC 2014. LNCS, vol. 8894, pp. 75–84. Springer, Heidelberg (2014)
3. Belmonte, R., Golovach, P.A., Hof, P., Paulusma, D.: Parameterized complexity of three edge contraction problems with degree constraints. Acta Informatica **51**(7), 473–497 (2014)
4. Bredereck, R., Hartung, S., Nichterlein, A., Woeginger, G.J.: The complexity of finding a large subgraph under anonymity constraints. In: Cai, L., Cheng, S.-W., Lam, T.-W. (eds.) Algorithms and Computation. LNCS, vol. 8283, pp. 152–162. Springer, Heidelberg (2013)
5. Bredereck, R., Froese, V., Hartung, S., Nichterlein, A., Niedermeier, R., Talmon, N.: The complexity of degree anonymization by vertex addition. In: Gu, Q., Hell, P., Yang, B. (eds.) AAIM 2014. LNCS, vol. 8546, pp. 44–55. Springer, Heidelberg (2014)
6. Cai, L.: Parameterized complexity of cardinality constrained optimization problems. Comput. J. **51**(1), 102–121 (2008)
7. Cai, L., Chan, S.M., Chan, S.O.: Random separation: a new method for solving fixed-cardinality optimization problems. In: Bodlaender, H.L., Langston, M.A. (eds.) IWPEC 2006. LNCS, vol. 4169, pp. 239–250. Springer, Heidelberg (2006)

8. Casas-Roma, J., Herrera-Joancomartí, J., Torra, V.: An algorithm for k-degree anonymity on large networks. In: Proceedings of ASONAM 2013, pp. 671–675. ACM Press (2013)
9. Chester, S., Kapron, B.M., Ramesh, G., Srivastava, G., Thomo, A., Venkatesh, S.: Why Waldo befriended the dummy? k-anonymization of social networks with pseudo-nodes. Soc. Netw. Analys. Min. **3**(3), 381–399 (2013)
10. Clarkson, K.L., Liu, K., Terzi, E.: Towards identity anonymization in social networks. In: Yu, P.S., Han, J., Faloutsos, C. (eds.) Link Mining: Models, Algorithms, and Applications, pp. 359–385. Springer, New York (2010)
11. Delling, D., Görke, R., Schulz, C., Wagner, D.: ORCA reduction and contraction graph clustering. In: Goldberg, A.V., Zhou, Y. (eds.) AAIM 2009. LNCS, vol. 5564, pp. 152–165. Springer, Heidelberg (2009)
12. Diestel, R.: Graph Theory. Graduate Texts in Mathematics, vol. 173, 4th edn. Springer, Heidelberg (2010)
13. Downey, R.G., Fellows, M.R.: Fundamentals of Parameterized Complexity. Springer, Heidelberg (2013)
14. Garey, M.R., Johnson, D.S.: Computers and Intractability: A Guide to the Theory of NP-Completeness. Freeman, New York (1979)
15. Golovach, P.A., van 't Hof, P., Paulusma, D.: Obtaining planarity by contracting few edges. In: Rovan, B., Sassone, V., Widmayer, P. (eds.) MFCS 2012. LNCS, vol. 7464, pp. 455–466. Springer, Heidelberg (2012)
16. Guillemot, S., Marx, D.: A faster FPT algorithm for bipartite contraction. Inf. Process. Lett. **113**(22), 906–912 (2013)
17. Hartung, S., Nichterlein, A., Niedermeier, R., Suchý, O.: A refined complexity analysis of degree anonymization in graphs. In: Fomin, F.V., Freivalds, R., Kwiatkowska, M., Peleg, D. (eds.) ICALP 2013, Part II. LNCS, vol. 7966, pp. 594–606. Springer, Heidelberg (2013)
18. Heggernes, P., Hof, P., Lévêque, B., Lokshtanov, D., Paul, C.: Contracting graphs to paths and trees. Algorithmica **68**(1), 109–132 (2014)
19. Hulett, H., Will, T.G., Woeginger, G.J.: Multigraph realizations of degree sequences: maximization is easy, minimization is hard. Oper. Res. Lett. **36**(5), 594–596 (2008)
20. Lu, X., Song, Y., Bressan, S.: Fast identity anonymization on graphs. In: Liddle, S.W., Schewe, K.-D., Tjoa, A.M., Zhou, X. (eds.) DEXA 2012, Part I. LNCS, vol. 7446, pp. 281–295. Springer, Heidelberg (2012)
21. Niedermeier, R.: Invitation to Fixed-Parameter Algorithms. Oxford University Press, Oxford (2006)
22. Oxley, J.G.: Matroid theory, vol. 3. Oxford University Press, Oxford (2006)
23. Wolle, T., Bodlaender, H.L.: A note on edge contraction. Technical report, Technical Report UU-CS-2004 (2004)

An Improved Exact Algorithm for Maximum Induced Matching

Mingyu Xiao[1]($\boxtimes$) and Huan Tan[2]

[1] School of Computer Science and Engineering,
University of Electronic Science and Technology of China, Chengdu, China
myxiao@gmail.com
[2] Library, University of Electronic Science and Technology of China,
Chengdu, China
huan1222@gmail.com

Abstract. This paper studies exact algorithms for the Maximum Induced Matching problem, in which an n-vertex graph is given and we are asked to find a set of maximum number of edges in the graph such that no pair of edges in the set have a common endpoint or are adjacent by another edge. This problem has applications in many different areas. We will give several structural properties of the problem and present an $O^*(1.4391^n)$-time algorithm, which improves previous exact algorithms for this problem.

Keywords: Exact algorithms · Graph algorithms · Maximum induced matching

1 Introduction

Recently, there has been an increasing interest in designing fast and nontrivial exact exponential algorithms for basic NP-hard graph problems. Many interesting exact algorithms have been developed for MAXIMUM INDEPENDENT SET (MIS) [6,20], 3-COLORING [1], FEEDBACK VERTEX SET [5,18], DOMINATING SET [6,17], EDGE DOMINATING SET [16,19] and many others. MAXIMUM INDEPENDENT SET is undoubtedly one of the most important problems in exact algorithms. There is a long list of contributions to the running-time bounds of exact algorithms and it can be solved in $O^*(1.1996^n)$ time and polynomial space now [20]. MAXIMUM INDEPENDENT SET is to find a maximum induced regular graph of degree 0. Gupta, Raman and Saurabh [10] studied exact algorithms for finding maximum induced regular graphs of degree r and presented an algorithm with running time $O^*((2 - \xi)^n)$, where $0 < \xi < 1$ depends on r. The special case that $r = 1$, i.e., the problem to find a maximum induced regular graph of degree 1, is known as MAXIMUM INDUCED MATCHING (MIM). In this paper, we will study structural properties and exact algorithms for MAXIMUM INDUCED MATCHING.

M. Xiao—Supported by NFSC of China under the Grant 61370071.

R. Jain et al. (Eds.): TAMC 2015, LNCS 9076, pp. 272–283, 2015.
DOI: 10.1007/978-3-319-17142-5_24

To find an *induced matching* (i.e., an induced regular graph of degree 1) of maximum size in a graph has received much attention because of the growing number of applications. Stockmeyer and Vazirani [15] showed that MIM has applications in the risk-free marriage problem – to find a maximum number of married couples such that each married person is compatible with no married person other than his/her spouse. Golumbic and Lewenstein [9] demonstrated some applications of induced matchings in secure communication channels, VLSI design and network flow problems. Golumbic and Laskar [8] gave the relations between the size of a maximum induced matching and the irredundancy number of a graph. MIM is also a subtask of the important problem of finding a strong edge coloring (i.e., a proper coloring of the edges such that no edge is adjacent to two edges of the same color) using a small number of colors (see [4,14] for more information).

It is unsurprised that MIM has been extensively studied on computational and algorithmic aspects. Although MIM is polynomial-time solvable in trees [9], chordal graphs [2], circular arc graphs [8], interval graphs [9] and many others, it has been known to be NP-hard in bipartite graphs with maximum degree 4 for more than 30 years [15]. In fact, it remains NP-hard even in planar 3-regular graphs or in planar bipartite graphs with degree-2 vertices in one part and degree-3 vertices in the other part [4,11]. Kobler and Rotics [12] also showed the NP-hardness of MIM in Hamiltonian graphs, claw-free graphs, chair-free graphs, line graphs and regular graphs.

MIM is hard to approximate or design parameterized algorithms. It is APX-complete even in d-regular graphs for each fixed $d \geq 3$ [4]. There is also an approximation algorithm with asymptotic performance ratio $d - 1$ for MIM in d-regular graphs [4]. In general graphs, MIM cannot be approximated within a factor of $n^{1/2-\epsilon}$ in polynomial time for any $\epsilon > 0$ unless $P = NP$ [14]. Take the size k of the induced matching as the parameter. To decide whether there is an induced matching of size at most k is W[1]-hard even in bipartite graphs, but fixed-parameter tractable in planar graphs, line graphs, graphs of bounded treewidth and graphs of girth at least 6 [13].

In terms of exact algorithms for MIM, Gupta, Raman and Saurabh [10] first gave an algorithm with running-time bound $O^*(1.6957^n)$ and then improved it to $O^*(1.4786^n)$. We also note another similar algorithm with the same running-time bound $O^*(1.4786^n)$ [3]. In this paper, we will improve the running-time bound to $O^*(1.4391^n)$ by presenting several new structural properties of MIM. Similar to the $O^*(1.4786^n)$-time algorithms in [10] and [3], our algorithm also uses fast algorithms for MAXIMUM INDEPENDENT SET as a subalgorithm to deal with graphs with maximum degree 4. Different from previous algorithms, the bottleneck cases in our algorithm are not to deal with vertices of degree ≥ 5 in the graph but to solve MIM in graphs with maximum degree 4. So with the results in this paper, faster algorithms for MAXIMUM INDEPENDENT SET or faster algorithms for MIM in graphs with maximum degree 4 directly imply faster algorithms for MIM in general graphs.

Proofs of some lemmas are omitted in this version due to space limitation.

2   Preliminaries

In this paper, a graph always means a simple and undirected graph. Let $G = (V, E)$ be a graph with $n = |V|$ vertices and $m = |E|$ edges. We may simply use v to denote the set $\{v\}$ of a singleton. The vertex set and edge set of a graph G are denoted by $V(G)$ and $E(G)$, respectively. The set of endpoints of edges in an edge set E' is also denoted by $V(E')$. For a subgraph (resp., a vertex subset) X, the subgraph induced by $V(X)$ (resp., X) is simply denoted by $G[X]$, and $G[V - V(X)]$ (resp., $G[V - X]$) is also written as $G - X$. For a vertex subset X, let $N(X)$ denote the *neighbors* of X, i.e., the vertices $y \in V - X$ adjacent to a vertex $x \in X$, and denote $N(X) \cup X$ by $N[X]$. Let $N_2(v)$ denote the set of vertices with distance exactly 2 from v. The *degree* of a vertex v in a graph G, denoted by $d(v)$, is defined to be the number of neighbors of v in G. A vertex v is *dominated* by a neighbor u of it if v is adjacent to all neighbors of u. A vertex $u \in N_2(v)$ is called a *satellite* of v if there is a neighbor p of v such that $N[p] - N[v] = \{u\}$. The vertex p is also called the *parent* of the satellite u at v. A vertex subset V' is called an *independent set* of a graph if there is no edge between any two vertices in V'. An edge subset E' is called an *induced matching* of a graph if the induced graph $G[V(E')]$ has maximum degree 1. We will use $\alpha(G)$ to denote the size of a maximum induced matching in G. For dominated vertices and satellites, we will use the following bounds on the number of vertices adjacent to an edge.

Lemma 1. *In a graph without dominated vertices, for any edge vu it holds*

$$|N[\{v, u\}]| \geq 2 + \max\{d(v), d(u)\}.$$

Lemma 2. *If a vertex v is not dominated by another vertex and has no satellites, then for any edge vu incident on v it holds*

$$|N[\{v, u\}]| \geq 3 + d(v).$$

2.1   Relations to MAXIMUM INDEPENDENT SET

MAXIMUM INDUCED MATCHING is to find an induced regular subgraph of degree 1 and MAXIMUM INDEPENDENT SET is to find an induced regular subgraph of degree 0. We show some relations between them.

There is a simple reduction from MAXIMUM INDEPENDENT SET to MAXIMUM INDUCED MATCHING. For a graph G, we construct a new graph G' from G by adding a new degree-1 vertex v' for each vertex v in G such that v' is adjacent to v. It is not hard to prove that the graph G has an independent set of size k if and only if the graph G' has an induced matching of size k.

There is also a method to reduce an instance of MAXIMUM INDUCED MATCHING to an instance of MAXIMUM INDEPENDENT SET. This technique seems to be firstly introduced by Cameron [2] and has been used in several previous algorithms to solve MAXIMUM INDUCED MATCHING in some special graph classes. For a graph $G = (V, E)$, let $L(G)$ denote the line graph of G and $G^{<i>}$ denote the graph with vertex set V and edge set $\{v_j v_k | v_j, v_k \in V$ & the distance between v_j and v_k in G is at most $i\}$. It is not hard to observe that

Proposition 1. [2] *A graph G has an induced matching of size k if and only if $L(G)^{<2>}$ has an independent set of size k.*

Note that the number of vertices in $L(G)^{<2>}$ equals to the number of edges in G. When the graph G has not many edges, we can use fast algorithms for MIS to solve MIM based on the above proposition. But when the graph is a dense graph, this method will not be effective. We will turn to branch-and-search algorithms.

2.2   Branch-and-Search Algorithms

We use the branch-and-search technique in our algorithm. We may search a maximum induced matching of a given graph by recursively branching on the current graph into several smaller graphs until the problem becomes polynomially solvable or satisfies some properties. To evaluate the size of the search tree generated by this paradigm, we need to evaluate all branches in the algorithm. In our algorithm, we will simply select the number n of vertices in the graph as the measure. Let $C(n)$ denote the maximum number of leaves in the search tree generated by the algorithm for any graph with at most n vertices. When we branch on a graph G with l branches such that in the i-th branch the number of vertices decreases by at least a_i, we obtain a recurrence

$$C(n) \leq C(n - a_1) + C(n - a_2) + \cdots + C(n - a_l).$$

The largest root of the function $f(x) = 1 - \sum_{i=1}^{l} x^{-a_i}$ is called the *branching factor* of the recurrence. Let γ be the maximum branching factor among all branching factors in the algorithm. The size of the search tree that represents the branching process of the algorithm applied to an input n-vertex graph is given by $O(\gamma^n)$. More details about the analysis and how to solve recurrences can be found in the monograph [7]. Note that for two recurrences $C(n) \leq \sum_{i=1}^{l} C(n-a_i)$ and $C(n) \leq \sum_{i=1}^{l} C(n - b_i)$ with $a_i \geq b_i$ $(i = 1, 2, \ldots, l)$, the branch factor of the first recurrence is not greater than this of the second one. This property will be frequently used in our analysis.

We can design branch-and-search algorithms for MIM mainly because of the following fact:

Fact 1. *Let G be a graph, if a vertex v is not in a maximum induced matching, then $\alpha(G) = \alpha(G - v)$; if there is a maximum induced matching containing an edge vu, then $\alpha(G) = \alpha(G - N[\{v, u\}]) + 1$.*

3   The Main Idea to Design Algorithms

By Proposition 1, we can reduce MIM to MIS and use fast algorithms for MIS to solve MIM when the number of edges in the input graph is not too large. When the maximum degree of the graph is at most 4, the number of edges in the graph is at most two times of the number of vertices. For this case, we can solve MIM in $O^*(1.1996^{2n}) = O^*(1.4391^n)$ time by using the $O^*(1.1996^n)$-time algorithm for MIS in [20]. When the graph has vertices of degree ≥ 5, we will use some branching rules to search a solution. There is a simple branching rule:

(B1) *Branching on a vertex v of degree d is to generate $d+1$ branches by either excluding v from the maximum induced matching or including each edge incident on v to the maximum induced matching.*

By Fact 1, we know that

$$\alpha(G) = \max\{\alpha(G-v), \max_{u_i \in N(v)} (\alpha(G - N[\{v, u_i\}]) + 1)\}.$$

Record that we use $C(n)$ to denote the number of leaves in the search tree generated by our algorithm when runs on a graph with at most n vertices. Let $d = d(v)$ be the degree of v. In the branch where edge vu_i is included to the maximum induced matching in the above branching operation, at least $d + 1$ vertices in $N[v] \subseteq N[\{v, u_i\}]$ are deleted. We can get the following recurrence

$$C(n) \leq C(n-1) + d \cdot C(n - (d+1)). \tag{1}$$

Note that it is possible $N[v] = N[\{v, u_i\}]$, since u_i may be adjacent to vertices only in $N[v]$. We use Rule (B1) only to deal with vertices of degree ≥ 5. Then we have $d \geq 5$ in (1). In fact, (1) has a largest branching factor when $d = 5$. For this case, we get $C(n) \leq C(n-1) + 5C(n-6)$ and the branching factor is 1.5532.

To improve the above result in (1), we can first use effective operations to deal with dominated vertices and then branch on vertices v of degree ≥ 5 with Rule (B1) only when each neighbor u of v is adjacent to at least one vertex in $V - N[v]$. By Lemma 1, we can improve (1) to

$$C(n) \leq C(n-1) + d \cdot C(n - (d+2)). \tag{2}$$

For the worst case that $d = 5$, the branching factor is 1.4786. This is the idea how the algorithms in [10] and [3] get the running time bound of $O^*(1.4786^n)$.

In this paper, we investigate more structural properties of MIM and deal with satellites of vertices of degree ≥ 5 in effective ways. Then we can guarantee that each neighbor u_i of v is adjacent to at least two vertices in $V - N[v]$ when branching on a vertex v of degree ≥ 5 with Rule (B1). By Lemma 2 we can improve (1) to

$$C(n) \leq C(n-1) + d \cdot C(n - (d+3)). \tag{3}$$

For the worst case that $d = 5$, the branching factor is 1.4231. The new running time bound will be $O^*(1.4391^n)$ to solve the problem in graphs with degree at most 4 by using algorithms for MIS in [20].

In the next section, we first give some structural properties that will be used to design our algorithm.

4   Structural Properties

Lemma 3. *Let G' be an induced subgraph of a graph G. If G has a maximum induced matching S such that $S \subseteq E(G')$, then any maximum induced matching of G' is also a maximum induced matching of G.*

The property in Lemma 3 is called the *induced subgraph property* of MIM. The induced subgraph property will be used to prove some lemmas.

Lemma 4. *If a graph G has two nonadjacent vertices v and u such that $N(v) = N(u)$, then $\alpha(G) = \alpha(G')$, where $G' = G - \{v\}$.*

By the induced subgraph property in Lemma 3, to prove this lemma we only need to show that there is a maximum induced matching S of G such that $S \subseteq E(G')$. More details can be found in the full version of this paper.

An edge vu is called *pendent* if one endpoint v of it is a degree-1 vertex and the neighbor set of the other endpoint u induces two cliques, i.e., $G[N(u)]$ consists of two cliques. Note that one clique in $G[N(u)]$ contains only one vertex v. We have the following lemma to deal with pendent edges.

Lemma 5. *If a graph has a pendent edge vu, then there is a maximum induced matching containing vu.*

Note that if a degree-1 vertex is adjacent to a degree-2 vertex or adjacent to a degree-3 vertex contained in a triangle, then the edge incident on the degree-1 vertex is a pendent edge.

A path of 5 vertices $v_1v_2v_3v_4v_5$ is called a *chain* if the three inner vertices v_2, v_3 and v_4 are degree-2 vertices and there is no edge between v_1 and v_5. We can use the following lemma to deal with chains.

Lemma 6. *Let G be a graph having a chain $v_1v_2v_3v_4v_5$, and G^* be the graph obtained from G by deleting $\{v_2, v_3, v_4\}$ and adding a new edge between v_1 and v_5. It holds that*

$$\alpha(G) = \alpha(G^*) + 1.$$

The operation to construct G^* from G in the above lemma is called *chain reduction*. We can apply chain reductions to eliminate all chains in the graph in polynomial time. Note that $v_1v_2v_3v_4v_5$ is not a chain when v_1 and v_5 are adjacent and the chain reduction is not applicable on it for this case.

Recall that for two vertices v and u, if $N[v] \supseteq N[u]$, we say that v is dominated by u. We also say that a vertex v is *strongly dominated* by two neighbors u_1 and u_2 if v is dominated by u_1 and u_2, respectively, and there is an edge between u_1 and u_2. A dominated vertex is *weakly dominated* if it is not strongly dominated by two neighbors. We use the following properties in our algorithms.

Lemma 7. *If a graph G contains a strongly dominated vertex v, then $\alpha(G) = \alpha(G')$, where $G' = G - \{v\}$.*

Lemma 8. *Let v be a vertex dominated by u. If there is a maximum induced matching S such that $v \in V(S)$, then there is a maximum induced matching S' containing edge vu.*

Proof. If $vu' \in S$ but $u' \neq u$, then we obtain another maximum induced matching $S' = (S - \{vu'\}) \cup \{vu\}$ containing vu. □

Lemma 8 has been used to design a branching rule to deal with dominated vertices in previous algorithms [10]. In our algorithms, we will use it to design a branching rule for some weakly dominated vertices.

(B2) *Branching on a weakly dominated vertex v means to generate two instances by deleting v from the graph or include an edge vu to the maximum induced matching, where u dominates v.*

Note that if v is weakly dominated by a degree-1 vertex u, then in the branch where v is removed, the vertex u becomes a degree-0 vertex, which can be removed directly. So for this case, we simply assume that v and u are removed in the first branch.

We also consider another special kind of weakly dominated vertices, which are dominated by a degree-2 vertex in a triangle.

Lemma 9. *If there is a degree-2 vertex v in a triangle vu_1u_2, then there is a maximum induced matching either containing one edge in $\{vu_1, vu_2\}$ or containing no edge incident on a vertex in $\{v, u_1, u_2\}$. Especially, if at least one vertex in $\{u_1, u_2\}$ is of degree ≤ 3, then there is a maximum induced matching containing one edge in $\{vu_1, vu_2\}$.*

Based on Lemma 9, we can design a good branching rule to deal with degree-2 vertices in triangles.

(B3) *Branching on a degree-2 vertex v with two adjacent neighbors u_1 and u_2 means*
   (i) *to generate three subbranches by either removing $\{v, u_1, u_2\}$ from the graph or including each of vu_1 and vu_2 to the maximum induced matching if both of u_1 and u_2 are of degree ≥ 4;*
   (ii) *to generate two subbranches by including either vu_1 or vu_2 to the maximum induced matching otherwise.*

5   The Algorithm and Its Analysis

We will use $\mathrm{MIM}(G)$ to denote our algorithm that is to compute the size of a maximum induced matching in the input graph G. For the purpose of presentation, our algorithm returns the size of a maximum induced matching instead of the maximum induced matching itself. Our algorithm $\mathrm{MIM}(G)$ is a recursive algorithm that consists of 12 steps, each of which except the last step will call the algorithm itself. We will analyze the correctness and running time of each step after describing it.

In the first four steps, we just reduce the graph without branching. Step 5 will branch on weakly dominated vertices with a branching factor at most 1.3803. After Step 5, the graph has no dominated vertices. Step 6 will branch on a graph having a degree-2 vertex with two degree-2 neighbors (the case not included in chains) with a branching factor at most 1.2168 and Step 7 will branch on a graph with two adjacent degree-2 vertices with a branching factor at most 1.4313. After

Step 7, each degree-2 vertex in the graph is adjacent to two vertices of degree ≥ 3. In Step 8, the algorithm will branch with a branching factor at most 1.4227 if there is degree-3 vertex in a triangle. Step 9 will deal with degree-2 vertices with at least a neighbor of degree ≥ 4 by branching with a branching factor at most 1.4231. After all the above steps, the algorithm can branch on satellites of vertices of degree ≥ 5 with a branching factor at most 1.4383. Finally, we can branch vertices of degree ≥ 5 with (3) in Step 11 and use fast algorithms for MIS to solve the problem in Step 12 when the graph has maximum degree 4.

Step 1 (Similar vertex pairs). If the graph has two nonadjacent vertices v and u such that $N(v) = N(u)$, return $\mathrm{MIM}(G - \{v\})$.

Step 2 (Pendent edges). If there is a pendent edge vu, return $\mathrm{MIM}(G - N[\{v, u\}]) + 1$.

Step 3 (Chains). If there is a chain $v_1 v_2 v_3 v_4 v_5$, then construct G^* from the current graph by deleting v_2, v_3 and v_4 and adding an edge between v_1 and v_5, and return $\mathrm{MIM}(G^*) + 1$.

Step 4 (Strong dominated vertices). If there is a strong dominated vertex v, return $\mathrm{MIM}(G - \{v\})$.

The correctness of the first four steps are based on Lemmas 4, 5, 6 and 7, respectively. Note that each application of the first four steps will decrease the number of vertices in the graph by at least 1.

Step 5 (Weakly dominated vertices). If there is a vertex v weakly dominated by u in the graph G, branch on v with Rule (B2) by returning

$$\max\{\mathrm{MIM}(G - \{v\}), \mathrm{MIM}(G - N[\{v, u\}]) + 1\}.$$

The correctness of this step is based on Lemma 8. Next we analyze the recurrence generated by this branch. Let $d = d(v)$ be the degree of v. We have that $d \geq 3$, otherwise there would be a pendent edge or a strong dominated vertex. In the branch where vu is included to the maximum induced matching, the $d + 1$ vertices in $N[v]$ are eliminated. This branch leads to the following recurrence

$$C(n) \leq C(n - 1) + C(n - (d + 1)) \quad \text{with} \quad d \geq 3. \tag{4}$$

For the worst case that $d = 3$, the recurrence has a maximum branching factor of 1.3803. If u is a degree-1 vertex, in the first branch of removing v the vertex u become a degree-0 vertex and will also be removed from the graph directly. For this case, we can get the recurrence $C(n) \leq C(n - 2) + C(n - (d + 1))$ with $d \geq 3$ indeed. After Step 5, the graph has no dominated vertex.

Step 6 (Pentagons with three adjacent degree-2 vertices). If there is a 5-cycle $v_1 v_2 v_3 v_4 v_5$ such that v_2, v_3 and v_4 are degree-2 vertices, then we branch into three branches by including each edge in $\{v_1 v_5, v_2 v_3, v_3 v_4\}$ to the maximum induced matching, i.e., return

$$1 + \max\{\mathrm{MIM}(G - N[\{v_1, v_5\}]), \mathrm{MIM}(G - N[\{v_2, v_3\}]), \mathrm{MIM}(G - N[\{v_3, v_4\}])\}.$$

The correctness of this step is based on the following observation. There are only three cases that either v_1v_5 is in the maximum induced matching or at least one of v_1 and v_5 is not in the maximum induced matching. When v_1 (resp., v_5) is not in a maximum induced matching, we delete it and v_2v_3 (resp., v_3v_4) becomes a pendent edge and can be included to the maximum induced matching directly. Then it is equivalent to including v_2v_3 (resp., v_3v_4) to the maximum induced matching.

Next we analyze the recurrence in this step. When v_1v_5 is included to the maximum induced matching, at least the 5 vertices in the 5-cycle are eliminated. When v_2v_3 (resp., v_3v_4) is included to the maximum induced matching, at least 4 vertices $\{v_1, v_2, v_3, v_4\}$ (resp., $\{v_2, v_3, v_4, v_5\}$) are eliminated. We get a recurrence

$$C(n) \le C(n-5) + 2C(n-4), \tag{5}$$

which is a branching factor of 1.2168.

Note that after Step 7 there is no degree-2 vertex adjacent to two degree-2 vertices, otherwise either Step 3 or Step 6 can be applied in the graph.

Step 7 (Pairs of adjacent degree-2 vertices). If there is a path $v_1v_2v_3v_4$ such that v_2 and v_3 are degree-2 vertices and v_1 and v_4 are of degree ≥ 3, then we first branch on v_1 with Rule (B1) and in the subbranch where v_1 is deleted we include the pendent edge v_2v_3 to the maximum induced matching directly, i.e., return

$$\max\{1 + \max_{u \in N(v_1)} \{MIM(G - N[\{v_1, u\}])\}, 1 + MIM(G - \{v_1, v_2, v_3, v_4\})\}.$$

Let $d = d(v_1)$ be the degree of v_1. There is no dominated vertex in this step. By Lemma 1, when v_1u is included to the maximum induced matching, at least $d+2$ vertices are removed. So we get a recurrence $C(n) \le dC(n - (d+2)) + C(n-4)$ with $d \ge 3$. For the worst case that $d = 3$, the recurrence has a maximum branching factor of 1.3413.

Step 8 (Degree-3 vertices in triangles). If there is a degree-3 vertex v with three neighbors u_1, u_2 and u_3 such that u_2 and u_3 are adjacent, then we first branch on u_1 with Rule (B1) and in the subbranch where u_1 is deleted we branch on the degree-2 vertex v in the triangle vu_2u_3 with Rule (B3), i.e., return

$$\max\{1 + \max_{u' \in N(u_1)} \{MIM(G - N[\{u_1, u'\}])\}, MIM(G - \{v, u_1, u_2, u_3\}),$$

$$1 + MIM(G - N[\{v, u_2\}]), 1 + MIM(G - N[\{v, u_3\}])\}$$

if both of u_2 and u_3 are of degree ≥ 4, and

$$1 + \max\{\max_{u' \in N(u_1)} \{MIM(G - N[\{u_1, u'\}])\}, MIM(G - N[\{v, u_2\}]), MIM(G - N[\{v, u_3\}])\}$$

if at least one of u_2 and u_3 is of degree ≤ 3.

Let $d \ge 2$ be the degree of u_1. By Lemma 1, when u_1u' is included to the maximum induced matching, at least $d + 2$ vertices are removed. When vu_i

$(i \in \{2,3\})$ is included to the maximum induced matching, at least 5 vertices are removed if the degree of u_i is 3 and at least 6 vertices are removed if the degree of u_i is at least 4. So we get recurrences

$$C(n) \leq dC(n - (d+2)) + C(n-4) + 2C(n-6) \quad \text{with} \quad d \geq 2, \qquad \text{or}$$

$$C(n) \leq dC(n - (d+2)) + 2C(n-5) \quad \text{with} \quad d \geq 2.$$

When $d = 3$, the above two recurrences have the maximum branching factors of 1.4227 and 1.3798, respectively.

Step 9 (Degree-2 vertices adjacent to a vertex of degree ≥ 4). If there is a degree-2 vertex v adjacent to a vertex u_1 of degree ≥ 4, then the other neighbor u_2 of v is a vertex of degree ≥ 3, since there are no two adjacent degree-2 vertices now. For this case, we first branch on u_1 with Rule (B1) and in the subbranch where u_1 is deleted we branch on the dominated vertex u_2 with Rule (B2), i.e., return

$$\max\{1 + \max_{u' \in N(u_1)} \{\mathrm{MIM}(G - N[\{u_1, u'\}])\}, \mathrm{MIM}(G - \{u_1, v, u_2\}), 1 + \mathrm{MIM}(G - N[\{v, u_2\}])\}.$$

Let $d_1 \geq 4$ and $d_2 \geq 3$ be the degree of u_1 and u_2, respectively. By Lemma 1, at least $d_1 + 2$ vertices are removed when $u_1 u'$ is included to the maximum induced matching, and at least $d_2 + 2$ vertices are removed when $v u_2$ is included to the maximum induced matching. We get a recurrence

$$C(n) \leq d_1 C(n - (d_1 + 2)) + C(n-3) + C(n - (d_2 + 2)) \quad \text{with} \quad d_1 \geq 4, d_2 \geq 3.$$

For the worst case that $d_1 = 4$ and $d_2 = 3$, the recurrence has a maximum branching factor of 1.4231.

Step 10 (Satellites of vertices of degree ≥ 5). Assume that there is a satellite v of a vertex u with $d(u) \geq 5$. Case (i): If u is also a satellite of v, we branch on u with Rule (B1) and in the subbranch where u is deleted we branch on the dominated vertex v with Rule (B2), i.e., return

$$\max\{1 + \max_{u' \in N(u)} \{\mathrm{MIM}(G - N[\{u, u'\}])\}, \mathrm{MIM}(G - \{u, v\}), 1 + \mathrm{MIM}(G - N[\{v, p'\}])\},$$

where p' dominates v in $G - \{u\}$.

Case (ii): If u is not a satellite of v, we branch on v with Rule (B1) and in the subbranch where v is deleted we branch on the dominated vertex u with Rule (B2), i.e., return

$$\max\{1 + \max_{v' \in N(v)} \{\mathrm{MIM}(G - N[\{v, v'\}])\}, \mathrm{MIM}(G - \{u, v\}), 1 + \mathrm{MIM}(G - N[\{u, p'\}])\},$$

where p' dominates u in $G - \{v\}$.

Let d_1 and d_2 be the degrees of u and v, respectively. We have that $d_1 \geq 5$ by the assumption. We can see that $d_2 \geq 3$. If v is a degree-2 vertex then v would be adjacent to a vertex of degree ≥ 4 or a degree-3 vertex in a triangle or a degree-2 vertex and then at least one of previous steps could be applied. We also let p denote the parent of v at u. Then $p \in N(v) \cap N(u)$.

First, we analyze Case (i). By Lemma 1, at least $d_1 + 2$ vertices are removed when an edge uu' incident on u is included to the maximum induced matching, and at least $d_2 + 2$ vertices are removed when vp is included to the maximum induced matching. We get a recurrence $C(n) \leq d_1 C(n - (d_1 + 2)) + C(n - 2) + C(n - (1 + d_2 + 2))$ with $d_1 \geq 5$ and $d_2 \geq 3$. When $d_1 = 5$ and $d_2 = 3$, the recurrence has a maximum branching factor of 1.4348.

Next, we consider Case (ii). By Lemma 1, at least $d_1 + 2$ vertices are removed when an edge vv' incident on v is included to the maximum induced matching, and at least $d_1 + 2$ vertices are removed when up is included to the maximum induced matching. Furthermore, we prove that at least $d_1 + 3$ vertices are removed when vp is included to the maximum induced matching. Since u is not a satellite of v, we know that there is a neighbor $u^* \neq u$ of p not adjacent to v. Then there are at least $d_1 + 3$ vertices in $N[v] \cup \{u, u^*\} \subseteq N[\{v, p\}]$. We get a recurrence $C(n) \leq (d_2 - 1)C(n - (d_2 + 2)) + C(n - (d_2 + 3)) + C(n - 2) + C(n - (d_1 + 2))$ with $d_1 \geq 5$ and $d_2 \geq 3$. When $d_1 = 5$ and $d_2 = 3$, the recurrence has a maximum branching factor of 1.4383.

Step 11 (Vertices of maximum degree ≥ 5). If there is a vertex v of degree ≥ 5, then branch on it with Rule (B1), i.e., return

$$\max\{1 + \max_{u \in N(v)} \{\text{MIM}(G - N[\{v, u\}])\}, \text{MIM}(G - v)\}.$$

Since in this step there is no dominated vertex and each vertices of degree ≥ 5 has no satellites, this branch leads to (3), i.e., $C(n) \leq C(n-1) + d \cdot C(n - (d+3))$ with $d = d(v) \geq 5$. For the worst case that $d = 5$, the branching factor is 1.4231.

Step 12 (Graphs with maximum degree at most 4). Construct the graph $L(G)^{<2>}$ in Proposition 1, use the $O^*(1.1996^n)$-time algorithm for MIS [20] to compute the size θ of a maximum independent set in $L(G)^{<2>}$, and return θ.

Since the graph in this step is of maximum degree 4, we know that there are at most $2n'$ vertices in $L(G)^{<2>}$, where n' is the number of vertices in the graph G in this step. This step can be executed in $O^*(1.4391^{n'})$ time. We can regard that the algorithm always branch with branching factor 1.4391 in this step.

Among all the steps above, Step 12 has the worst performance. In the worst case, for example the input graph is a graph with maximum degree 4, the algorithm may only excuse Step 12 and the running time will be $O^*(1.4391^n)$. Note that the algorithm for MIS in [20] uses only polynomial space and each of the first eleven steps uses polynomial space. Then our algorithm uses only polynomial space.

Theorem 2. *A maximum induced matching in a graph with n vertices can be computed in $O^*(1.4391^n)$ time and polynomial space.*

References

1. Beigel, R., Eppstein, D.: 3-coloring in time $o(1.3289^n)$. J. Algorithms **54**(2), 168–204 (2005)
2. Cameron, K.: Induced matchings. Discret. Appl. Math. **24**, 97–102 (1989)

3. Chang, M.-S., Hung, L.-J., Miau, C.-A.: An $O^*(1.4786^n)$-time algorithm for the maximum induced matching problem. In: Chang, R.-S. et al. (eds.): Advances in Intelligent Systems and Applications, SIST 20, pp. 49–58 (2013)
4. Duckwortha, W., Manloveb, D.F., Zito, M.: On the approximability of the maximum induced matching problem. J. Discret. Algorithms 3(1), 79–91 (2005)
5. Fomin, F.V., Gaspers, S., Pyatkin, A.V., Razgon, I.: On the minimum feedback vertex set problem: exact and enumeration algorithms. Algorithmica 52(2), 293–307 (2008)
6. Fomin, F.V., Grandoni, F., Kratsch, D.: A measure & conquer approach for the analysis of exact algorithms. J. ACM 56(5), 1–32 (2009)
7. Fomin, F.V., Kratsch, D.: Exact Exponential Algorithms. Springer, New York (2010)
8. Golumbic, M.C., Laskar, R.: Irredundancy in circular arc graphs. Discret. Appl. Math. 44, 79–89 (1993)
9. Golumbic, M.C., Lewenstein, M.: New results on induced matchings. Discret. Appl. Math. 101, 157–165 (2000)
10. Gupta, S., Raman, V., Saurabh, S.: Maximum r-regular induced subgraph problem: fast exponential algorithms and combinatorial bounds. SIAM J. Discret. Math. 26(4), 1758–1780 (2012)
11. Ko, C.W., Shepherd, F.B.: Bipartite domination and simultaneous matroid covers. SIAM J. Discret. Math. 16(4), 517–523 (2003)
12. Kobler, D., Rotics, U.: Finding maximum induced matchings in subclasses of clawfree and p5-free graphs, and in graphs with matching and induced matching of equal maximum size. Algorithmica 37, 327–346 (2003)
13. Mosera, H., Sikdar, S.: The parameterized complexity of the induced matching problem. Discret. Appl. Math. 157, 715–727 (2009)
14. Orlovicha, Y., Finkeb, G., Gordonc, V., Zverovichd, I.: Approximability results for the maximum and minimum maximal induced matching problems. Discret. Optim. 5, 584–593 (2008)
15. Stockmeyer, L.J., Vazirani, V.V.: NP-completeness of some generalizations of the maximum matching problem. Inf. Process. Lett. 15(1), 14–19 (1982)
16. Van Rooij, J.M., Bodlaender, H.L.: Exact algorithms for edge domination. Algorithmica 64(4), 535–563 (2012)
17. Van Rooij, J.M., Bodlaender, H.L.: Exact algorithms for dominating set. Discret. Appl. Math. 159(17), 2147–2164 (2011)
18. Xiao, M., Nagamochi, H.: An improved exact algorithm for undirected feedback vertex set. J.Comb. Optim. (2014). doi:10.1007/s10878-014-9737-x
19. Xiao, M., Nagamochi, A.: A refined exact algorithm for edge dominating. Theoret. Comput. Sci. 560, 207–216 (2014)
20. Xiao, M., Nagamochi, H.: Exact algorithms for maximum independent set. In: Cai, L., Cheng, S.-W., Lam, T.-W. (eds.) Algorithms and Computation. LNCS, vol. 8283, pp. 328–338. Springer, Heidelberg (2013)

Completion of the Mixed Unit Interval Graphs Hierarchy

Alexandre Talon[1(✉)] and Jan Kratochvil[2]

[1] ENS Lyon, Lyon, France
alexandre.talon@ens-lyon.fr
[2] Department of Applied Mathematics, Faculty of Mathematics and Physics,
Charles University, Praha, Czech Republic
honza@kam.mff.cuni.cz

Abstract. We describe the missing class of the hierarchy of mixed unit interval graphs, generated by the intersection graphs of closed, open and one type of half-open intervals of the real line. This class lies strictly between unit interval graphs and mixed unit interval graphs. We give a complete characterization of this new class, as well as a polynomial time algorithm to recognize graphs from this class and to produce a corresponding interval representation if one exists.

Keywords: Unit interval graph · Mixed unit interval graph · Proper interval graph · Intersection graph

1 Introduction

A graph is an interval graph if one can associate with each of its vertices an interval of the real line such that two vertices are adjacent if and only if the corresponding intervals intersect. A well-studied subclass of the class of interval graphs is the one of proper interval graphs where it is required that no interval properly contains another one. This class coincides with the class of unit interval graphs where all intervals have length one [6].

However, in this description no particular attention is paid to the types of intervals we use: are they open, closed, or semi-closed? Dourado and al. proved in [1] that this is of no importance as far as interval graphs are concerned. This is no longer the case, though, for unit interval graphs: deciding which types of intervals are allowed to represent the vertices of a graph is crucial. This fact was notably studied in [1–3,5–7]. In these papers one can find results about the classes of graphs we can get depending on the types of unit intervals we allow for their representations. In particular it is shown that if all intervals in a representation are required to be of the same type (all closed, all open, all left-closed-right-open, or all left-open-right-closed), one gets the same

Supported by CE-ITI project GACR P202/12/G061.

A full version can be found at http://arxiv.org/abs/1412.0540.

R. Jain et al. (Eds.): TAMC 2015, LNCS 9076, pp. 284–296, 2015.
DOI: 10.1007/978-3-319-17142-5_25

class of *unit interval graphs* which is a proper subclass of *mixed unit interval graphs*, i.e., graphs obtained if no restriction – apart from the unit length – on the intervals is imposed. Recently, Joos [3] gave a characterization of mixed unit interval graphs by an infinite class of forbidden induced subgraphs, and Shuchat et al. [7] complemented it by a polynomial-time recognition algorithm. In [4], Le and Rautenbach take a different approach and study the graphs which are representable by intervals beginning at integer positions.

The aim of this paper is to complete this hierarchy of classes. We consider all subsets of the four types of unit intervals, show that several of them lead to the classic unit interval graphs (where all intervals are closed), recall the previously studied and characterized class determined by open and closed unit intervals, and then show that – with respect to this parametrization – there exists exactly one other proper subclass of the class of mixed unit interval graphs. We characterize this class by an infinite list of forbidden induced subgraphs, give a polynomial-time algorithm to check whether a graph belongs to this class, as well as an algorithm to produce an appropriate interval representation of any graph of this class.

2 Preliminaries

2.1 First Definitions and Notations

All the graphs we consider here are finite, undirected, and simple. Let G be a graph. We denote the vertex and edge set of G by $V(G)$ and $E(G)$, respectively, or V and E if there is no ambiguity. We say that two vertices u and v are neighbors, adjacent, or connected if $\{u, v\} \in E(G)$. For a vertex $v \in V(G)$, let the *neighborhood* $N_G(v)$ of v be the set of all vertices which are adjacent to v and let the *closed neighborhood* $N_G[v]$ be defined by $N_G(v) \cup \{v\}$. Two distinct vertices u and v are *twins* (in G) if $N_G[u] = N_G[v]$. If G contains no twins, then G is *twin-free*. If C is a set of vertices, then we denote by $G[C]$ the subgraph of G induced by C. Let $\mathcal{M}$ be a set of graphs. We say that G is $\mathcal{M}$-*free* if for every $H \in \mathcal{M}$, the graph H is not an induced subgraph of G. Let $\mathcal{N}$ be a family of intervals. We say that a graph G has an $\mathcal{N}$-*representation* if there is a function $I : V(G) \rightarrow \mathcal{N}$ such that for any two distinct vertices u and v, there is an edge joining u and v if and only if $I(u) \cap I(v) \neq \emptyset$. We say that G is an $\mathcal{N}$-*graph* if there is an $\mathcal{N}$-representation of G. Let $x, y \in \mathbb{R}$. We define the *closed interval* $[x, y] = \{z \in \mathbb{R} : x \leq z \leq y\}$, the *open interval* $(x, y) = \{z \in \mathbb{R} : x < z < y\}$, the *open-closed interval* $(x, y] = \{z \in \mathbb{R} : x < z \leq y\}$ and the *closed-open interval* $[x, y) = \{z \in \mathbb{R} : x \leq z < y\}$. We will draw the different types of intervals as follows (Fig. 1):

Fig. 1. The closed, open, closed-open, and open-closed intervals.

For an interval A, let $\ell(A) = \inf(\{x \in \mathbb{R} \mid x \in A\})$ and $r(A) = \sup(\{x \in \mathbb{R} \mid x \in A\})$. If I is an interval representation of G and $v \in V(G)$, then we write $\ell(v)$ and $r(v)$ instead of $\ell(I(v))$ and $r(I(v))$, if there are no ambiguities.

Let $\mathcal{U}^{++}$ be the set of all closed unit intervals, $\mathcal{U}^{--}$ be the set of all open unit intervals, $\mathcal{U}^{-+}$ be the set of all open-closed unit intervals, $\mathcal{U}^{+-}$ be the set of all closed-open unit intervals, and $\mathcal{U}$ be the set of all unit intervals. We also define $\mathcal{U}^{\pm} = \mathcal{U}^{++} \cup \mathcal{U}^{--}$ and $\mathcal{U}^X = \bigcup_{x \in \{X\}} \mathcal{U}^x$ for every $\{X\} \subset \mathcal{P}(\{++, --, -+, +-, \pm\})$. For instance, $\mathcal{U} = \mathcal{U}^{\pm, +-, -+}$. In this terminology, $\mathcal{U}$-graphs are *mixed unit interval graphs*. Let us call a $\mathcal{U}^{\pm, +-}$-graph an *almost-mixed unit interval graph*.

2.2 Previous Results

First we can see that if a graph contains twins, then they can be assigned the same intervals, so in what follows we will mostly consider twin-free graphs. We will denote by $\mathcal{G}^X$ the set of all twin-free $\mathcal{U}^X$-graphs. We begin by recalling the results about classifying the unit interval classes and characterizing them.

Theorem 1 (Roberts [6]). *A graph G is a $\mathcal{U}^{++}$-graph if and only if it is a $K_{1,3}$-free interval graph.*

Theorem 2 (Dourado et al., Frankl and Maehara [1,2]). *The classes of $\mathcal{U}^{++}$-graphs, $\mathcal{U}^{--}$-graphs, $\mathcal{U}^{+-}$-graphs, $\mathcal{U}^{-+}$-graphs, and $\mathcal{U}^{+-,-+}$-graphs are the same.*

Theorem 3 (Rautenbach and Szwarcfiter [5]). *A graph G is in $\mathcal{G}^{\pm}$ if and only if G is a $\{K_{1,4}, K_{1,4}^*, K_{2,3}^*, K_{2,4}^*\}$-free interval graph (Fig. 2).*

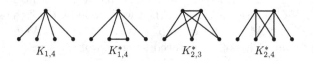

$$K_{1,4} \qquad K_{1,4}^* \qquad K_{2,3}^* \qquad K_{2,4}^*$$

Fig. 2. Forbidden induced subgraphs for twin-free $\mathcal{U}^{\pm}$-graphs

It is easy to see that these three classes of interval graphs are not the same. Indeed, $K_{1,3}$ is a $\mathcal{U}^{\pm}$-graph but not a $\mathcal{U}^{++}$-graph. Also, the graph of Fig. 3 is a $\mathcal{U}$-graph but not a $\mathcal{U}^{\pm}$-graph. A characterization of twin-free $\mathcal{U}$-graphs was recently given by Joos (the classes $\mathcal{R}$, $\mathcal{S}$, $\mathcal{S}'$, and $\mathcal{T}$ of forbidden induced subgraphs are depicted in Figs. 4, 5, 6 and 7):

Fig. 3. A graph, which is a $\mathcal{U}$-graph, but not a $\mathcal{U}^{\pm}$-graph

Fig. 4. The class $\mathcal{R}$

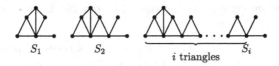

Fig. 5. The class $\mathcal{S}$

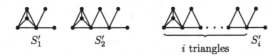

Fig. 6. The class $\mathcal{S}'$

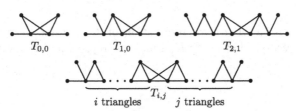

Fig. 7. The class $\mathcal{T}$

Theorem 4 (Joos [3]). *A graph G is in $\mathcal{G}$ if and only if G is a $\{K_{2,3}^*\} \cup \mathcal{R} \cup \mathcal{S} \cup \mathcal{S}' \cup \mathcal{T}$-free interval graph.*

To summarize, so far we have the following inclusions, all being proper:
$\{(\varnothing,\varnothing)\} \subsetneq \{\mathcal{U}^{++},\mathcal{U}^{--},\mathcal{U}^{+-},\mathcal{U}^{-+}, \text{or } \mathcal{U}^{+-,-+}\}$-graphs $\subsetneq \mathcal{U}^{\pm}$-graphs $\subsetneq$ $\mathcal{U}$-graphs.

However so far we have seen only 9 different sets of unit interval types, out of the 16 which exist. In the next section we will complete the picture.

3 Our Results

In this part we take care of each of the 7 missing subsets for the unit interval representations of graphs. We first consider the subsets which lead to the class of $\mathcal{U}^{++}$-graph, and then introduce the new one.

3.1 Completion of the Unit Interval Graphs Hierarchy

Theorem 5. *The classes of $\mathcal{U}^{++}$-graphs, $\mathcal{U}^{++,+-}$-graphs, $\mathcal{U}^{++,-+}$-graphs, $\mathcal{U}^{--,+-}$-graphs, $\mathcal{U}^{--,-+}$-graphs, $\mathcal{U}^{++,+-,-+}$ and $\mathcal{U}^{--,+-,-+}$ are the same.*

Proof. The proof is straightforward and appears in the full version.   ■

We now deal with the remaining two subsets of intervals: $\mathcal{U}^{\pm,+-}$ and $\mathcal{U}^{\pm,-+}$ which lead, by symmetry, to the same class of graphs. We first show that this is a proper new class. In order to do so, we introduce a lemma about the essence of the $\mathcal{U}^{\pm,+-}$ class: the existence of an induced $K_{1,4}^*$. We call a representation *injective* if no two vertices are represented by the same interval. Every representation of a twin-free graph is injective.

Lemma 1. *Up to symmetry, there are only two injective $\mathcal{U}$-representations of $K_{1,4}^*$, shown in Fig. 8 (the leftmost interval is either open-closed or closed).*

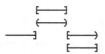

Fig. 8. The unique representations of $K_{1,4}^*$

Proof. Let us consider I a $\mathcal{U}$-representation of $K_{1,4}^*$. First from the proof of Theorem 5, we can see that every $K_{1,3}$ must be represented this way (Fig. 9) :

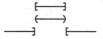

Fig. 9. The unique injective $\mathcal{U}$-representation of $K_{1,3}$

Let us denote the two leaves of $K_{1,4}^*$ by a and b, the vertex of maximum degree by c, and the other two nodes by d and e. We have the following claws: *cabd* and *cabe*. Since c is connected to all the other vertices, I(c) must be the middle closed interval. Then I(a) and I(b) must only intersect I(c), so one must be the middle open interval, and the other one for instance the leftmost one.

Once this is done, the positions of the intervals representing d and e are uniquely determined, but the two intervals cannot be equal since I is injective. They must also intersect I(c), and must not intersect I(b), so they are uniquely determined as in Fig. 8. Note that the left end of the leftmost interval is free.   ■

Theorem 6. *The following strict inclusions hold: $\mathcal{U}^{\pm}$-graphs $\subsetneq \mathcal{U}^{\pm,+-}$-graphs $\subsetneq \mathcal{U}$-graphs.*

Proof. The inclusions are immediate, we only need to show that they are strict. The proof can be found in the full version.   ■

To conclude this part, we now have a complete picture of the different subclasses of the mixed unit interval class. In the schematic figure below, $\mathcal{U}^X \subsetneq \mathcal{U}^Y$

is a shorthand notation for $\mathcal{U}^X$-graphs $\subsetneq \mathcal{U}^Y$-graphs. Sets separated by commas define the same classes of graphs (Fig. 10).

$$\varnothing$$

$$\text{\rotatebox{90}{$\subseteq$}}$$

$$\mathcal{U}^{++}, \quad \mathcal{U}^{--}, \quad \mathcal{U}^{+-}, \quad \mathcal{U}^{-+}, \quad \mathcal{U}^{+-,-+}, \quad \mathcal{U}^{++,+-},$$
$$\mathcal{U}^{++,-+}, \quad \mathcal{U}^{--,+-}, \quad \mathcal{U}^{--,-+}, \quad \mathcal{U}^{++,+-,-+}, \quad \mathcal{U}^{--,+-,-+}$$

$$\text{\rotatebox{90}{$\subseteq$}}$$

$$\mathcal{U}^{\pm}$$

$$\text{\rotatebox{90}{$\subseteq$}}$$

$$\mathcal{U}^{\pm,+-}, \quad \mathcal{U}^{\pm,-+}$$

$$\text{\rotatebox{90}{$\subseteq$}}$$

$$\mathcal{U}$$

Fig. 10. Classification of the mixed unit interval graphs subclasses

3.2 Characterization of the New Class

In this part, we characterize the new $\mathcal{G}^{\pm,+-}$ class by a list of minimal forbidden induced subgraphs. We begin by finding this list through a reasoning by inference, and afterwards check that all these graphs are indeed forbidden, and minimal.

We begin by a very important lemma for what follows. It guarantees that any graph belonging to $\mathcal{G} \setminus \mathcal{G}^{\pm,+-}$ has a "good" interval representation in which almost each half-closed interval is surrounded by a certain neighborhood.

Lemma 2. *Let $G \in \mathcal{G} \setminus \mathcal{G}^{\pm}$ and I a $\mathcal{U}$-representation of it. Then one of the following statements is true:*

(i) There exists a $\mathcal{U}$-representation I' of G with fewer open-closed (resp. closed-open) intervals

(ii) For every vertex u' (resp. d') such that $I(u')$ (resp. $I(d')$) is an open-closed (resp. closed-open) interval there exist vertices u, v, w, x, y (resp. a, b, c, d, e) in the same connected component as u' (resp. d') such that their intervals are the following:

Proof. We prove the lemma only for the case with u', the other one being completely symmetrical. Also, up to translation, we will assume that $\ell(u) = 0$. We assume that (i) is false, and show that in this case (ii) is true. We first set $u = u'$.

The overall idea of the proof is that, if one of the mentioned intervals is missing, then we can shift some intervals and close the left end of I(u) so as to get a representation I$'$, equivalent to I, with the same number of closed-open intervals but with one fewer open-closed intervals, hence a contradiction. To do so, we first define

$$\epsilon = \min(\{1\} \cup \{|x - y| \mid x, y \in \bigcup_{t \in V(G)} \{l(t), r(t)\} \wedge x \neq y\}).$$

This quantity equals the smallest non-zero distance between any two distinct ends of any two (non necessarily different) intervals, or 1 if such a distance does not exist.

Remark 1. Let $0 < \epsilon' < \epsilon$. If a vertex x is such that I(x) has an open left (resp. right) end, we can either shift it by ϵ' (resp. $-\epsilon'$) or shift any other set of intervals by $-\epsilon'$ (resp. ϵ') without loosing any intersection involving I(x).

Definition 1. *We say that the interval of a vertex x is left-free (resp. right-free) if there is no other vertex t such that $r(t) = \ell(x)$ (resp. $\ell(t) = r(x)$).*

Remark 2. Let $0 < \epsilon' < \epsilon$ and I(x) be a left-free (resp. right-free) interval. Closing its left (resp. right) end does not create any intersection.

Definition 2. *We say that a vertex x has an integer interval if $\ell(x) \in \mathbb{Z}$.*

Claim 1. If I(u) is open-closed, then there exists some closed I(v) at the same position.

Proof (of Claim 1). We assume for contradiction that there is no such I(v). We would like to close the left end of I(u). To do so, let us define I$'$ the following way:

- I$'(t) = $ I(t) $- \epsilon/2$ if $\ell(I(t)) \in \mathbb{Z}$, $\ell(I(t)) \leq 0$ and $t \neq u$
- I$'(u) = [0, 1]$ (now it is closed)
- I$'(t) = $ I(t) otherwise

We now show that I and I$'$ are equivalent.

By the definition of ϵ, we modify no intersection involving any non-integer interval. Since we do not shift the intervals beginning from 1 on, and we shift all integer intervals J such that $\ell(J) \leq 0$ by the same quantity, the only intersections we can change involve I(u) or an interval at the same position as I(u). Since I is injective and there is no $[0, 1]$ interval, any interval sharing the position of I(u) must have an open right end. Therefore it had no intersection at 1, and shifting it does not remove any intersection. The same applies for I(u): since its left end is open, it does not loose any intersection. Moreover, since we shifted all other integer intervals, we can close it without creating any new intersection.

This shows the equivalence between I and I$'$, so (i) is true, which is a contradiction. □

Claim 2. If $I(u)$ is open-closed, then there exists some closed $I(w)$ like in (i).

Claim 3. If $I(u')$ is open-closed, then there exist, in the same connected component as u', some vertices u, v, w and y with intervals like in (i) and such that there is no open-closed interval at the same position.

The proofs of Claims 2 and 3 are similar to the one of Claim 1 and can be found, as well as the proof of the existence of $I(z)$, in the full version. ∎

Now we look for all possible forbidden induced minimal subgraphs of any $G \in \mathcal{G} \setminus \mathcal{G}^{\pm,+-}$. Let us take such a graph G and consider I a $\mathcal{U}$-representation of G *with minimum number of open-closed intervals, and subject to this condition, minimum number of closed-open intervals.*

First, since $G \notin \mathcal{G}^{\pm,+-}$, there exist one open-closed interval $I(u)$ and one closed-open interval $I(d)$. By Lemma 2, they come with some neighbors $a, b, c, e, v,$ w, y, z represented by intervals exactly like in the lemma.

Remark 3. We may assume that $I(u)$ and $I(d)$ are connected through a succession of intervals.

Proof. We proceed by contradiction. If every such pair (u, d) was composed of "disconnected" vertices, then by symmetrizing all components containing (only) open-closed intervals we would get an interval representation I' with intervals in $\mathcal{U}^{\pm,+-}$. ∎

So from now on, we assume that u and d are in a same component. We also assume, translating the whole interval representation if necessary, that the intervals for a, b, c, d, e are fixed and that $\ell(a) = 0$. We now explore all the possible values for $\ell(u)$:

– $\ell(u) < -2$: This leads (see the full version for details) to class $\mathcal{A}$:

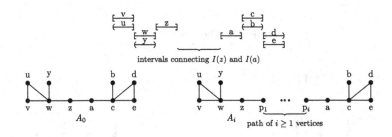

Fig. 11. The class $\mathcal{A}$ and its interval representation

– $\ell(u) \geq 3$: This leads (see the full version for details) to classes $\mathcal{B}, \mathcal{B}'$ and $\mathcal{B}''$:

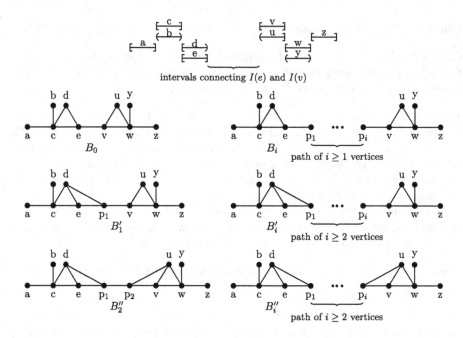

intervals connecting $I(e)$ and $I(v)$

path of $i \geq 1$ vertices

path of $i \geq 2$ vertices

path of $i \geq 2$ vertices

Fig. 12. The classes $\mathcal{B}$, $\mathcal{B}'$, $\mathcal{B}''$ and their interval representations

- $\ell(u) \in \mathbb{Z}$ and $-2 \leq \ell(u) < 3$:
  - $\ell(u) = -2$ (Fig. 13):

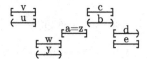   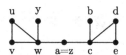

Fig. 13. The graph C_{-2}

- $\ell(u) = -1$ (Fig. 14):

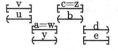

Fig. 14. The graph C_{-1}

- $\ell(u) = 0$ (Fig. 15):

Fig. 15. The graph C_0

- $\ell(u) = 1$ (Fig. 16):

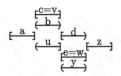 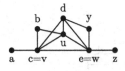

Fig. 16. The graph C_1

- $\ell(u) = 2$ (Fig. 17):

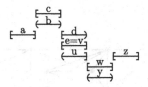 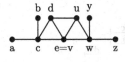

Fig. 17. The graph C_2

- $-2 < \ell(u) < 3$ and $\ell(u) \notin \mathbb{Z}$:
  - $-2 < \ell(u) < -1$ (Fig. 18):

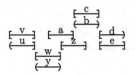 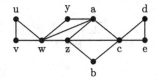

Fig. 18. The graph C'_{-2}

- $-1 < \ell(u) < 0$ (Fig. 19):

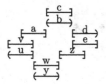 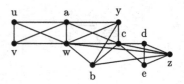

Fig. 19. The graph C'_{-1}

- $0 < \ell(u) < 1$ (Fig. 20):

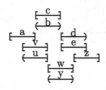 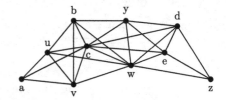

Fig. 20. The graph C'_0

- $1 < \ell(u) < 2$ (Fig. 21):

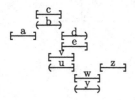 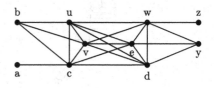

Fig. 21. The graph C'_1

- $2 < \ell(u) < 3$ (Fig. 22):

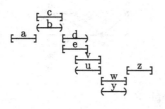 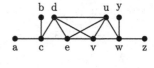

Fig. 22. The graph C'_2

We have to add the graphs which are forbidden even for $\mathcal{G}$. From the class $\mathcal{R}$ we only need R_0 and R_1 since the other ones are supergraphs of graphs in $\mathcal{B}$. We need all the graphs in $\mathcal{S}$ and $\mathcal{S}'$. We only have to add the graphs $T_{0,j}$ for $j \geq 0$ and $T_{1,1}$ because the $T_{i,j}$ with $i > 0$ or $j > 0$ are supergraphs of graphs in $\mathcal{B}$ and because for every $i, j \geq 0$, $T_{i,j} = T_{j,i}$.

Now we check that all these graphs are indeed forbidden. Since $\mathcal{G}^{\pm,+-} \subset \mathcal{G}$, we only need to check the classes we introduce in this article: $\mathcal{A}$, $\mathcal{B}$, $\mathcal{B}'$, $\mathcal{B}''$, $\mathcal{C}$ and $\mathcal{C}'$.

First, we justify the fact that the classes $\mathcal{B}$, $\mathcal{B}'$ and $\mathcal{B}''$ are forbidden. This is because they contain the following pattern:

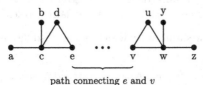

path connecting e and v

Indeed, Lemma 1 specifies that the two copies of $K^*_{1,4}$ must be represented, up to symmetry, as in Fig. 8. Since there is a path between e and v, which is vertex-disjoint from the vertices of the copies of $K^*_{1,4}$, the two interval representations must be symmetrical, hence the need for the two types of semi-closed intervals. For the class $\mathcal{A}$, we have the following pattern:

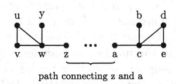

path connecting z and a

Here again we must have two occurrences of Fig. 8, but here vertices a and z are connected by a path which is vertex-disjoint from the two $K^*_{1,4}$, so these two occurrences must be symmetrical, hence the fact that these graphs are forbidden.

For the graphs C'_{-2}, C'_{-1}, C'_0, C'_1 and C'_2 the point is that we have two vertex-disjoint $K^*_{1,4}$ ($decba$ and $uvwyz$). By Lemma 1 we know that they can be represented by only two sets of intervals. However if we begin to draw the intervals for $decba$, then there is only one choice for $uvwyz$, up to a small translation.

For the graphs C_{-2}, C_{-1}, C_0, C_1 and C_2 the argument is the same, except that the two $K^*_{1,4}$ share some vertices. We first begin to draw $decba$, and then realize that the other intervals must be exactly like in the above figures.

From what precedes we can state:

Theorem 7. *A graph G is in $\mathcal{G}^{\pm,+-}$ if and only if it is a $\mathcal{A} \cup \mathcal{B} \cup \mathcal{B}' \cup \mathcal{B}'' \cup \mathcal{C} \cup$*
$\mathcal{C}' \cup \mathcal{S} \cup \mathcal{S}' \cup \{T_{0,j} \mid j \geq 0\} \cup \{T_{1,1}\} \cup \{R_0, R_1\}$-free interval graph.

Furthermore:

Theorem 8. *The graphs of Theorem 7 are minimal forbidden induced subgraphs for the class $\mathcal{G}^{\pm,+-}$.*

Proof. The proof is not difficult and can be found in the full version.   ∎

From Theorem 7 we can design an algorithm to recognize an almost-mixed unit interval graph: first check whether it is a mixed unit interval graph. After this, prune it to make it twin-free and search the result for the graphs in $\mathcal{C} \cup \mathcal{C}'$. To check for the infinity of graphs in the remaining classes we look for all the copies of $K_{1,4}^*$ and how they connect to one another, to find two of them connected like in Fig. 11 to Fig. 12. In fact, it is sufficient to check for every vertex-disjoint copies of $K1,4^*$ that there exist two vertices u and v such that:

- u and v are not in the same copy of $K_{1,4}^*$
- none of them is of maximum degree (*ie* degree 4) in $K_{1,4}^*$
- they are of equal degrees (*ie* they have the same "roles" in $K_{1,4}^*$)

and look for a path between u and v which is vertex-disjoint from the two $K_{1,4}^*$'s. This leads to:

Theorem 9. *The class of almost-mixed unit interval graphs can be recognized in polynomial time.*

In addition to that:

Theorem 10. *There exists a polynomial-time algorithm which, given a $\mathcal{U}^{\pm,+-}$-graph G, produces a $\mathcal{U}^{\pm,+-}$-representation of G.*

Proof. Let G be a $\mathcal{U}^{\pm,+-}$-graph. We first use prune G into G' which is twin-free. We then use the algorithm of [7] to get a $\mathcal{U}$-representation of G' in quatratic time. After this, we use the arguments of the proof of Lemma 2 to try to close first all open-closed intervals, and then all closed-open intervals. This can be done in time polynomial in the number of intervals, hence the result. We get a representation of G by assigning to each vertex the same interval as its twin which is in G'. At the end, we get a representation of G with at most one type of semi-closed intervals.   ∎

References

1. Dourado, M.C., Le, V.B., Protti, F., Rautenbach, D., Szwarcfiter, J.L.: Mixed unit interval graphs. Discrete Math. **312**, 3357–3363 (2012)
2. Frankl, P., Maehara, H.: Open interval-graphs versus closed interval-graphs. Discrete Math. **63**, 97–100 (1987)
3. Joos, F.: A characterization of mixed unit interval graphs. In: Kratsch, D., Todinca, I. (eds.) WG 2014. LNCS, vol. 8747, pp. 324–335. Springer, Heidelberg (2014)
4. Le, V.B., Rautenbach, D.: Integral mixed unit interval graphs. In: Gudmundsson, J., Mestre, J., Viglas, T. (eds.) COCOON 2012. LNCS, vol. 7434, pp. 495–506. Springer, Heidelberg (2012)
5. Rautenbach, D., Szwarcfiter, J.L.: Unit interval graphs of open and closed intervals. J. Graph Theory **72**(4), 418–429 (2013)
6. Roberts, F.S.: Indifference graphs. In: Harary, F. (Ed.) Proof Techniques in Graph Theory, pp. 139–146. Academic Press, New York (1969)
7. Shuchat, A., Shull, R., Trenk, A.N., West, L.C. Unit Mixed Interval Graphs, arXiv preprint arXiv:1405.4247 (2014)

Bounded Treewidth and Space-Efficient Linear Algebra

Nikhil Balaji$^{(\boxtimes)}$ and Samir Datta

Chennai Mathematical Institute (CMI), Chennai, India
{nikhil,sdatta}@cmi.ac.in

Abstract. Motivated by a recent result of Elberfeld, Jakoby and Tantau [EJT10] showing that MSO properties are Logspace computable on graphs of bounded treewidth, we consider the complexity of computing the determinant of the adjacency matrix of a bounded treewidth graph and as our main result prove that it is in Logspace. It is important to notice that the determinant is neither an MSO-property nor counts the number of solutions of an MSO-predicate. This technique yields Logspace algorithms for counting the number of spanning arborescences and directed Euler tours in bounded treewidth digraphs.

We demonstrate some linear algebraic applications of the determinant algorithm by describing Logspace procedures for the characteristic polynomial, the powers of a weighted bounded treewidth graph and feasibility of a system of linear equations where the underlying bipartite graph has bounded treewidth.

Finally, we complement our upper bounds by proving L-hardness of the problems of computing the determinant, and of powering a bounded treewidth matrix. We also show the GapL-hardness of Iterated Matrix Multiplication where each matrix has bounded treewidth.

1 Introduction

The determinant is a fundamental algebraic invariant of a matrix. For an $n \times n$ matrix A the determinant is given by the expression $\mathrm{Det}(A) = \sum_{\sigma \in S_n} \mathrm{sign}(\sigma) \prod_{i \in [n]} a_{i,\sigma(i)}$ where S_n is the symmetric group on n elements, σ is a permutation from S_n and $\mathrm{sign}(\sigma)$ is the parity of the number of inversions in σ ($\mathrm{sign}(\sigma) = 1$ if the number of inversions in σ is even and 0 if it is odd). Even though the summation in the definition runs over $n!$ many terms, there are many efficient sequential [vzGG13] and parallel [Ber84] algorithms for computing the determinant.

Apart from the inherently algebraic methods to compute the determinant there are also combinatorial algorithms (see, for instance, Mahajan and Vinay [MV97]) which extend the definition of determinant as a signed sum of cycle covers in the weighted adjacency matrix of a graph. Mahajan and Vinay [MV97] are thus able to give another proof of the GapL-completeness of the determinant, a result first proved by Toda [Tod91]. For a more complete discussion on the known algorithms for the determinant, see [MV97].

S. Datta—Part of the work was done on a visit to the Institute for Theoretical Computer Science at Leibniz University Hannover.

© Springer International Publishing Switzerland 2015
R. Jain et al. (Eds.): TAMC 2015, LNCS 9076, pp. 297–308, 2015.
DOI: 10.1007/978-3-319-17142-5_26

Armed with this combinatorial interpretation of the determinant and faced with its GapL-hardness, one can ask if the determinant is any easier when the underlying matrix represents simpler classes of graphs. Datta, Kulkarni, Limaye, Mahajan [DKLM10] study the complexity of the determinant and permanent, when the underlying directed graph is planar and show that they are as hard as the general case - GapL and #P-hard, respectively. We revisit these questions in the context of bounded treewidth graphs.

Many NP-complete graph problems become tractable when restricted to graphs of bounded treewidth. In an influential paper, Courcelle [Cou90] proved that any property of graphs expressible in Monadic Second Order MSO logic can be decided in linear time on bounded treewidth graphs. For example, Hamiltonicity is an MSO property and hence deciding if a bounded treewidth graph has a Hamiltonian cycle can be done in linear time. More recently Elberfeld, Jakoby, Tantau [EJT10] showed that in fact, MSO properties on bounded treewidth graphs can be decided in L.

We study the Determinant problem when the underlying directed graph has bounded treewidth and show a Logspace upper bound. In the same vein we also compute other linear algebraic invariants of a bounded treewidth matrix, such as the characteristic polynomial, rank and powers of a matrix in Logspace. Interpreting rectangular matrices as (weighted) bipartite graphs, we are also able to show that checking for the feasibility of a system of linear equations for such matrices arising from bounded treewidth bipartite graphs is in L. FSLE has previously been studied for general graphs in [ABO99] where it is shown to be complete for the first level of the Logspace counting hierarchy: $L^{C=L}$.

We give a tight bound on the complexity of the determinant by showing that it is L-hard via a reduction from directed reachability in paths. We also show that it is unreasonable to attempt to extend the Logspace upper bound of determinant and powering to Iterated Matrix Multiplication (IMM) of bounded treewidth matrices, by showing GapL-hardness for IMM. It is worthwhile to contrast this with the case of general graphs, where the Determinant, IMM and Matrix Powering are known to be inter-reducible to each other and hence complete for GapL.

1.1   Our Results and Techniques

Throughout this paper, we work with matrices with entries from $\mathbb{Q}$, unless stated otherwise. We show that the following can be computed/tested in L:

1. (Main Result) The Determinant of an $(n \times n)$ matrix A whose underlying undirected graph has bounded treewidth. As a corollary we can also compute the coefficients of the characteristic polynomial of a matrix.
2. The inverse of an $(n \times n)$ matrix A whose underlying undirected graph has bounded treewidth. As a corollary we get a Logspace algorithm to compute the powers A^k of a matrix A (with rational entries) whose support is a bounded treewidth digraph.

3. Testing if a system of rational linear equations $Ax = b$ is feasible where A is (a not necessarily square) matrix whose support is the biadjacency matrix of an undirected bipartite graph of bounded treewidth.
4. The number of Spanning Trees in graphs of bounded treewidth.
5. The number of Euler tours in a bounded treewidth directed graph.

We also show hardness results to complement the above easiness results:

1. Computing the determinant of a bounded treewidth matrix is L-hard which precludes further improvement in the Logspace upper bound.
2. Computing the iterated matrix multiplication of bounded treewidth matrices is GapL-hard which precludes attempts to extend the L-bound on powering matrices of bounded treewidth to iterated matrix multiplication.
3. Powering matrices are however L-hard which prevents attempts to further improve the L-bound on matrix powering.

At the core of the upper bound results is our algorithm to compute the determinant by writing down an MSO formula that evaluates to true on every valid cycle cover of the bounded treewidth graph underlying A. The crucial point being that the cycle covers are parameterised by the number of cycles in the cycle cover, a quantity closely related to the sign of the cycle covers. This makes it possible to invoke the cardinality version of Courcelle's theorem (for Logspace) due to [EJT10] to compute the determinant. A more subtle point is that in order to keep track of the number of cycles as the size of a set of vertices, we need to pick one vertex per cycle. Picking one vertex per cycle is done by choosing the "smallest" vertex in the cycle. In order to pick a vertex in a cycle cover, we need to define a total order on the vertices which makes this part of the proof technically challenging.

We use this determinant algorithm and the Kirchoff matrix tree theorem along with the BEST theorem to count directed Euler tours.

1.2   Organization of the Paper

Section 2 introduces some notation and terminology required for the rest of the paper. In Sect. 3, we give a Logspace algorithm to compute the Determinant of matrices of bounded treewidth and give some linear algebraic and graph theoretic applications. In Sect. 4, we give some L-hardness results to complement our Logspace upperbounds. In Sect. 5, we mention some problems that remain open.

2   Preliminaries

2.1   Background on Graph Theory

Definition 1. *Given an undirected graph $G = (V_G, E_G)$ a tree decomposition of G is a tree $T = (V_T, E_T)$ (the vertices in $V_T \subseteq 2^{V_G}$ are called bags), such that*

1. *Every vertex $v \in V_G$ is present in at least one bag, i.e., $\cup_{X \in V_T} X = V_G$.*
2. *If $v \in V_G$ is present in bags $X_i, X_j \in V_T$, then v is present in every bag X_k in the unique path between X_i and X_j in the tree T.*
3. *For every edge $(u, v) \in E_G$, there is a bag $X_r \in V_T$ such that $u, v \in X_r$.*

The width of a tree decomposition is the $\max_{X \in V_T}(|X| - 1)$. The treewidth of a graph is the minimum width over all possible tree decomposition of the graph.

Definition 2. *Given a weighted directed graph $G = (V, E)$ by its adjacency matrix $[a_{ij}]_{i,j \in [n]}$, a cycle cover C of G is a set of vertex-disjoint cycles that cover the vertices of G. i.e., $C = \{C_1, C_2, \ldots, C_k\}$, where $V(C_i) = \{c_{i_1}, \ldots, c_{i_r}\} \subseteq V$ such that $(c_{i_1}, c_{i_2}), (c_{i_2}, c_{i_3}), \ldots, (c_{i_{r-1}}, c_{i_r}), (c_{i_r}, c_{i_1}) \in E(C_i) \subseteq E$ and $\sqcup_{i=1}^{k} V(C_i) = V$. The least numbered vertex in cycle C_i, denoted h_i, is called the head of the cycle.*

Fact 1. *The weight of the cycle $C_i = \prod_{j \in [r]} wt(a_{ij})$ and the weight of the cycle cover $wt(C) = \prod_{i \in [k]} wt(C_i)$. The sign of the cycle cover C, $sign(C)$ is $(-1)^{n+k}$.*

Every permutation $\sigma \in S_n$ can be written as a union of vertex disjoint cycles. Hence a permutation corresponds to a cycle cover of a graph on n vertices. In this light, the determinant of an $(n \times n)$ matrix A can be seen as a signed sum of cycle covers: $\det(A) = \sum_C sign(C) wt(C)$

2.2   Background on MSO-Logic

Definition 3 (Monadic Second Order Logic). *Let the variables $V = \{v_1, v_2, \ldots, v_n\}$ denote the vertices of a graph $G = (V, E)$. Let X, Y denote subsets[1] of V or E. Let $E(x, y)$ be the predicate that evaluates to 1 when there is an edge between x and y in G. A logical formula ϕ is called an MSO-formula if it can be constructed using the following: (1) $v \in X$ (2) $v_1 = v_2$ (3) $E(v_1, v_2)$ (4) $\phi_1 \vee \phi_2$, $\phi_1 \wedge \phi_2$, $\neg \phi$ (5) $\exists x \phi, \forall x \phi$ (6) $\exists X \phi, \forall X \phi$*

In addition, if the Gaifman graph[2] of a relation $R(x_1, \ldots, x_n)$ is bounded treewidth, then we can use R in item (3) above. A property Π of graphs is MSO-definable, if it can be expressed as a MSO formula ϕ such that ϕ evaluates to TRUE on a graph G if and only if G has property Π. (See [FG06] for more background on MSO)

Definition 4 (Solution Histogram). *Given a graph $G = (V, E)$ and an MSO formula $\phi(X_1, \ldots, X_d)$ in free variables $X_1, \ldots, X_d$, where $X_i \subseteq V$ (or E), the $(i_1, \ldots, i_d)$-th entry of $histogram(G, \phi)$ gives the number of subsets $S_1, \ldots, S_d$ such that $|S_j| = i_j$ for which $\phi(S_1, \ldots, S_d)$ is true.*

---

[1] The case when quantification over subset of edges is not allowed is referred to as MSO₁ which is known to be strictly less powerful than MSO₂, the case when edge set quantification is allowed. Throughtout our paper, we will work with MSO₂ and hence we will just refer to it as MSO.

[2] The Gaifman graph (also called the *Primal Graph*) of a binary relation $R \subseteq A \times A$ is the graph whose nodes are elements of A and an edge joins a pair of variables x, y if $(x, y) \in R$.

We need the following results from [EJT10]:

Theorem 1 (Logspace Version of Bodlaender's Theorem). *For every* $k \geq 1$, *there is a Logspace machine that on input of any graph G of treewidth at most k outputs a width-k tree decomposition of G.*

Theorem 2 (Logspace Version of Courcelle's Theorem). *For every $k \geq 1$ and every MSO-formula ϕ, there is a Logspace machine that on input of any logical structure $\mathcal{A}$ of treewidth at most k decides whether $A \vDash \phi$ holds.*

Theorem 3 (Cardinality Version of Courcelle's Theorem). *Let $k \geq 1$ and let $\phi(X_1, \ldots, X_d)$ be an MSO-formula on free variables $X_1, \ldots, X_d$. Then there is a Logspace machine that on input of the tree decomposition of a graph G of treewidth at most k, MSO-formula ϕ and $(i_1, \ldots, i_d)$, outputs the value of histogram(G, ϕ) at $|X_1| = i_1, \ldots, |X_d| = i_d$.*

3 Determinant Computation

Given a square $\{0, 1\}$-matrix A, we can view it as the bipartite adjacency matrix of a bipartite graph H_A. The permanent of this matrix A counts the number of perfect matchings in H_A, while the determinant counts the signed sum of perfect matchings in H_A.

If G is a bounded treewidth graph then we can count the number of perfect matchings in G in L[EJT10] (see also [DDN13]). Hence the complexity of the permanent of A, above is well understood in this case while the complexity of computing the determinant is not clear.

On the other hand the determinant of a $\{0, 1\}$-matrix reduces to counting the number of paths in another graph (see e.g. [MV97]). Also counting s, t-paths in a bounded treewidth graph is again in L via [EJT10] (see also [DDN13]). But the problem with this approach is that that the graph obtained by reducing a bounded treewidth G is not of bounded treewidth.

However, we can also view A as the adjacency matrix of a directed graph G_A. If G_A has bounded treewidth (which implies that H_A also has bounded treewidth), then we have a way of computing the determinant of A. The following lemma will be a useful preprocessing step:

Let G be the input graph of bounded treewidth. We will augment G with some new vertices and edges to yield a graph G' again with a tree decomposition T' of bounded treewidth. Then we have:

Lemma 1. *There exists a relation NXT on vertices of G' which satisfies the following:*

1. *NXT is compatible[3] with the tree decomposition T'*
2. *NXT is a partial order on the vertices of G'*

---
[3] Binary relation R is said to be compatible with the tree decomposition T' of G if the Gaifman graph of R has T' as its tree decomposition.

3. NXT *is computable in* L

4. *The transitive closure* NXT* *is a total order when restricted to the vertices of* G

5. NXT* *is expressible as an* MSO-*formula over the vocabulary of* G′ *along with* NXT.

The construction of such a relation is fairly straight forward and considered folklore in the Finite Model Theory literature (See for example Theorem VI.4 in [CF12]).

Lemma 2. *There is an* MSO-*formula* $\phi(X, Y)$ *with free variables* X, Y *that take values from the set of subsets of vertices and edges respectively, such that* $\phi(X, Y)$ *is true exactly when* X *is the set of heads of a cycle cover* Y *of the given graph.*

Proof. We write an MSO formula ϕ on free variables X, Y, such that $Y \subseteq E$ and $X \subseteq V$, such that ϕ evaluates to true on any set of heads of a cycle cover S. The MSO predicate essentially verifies that the subgraph induced by Y indeed forms a cycle cover of G. Our MSO formula is of the form[4]:

$$\phi(X, Y) \equiv (\forall v \in V)(\exists! h \in X)[\mathsf{DEG}(v, Y) \wedge \mathsf{PATH}(h, v, Y) \wedge (\mathsf{NXT}^*(h, v) \vee (h = v))]$$

where,

1. $\mathsf{DEG}(v, Y)$ is the predicate that says that the in-degree and out-degree of v (in the subgraph induced by the edges in Y) is 1.
2. $\mathsf{PATH}(x, y, Y)$ is the predicate that says that there is a path from x to y in the graph induced by edges of Y.

One can check that all the predicates above are MSO-definable.

Lemma 2 along with the Fact 1 yields:

Lemma 3. *For any matrix* $A_{n \times n}$ *of treewidth* $k \geq 2$ *and having integer entries, there is a Logspace algorithm that constructs an* $(m \times m)$ *(where* $m = \mathsf{poly}(n)$*) matrix* B *with entries from* $\{0, 1\}$, *such that* $det(A) = det(B)$ *and the treewidth of* B *is the same as the treewidth of* A.

Thus, using the histogram version of Courcelle's theorem from [EJT10] and Lemma 3, we get:

Theorem 4. *The determinant of a matrix* A *with integer entries, which can be viewed as the adjacency matrix of a weighted directed graph of bounded treewidth, is in* L.

---

[4] Note that since we require that for a given X, Y, every $v \in V$ has a unique $h \in X$, our formula is not monotone, i.e., If $X \subseteq X'$ are two sets of heads then if $\phi(X, Y)$ is true doesn't imply $\phi(X', Y)$ is also true (consider vertices in $X' \setminus X$, since $X' \subseteq X$, they will have two different h, h' such that the PATH and NXT* predicates are true contradicting uniqueness of h.

Proof. Firstly, obtain the matrix B from A using Lemma 3. The histogram version of Courcelle's theorem as described in [EJT10] when applied to the formula $\phi(X, Y)$ above yields the number of cycle covers of G_B parametrized on $|X|, |Y|$. But in the notation of Fact 1 above, $|X| = k$ and $|Y| = n$, so we can easily compute the determinant as the alternating sum of these counts. $\square$

Corollary 1. *There is a Logspace algorithm that takes as input a $(n \times n)$ bounded treewidth matrix A, 1^m, where $1 \leq m \leq n$ and computes the coefficient of x^m in the characteristic polynomial $(\chi_A(x) = det(xI - A))$ of A.*

The characteristic polynomial of an $(n \times n)$ matrix A is the determinant of the matrix $A(x) = xI - A$. We could use Theorem 4 to compute this quantity (since $A(x)$ is bounded treewidth, if A is bounded treewidth). However, Theorem 4 holds only for matrices with integer entries while the matrix $A(x)$ contains entries in the diagonal involving the indeterminate x.

We proceed as follows: In the directed graph corresponding to A, replace a self loop on a vertex of weight $x - d$ by a gadget of weight $-d$ in parallel with a self loop of weight x (In the event that there is no self loop on a vertex in A, add a self loop of weight x on the vertex). Replace the weights on the other edges according to the gadget in Lemma 3. We have added exactly n self loops, each of weight x (for the original vertices of A).

We first consider a generalisation of the determinant of $\{0, 1\}$-matrices of bounded treewidth viz. the determinant of matrices where the entries are from a set whose size is a fixed universal constant and the underlying graph consisting of the non-zero entries of A is of bounded treewidth.

Lemma 4. *Let A be a matrix whose entries belong to a set S of fixed size independent of the input or its length. If the underlying digraph with adjacency matrix A', where $A'_{ij} = 1$ iff $A_{ij} \neq 0$, is of bounded treewidth then the determinant of A can be computed in L.*

Proof. Let $s = |S|$ be a universal constant, $S = \{c_1, \ldots, c_s\}$ and let val_i be the predicates that partitions the edges of G according to their values i.e., $val_i(e)$ is true iff the edge e has value $c_i \in S$. Our modified formula $\psi(X, Y_1, \ldots, Y_s)$ will contain s unquantified new edge-set variables $Y_1, \ldots, Y_s$ along with the old vertex variable X, and is given by:

$$\forall e \in E((e \in Y_i \rightarrow val_i(e)) \wedge (e \in Y \leftrightarrow \vee_{i=1}^s (e \in Y_i) \wedge \phi(X, Y))$$

Notice that we verify that the edges in the set Y_i belong to the ith partition and each eadge in Y is in one of the Y_i's. The fact that the Y_i's form a partition of Y follows from the assumption that $val_i(e)$ is true for exactly one $i \in [s]$ for any edge e.

To obtain the determinant we consider the histogram parameterised on the s variables $Y_1, \ldots, Y_s$ and the heads X. For an entry indexed by $x, y_1, \ldots, y_s$, we multiply the entry by $(-1)^{n+x} \prod_{i=1}^s c_i{}^{y_i}$ and take a sum over all entries. $\square$

In light of the Lemma above, we can compute the characteristic polynomial as follows:

Proof (of Corollary 1). While counting the number of cycle covers with k cycles, we can keep track of the number of self-loops occurring in a cycle cover. It is easy to see that we can obtain the coefficient of x^r in the characteristic polynomial from the histogram outlined in Lemma 4. Hence we can also compute the characteristic polynomial in L. $\square$

Corollary 2. *Given a bounded treewidth matrix* $A_{m \times n}$ *the rank of* A *can be computed in* L.

Proof. A can be interpreted as the biadjacency matrix of a bipartite graph. Now, consider the matrix $B = \begin{pmatrix} 0 & A \\ A^T & 0 \end{pmatrix}$ – this is a matrix of dimension $(m + n) \times (m + n)$. It is easy to see that B corresponds to the adjacency matrix of A. Let row-rank(A) = column-rank(A) = r. Since A and A^T have the same rank, rank(A) + rank(A^T) = $2r$ = rank(B). Now we use Mulmuley's method [Mul87]: Let $Z_{ii} = z^{i-1}$ be a diagonal matrix in the indeterminate z. Compute the characteristic polynomial of $ZB = \det(xI - ZB)$ and use the fact that the rank of ZB is a number r such that x^{n-r} is the smallest power of x with a non-zero coefficient. We are now done with the help of Corollary 1. $\square$

FSLE(A, b) is the following problem: Given a system of m linear equations (with integer coefficients, w.l.o.g) in variables $z_1, \ldots, z_n$ and a target vector b, we want to check if there is a feasible solution to $Az = b$. That is, we want to decide if there is a setting of the variable vector $z \in \mathbb{Q}^n$ such that, $Az = b$ holds for a bounded treewidth matrix $A \in \mathbb{Z}^{m \times n}$ (when we say a rectangular matrix is bounded treewidth, we mean the underlying bipartite graph on $(m+n)$ vertices has bounded treewidth).

Corollary 3. *For a bounded treewidth matrix* $A_{m \times n}$ *and vector* $b_{n \times 1}$, FSLE(A, b) *is in* L.

Proof. We know that the system of linear equations given by A, b is feasible if and only if rank(A) = rank$([A : b])$. Therefore, we can use the Logspace procedure for matrix rank given by Corollary 2 to decide FSLE. $\square$

Corollary 4. *There is a Logspace algorithm that takes as input a* $(n \times n)$ *bounded treewidth matrix* A, $1^i, 1^j, 1^k$ *and computes the* k*-th bit of* A^{-1}_{ij}.

Proof. The inverse of a matrix A is the matrix $B = \frac{\mathbf{C}^T}{\det(A)}$ where $\mathbf{C} = (C_{ij})_{1 \le i,j \le n}$ is the cofactor matrix, whose (i, j)-th entry $C_{ij} = (-1)^{i+j}\det(A_{ij})$ is the determinant of the $(n - 1) \times (n - 1)$ matrix obtained from A by deleting the i-th row and j-th column. If we can compute C_{ij} in L, we can compute the entries of B via integer division which is known to be in L from [HAB02]. To this end, consider the directed graph G_A represented by A. To compute $\det(A_{ij})$, swap the columns of A such that the j-th column becomes the i-th column. The graph so obtained is of bounded treewidth (To see this, notice that the swapping operation just re-routes all incoming edges of j to i and those of i to j. The tree decomposition of this graph is just obtained by adding vertices (i, j) to every bag in the tree decomposition of G_A and also the edges rerouted to the respective bags. This increases the treewidth by 2). Now, remove the i-th vertex in G_A and all edges incident to it to get a graph $G_{A'_{ij}}$ on $(n - 1)$ vertices. The swapping operation changes the determinant of A_{ij} by a sign that is $(-1)^{i-j} = (-1)^{i+j}$. Computing the determinant of this modified matrix A'_{ij} yields C_{ij} as required. Since A'_{ij} is obtained from A by removing a vertex and all the edges incident on it, the treewidth of A'_{ij} is at most the treewidth of A. By Theorem 4, C_{ij} is in FL. □

Corollary 5. *There is a Logspace algorithm that on input an $(n \times n)$ bounded treewidth matrix A, $1^m, 1^i, 1^j, 1^k$ gives the k-th bit of (i, j)-th entry of A^m.*

Proof. Consider $A' = (I - tA)^{-1}$ where I is the $(n \times n)$ identity matrix and t is a small constant to be chosen later. Notice that $A' = (I - tA)^{-1} = \sum_{j \geq 0} t^j A^j$. By choosing t as a suitably small power of 2 (say $2^{-p} = t$ such that $2^p > \|A\|$) and computing A' to a suitable accuracy, we can read the (i, j)-th entry of A^m off the appropriate bit positions of the (i, j)-th entry of A'. So, in essence the problem of powering bounded treewidth matrix A reduces to the problem of computing the inverse of a related matrix which is known to be in L via Corollary 4. □

3.1 Spanning Trees and Directed Euler Tours

Fact 2. *The number of arborescences[5] of a digraph equals any cofactor of its Laplacian.*

where the Laplacian of a directed graph G is $D - A$ where D is the diagonal matrix with the D_{ii} being the out-degree of vertex i and A is the adjacency matrix of the underlying undirected graph. The BEST Theorem states:

Fact 3 ([AEB87, TS41]). *The number of Euler Tours in a directed Eulerian graph K is exactly: $t(K) \prod_{v \in V} (deg(v) - 1)!$ where $t(K)$ is the number of arborescences in K rooted at an arbitrary vertex of K and $deg(v)$ is the indegree as well as the outdegree of the vertex v.*

We combine Facts 2 and 3 with Theorem 4 to compute the number of directed Euler Tours in a directed Eulerian graph in L. Use the Kirchoff Matrix Tree theorem [Sta13] and Fact 3:

Corollary 6. *Counting arborescences and directed Euler Tours in a directed Eulerian graph G (where the underlying undirected graph is bounded treewidth) is in L.*

---

[5] Alternately, arborescences are MSO_2-definable and, thus, counting them in bounded treewidth graphs can be implemented in L via [EJT10].

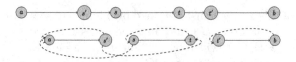

Fig. 1. s occurs before t

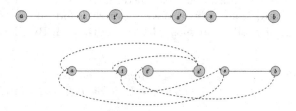

Fig. 2. t occurs before s

4   Hardness Results

Proposition 1 (Hardness of Bounded Treewidth Determinant). *For all constant $k \geq 2$, computing the determinant of an $(n \times n)$ matrix A whose underlying undirected graph has treewidth at most k is L-hard.*

Proof. We reduce the problem ORD of deciding for a directed path P and two vertices $s, t \in V(P)$ if there is a path from s to t (known to be L-complete via [Ete97]) to computing the determinant of bounded treewidth matrices (Note that P is a path and hence it has treewidth 1). Our reduction is as follows: Given a directed path P with source a, sink b and distinguished vertices s and t, we construct a new graph P' as follows: Add edges $(s', a), (s', t), (t, s), (s, a)$ and (b, t') and remove edges $(s', s), (t, t')$ where s' and t' are vertices in P such that $(s', s), (t, t') \in E(P)$ (See Fig. 1).

We claim that there is a directed path between s and t if and only if the determinant of the adjacency matrix of P' is zero. If there is a directed path from s to t in P, then there are two cycle covers in P' : $(a, s')(s, t)(t', b)$, with three cycles and $(a, s', t, s), (t', b)$, with two cycles. Using Fact 1, the signed sum of these cycle covers is $(-1)^{n+3} + (-1)^{n+2} = 0$, which is the determinant of P'.

In the case that P has a directed path from t to s (see Fig. 2), then there is one cycle namely (a, t, s, b, t', s'). We argue as follows: The edges $(t, s), (s, b), (b, t')$, (t', s') are in the cycle cover since they are the only incoming edges to s, b, t', s' respectively. So (t, s, b, t', s') is a part of any cycle cover of the graph. This forces one to pick the edge (s', a) and hence we have one cycle in the cycle cover for P'. □

Proposition 2 (Hardness of Bounded Treewidth Matrix Powering). *Bounded Treewidth Powering is L-hard under disjunctive truth table reductions.*

Proof. We reduce ORD to matrix powering. Given an directed path P on n vertices and distinguished vertices s and t, we argue as follows: There is a directed path between s and t, then it must be of length i for an unique $i \in [n]$. Consider the matrix $(I + A_P)^n$: s and t are connected by a path if and only if $(I + A_P)^n_{s,t} \neq 0$. This is because $(I + A_P)^n_{s,t}$ gives the walks from s to t, and if at all there is a path from s to t, then there is definitely a walk of length at most n between them. Checking if this entry is zero can be done by a DNF which takes as input the bits of $(I + A_P)^n_{s,t}$. $\qquad\square$

Proposition 3 (Hardness of Bounded Treewidth IMM). *Given a sequence of bounded treewidth matrices with rational entries $M_1, M_2, \ldots, M_n$ and $1^i, 1^j, 1^k$ as input, computing the k-th bit of (i,j)-th entry of $\prod_{l=1}^{n} M_l$ is GapL-hard.*

5   Open Problems

What is the complexity of other linear algebraic invariants such as minimal polynomial of a bounded treewidth matrix? What is the complexity of counting Euler Tours in undirected tours in bounded treewidth graphs? On general graphs, this problem is known to be #P-complete [BW05]. See [CCM12, CCM13] for some recent progress on this problem.

Acknowledgement. We would like to thank Abhishek Bhrushundi, Arne Meier, Rohith Varma and Heribert Vollmer for illuminating discussions regarding this paper. Special thanks are due to Johannes Köbler and Sebastian Kuhnert who were involved in the initial discussions on the proof of Theorems 4, 6; to Stefan Mengel for proof reading the paper and finding a gap in a previous "proof" of Theorem 4; and to Raghav Kulkarni for suggesting proof strategies for Corollary 5 and Lemma 3; and to Sebastian Kuhnert for the proof of Proposition 2. Thanks are also due to anonymous referees for pointing out errors in a previous version of the paper and for greatly simplifying the proof of Corollary 3. This work is partially funded by a grant from Infosys Foundation.

References

[ABO99] Allender, E., Beals, R., Ogihara, M.: The complexity of matrix rank and feasible systems of linear equations. Comput. Complex. **8**(2), 99–126 (1999)

[AEB87] Aardenne-Ehrenfest, T., Bruijn, N.G.: Circuits and trees in oriented linear graphs. In: Gessel, I., Rota, G.-C. (eds.) Classic Papers in Combinatorics. Modern Birkhäuser Classics, pp. 149–163. Birkhäuser, Boston (1987)

[Ber84] Berkowitz, S.J.: On computing the determinant in small parallel time using a small number of processors. Inf. Process. Lett. **18**(3), 147–150 (1984)

[BW05] Brightwell, G., Winkler, P.: Counting eulerian circuits is# p-complete. In: ALENEX/ANALCO, pp. 259–262. Citeseer (2005)

[CCM12] Chebolu, P., Cryan, M., Martin, R.: Exact counting of euler tours for generalized series-parallel graphs. J. Discrete Algorithms **10**, 110–122 (2012)

[CCM13] Chebolu, P., Cryan, M., Martin, R.: Exact counting of euler tours for graphs of bounded treewidth. In: CoRR, abs/1310.0185 (2013)

[CF12]   Chen, Y., Flum, J.: On the ordered conjecture. In: Proceedings of the 2012 27th Annual IEEE/ACM Symposium on Logic in Computer Science, pp. 225–234. IEEE Computer Society (2012)

[Cou90]   Courcelle, B.: The monadic second-order logic of graphs. i. recognizable sets of finite graphs. Inf. comput. **85**(1), 12–75 (1990)

[DDN13]   Das, B., Datta, S., Nimbhorkar, P.: Log-space algorithms for paths and matchings in k-trees. Theor. Comput. Syst. **53**(4), 669–689 (2013)

[DKLM10]   Datta, S., Kulkarni, R., Limaye, N., Mahajan, M.: Planarity, determinants, permanents, and (unique) matchings. TOCT 1(3) (2010)

[EJT10]   Elberfeld, M., Jakoby, A., Tantau, T.: Logspace versions of the theorems of bodlaender and courcelle. In: FOCS, pp. 143–152 (2010)

[Ete97]   Etessami, K.: Counting quantifiers, successor relations, and logarithmic space. J. Comput. Syst. Sci. **54**(3), 400–411 (1997)

[FG06]   Flum, J., Grohe, M.: Parameterized Complexity Theory, vol. 3. Springer, Heidelberg (2006)

[HAB02]   Hesse, W., Allender, E., Barrington, D.A.M.: Uniform constant-depth threshold circuits for division and iterated multiplication. J. Comput. Syst. Sci. **65**, 695–716 (2002)

[Mul87]   Mulmuley, K.: A fast parallel algorithm to compute the rank of a matrix over an arbitrary field. Combinatorica **7**(1), 101–104 (1987)

[MV97]   Mahajan, M., Vinay, V.: Determinant: combinatorics, algorithms, and complexity. Chicago J. Theor. Comput. Sci. **1997**, 26 (1997)

[Sta13]   Stanley, R.P.: Algebraic Combinatorics. Springer-Verlag, New York (2013)

[Tod91]   Toda, S.: Counting problems computationally equivalent to the determinant. Technical report CSIM 91–07, Dept of Comp Sc & Information Mathematics, Univ of Electro-Communications, Chofu-shi, Tokyo (1991)

[TS41]   Tutte, W.T., Smith, C.A.B.: On unicursal paths in a network of degree 4. Am. Math. Monthly **48**(4), 233–237 (1941)

[vzGG13]   von zur Gathen, J., Gerhard, J.: Modern Computer Algebra (3. ed.). Cambridge University Press, New York (2013)

Quantum Computing

Quantum Game Players Can Have Advantage Without Discord

Zhaohui Wei[1]([✉]) and Shengyu Zhang[2]

[1] School of Physics and Mathematical Sciences, Nanyang Technological University
and Centre for Quantum Technologies, Singapore, Singapore
weizhaohui@gmail.com
[2] Department of Computer Science and Engineering,
The Chinese University of Hong Kong, Hong Kong, China
syzhang@cse.cuhk.edu.hk

Abstract. The last two decades have witnessed a rapid development
of quantum information processing, a new paradigm which studies the
power and limit of "quantum advantages" in various information processing tasks. Problems such as when quantum advantage exists, and if existing, how much it could be, are at a central position of these studies. In
a broad class of scenarios, there are, implicitly or explicitly, at least two
parties involved, who share a state, and the correlation in this shared
state is the key factor to the efficiency under concern. In these scenarios,
the shared *entanglement* or *discord* is usually what accounts for quantum advantage. In this paper, we examine a fundamental problem of this
nature from the perspective of game theory, a branch of applied mathematics studying selfish behaviors of two or more players. We exhibit a
natural zero-sum game, in which the chance for any player to win the
game depends only on the ending correlation. We show that in a certain
classical equilibrium, a situation in which no player can further increase
her payoff by any local classical operation, whoever first uses a quantum
computer has a big advantage over its classical opponent. The equilibrium is fair to both players and, as a shared correlation, it does not
contain any discord, yet a quantum advantage still exists. This indicates
that at least in game theory, the previous notion of discord as a measure
of non-classical correlation needs to be reexamined, when there are two
players with different objectives.

1 Introduction

Quantum computers have exhibited tremendous power in algorithmic, cryptographic, information theoretic, and many other information processing tasks,
compared with their classical counterparts. Meanwhile, for a large number of
problems, quantum computers are not able to offer much advantage over classical ones. When and why quantum computers are more powerful are always
at a central position in studies on quantum computation and quantum information processing. A particularly interesting class of scenarios is when there
are, implicitly or explicitly, at least two parties involved who share a state, the

R. Jain et al. (Eds.): TAMC 2015, LNCS 9076, pp. 311–323, 2015.
DOI: 10.1007/978-3-319-17142-5_27

correlation in this state is the key factor. What accounts for the quantum advantage is often *entanglement*, one of the most distinctive characters of quantum information. Indeed, it has been showed that a quantum algorithm with only slight entanglement can be simulated efficiently by a classical computer [Vid03]. In certain potential applications of quantum algorithms, it is also shown that entangled measurement is necessary for the existence of efficient quantum algorithms [HMR+10].

Recently people started to realize that entanglement is not always a necessary resource needed for generating quantum correlations. It has been found that *discord*, another unique character of quantum states, also plays an important role in quantum information processing [OZ01]. Discord is a relaxed version of entanglement—states with positive entanglement must also have positive discord, but there are states with positive discord but zero entanglement. People has discovered cases where quantum speed-up exists without entanglement involved, and discord is considered to be responsible for the quantum advantage [DSC08]. Till today, discord is widely considered as necessary for the existence of quantum advantages.

In this paper, we reexamine this notion from the perspective of game theory [OR94]. Game theory studies the situation in which there are two or more players with possibly different goals. There are two broad classes of games, one is strategic-form (or normal-form) games, in which all players make their choice simultaneously; a typical example is Rock-Paper-Scissors. The other class is extensive-form games, in which players make their moves in turn; a typical example is chess.

The research on quantum games began about one decade ago, starting with two pioneering papers.[1] The first one [EWL99] aimed to quantize a specific strategic-form game called *Prisoners' Dilemma* [EWL99], and it unleashed a long sequence of follow-up works in the same model. Despite the rapid growth of literature, controversy also largely exists [BH01,vEP02,CT06], which questioned the meaning of the claimed quantum solution, the ad hoc assumptions in the model, and the inconsistency with standard settings of classical strategic games. Recently a new model was proposed for quantizing general strategic-form games [Zha12]. Compared with [EWL99], the new model corresponds to the classical games more precisely, and has rich mathematical structures and game-theoretic questions; also see later theoretical developments [KZ12, WZ13, JSWZ13, PKL+15].

Back to the early stage of the development of quantum game theory, the other pioneering paper was [Mey99], which demonstrated the power of using quantum strategies in an extensive-form game. More specifically, Meyer considered the quantum version of the classical Penny Matching game. The basic setting is as follows. There are two players, and each has two possible actions on one bit: Flip it or not. Starting with the bit being 0, Player 1 first takes an action, and

---

[1] Note that there is also a class of "nonlocal games", such as CHSH or GHZ games [BCMdW10], where all the players have the *same* objective. But general game theory focuses more on situation that the players have *different* objective functions, and the players are selfish, each aiming to optimize her own objective function only.

then Player 2 takes an action, and finally Player 1 takes another action, and the game is finished. If the bit is finally 0, then Player 1 wins; otherwise Player 2 wins. It is not hard to see that if Player 2 flips the bit with half probability, then no matter what Player 1 does, each player wins the game with half probability. Now consider the following change of setting: The bit becomes a qubit; the first player uses a quantum computer in the sense that she can perform any quantum admissible operation on the bit; the second player uses a classical computer in the sense that she can perform either Identity or the flip operation $\begin{bmatrix} 0 & 1 \\ 1 & 0 \end{bmatrix}$. In this new setting, Player 1 can win the game with certainty! Her winning strategy is simple: she first applies a Hadamard gate to change the state to $|+\rangle = (|0\rangle + |1\rangle)/\sqrt{2}$, and then no matter whether Player 2 applies the flip operation or not, the state remains the same $|+\rangle$, thus in the third step Player 1 can simply apply a Hadamard gate again to rotate the state back to $|0\rangle$. This shows that a player using a quantum computer can have big advantage over one using a classical computer.

Despite a very interesting phenomena it exhibits, the quantum advantage is not the most convincing due to a fairness issue. After all, the quantum player takes two actions and the classical player takes just one. And the order of "Player 1 $\rightarrow$ Player 2 $\rightarrow$ Player 1" is also crucial for the quantum advantage. One remedy is to consider normal-form games, in which the players give their strategies *simultaneously*, thus there is no longer the issue of the action order. Taking the model in [Zha12], two players play a complete-information normal-form game, with a starting state ρ in systems (A_1, A_2), and A_i being given to Player i. A classical player can only measure her part of the state in the computational basis, followed by whatever classical operation C (on the computational basis). In previous works [EWL99, Mey99, ZWC+12] the classical player is usually assumed to be able to apply any classical operation on computational basis (such as X-gate), followed by a measurement in the computational basis. A classical operations there is implicitly assumed to be unitary, so the operation in the matrix form is a permutation matrix. Here we allow classical player to measure first and then perform any classical operation, which gives her more power since the second-step classical operation does need to be unitary. Indeed, in Meyer's Penny Matching game, in the second step Player 2 could measure the state first and then randomly set it to be $|0\rangle$ or $|1\rangle$ each with half probability. Then in the third step, Player 1's Hadamard gate will change the state to $|+\rangle$ or $|-\rangle$, in either case, Player 1 could win with only half probability.

Even if we now enlarge the space of possible operations of the classical player, we will show examples where the quantum player has advantage of winning the game. Furthermore, the examples have the following nice properties respecting the fairness of the game:

1. If both players are classical, then both get expected payoff 0, and ρ is a correlated equilibrium in the sense that any classical operation C by one player cannot increase her expected payoff.

2. Suppose that one player remains classical and the other player uses a quantum computer. To illustrate the power of using quantum strategies, we cut the classical player some slack as follows. The classical player can (1) pick one subsystem, A_1 or A_2, of ρ, leaving the other subsystem to the quantum player, and (2) "take side" by picking one of the two payoff matrices, leaving the other to the quantum player.

Examples were found that even with the advantage of taking side and taking part of the shared state, the classical player still has a disadvantage compared to the quantum player. Consider the canonical 2×2 zero-sum game with the payoff matrices being

$$U_1 = \begin{pmatrix} 1 & -1 \\ -1 & 1 \end{pmatrix} \quad \text{and} \quad U_2 = \begin{pmatrix} -1 & 1 \\ 1 & -1 \end{pmatrix}. \tag{1}$$

Quantum game with entanglement. Each player i owns a 2-dimensional Hilbert space, and they share the quantum state

$$|\psi\rangle = \frac{1}{\sqrt{2}}(|+0\rangle + |-1\rangle) = \frac{1}{\sqrt{2}}(|0+\rangle + |1-\rangle), \tag{2}$$

where $|+\rangle = \frac{1}{\sqrt{2}}(|0\rangle + |1\rangle)$, and $|-\rangle = \frac{1}{\sqrt{2}}(|0\rangle - |1\rangle)$. It is not difficult to verify that if both players measure their parts in the computational basis, then each gets payoff 1 and -1 with equal probability, resulting an average payoff of zero for both players. This is a correlated equilibrium for classical operations.

Now suppose that Player 1 employs a quantum computer. Since the state is symmetric, it does not matter which part Player 2, the classical player, chooses. Let us assume that Player 2 chooses part 2, and the payoff matrix U_2. Then Player 1 can apply the Hadamard transformation on her qubit, followed by the measurement in computational basis. The state immediately before the measurement is $|\psi'\rangle = (|00\rangle + |11\rangle)/\sqrt{2}$. Therefore the measurement in the computational basis gives Player 1 and Player 2 payoff 1 and -1, respectively, with certainty. In other words, Player 1 wins with certainty, whereas she could only win with half probability when using a classical computer.

In this example where the quantum player has an advantage, the state shared by players is highly entangled, which motivates the following natural question: Is entanglement necessary for quantum advantage in the game? It turns out that the answer is no. Consider the example below.

Quantum Game with Discord. The payoff matrices are the same as before, but the quantum state shared by players is the following.

$$\rho = \frac{1}{4}(|+\rangle\langle+| \otimes |0\rangle\langle0| + |0\rangle\langle0| \otimes |+\rangle\langle+| + |-\rangle\langle-| \otimes |1\rangle\langle1| + |1\rangle\langle1| \otimes |-\rangle\langle-|). \tag{3}$$

This state is separable and thus does not have any entanglement. It can be checked that if the players measure this state in computational basis, the probability of getting each of the four possible outcomes is $1/4$. Thus the overall payoff of each player is zero, and it can be verified that it is a classical correlated equilibrium.

In the quantum setting, again without loss of generality assume that the classical computer picks the second part of ρ and the second payoff matrix. The quantum player can again perform a Hadamard operation on her system, resulting in a new state

$$\rho' = \frac{1}{4}(|0\rangle\langle 0| \otimes |0\rangle\langle 0| + |+\rangle\langle +| \otimes |+\rangle\langle +| + |1\rangle\langle 1| \otimes |1\rangle\langle 1| + |-\rangle\langle -| \otimes |-\rangle\langle -|). \quad (4)$$

Measuring the new state, the quantum player gets state $|00\rangle, |01\rangle, |10\rangle, |11\rangle$ with probability 3/8, 1/8, 1/8, 3/8 respectively. As a result, her winning probability increases from 1/2 to 3/4; in other words, she gets an expected payoff of 1/2.

Note that the quantum state in Eq. (4) is separable, and there is no any entanglement, but the quantum player still gets a quantum advantage. Thus, entanglement is not necessary for quantum advantage to exist in this game. Note that, however, the state in Eq. (4) has a positive discord. As we have mentioned, it was known that in some scenarios, it is discord, rather than entanglement, that produces non-classical correlations. So the above example confirms this traditional notion in the new game-theoretic setting.

These two examples were also experimentally verified recently [ZWC+12]. The present paper makes further studies on the foregoing notion by asking the following fundamental question.

Is discord necessary for quantum advantage to exist in games where play-ers share a symmetric state?

It is tempting to conjecture that the answer is Yes. In the rest of the paper, we will show that, first, discord is indeed necessary for any quantum advantage to exist in a 2-player games where each player has $n = 2$ strategies. We will then show that when $n \geq 3$, however, there are games where the quantum player has a positive advantage even when the shared symmetric state has zero discord.

2   Preliminaries

Suppose that in a classical game there are k players, labeled by $\{1, 2, \ldots, k\}$. Each player i has a set S_i of strategies. To play the game, each player i selects a strategy s_i from S_i. We use $s = (s_1, \ldots, s_k)$ to denote the *joint strategy* selected by the players and $S = S_1 \times \ldots \times S_k$ to denote the set of all possible joint strategies. Each player i has a utility function $u_i : S \to \mathbb{R}$, specifying the *payoff* or *utility* $u_i(s)$ of Player i on the joint strategy s. For simplicity of notation, we use subscript $-i$ to denote the set $[k] - \{i\}$, so s_{-i} is $(s_1, \ldots, s_{i-1}, s_{i+1}, \ldots, s_k)$, and similarly for S_{-i}, p_{-i}, etc. In this paper, we will mainly consider 2-player games.

Nash equilibrium is a fundamental solution concept in game theory. Roughly, it says that in a joint strategy, no player can gain more by changing her strategy, provided that all other players keep their current strategies unchanged. The precise definition is as follows.

Definition 1. *A pure Nash equilibrium is a joint strategy* $s = (s_1, \ldots, s_k) \in S$ *satisfying that*

$$u_i(s_i, s_{-i}) \geq u_i(s'_i, s_{-i}), \qquad \forall i \in [k], \forall s'_i \in S_i.$$

Pure Nash equilibria can be generalized by allowing each player to independently select her strategy according to some probability distribution, leading to the following concept of *mixed Nash equilibrium*.

Definition 2. *A (mixed) Nash equilibrium (NE) is a product probability distribution* $p = p_1 \times \ldots \times p_k$, *where each* p_i *is a probability distributions over* S_i, *satisfying that*

$$\sum_{s_{-i}} p_{-i}(s_{-i}) u_i(s_i, s_{-i}) \geq \sum_{s_{-i}} p_{-i}(s_{-i}) u_i(s'_i, s_{-i}), \; \forall i \in [k], \; \forall s_i, s'_i \in S_i \text{ with } p_i(s_i) > 0.$$

A fundamental fact proved by Nash [Nas51] is that every game with a finite number of players and a finite set of strategies for each player has at least one mixed Nash equilibrium.

There are various further extensions of mixed Nash equilibria. Aumann [Aum74] introduced a relaxation called *correlated equilibrium*. This notion assumes an external party, called Referee, to draw a joint strategy $s = (s_1, \ldots, s_k)$ from some probability distribution p over S, possibly correlated in an arbitrary way, and to suggest s_i to Player i. Note that Player i only sees s_i, thus the rest strategy s_{-i} is a random variable over S_{-i} distributed according to the conditional distribution $p|_{s_i}$, the distribution p conditioned on the i-th part being s_i. Now p is a correlated equilibrium if any Player i, upon receiving a suggested strategy s_i, has no incentive to change her strategy to a different $s'_i \in S_i$, assuming that all other players stick to their received suggestion s_{-i}.

Definition 3. *A correlated equilibrium (CE) is a probability distribution* p *over* S *satisfying that*

$$\sum_{s_{-i}} p(s_i, s_{-i}) u_i(s_i, s_{-i}) \geq \sum_{s_{-i}} p(s_i, s_{-i}) u_i(s'_i, s_{-i}), \; \forall i \in [k], \forall s_i, s'_i \in S_i.$$

The above statement can also be restated as

$$\mathbb{E}_{s_{-i} \leftarrow \mu|s_i}[u_i(s_i, s_{-i})] \geq \mathbb{E}_{s_{-i} \leftarrow \mu|s_i}[u_i(s'_i, s_{-i})]. \tag{5}$$

where $\mu|s_i$ is the distribution μ conditioned on the i-th component being s_i. Notice that a classical correlated equilibrium p is a classical Nash equilibrium if p is a product distribution.

Correlated equilibria captures natural games such as the Traffic Light and the Battle of the Sexes ([VNRET97], Chap. 1). The set of CE also has good mathematical properties such as being convex (with Nash equilibria being some of the vertices of the polytope). Algorithmically, it is computationally benign for finding the best CE, measured by any linear function of payoffs, simply by solving a linear program (of polynomial size for games of constant players). A natural learning dynamics also leads to an approximate CE ([VNRET97], Chap. 4) which we will define next, and all CE in a graphical game with n players and with $\log(n)$ degree can be found in polynomial time ([VNRET97], Chap. 7).

3   Quantum Game Without Discord

In this section, we will address the question proposed at the end of the first section. Suppose that a game has two players and both of them have n strategies. In other words, each player holds an n-dimensional quantum system. Recall that we also require the shared quantum state $\rho \in H \otimes H$ be symmetric, so that swapping the two systems does not change the state. It is not hard to derive from the general criteria of zero-discord state [DVB10] that these quantum states ρ have the form of

$$\rho = \sum_{i,j=0}^{n-1} p(i,j)|\psi_i\rangle\langle\psi_i| \otimes |\psi_j\rangle\langle\psi_j|, \tag{6}$$

where $\{|\psi_i\rangle\}$ is a set of orthogonal basis of the n-dimensional Hilbert space H, and $P = [p(i,j)]_{ij} \in \mathbb{R}_+^{n \times n}$ is a symmetric matrix with nonnegative entries satisfying that $\sum_{ij} p(i,j) = 1$. (In general, we use the upper case letter P to denote the matrix and the lower case letter p to denote the corresponding two-variate distribution $p(i,j)$.) We sometimes also write the state as

$$\rho = \sum_i p_1(i)|\psi_i\rangle\langle\psi_i| \otimes \sigma_i \tag{7}$$

where $p_1(i) = \sum_j p(i,j)$ is the marginal distribution on the first system, and $\sigma_i = \sum_j \frac{p(i,j)}{p_1(i)}|\psi_j\rangle\langle\psi_j|$ (if $p_1(i) = 0$ then let $\sigma_i = |0\rangle\langle 0|$).

Consider the following game as a natural extension of the Penny Matching game in Sect. 1. The payoff matrices are

$$U_1 = nI - J \quad and \quad U_2 = -U_1, \tag{8}$$

where J is the all-one matrix. Intuitively, whoever takes the first matrix bets that the two n-sided dice give the same side, and the other player bets that the two dice give different sides. We first show that there is a unique correlated equilibrium in the game.

Lemma 1. *The game given by Eq. (8) has only one classical correlated equilibrium $Q = J/n^2$.*

Proof. According to the definition of correlated equilibrium, if a distribution q on $[n] \times [n]$ is a classical correlated equilibrium, then the following relationships hold:

$$\sum_j q(i,j)U_1(i,j) \geq \sum_j q(i,j)U_1(i',j), \qquad \forall i, i' \in \{0,1,...,n-1\}, \tag{9}$$

and

$$\sum_i q(i,j)U_2(i,j) \geq \sum_i q(i,j)U_2(i,j'), \qquad \forall j, j' \in \{0,1,...,n-1\}. \tag{10}$$

Plugging the definition of U_1 and U_2 into the above inequalities, one can verify that $Q = J/n^2$ is the only solution.

Recall that $\rho = \sum_i p_1(i)|\psi_i\rangle\langle\psi_i|\otimes\sigma_i$. Since ρ is symmetric, it does not matter which part the classical player, Player 2, chooses to hold. For the convenience of discussions, let us assume that the classical player takes the second part. We use $\mathsf{supp}(p)$ to denote the support of a distribution p, i.e., the set of elements with non-zero probability. The next lemma gives a sufficient and necessary condition for the existence of quantum advantage.

Lemma 2. *Suppose that measuring the state ρ gives a classical correlated equilibrium for the game given in Eq. (8). Then Player 1 (who is quantum) does not have any advantage if and only if*

$$\langle i|\sigma_j|i\rangle = 1/n, \quad \forall i \in \{0, 1, ..., n-1\} \text{ and } j \in \mathsf{supp}(p_1). \tag{11}$$

Proof. "*Only if*": Assume that Player 1 first measures her part in the orthonormal basis $\{|\psi_i\rangle\}$. Note that this does not affect the state. If outcome j occurs, then Player 1 knows that the state of Player 2 is σ_j. We consider which utility matrix in Eq.(8) Player 1 has. In the first case, Player 1 takes the utility matrix U_1. It is not hard to see that her optimal strategy is to replace her part $|\psi_j\rangle$ by $|i\rangle$, where i is a maximizer of $\max_i\langle i|\sigma_j|i\rangle$. Thus Player 1 has a strict positive advantage if and only if there is some i and j, where $j \in \mathsf{supp}(p_1)$, with $\langle i|\sigma_j|i\rangle > 1/n$, which is equivalent to saying that there is some i and $j \in \mathsf{supp}(p_1)$ with $\langle i|\sigma_j|i\rangle \neq 1/n$.

Similarly, if Player 1 takes the utility matrix U_2, then her optimal strategy is to replace $|\psi_j\rangle$ with $|i\rangle$, where i is a minimizer of $\min_i\langle i|\sigma_j|i\rangle$. Thus Player 1 has a strict positive advantage if and only if there is some i and j with $\langle i|\sigma_j|i\rangle < 1/n$, which is again equivalent to saying that there is some i and j with $\langle i|\sigma_j|i\rangle \neq 1/n$.

"*If*": Player 2 measures her part in the computational basis, yielding the state

$$\frac{1}{n}\sum_{i,j} p_1(j)|\psi_j\rangle\langle\psi_j| \otimes |i\rangle\langle i|.$$

Now whatever quantum operation Player 1 applies, the probability of observing the same bits (i.e., the state after the measurement is $|ii\rangle$ for some i) is $1/n$, with the expected payoff of 0 for both players.

Though the above lemma gives a sufficient and necessary condition, it is still not always clear whether quantum advantage could exist for any symmetric state ρ with zero discord. Next we will further the study by considering a related matrix $M \in \mathbb{R}_+^{n\times n}$, whose (i,j)-th entry is defined to be

$$M(i,j) = |\langle i|\psi_j\rangle|^2. \tag{12}$$

It turns out that the rank of M is an important criteria to our question. In the rest of this section, we will consider two cases, depending on whether M is full rank or not.

3.1   Case 1: M Is Full-Rank

We will first show that if M is full-rank, then the quantum player cannot have any advantage.

Theorem 3. *Suppose that the two players of the game Eq.(8) share a symmetric state ρ, measuring which gives a classical correlated equilibrium. Then Player 1 (who is quantum) does not have any advantage if M in Eq.(12) is full-rank.*

Proof. By Lemma 1, for any $0 \leq k, j \leq n-1$ we have

$$\sum_{i=0}^{n-1} p_1(i)|\langle k|\psi_i\rangle|^2 \cdot \langle j|\sigma_i|j\rangle = \frac{1}{n^2}$$

Summing over j, we obtain another equality

$$\sum_{i=0}^{n-1} p_1(i)|\langle k|\psi_i\rangle|^2 = \frac{1}{n}.$$

Combining these two equalities, we have

$$\sum_{i=0}^{n-1} |\langle k|\psi_i\rangle|^2 \cdot p_1(i)\left(\langle j|\sigma_i|j\rangle - \frac{1}{n}\right) = 0.$$

Define a matrix $A = [a(i,j)]_{ij} \in \mathbb{R}^{n \times n}$ by $a(i,j) = p_1(i)\left(\langle j|\sigma_i|j\rangle - \frac{1}{n}\right)$. Then the above equality is just $\sum_i M(k,i)a(i,j) = 0$ for all k, j. In other words, we have $M \cdot A = 0$. Since the matrix M is assumed to be full-rank, we have $A = M^{-1}0 = 0$. The conclusion thus follows by Lemma 2.

Two corollaries are in order. First, note that M is full-rank for a generic orthogonal basis $\{|\psi_i\rangle\}$, it is generically true that no discord implies no quantum advantage.

Corollary 4. *If a set of orthonormal basis $\{|\psi_i\rangle\}$ is picked uniformly at random, then with probability 1, the quantum player does not have any advantage.*

The second corollary considers the case of $n = 2$, which is settled by the above theorem completely. Indeed, when $n = 2$, the rank of M is either 1 or 2. The rank-2 case is handled by the above theorem. If the rank is 1, it is not hard to see that the only possible M is $M = \begin{bmatrix} 1/2 & 1/2 \\ 1/2 & 1/2 \end{bmatrix}$. In this case, for any i and any k it holds that

$$\langle k|\sigma_i|k\rangle = \langle k|\left(\sum_j p(j|i)|\psi_j\rangle\langle\psi_j|\right)|k\rangle = \sum_j p(j|i)|\langle k|\psi_j\rangle|^2 = \frac{1}{2}\sum_j p(j|i) = \frac{1}{2}.$$

Applying Lemma 2, we thus get the following corollary.

Corollary 5. *There is no quantum advantage for the game defined in Eq.(1) on any symmetric state ρ with zero discord.*

3.2   Case 2: M Is Not Full Rank

Somewhat surprisingly, the quantum player *can* have an advantage when M is not full-rank. In this section we exhibit a counterexample for $n = 3$. In this case, recall that the payoff matrices are

$$U_1 = \begin{pmatrix} 2 & -1 & -1 \\ -1 & 2 & -1 \\ -1 & -1 & 2 \end{pmatrix} \quad \text{and} \quad U_2 = \begin{pmatrix} -2 & 1 & 1 \\ 1 & -2 & 1 \\ 1 & 1 & -2 \end{pmatrix}. \tag{13}$$

We consider the following quantum state,

$$\rho = \sum_{i,j=0}^{2} p(i,j)|\psi_i\rangle\langle\psi_i| \otimes |\psi_j\rangle\langle\psi_j|, \tag{14}$$

where

$$|\psi_0\rangle = \frac{1}{\sqrt{2}}(|0\rangle + |1\rangle), \quad |\psi_1\rangle = \frac{1}{\sqrt{2}}(|0\rangle - |1\rangle), \quad |\psi_2\rangle = |2\rangle. \tag{15}$$

It is not hard to calculate M:

$$M = \begin{pmatrix} 1/2 & 1/2 & 0 \\ 1/2 & 1/2 & 0 \\ 0 & 0 & 1 \end{pmatrix}. \tag{16}$$

which has rank 2. Define

$$P = \begin{pmatrix} 4/9 & 0 & 0 \\ 0 & 0 & 2/9 \\ 0 & 2/9 & 1/9 \end{pmatrix}. \tag{17}$$

It can be easily verified that if the two players measure the state in computational basis, the probability distribution yielded is uniform, which is a classical Nash equilibrium.

Now suppose that Player 1 uses a quantum computer. One can verify that the condition in Lemma 2 does not hold. For a concrete illustration, let us consider the protocol in Lemma 2 again. Player 1 first measures in the basis $\{|+\rangle, |-\rangle, |2\rangle\}$. With probability 4/9, she observes $|+\rangle$, then changes it to $|0\rangle$. Player 2's state is also $|+\rangle$ in this case, thus a measurement in the computational basis gives the $|00\rangle$ and $|01\rangle$ each with half probability. Thus Player 1's payoff in this case is $2 \cdot \frac{1}{2} - 1 \cdot \frac{1}{2} = \frac{1}{2}$. The second case is that Player 1 observes $|-\rangle$, which happens with probability 2/9, and Player 2's state is $|2\rangle$ for sure. Player 1 changes her part to $|2\rangle$, and gets payoff 2. The third case is that Player 1 observes $|2\rangle$, which happens with probability 1/3, leaving Player 2 $\sigma_3 = (2/3)|1\rangle\langle1| + (1/3)|2\rangle\langle2|$. Player 1 then changes her qubit to $|1\rangle$, collides with Player 2's outcome with probability 1/3, thus Player 1's payoff is $2 \cdot \frac{1}{3} - 1 \cdot \frac{2}{3} = 0$. On average, the quantum player has a payoff of $(4/9)(1/2) + (2/9) \cdot 2 + (1/3) \cdot 0 = 2/3$.

It should be pointed out that the matrix P achieving the quantum advantage of $2/3$ is not unique. For example, the following matrix also works with the same effect:

$$P = \begin{pmatrix} 2/9 & 2/9 & 0 \\ 0 & 0 & 2/9 \\ 1/9 & 1/9 & 1/9 \end{pmatrix}. \qquad (18)$$

3.3 Optimization

In this subsection, we show that the 3-dimensional example in the above subsection is actually optimal for M defined in Eq. (16). Actually the theorem below shows more. Note that if the rank of M is 1, it is easy to prove that M must be the uniform matrix, and the quantum advantage must be zero, thus in the following we suppose the rank of M to be 2.

Theorem 6. *Suppose that measuring the state ρ gives a classical correlated equilibrium. Suppose the columns of M are M^0, M^1 and M^2. Without loss of generality, suppose $M^0 = xM^1 + (1-x)M^2$, where $0 \le x \le 1$. Then the quantum advantage*

$$QA \le \frac{1}{3} + \frac{1}{3x_b}, \quad \text{where } x_b = \max\{x, 1-x\}. \qquad (19)$$

Proof. By Lemma 1, for any $0 \le k,l \le 2$ we have $\sum_{i,j=0}^{2} p(i,j)|\langle k|\psi_i\rangle|^2 \cdot |\langle l|\psi_j\rangle|^2 = \frac{1}{9}$., which turns out to be equivalent to $M \cdot P \cdot M^T = \frac{J}{9}$. Noting that $M \cdot (J/9) \cdot M^T = J/9$, we know that P can be expressed as

$$P = \frac{J}{9} + \bar{P}, \qquad (20)$$

where $M \cdot \bar{P} \cdot M^T = 0$. By straightforward calculation, one can show that Eq. (20) indicates $M \cdot \bar{P} = 0$. Considering the form of M, $\bar{P}$ can now be expressed as

$$\bar{P} = \begin{pmatrix} k_0 & k_1 & k_2 \\ -k_0 x & -k_1 x & -k_2 x \\ -k_0(1-x) & -k_1(1-x) & -k_2(1-x) \end{pmatrix}, \qquad (21)$$

where k_0, k_1 and k_2 are real numbers.

According to the discussion above, we know that the maximal quantum advantage is

$$QA = \sum_{i=0}^{2} p_1(i)[2 \cdot \langle l_i|\sigma_i|l_i\rangle - 1 \cdot (1 - \langle l_i|\sigma_i|l_i\rangle)], \qquad (22)$$

where $l_i = \max_l \langle l|\sigma_i|l\rangle$. Then it holds that

$$QA = 3\sum_{i=0}^{2} p_1(i) \cdot \langle l_i|\sigma_i|l_i\rangle - 1 = 3\sum_{i,j=0}^{2} p(i,j)|\langle l_i|\psi_j\rangle|^2 - 1$$

$$= 3\sum_{i,j=0}^{2} \left(\frac{1}{9} + \bar{p}(i,j)\right)|\langle l_i|\psi_j\rangle|^2 - 1 = 3\sum_{i=0}^{2}\left(\sum_{j=0}^{2} \bar{p}(i,j)|\langle l_i|\psi_j\rangle|^2\right),$$

where $\bar{p}(i,j)$ is the element of $\bar{P}$. At the same time, it can be obtained that $l_i = \max_l \sum_j \bar{p}(i,j)|\langle l|\psi_j\rangle|^2$. Besides, recall that the rank of M is 2, then there must be one row of M, say M_2, has the form of $aM_0 + (1-a)M_1$, where M_0 and M_1 are the other two rows of M, and $0 \le a \le 1$. Then it can be known that every l_i must be 0 or 1. Based on the form of $\bar{P}$, we have that $l_0 \neq l_1 = l_2$. Without loss of generality, we suppose $l_0 = 0$, and $l_1 = l_2 = 1$. Then

$$QA = 3\sum_{j=0}^{2} \bar{p}(0,j)|\langle 0|\psi_j\rangle|^2 + 3\sum_{j=0}^{2}(\bar{p}(1,j) + \bar{p}(2,j))|\langle 1|\psi_j\rangle|^2$$

$$= 3\sum_{j=0}^{2} \bar{p}(0,j)|\langle 0|\psi_j\rangle|^2 - 3\sum_{j=0}^{2} \bar{p}(0,j)|\langle 1|\psi_j\rangle|^2.$$

Note that $\bar{P}+J/9$ is a matrix with nonnegative elements. Thus, for any $0 \le i \le 2$, if $k_i \ge 0$ we have $-k_i x \ge -\frac{1}{9}$ and $-k_i(1-x) \ge -\frac{1}{9}$, and if $k_i < 0$, we have $-k_i \le \frac{1}{9}$. And the above inequality indicates that if $0 < x < 1$, $k_i \le \frac{1}{9x}$ and $k_i \le \frac{1}{9(1-x)}$, which is equivalent to $k_i \le \frac{1}{9x_b}$. Actually, this also holds when $x = 0$ or $x = 1$. Therefore, we obtain that

$$QA = 3\sum_{j=0}^{2} \bar{p}(0,j)|\langle 0|\psi_j\rangle|^2 - 3\sum_{j=0}^{2} \bar{p}(0,j)|\langle 1|\psi_j\rangle|^2$$

$$= 3\sum_{j=0}^{2} k_j|\langle 0|\psi_j\rangle|^2 - 3\sum_{j=0}^{2} k_j|\langle 1|\psi_j\rangle|^2 \le 3 \cdot \frac{1}{9x_b} + 3 \cdot \frac{1}{9} = \frac{1}{3} + \frac{1}{3x_b},$$

where the relationship $\sum_j |\langle 0|\psi_j\rangle|^2 = \sum_j |\langle 1|\psi_j\rangle|^2 = 1$ is utilized.

Go back to the example in the above subsection. Note that for M in Eq. (16) we have $M^0 = 1 \cdot M^1 + 0 \cdot M^2$ (thus in order to utilize Theorem 6, we need to adjust the order of the columns). Thus we can choose $x = 0$, and then $x_b = 1$. As a result, the discussion above shows that $QA \le 2/3$, which means the choice of P in Eq. (17) is optimal for M in Eq. (16).

Open Problems. From the mathematical perspective, some questions remain open. Two of them are listed as below: (1) What is the maximum gain in a zero-sum $[-1,1]$-normalized game[2] on a state in symmetric subspace without entanglement? (2) What is the maximum gain in a zero-sum $[-1,1]$-normalized game on a state in symmetric subspace without discord?

Acknowledgments. Z.W. thanks Leong Chuan Kwek and Luming Duan for helpful comments. Z.W. was supported by the Singapore National Research Foundation under NRF RF Award No. NRF-NRFF2013-13 and the WBS grant under contract no. R-710-000-007-271. S.Z. was supported by Research Grants Council of the Hong Kong (Project no. CUHK419011, CUHK419413), and this research benefited from visits to Tsinghua

---

[2] A game is $[-1,1]$-normalized if all utility functions have ranges within $[-1,1]$.

University partially supported by China Basic Research Grant 2011CBA00300 (subproject 2011CBA00301) and to Centre for Quantum Technologies partially under their support.

References

[Aum74] Aumann, R.: Subjectivity and correlation in randomized strategies. J. Math. Econ. **1**, 67–96 (1974)

[BCMdW10] Buhrman, H., Cleve, R., Massar, S., de Wolf, R.: Nonlocality and communication complexity. Rev. Mod. Phys. **82**, 665–698 (2010)

[BH01] Benjamin, S., Hayden, P.: Comment on "quantum games and quantum strategies". Phys. Rev. Lett. **87**(6), 069801 (2001)

[CT06] Cheon, T., Tsutsui, I.: Classical and quantum contents of solvable game theory on hilbert space. Phys. Lett. A **348**, 147–152 (2006)

[DSC08] Datta, A., Shaji, A., Caves, C.M.: Quantum discord and the power of one qubit. Phys. Rev. Lett. **100**, 050502 (2008)

[DVB10] Dakic, B., Vedral, V., Brukner, C.: Necessary and sufficient condition for nonzero quantum discord. Phys. Rev. Lett. **105**, 190502 (2010)

[EWL99] Eisert, J., Wilkens, M., Lewenstein, M.: Quantum games and quantum strategies. Phys. Rev. Lett. **83**(15), 3077–3080 (1999)

[HMR+10] Hallgren, S., Moore, C., Roetteler, M., Russell, A., Sen, P.: Limitations of quantum coset states for graph isomorphism. J. ACM **57**(6), 1–33 (2010)

[JSWZ13] Jain, R., Shi, Y., Wei, Z., Zhang, S.: Efficient protocols for generating bipartite classical distributions and quantum states. IEEE Trans. Inf. Theor. **59**(8), 5171–5178 (2013)

[KZ12] Kerenidis, I., Zhang, S.: A quantum protocol for sampling correlated equilibria unconditionally and without a mediator. In: Kawano, Y. (ed.) TQC 2012. LNCS, vol. 7582, pp. 13–28. Springer, Heidelberg (2012)

[Mey99] Meyer, D.: Quantum strategies. Phys. Rev. Lett. **82**(5), 1052–1055 (1999)

[Nas51] Nash, J.: Non-cooperative games. The Ann. Math. **54**(2), 286–295 (1951)

[OR94] Osborne, M.J., Rubinstein, A.: A Course in Game Theory. MIT Press, Cambridge (1994)

[OZ01] Ollivier, H., Zurek, W.H.: Quantum discord: a measure of the quantumness of correlations. Phys. Rev. Lett. **88**, 017901 (2001)

[PKL+15] Pappa, A., Kumar, N., Lawson, T., Santha, M., Zhang, S., Diamanti, E., Kerenidis, I.: Nonlocality and conflicting interest games. Phys. Rev. Lett. **114**, 020401 (2015)

[vEP02] van Enk, S.J., Pike, R.: Classical rules in quantum games. Phys. Rev. A **66**, 024306 (2002)

[Vid03] Vidal, G.: Efficient classical simulation of slightly entangled quantum computations. Phys. Rev. Lett. **91**, 147902 (2003)

[VNRET97] Vazirani, V., Nisan, N., Roughgarden, T., Éva, T.: Algorithmic Game Theory. Cambridge University Press, New York (1997)

[WZ13] Wei, Z., Zhang, S.: Full characterizing quantum correlated equilibria. Quantum Inf. Comput. **13**(9–10), 0846–0860 (2013)

[Zha12] Zhang, S.: Quantum strategic game theory. In: Proceedings of the 3rd Innovations in Theoretical Computer Science, pp. 39–59 (2012)

[ZWC+12] Zu, C., Wang, Y., Chang, X., Wei, Z., Zhang, S., Duan, L.: Experimental demonstration of quantum gain in a zero-sum game. New J. Phys. **14**(033002), 39–59 (2012)

Quantum Circuits for the
Unitary Permutation Problem

Stefano Facchini[1] and Simon Perdrix[2]([✉])

[1] LIG, University of Grenoble Alpes, 38000 Grenoble, France
stefano.facchini@imag.fr
[2] CNRS, LORIA, Inria Project Team Carte, Nancy, France
simon.perdrix@loria.fr

Abstract. We consider the *Unitary Permutation* problem which consists, given n unitary gates $U_1, \ldots, U_n$ and a permutation σ of $\{1, \ldots, n\}$, in applying the unitary gates in the order specified by σ, i.e. in performing $U_{\sigma(n)} \circ \ldots \circ U_{\sigma(1)}$.

This problem has been introduced and investigated in [6] where two models of computations are considered. The first is the (standard) model of query complexity: the complexity measure is the number of calls to any of the unitary gates U_i in a quantum circuit which solves the problem. The second model provides *quantum switches* and treats unitary transformations as inputs of second order. In that case the complexity measure is the number of quantum switches. In their paper, Colnaghi et al. [6] have shown that the problem can be solved within n^2 calls in the query model and $\frac{n(n-1)}{2}$ quantum switches in the new model, moreover both results was claimed to be optimal.

We refine these results and contradict their optimality, by proving that $n \log_2(n) + \Theta(n)$ quantum switches are necessary and sufficient to solve this problem, whereas $n^2 - 2n + 4$ calls are sufficient to solve this problem in the standard quantum circuit model. We prove, with an additional assumption on the family of gates used in the circuits, that $n^2 - o(n^{7/4+\epsilon})$ queries are required, for any $\epsilon > 0$. The upper and lower bounds for the standard quantum circuit model are established by pointing out connections with the *permutation as substring* problem introduced by Karp.

1 Introduction

The problem of applying two unitary gates U and V in an order specified by a control bit x is a natural problem: one wants to apply VU if $x = 0$ and UV if $x = 1$. Surprisingly, Chiribella et al. [4] showed that in the standard model of quantum circuits, this task cannot be realised using a single call to U and a single call to V, whereas, in the lab, a simple procedure can be implemented – using standard tools in quantum optics for instance – that performs this task using a single call to U and single call to V (see [4] for details). To model such procedure, they introduced the notion of quantum switch (QS) which is a gate that inputs two unitary transformations and a control bit and performs a switch or not of the unitary transformations depending on the control bit: $QS(x, U_0, U_1) = (U_x, U_{\bar{x}})$.

R. Jain et al. (Eds.): TAMC 2015, LNCS 9076, pp. 324–331, 2015.
DOI: 10.1007/978-3-319-17142-5_28

To point out a separation between the standard model of quantum circuits and the quantum switch model, Colnaghi et al. [6] considered the generalisation of the previous problem: given n unitary transformations $U_1, \ldots, U_n$, and a permutation σ of $[n]$, the task consists in performing $U_{\sigma(n)} \circ \ldots U_{\sigma(1)}$, where $[n]$ denotes the interval $\{1, \ldots, n\}$. This problem is called the *Unitary Permutation* problem of size n or UP_n problem for short. They proved that the UP_n problem can be solved within n^2 queries in the standard model of quantum circuits whereas $\frac{n(n-1)}{2}$ quantum switches suffice to solves this problem. Moreover, they claimed the optimality of their constructions in both cases. However, we improve both constructions: in the standard model of quantum circuits we show that $n^2 - 2n + 4$ queries are sufficient. To this end we reduce the problem to the existence of a *complete* sequence over the set $\{1, \ldots, n\}$, i.e. a sequence which contains all permutations of $[n]$ as subsequences. This problem has been originally introduced by Karp [5, Problem 36] and bounds on the size of the minimal complete sequences are known [1,7,11].

Moreover, we show that the complexity of the problem in the quantum switch model is $n \log_2(n) + \Theta(n)$. This problem is actually strongly related to the classical problem of permutation networks [14], i.e. the problem of implementing a permutation in the classical binary circuit model.

2   Bounds for the Standard Model

In [6] a simple circuit is provided which solves UP_n within n^2 calls to the unitary gates. The circuit is made of $n + 1$ layers, each composed of controlled-swaps, interspersed by n U_i's in parallel (see Fig. 1).

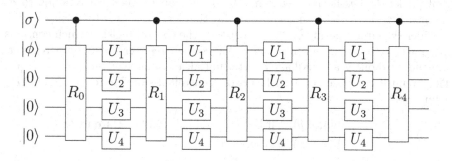

Fig. 1. Circuit introduced by Colnaghi et al. [6] for the UP_n problem.

Each layer of generalised controlled-swap (ΛR_i) performs a rewiring of the qubits: the first layer maps qubit 1 (the data qubit) to qubit $\sigma(1)$, on which $U_{\sigma(1)}$ is applied, then the second layer of controlled-swap maps qubit $\sigma(1)$ to qubit $\sigma(2)$ and so on.

In this section we provide a more efficient circuit to solve UP_n using the *permutation as substrings* problem introduced by Karp (see [5], Problem 36).

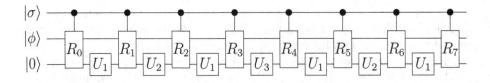

Fig. 2. Circuit based on complete sequences for the UP_n problem.

A sequence over the finite set $[n]$ is complete if it contains all permutations of $[n]$ as a (not necessarily consecutive) subsequence. For instance $w = 1213121$ is a complete sequence for $n = 3$. Finding the shortest complete sequence is known as the *permutation as substring* problem.

Definition 2.1. *Given $n \geq 1$, let $S(n)$ be the size of the shortest sequence over $[n]$ which contains each permutation of $[n]$ as a subsequence.*

Sequences of size $n^2 - 2n + 4$ are known to be complete [1,7,9,11,13], and $n^2 - 2n + 4$ is actually the size of the shortest complete subsequences for $3 \leq n \leq 7$. When $n \geq 10$, the size of the shortest complete sequence is upper bounded by $n^2 - 2n + 3$ [15], whereas the best known upper bound for $n \geq 13$ is $\lceil n^2 - \frac{7}{3}n + \frac{19}{3} \rceil$ [12]. Regarding the lower bound, Kleitman and Kwiatkowski [8] proved that $S(n) \geq n^2 - C_\epsilon n^{7/4+\epsilon}$ for any $\epsilon > 0$ where C_ϵ is a constant depending on ϵ.

Theorem 2.2. *There exists a quantum circuit which solves the UP_n problem with $S(n)$ calls.*

Proof. Given a complete sequence w of $[n]$ of size $S(n)$, for any permutation σ of $[n]$ let f_σ be the indices of a subsequence of w which corresponds to σ, i.e. $f_\sigma : [n] \to [S(n)]$ is an increasing function s.t. $\forall i \in [n], \sigma(i) = w_{f_\sigma(i)}$. We consider the circuit acting on 3 sub-registers: the control register which contains the description of the permutation σ, the second register is the data register, and the third one is an auxiliary register initialised in an arbitrary state, say $|0\rangle$. The circuit is composed of $n + 1$ layers of "controlled-swap" gates $\Lambda R_0, \ldots, \Lambda R_n$ defined as

$$\Lambda R_i \left| \sigma, x, y \right\rangle = \begin{cases} |\sigma, x, y\rangle & \text{if both or none of } i \text{ and } i{+}1 \text{ are in } Im(f_\sigma) \\ |\sigma, y, x\rangle & \text{otherwise} \end{cases}$$

where $Im(f_\sigma) = \{ f_\sigma(i) \mid i \in [n] \}$ is the image of f_σ. The unitary transformations ΛR_{i-1} and ΛR_i are interspersed by a call to U_{w_i} on the auxiliary register and the identity on the data register. Given a permutation σ, the R_i act either as a swap or as the identity in such a way that the unitary transformations applied on the data register are $U_{\sigma(1)}$ then $U_{\sigma(2)}$ and so on. An example of complete sequence for $n = 3$ is $w = 1213121$ which leads to the circuit described in Fig. 2. $\square$

Corollary 2.3. *The UP_n problem can be solved within $n^2 - 2n + 4$ calls in the standard model.*

Solving the UP_n problem within $n^2 - 2n + 4$ calls contradicts Theorem 2 in [6] which claims that Colnaghi et al. construction *is the most efficient implementation of* a circuit which solves the unitary permutation problem. Indeed, Corollary 2.3 implies that construction in [6] is not optimal in terms of number of calls, however in terms of depth or total number of quantum operations our result does not disprove the optimality of the construction proposed by Colnaghi et al.

We conjecture that $S(n)$ is also a lower bound on the number of calls necessary to solve the unitary permutation problem in the standard model, which would imply that any circuit which solves the unitary permutation problem uses at least $n^2 - o(n^{7/4+\epsilon})$ queries for any $\epsilon > 0$.

Conjecture 2.4. *The UP_n problem requires $n^2 - o(n^{7/4+\epsilon})$ calls in the standard quantum circuit model, for any $\epsilon > 0$.*

We prove the conjecture in a particular setting where only rewiring gates – like controlled swaps – are allowed.

Definition 2.5 (Rewiring gates). *A rewiring gate R is a unitary gate acting on a control register and a k-qubit target register as follows: for any permutation σ of $[n]$, $R |\sigma, x_1, \ldots x_k\rangle = |\sigma, x_{\tau_1}, \ldots x_{\tau_k}\rangle$ where τ is a permutation of $[k]$ which depends on σ.*

Lemma 2.6. *Any circuit composed of rewiring gates and calls to the U_i which solves the UP_n problem is composed of at least $n^2 - o(n^{7/4+\epsilon})$ calls to the oracle for any $\epsilon > 0$.*

Proof. The circuit that solves the UP_n problem has 3 inputs: the permutation $|\sigma\rangle$, the input state $|\phi\rangle$ and some ancillary qubits that we assume w.l.o.g. in the state $|0\rangle$. The calls to the oracle are performed in a certain order, independent of the permutation σ, which can be represented by a sequence w over the set $\{1, \ldots, n\}$: the first call is to U_{w_1}, the second to U_{w_2} and so on. If two or more calls are made in parallel we arbitrarily sequentialise the calls. Each call to the oracle is preceded and followed by a rewiring gate (or by the identity which is a particular rewiring gate). Thus for any fixed input permutation σ, the input state goes through some of the U-gates. So the applied unitary is $U_{\tau(m)} \ldots U_{\tau(1)}$ for some sub sequence τ of w of size m, such that $U_{\sigma(n)} \ldots U_{\sigma(1)} = \alpha U_{\tau(m)} \ldots U_{\tau(1)}$ for some $\alpha \in \mathbb{C}$ (unitary transformations are usually defined up to a global phase). We consider the following particular family of unitary gates U_j acting on 2 registers as follows $\forall d \in \mathbb{N}, \forall x \in \{0,1\}$, $U_j |d, x\rangle = e^{ixjn^d} |d+1, x\rangle$. $U_{\sigma(n)} \ldots U_{\sigma(1)} = \alpha U_{\tau(m)} \ldots U_{\tau(1)}$ implies $U_{\sigma(n)} \ldots U_{\sigma(1)} |0,0\rangle = \alpha U_{\tau(m)} \ldots U_{\tau(1)} |0,0\rangle$ so $|n,0\rangle = \alpha |m,0\rangle$, as a consequence $\alpha = 1$ and $n = m$. Moreover $U_{\sigma(n)} \ldots U_{\sigma(1)} |0,1\rangle = U_{\tau(n)} \ldots U_{\tau(1)} |0,1\rangle$ $\implies e^{i \sum_{d=0}^{n-1} \sigma(d) n^d} |n,1\rangle = e^{i \sum_{d=0}^{n-1} \tau(d) n^d} |n,1\rangle$, thus $\tau = \sigma$. As a consequence any permutation is a subsequence of w, so w is complete and its size, i.e. the number of calls, is at least $n^2 - C_\epsilon n^{7/4+\epsilon}$, for any $\epsilon > 0$ [8]. So, for any $\epsilon > 0$, the minimal number of calls is at least $n^2 - C_{\frac{\epsilon}{2}} n^{7/4+\frac{\epsilon}{2}} = n^2 - o(n^{7/4+\epsilon})$. $\square$

Although the unitary permutation problem seems to be strongly related to the permutation as substring problem, Conjecture 2.4 is false if one considers a slightly different model. For instance, the UP_n problem can be solved with a non zero probability using n calls only. Such circuit is based on the quantum teleportation and generalises the construction given for the particular case $n = 2$ in [4]. In a post-selected quantum circuit model where one can choose the outcome of each measurement among those which occur with a non zero probability, the UP_n problem can be solved within n calls. The probabilistic and postselection settings point out that the proof of Conjecture 2.4 should rely on some fundamental properties of the quantum circuits, like causality. In [4], the case $n = 2$ of the conjecture is proved. The proof is based on the fact that time loops are forbidden in quantum circuits.

3   Bounds for the Quantum Switch Circuit Model

A quantum switch (QS) is a gate that inputs two unitary transformations and a control bit and performs a switch or not of the unitary transformations depending on the control qubit: $\mathsf{QS}(x, U_0, U_1) = (U_x, U_{\bar{x}})$. Following [6], QS gate are represented as follows.

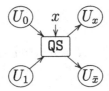

In [6], it is proved that the following network (for $n = 4$) solves the unitary permutation problem:

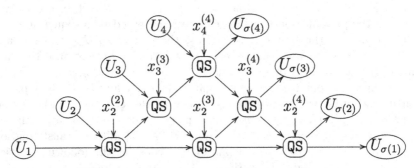

More generally, the UP_n problems can be solved using $\frac{n(n-1)}{2}$ quantum switches as described in Fig. 3.

Even if this network is claimed to solve the UP_n problem in the most efficient way, minimising the number of QS (Theorem 3 in [6]), we show that the problem can be solved much more efficiently using $n \log_2(n)$ QS only. The flaw in Theorem 3 in [6] comes from the fact that the authors claim that a permutation on $[n]$ needs to be specified by the relative ordering of each pair of different unitaries, whereas any permutation is characterised by $n - 1$ pairs of unitaries.

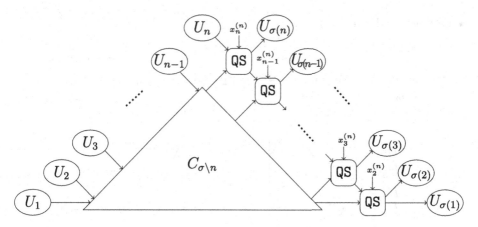

Fig. 3. Inductive definition of C_σ, where $\sigma\backslash n$ is the permutation over $[n-1]$ and $x_i^{(n)}$ are control bits defined as follows: $\forall i \in [1, \sigma^{-1}(n))$, $\sigma\backslash n(i) := \sigma(i)$ and $x_i^{(n)} := 0$, $\forall i \in [\sigma^{-1}(n), n)$, $\sigma\backslash n(i) := \sigma(i+1)$ and $x_i^{(n)} := 1$. Intuitively, $C_{\sigma\backslash n}$ produces the permutation of the $\{U_i, 1 \leq i < n\}$ according to $\sigma\backslash n$, then U_n is added at the appropriate position. The number s_n of QS in the circuit C_σ which solves the UP_n problem satisfies $s_{n+1} = s_n + n$ and $s_1 = 0$, thus $s_n = \frac{n(n-1)}{2}$.

To prove that $n \log_2(n)$ QS are sufficient to solve the UP_n problem, we use the Beneš network that solves the classical permutation network problem [2,10,14].

Lemma 3.1. *For any $n \geq 1$ there exists a circuit composed of less than $n \log_2(n)$ QS which solves the UP_n problem.*

Proof. Following the definition of the Beneš networks used for implementing arbitrary permutations (see [14] when n is a power of two and [3] in the general case), let $B(n)$ be a QS-circuit inductively defined as follows – where the control bits are omitted – depending on the parity of the number of inputs. For any $n > 0$,

- $B(2n)$ is defined as follows:

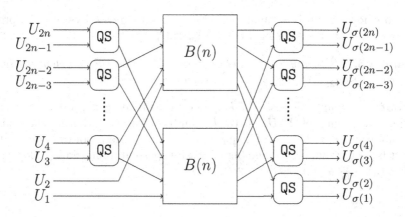

- $B(2n{+}1)$ is defined as follows:

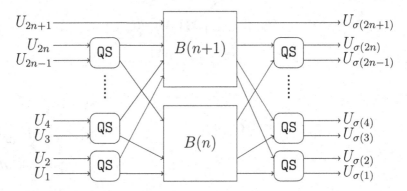

Moreover we define $B(1)$ to be the identity. Notice that according to this definition, $B(2)$ is, as expected, a single QS. Beneš networks are known to implement any permutation, see for instance [14] for a proof. Roughly speaking, given a permutation σ, the control bits can be assigned as follows: assume n is even, the first input i.e., the input which is associated with U_1, has to be connected to the output $\sigma^{-1}(1)$ through the bottom subnetwork. Thus the control bit of the right QS connected to the output $\sigma^{-1}(1)$ is forced by this constraint. Moreover, for the output $u = \sigma^{-1}(1) \pm 1$ which is connected to the same right QS as the output $\sigma^{-1}(1)$, the path from its corresponding input $\sigma(u)$ has to go through the upper subnetwork, which forces the control bit of the left QS connected to $\sigma(u)$. Then the constraint is propagated to the input $u \pm 1$ which is connected to the same left QS as u and so on. These constraints propagate until a loop is created. At that step if not all the control bits have been assigned, a control bit of a left QS which has not been assigned yet is assigned to 0, and the constraints generated by this assignment are propagated, and so on, until all control bits have been assigned.

Regarding the number of QS, the size b_n of the Beneš network $B(n)$ satisfies $b_{2n} = 2n - 1 + 2b_n$, $b_{2n+1} = 2n + b_n + b_{n+1}$, and $b(1) = 0$. As a consequence, for any $n > 0$, $b_n \leq n\log_2(n)$.   □

The previous circuit is optimal, indeed any circuit which solves the UP_n problem is composed of at least $n\log_2(n) - 2n$ QS gates:

Lemma 3.2. *For any $n \geq 1$, a QS circuit solving the UP_n problem is composed of at least $\lceil \log_2(n!) \rceil \geq n\log_2(n) - 2n$ quantum switches.*

Proof. A circuit composed of k quantum switches can produce at most 2^k different orderings of the U_i. As the U_i can be chosen such that for every distinct permutations σ, τ, $U_{\sigma(n)} \ldots U_{\sigma(1)} \neq U_{\tau(n)} \ldots U_{\tau(1)}$ (see proof of Lemma 2.6), 2^k must be larger than $n!$ the number of possible permutations. So $k \geq \lceil \log_2(n!) \rceil \geq n\log_2(n) - \frac{n-1}{\ln(2)} \geq n\log_2(n) - 2n$.   □

Acknowledgements. The authors want to thank G. Chiribella, P. Perinotti, and B. Valiron for fruitful discutions. This work has been funded by the ANR-10-JCJC-0208 CausaQ grant.

References

1. Adleman, L.: Short permutation strings. Discrete Math. **10**, 197 (1974)
2. Beneš, V.: Mathematical Theory of Connecting Networks and Telephone Traffic. Academic Press, New York (1965)
3. Chang, C., Melhem, R.: Arbitrary size benes networks. Parallel Process. Lett. **7**, 279–284 (1997)
4. Chiribella, G., D'Ariano, G.M., Perinotti, P., Valiron, B.: Quantum computations without definite causal structure. Phys. Rev. A **88**, 022318 (2013)
5. Chvátal, V., Klarner, D.A., Knuth, D.E.: Selected combinatorial research problems. Technical report 292 (1972)
6. Colnaghi, T., D'Ariano, G.M., Facchini, S., Perinotti, P.: Quantum computation with programmable connections between gates. Phys. Lett. A **376**, 2940 (2012)
7. Galbiati, G., Preparata, F.P.: On permutation embedding sequences. SIAM J. Appl. Math. **30**, 421 (1976)
8. Kleitman, D., Kwiatkowsky, D.: A lower bound on the length of a sequence containing all permutations as subsequences. J. Comb. Theory Ser. A **21**, 129 (1976)
9. Mohanty, S.P.: Shortest string containing all permutations. Discrete Math. **31**, 91 (1980)
10. Nassimi, D., Sahni, S.: A self-routing benes network and parallel permutation algorithms. IEEE Trans. Comput. **30**, 332 (1981)
11. Newey, M.: Notes on a problem involving permutations as subsequences. Technical report 340 (1973)
12. Radomirović, S.: A construction of short sequences containing all permutations of a set as subsequences. Electron. J. Comb. **19**, 31 (2012)
13. Savage, C.: Short strings containing all k-element permutations. Discrete Math. **42**, 281 (1982)
14. Waksman, A.: A permutation network. J. Ass. Comput. Mach. **15**, 159 (1968)
15. Zălinescu, E.: Shorter strings containing all k-element permutations. Inf. Process. Lett. **111**, 605 (2011)

Parallelism and Statistics

Algorithms in the Ultra-Wide Word Model

Arash Farzan[1], Alejandro López-Ortiz[2], Patrick K. Nicholson[3],
and Alejandro Salinger[4](✉)

[1] Facebook Inc., New York, NY, USA
afarzan@fb.com
[2] David R. Cheriton School of Computer Science, University of Waterloo,
Waterloo, ON, Canada
alopez-o@uwaterloo.ca
[3] Max-Planck-Institut Für Informatik, Saarbrücken, Germany
pnichols@mpi-inf.mpg.de
[4] SAP SE, Walldorf, Germany
alejandro.salinger@sap.com

Abstract. The effective use of parallel computing resources to speed up algorithms in current multi-core parallel architectures remains a difficult challenge, with ease of programming playing a key role in the eventual success of various parallel architectures. In this paper we consider an alternative view of parallelism in the form of an ultra-wide word processor. We introduce the Ultra-Wide Word architecture and model, an extension of the word-RAM model that allows for constant time operations on thousands of bits in parallel. Word parallelism as exploited by the word-RAM model does not suffer from the more difficult aspects of parallel programming, namely synchronization and concurrency. For the standard word-RAM algorithms, the speedups obtained are moderate, as they are limited by the word size. We argue that a large class of word-RAM algorithms can be implemented in the Ultra-Wide Word model, obtaining speedups comparable to multi-threaded computations while keeping the simplicity of programming of the sequential RAM model. We show that this is the case by describing implementations of Ultra-Wide Word algorithms for dynamic programming and string searching. In addition, we show that the Ultra-Wide Word model can be used to implement a non-standard memory architecture, which enables the side-stepping of lower bounds of important data structure problems such as priority queues and dynamic prefix sums. While similar ideas about operating on large words have been mentioned before in the context of multimedia processors [27], it is only recently that an architecture like the one we propose has become feasible and that details can be worked out.

1 Introduction

In the last few years, multi-core architectures have become the dominant commercial hardware platform. The potential of these architectures to improve performance through parallelism remains to be fully attained, as effectively using all cores on a single application has proven to be a difficult challenge. In this

© Springer International Publishing Switzerland 2015
R. Jain et al. (Eds.): TAMC 2015, LNCS 9076, pp. 335–346, 2015.
DOI: 10.1007/978-3-319-17142-5_29

paper we introduce the Ultra-Wide Word architecture and model of computation, an alternate view of parallelism for a modern architecture in the form of an ultra-wide word processor. This can be implemented by replacing one or more cores of a multi-core chip with a very wide word Arithmetic Logic Unit (ALU) that can perform operations on a very large number of bits in parallel.

The idea of executing operations on a large number of bits simultaneously has been successfully exploited in different forms. In Very Long Instruction Word (VLIW) architectures [14], several instructions can be encoded in one wide word and executed in one single parallel instruction. Vector processors allow the execution of one instruction on multiple elements simultaneously, implementing Single-Instruction-Multiple-Data (SIMD) parallelism. This form of parallelism led to the design of supercomputers such as the Cray architecture family [26] and is now present in Graphics Processing Units (GPUs) as well as in Streaming SIMD Extensions (SSE) to scalar processors.

In 2003, Thorup [27] observed that certain instructions present in some SSE implementations were particularly useful for operating on large integers and speeding up algorithms for combinatorial problems. To a certain extent, some of the ideas in the Ultra Wide Word architecture are presaged in the paper by Thorup, which was proposed in the context of multimedia processors. Our architecture developed independently and differs on several aspects (see discussion in full version [15]) but it is motivated by similar considerations.

As CPU hardware advances, so does the model used in theory to analyze it. The increase in word size was reflected in the word-RAM model in which algorithm performance is given as a function of the input size n and the word size w, with the common assumption that $w = \Theta(\log n)$. In its simplest version, the word-RAM model allows the same operations as the traditional RAM model. Algorithms in this model take advantage of bit-level parallelism through packing various elements in one word and operating on them simultaneously. Although similar to vector processing, the word-RAM provides more flexibility in that the layout of data in a word depends on the algorithm and data elements can be packed in an arbitrary way. Unlike VLIW architectures, the Ultra-Wide Word model we propose is not concerned with the compiler identifying operations which can be done in parallel but rather with achieving large speedups in implementations of word-RAM algorithms through operations on thousands of bits in parallel.

As multi-core chip designs evolve, chip vendors try to determine the best way to use the available area on the chip, and the options traditionally are an increased number of cores or larger caches. We believe that the current stage in processor design allows for the inclusion of an architecture such as the one we propose. In addition, ease of programming is a major hurdle to the eventual success of parallel and multi-core architectures. In contrast, bit parallelism as exploited by the word-RAM model does not suffer from this drawback: there is a large selection of word-RAM algorithms (see, e.g., [2,11,19,21]) that readily benefit from bit parallelism without having to deal with the more difficult aspects of concurrency such as mutual exclusion, synchronization, and resource contention. In this sense, the advantage of an on-chip ultra-wide word architecture is that it can enable word-RAM algorithms to achieve speedups comparable to those of multi-threaded computations, while at the same time keeping the simplicity of sequential programming that is inherent to the RAM model. We argue that this is the case by showing several examples of implementations of word-RAM algorithms using the wide word, usually with simple modifications to existing algorithms, and extending the ideas and techniques from the word-RAM model.

In terms of the actual architecture, we envision the ultra-wide ALU together with multi-cores on the same chip. Thus, the Ultra-Wide Word architecture adds to the computing power of current architectures. The results we present in this paper, however, do not use multi-core parallelism.

Summary of Results. We introduce the Ultra-Wide Word architecture and model, which extends the w-bit word-RAM model by adding an ALU that operates on w^2-bit words. We show that several broad classes of algorithms can be implemented in this model. In particular:

- We describe Ultra-Wide Word implementations of dynamic programming algorithms for the subset sum problem, the knapsack problem, the longest common subsequence problem, as well as many generalizations of these problems. Each of these algorithms illustrates a different technique (or combination of techniques) for translating an implementation of an algorithm in the word-RAM model to the Ultra-Wide Word model. In all these cases we obtain a w-fold speedup over word-RAM algorithms.
- We also describe Ultra-Wide Word implementations of popular string searching algorithms: the Shift-And/Shift-Or algorithms [3, 28] and the Boyer-Moore-Horspool algorithm [22]. Again, we obtain a w-fold speedup over the original algorithms.
- Finally, we show that the Ultra-Wide Word model is powerful enough to simulate a non-standard memory architecture in which bytes can overlap, which we shall call FS-RAM [16]. This allows us to implement data structures and algorithms that circumvent known lower bounds for the word-RAM model.

Due to space constraints, we only present a high-level description of our results. The full details can be found in the full version of this paper [15].

2   The Ultra-Wide Word-RAM Model

The Ultra-Wide word-RAM model (UW-RAM) we propose is an extension of the word-RAM model. The word-RAM is a variant of the RAM model in which a word has length w bits, and the contents of memory are integers in the range $\{0, \ldots, 2^w - 1\}$ [19]. This implies that $w \geq \log n$, where n is the size of the input, and a common assumption is $w = \Theta(\log n)$ (see, e.g., [7, 24]). Algorithms in this model take advantage of the intrinsic parallelism of operations on w-bit words. We provide a more detailed description of the word-RAM in the full version [15].

The Ultra-Wide word-RAM model extends the word-RAM model by introducing an ultra-wide ALU with w^2-bit *wide words*. The ultra-wide ALU supports the basic operations available in a word-RAM on the entire word at once. As in the word-RAM model, the available set of instructions can be assumed to be those of the restricted, multiplication, or the AC^0 models. For the results in this paper we assume the instructions of the restricted model (addition, subtraction, left and right shift, and bitwise boolean operations), plus two non-standard straightforward AC^0 operations that we describe at the end of this section.

The model maintains the standard w-bit ALU as well as w-bit memory addressing. In general, we use the parameter w for the word size in the description and analysis of algorithms, although in some cases we explicitly assume $w = \Theta(\log n)$. In terms of real world parameters, the wide word in the ultra-wide ALU would presently have between 1,000 and 10,000 bits and could increase even further in the future. In reality, the addition of an ALU that supports operations on thousands on bits would require appropriate adjustments to the data and instruction caches of a processor as well as to the instruction pipeline implementation. Similarly to the abstractions made by the RAM and word-RAM models, the UW-RAM model ignores the effects of these and other architectural features and assumes that the execution of instructions on ultra-wide words is as efficient as the execution of operations on regular w-bit words, up to constant factors.

Provided that the UW-RAM supports the same operations as the word-RAM, the techniques to achieve bit-level parallelism in the word-RAM extend directly to the UW-RAM. However, since the word-RAM assumes that a word can be read from memory in constant time, many operations in word-RAM algorithms can be implemented through constant time table lookups. With words of w^2 bits, we cannot expect to achieve constant time lookups since the size of the tables would be prohibitive. However, the memory access operations of our model allow for the implementation of simultaneous table lookups of several w-bit words within a wide word, as we shall explain below.

We first introduce some notation. Let W denote a w^2-bit word. Let $W[i]$ denote the i-th bit of W, and let $W[i..j]$ denote the contiguous subword of W from bit i to bit j, inclusive. The least significant bit of W is $W[0]$, and thus $W = \sum_{i=0}^{w^2-1} W[i] \times 2^i$. For the sake of memory access operations, we divide W into w-bit blocks. Let W_j denote the j-th contiguous block of w bits in W, for $0 \leq j \leq w - 1$, and let $W_j[i]$ denote the i-th bit within W_j. Thus, $W_j = W[jw..(j+1)w-1]$. The division of a wide word in blocks is solely intended for certain memory access operations, but basic operations of the model have no notion of block boundaries. Figure 1 shows a representation of a wide word, depicting bits with increasing significance from left to right. In the description of operations with wide words we generally refer to variables with uppercase letters, whereas we use lowercase to refer to regular variables that use one w-bit word. Thus, shifts to the left (right) by i are equivalent to division (multiplication) by 2^i. In addition, we use $\mathbf{0}$ to denote a wide word with value 0. We use standard C-like notation for operations AND ('&'), OR ('|'), NOT ('~') and shifts ('<<','>>').

Memory Access Operations. In this architecture w (not necessarily contiguous) words from memory can be transferred into the w blocks of a wide word W in constant time. These blocks can be written to memory in parallel as well.

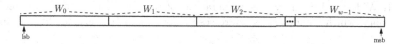

Fig. 1. A wide word in the Ultra-Wide Word architecture. The wide word is divided in w blocks of w bits each, shown here in increasing number of block from left to right.

As with PRAM algorithms, the memory access type of the model can be assumed to allow or disallow concurrent reads and writes. For the results in this paper we assume the Concurrent-Read-Exclusive-Write (CREW) model.

The memory access operations that involve wide words are of three types: block, word, and content. We describe read accesses (write accesses are analogous). A *block access* loads a single w-bit word from memory into a given block of a wide word. A *word access* loads w contiguous w-bit words from memory into an entire wide word in constant time. Finally, a *content access* uses the contents of a wide word W as addresses to load (possibly non-contiguous) words of memory simultaneously: for each block j within W, this operation loads from memory the w-bit word whose address is W_j (plus possibly a base address). The specifics of read and write operations are shown in Table 1.

Note that accessing several (possibly non-contiguous) words from memory simultaneously is an assumption that is already made by any shared memory multiprocessing model. While, in reality, simultaneous access to all addresses in actual physical memory (e.g., DRAM) might not be possible, in shared memory systems, such as multi-core processors, the slowdown is mitigated by truly parallel access to private and shared caches, and thus the assumption is reasonable. We therefore follow this assumption in the same spirit.

In fact, for w equal to the regular word size (32 or 64 bits), the choice of w blocks of w bits each for the wide word ALU was judiciously made to provide the model with a feasible memory access implementation. w^2 lines to memory are well within the realm of the possible, as they are of the same order of magnitude (a factor of 2 or 8) as modern GPUs, some of which feature bus widths of 512 bits (see, e.g., [1,18]). We note that a more general model could feature a wide word with k blocks of w bits each, where k is a parameter, which can be adjusted in reality according to the feasibility of implementation of parallel memory accesses. Although described for w blocks, the algorithms presented in this paper can easily be adapted to work with k blocks instead. Naturally, the speedups obtained would depend on the number of blocks assumed, but also on the memory bandwidth of the architecture. A practical implementation with a large number of blocks would likely suffer slowdowns due to congestion in the memory bus. We believe that an implementation with k equal to 32 or 64 can be realized with truly parallel memory access, leading to significant speedups.

UW-RAM Subroutines. A procedure called *compress* serves to bring together bits from all blocks into one block in constant time, while a procedure called *spread* is the inverse function[1]. Both operations can be implemented by straight-

---

[1] These operations are also known as PackSignBits and UnPackSignBits [27].

Table 1. Wide word memory access operations of the UW-RAM. MEM denotes regular RAM memory, which is indexed by addresses to words, and *base* is some base address.

Name	Input	Semantics
read_block	W, j, base	$W_j \leftarrow \text{MEM}[\text{base}+j]$
read_word	W, base	for all j in parallel: $W_j \leftarrow \text{MEM}[\text{base}+j]$
read_content	W, base	for all j in parallel: $W_j \leftarrow \text{MEM}[\text{base}+W_j]$
write_block	W, j, base	$\text{MEM}[\text{base}+j] \leftarrow W_j$
write_word	W, base	for all j in parallel: $\text{MEM}[\text{base}+j] \leftarrow W_j$
write_content	W, V, base	for all j in parallel: $\text{MEM}[\text{base}+V_j] \leftarrow W_j$

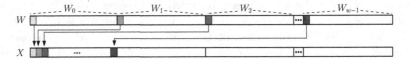

Fig. 2. The *compress* operation takes a wide word W whose set bits are restricted to the first bit of each block and compresses them to the first block of a wide word.

forward constant-depth circuits. We will also use parallel comparators, a standard technique used in word-RAM algorithms [19] (see details in full version [15]). Although these are all the subroutines that we need for the results in this paper, other operations of similar complexity could be defined if proved useful.

- **Compress:** Let W be a wide word in which all bits are zero except possibly for the first bit of each block. The compress operation copies the first bit of each block of W to the first block of a word X. I.e., if $X = \text{compress}(W)$, then $X[j] \leftarrow W_j[0]$ for $0 \le j < w$, and $X[j] = 0$ for $j \ge w$ (see Fig. 2).
- **Spread:** This operation is the inverse of the compress operation. It takes a word W whose set bits are all in the first block and spreads them across blocks of a word X so that $X_j[0] \leftarrow W[j]$ for $0 \le j < w$.

Relation to Other Models. We provide a discussion of similarities and differences between the UW-RAM and other existing models in the full version [15].

3   Simulation of FS-RAM

In the standard RAM model of computation memory is organized in registers or words, each word containing a set of bits. Any bit in a word belongs to that word only. In contrast, in the FS-RAM model [16]—also known as Random Access Machine with Byte Overlap (RAMBO)—words can overlap, that is, a single bit of memory can belong to several words. The topology of the memory, i.e., a specification of which bits are contained in which words, defines a particular variant of the FS-RAM model. Variants of this model have been used to sidestep lower bounds for important data structure problems [9,10].

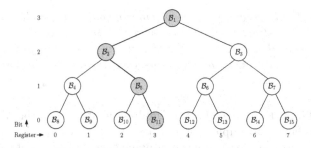

Fig. 3. Yggdrasil memory layout [9]: each node in a complete binary tree is an FS-RAM bit and registers are defined as paths from a leaf to the root. For example, register 3 contains bits $\mathcal{B}_{11}, \mathcal{B}_5, \mathcal{B}_2$, and $\mathcal{B}_1$ (shaded nodes).

We show how the UW-RAM can be used to implement memory access operations for any given FS-RAM of word size at most w bits in constant time. Thus, the time bounds of any algorithm in the FS-RAM model carry over directly to the UW-RAM. Note that each FS-RAM layout requires a different specialized hardware implementation, whereas a UW-RAM architecture can simulate any FS-RAM layout without further changes to its memory architecture.

Let $\mathcal{B}_1, \ldots, \mathcal{B}_B$ denote the bits of FS-RAM memory. A particular FS-RAM memory layout can be defined by the registers and the bits contained in them [8]. For example, in the *Yggdrasil* model in Fig. 3, reg[0]=$\mathcal{B}_8\mathcal{B}_4\mathcal{B}_2\mathcal{B}_1$, and in general reg[$i$].bit[$j$]= $\mathcal{B}_k$, where $k = \lfloor i/2^j \rfloor + 2^{m-j-1}$ ($m = 4$ in the example) [9].

In order to implement memory access operations on a given FS-RAM using the UW-RAM, we need to represent the memory layout of FS-RAM in standard RAM. Assume an FS-RAM memory of r registers of $b \leq w$ bits each and $B \leq br$ distinct FS-RAM bits. We assume that the FS-RAM layout is given as a table $\mathcal{R}$ that stores, for each register and bit within the register, the number of the corresponding FS-RAM bit. Thus, if reg[i].bit[j]= $\mathcal{B}_k$, for some k, then $\mathcal{R}[i, j] = k$. We assume $\mathcal{R}$ is stored in row major order. We simply store the value of each FS-RAM bit $\mathcal{B}_i$ in a different w-bit entry of an array A in RAM, i.e., $A[i] = \mathcal{B}_i$.

Given an index t of a register of an FS-RAM represented by $\mathcal{R}$, we can read the values of each bit of reg[t] from RAM and return the b bits in a word in constant time using the parallel reading and compress operations. Let reg[t]= $\mathcal{B}_{i_0} \ldots \mathcal{B}_{i_{b-1}}$. The read operation first obtains the address in A of each bit of register t from $\mathcal{R}$. Then, it uses a content access to read the value of each bit $\mathcal{B}_{i_j}$ into block W_j of W, thus assigning $W_j \leftarrow A[\mathcal{R}[t, j]]$. Finally, it applies one compress operation, after which the b bits are stored in W_0. In order to implement the write operation reg[t]$\leftarrow \mathcal{B}_{i_0} \ldots \mathcal{B}_{i_{b-1}}$ of FS-RAM, we first set $W_0 \leftarrow \mathcal{B}_{i_0} \ldots \mathcal{B}_{i_{b-1}}$ and perform a spread operation to place each bit $\mathcal{B}_j$ in block W_j. We then write the contents of each W_j in $A[\mathcal{R}[t, j]]$. Both read and write take constant time. We describe these operations in pseudocode in the full version [15].

Since the read and write operations described above are sufficient to implement any operation that uses FS-RAM memory (any other operation is implemented in RAM), we have the following result (see [15] for the proof).

Theorem 1. *Let $\mathcal{R}$ be any* FS-RAM *memory layout of r registers of at most b bits each and B distinct* FS-RAM *bits, with $b \leq w$ and $\log B \leq w$. Let A be any* FS-RAM *algorithm that uses $\mathcal{R}$ and runs in time T. Algorithm A can be implemented in the* UW-RAM *to run in time $O(T)$, using $rb + B$ additional words of* RAM.

Constant Time Priority Queue. Brodnik et al. [9] use the Yggdrasil FS-RAM memory layout to implement priority queue operations in constant time using $3M - 1$ bits of space ($2M$ of ordinary memory and $M - 1$ of FS-RAM memory), where M is the size of the universe. This problem has non-constant lower bounds for several models, including the RAM model [5]. For a universe of size $M = 2^m$, for some m, the Yggdrasil FS-RAM layout consists of $r = M/2$ registers of $b = \log M$ bits each and $B = M - 1$ distinct FS-RAM bits (Figure 3 is an example with $M = 16$). Thus, by Theorem 1 we obtain the following result:

Corollary 1. *The discrete extended priority queue problem can be solved in the* UW-RAM *in $O(1)$ time per operation using $2M + w(M/2) \log M + w(M - 1)$ bits, thus in $O(M \log M)$ words of* RAM.

Constant Time Dynamic Prefix Sums. Brodnik et al. [10] use a modified version of the Yggdrasil FS-RAM to solve the dynamic prefix sums problem in constant time. This problem consists of maintaining an array A of size N over a universe of size M that supports the operations $update(j, d)$, which sets $A[j]$ to $A[j] \oplus d$, and $retrieve(j)$, which returns $\oplus_{i=0}^{j} A[i]$ [10,17], where $\oplus$ is any associative binary operation. This FS-RAM implementation sidesteps lower bounds on various models [17,20]. See the full version [15] for more details.

Corollary 2. *The operations of the dynamic prefix sums problem can be supported in $O(1)$ time in the* UW-RAM *with $O(M^{\sqrt{\log N}})$ bits of* RAM.

4 Dynamic Programming

In this section we describe UW-RAM implementations of dynamic programming algorithms for the subset sum, knapsack, and longest common subsequence problems. A word-RAM algorithm that only uses bit parallelism can be translated directly to the UW-RAM. The algorithm for subset sum is an example of this. In general, however, word-RAM algorithms that use lookup tables cannot be directly extended to w^2 bits, as this would require a mechanism to address $\Theta(w^2)$-bit words in memory as well as lookup tables of prohibitively large size. Hence, extra work is required to simulate table lookup operations. The knapsack implementation that we present is a good example of such case. We note that these problems have many generalizations that can be solved using the same techniques and describe them further in the full version [15].

Subset Sum. Given a set $S = \{a_1, a_2, \ldots, a_n\}$ of nonnegative integers (weights) and an integer t (capacity), the subset sum problem is to find $S' \subseteq S$ such that

$\sum_{a_i \in S'} a_i = t$ [12]. This problem is NP-hard, but it can solved in pseudopoly-nomial time via dynamic programming in $O(nt)$ time, using the following recurrence [6]: for each $0 \leq i \leq n$ and $0 \leq j \leq t$, $C_{i,j} = 1$ if and only if there is a subset of elements $\{a_1, \ldots, a_i\}$ that adds up to j. Thus, $C_{0,0} = 1$, $C_{0,j} = 0$ for all $j > 0$, and $C_{i,j} = 1$ if $C_{i-1,j} = 1$ or $C_{i-1,j-a_i} = 1$ ($C_{i,j} = 0$ for any $j < 0$). The problem admits a solution if $C_{n,t} = 1$.

Pisinger [25] gives an algorithm that implements this recursion in the word-RAM with word size w by representing up to w entries of a row of C. Using bit parallelism, w bits of a row can be updated simultaneously in constant time from the entries of the previous row: C_i is updated by computing $C_i = (C_{i-1} \mid (C_{i-1} >> a_i))$ (which might require shifting words containing C_{i-1} first by $\lfloor a_i/w \rfloor$ words and then by $a_i - \lfloor a_i/w \rfloor$) [25]. Assuming $w = \Theta(\log t)$, this approach leads to an $O(nt/\log t)$ time solution in $O(t/\log t)$ space.

This algorithm can be implemented directly in the UW-RAM: entries of row C_i are stored contiguously in memory; thus, we can load and operate on w^2 bits in $O(1)$ time when updating each row. Hence, the UW-RAM implementation runs in $O(nt/\log^2 t)$ time using the same $O(t/\log t)$ space (number of w-bit words).

Knapsack. Given a set S of n elements with weights and values, the knapsack problem asks for a subset of S of maximum value such that the total weight is below a given capacity bound b. Let $S = \{(w_i, v_i)\}_{i=1}^n$, where w_i and v_i are the weight and value of the i-th element. Like subset sum, this problem is NP-hard but can be solved in pseudopolynomial time using the following recurrence [6]: let $C_{i,j}$ be the maximum value of a solution containing elements in the subset $S_i = \{(w_k, v_k)\}_{k=1}^i$ with maximum capacity j. Then, $C_{0,j} = 0$ for all $0 \leq j \leq b$, and $C_{i,j} = \max\{C_{i-1,j}, C_{i-1,j-w_i} + v_i\}$. The value of the optimal solution is $C_{n,b}$. This leads to a dynamic program that runs in $O(nb)$ time.

The word-RAM algorithm by Pisinger [25] represents partial solutions of the dynamic programming table with two binary tables g and h and operates on $O(w)$ entries at a time. More specifically, $g_{i,u} = 1$ and $h_{i,v} = 1$ if and only if there is a solution with weight u and value v that is not dominated by another solution in $C_{i,*}$ (i.e., there is no entry $C_{i,u'}$ such that $u' < u$ and $C_{i,u'} \geq v$). Pisinger shows how to update each entry of g and h with a constant time procedure, which can be encoded as a constant size lookup table T. A new lookup table T^α is obtained as the product of α times the original table T. Thus, α entries of g and h can be computed in constant time. Setting $\alpha = w/10$, an entire row of g and h can be computed in $O(m/w)$ time and $O(m/w)$ space [25], where m is the maximum of the capacity b and the value of the optimal solution. The optimal solution can then be computed in $O(nm/w)$ time.

Compared to the subset sum algorithm, which relies mainly on bit-parallel operations, this word-RAM algorithm for knapsack relies on precomputation and use of lookup tables to achieve a w-fold speedup. While we cannot precompute a composition of $\Theta(w^2)$ lookup tables to compute $\Theta(w^2)$ entries of g and h at a time, we can use the same tables with $\alpha = w/10$ as in Pisinger's algorithm and use the *read_content* operation of the UW-RAM to make w simultaneous lookups to the table. Since the entries in a row i of h and g depend only on entries in row $i-1$, then there are no dependencies between entries in the same row.

One difficulty is that in order to compute the entries in row i in parallel we must first preprocess row $i - 1$ in both h and g, such that we can return the number of one bits in both $g_{i-1,0}, ..., g_{i-1,j}$ and $h_{i-1,0}, ..., h_{i-1,j}$ in $O(1)$ time for any column $j \in \{0, m-1\}$. That is, the prefix sums of the one bits in row $i - 1$. Note that this is *not* the same as the dynamic problem described in Sect. 3, but it is a static prefix sums problem. We describe how to compute the prefix sums of a row of g and h in $O(m/w^2)$ time in the full version [15]. Then, each row of g and h takes $O(m/w^2)$ time to compute, and since there are n rows, the total time to compute g and h (and hence the optimal solution) on the UW-RAM is $O(nm/w^2)$. This achieves a w-fold speedup over Pisinger's word-RAM solution.

Longest Common Subsequence. The final dynamic programming problem we examine is that of computing the longest common subsequence (LCS) of two string sequences (see the full version [15] for a definition). This problem can be solved via a classic dynamic programming algorithm in $O(nm)$ time [12]. In [15] we describe a UW-RAM algorithm for LCS based on an algorithm by Masek and Paterson [23]. We note that there exist other approaches to solving the LCS problem with bit-parallelism (e.g., [13]) that could also be adapted to work in the UW-RAM. The approach we show here is a good example of bit parallelism combined with the parallel lookup power of the model, which we use to implement the Four Russians technique. We obtain the following results:

Theorem 2. *The length of the LCS of two strings X and Y over an alphabet of size σ, with $|X| = m$ and $|Y| = n$, can be computed in the UW-RAM in $O(\frac{nm}{w^2} \log \sigma + m + n)$ time and $O(\frac{\min(n,m)}{w} \log \sigma)$ words in addition to the input.*

Theorem 3. *The length of the LCS of two strings X and Y of length n over an alphabet of size σ can be computed in the UW-RAM in $O(n^2 \log^2(\sigma)/w^3 + n \log(\sigma)/w)$ time. For $\sigma = O(1)$ and $w = \Theta(\log n)$ this time is $O(n^2/\log^3 n)$.*

5   String Searching

Another example of a problem where a large class of algorithms can be sped up in the UW-RAM is string searching. Given a text T of length n and a pattern P of length m, both over an alphabet Σ, string searching consists of reporting all the occurrences of P in T. We assume in general that $n \gg m$. We use two classic algorithms for this problem to illustrate different ways of obtaining speedups via parallel operations in the wide word. More specifically, we obtain speedups of $w = \Omega(\log n)$ for UW-RAM implementations of the Shift-And and Shift-Or algorithms [3,28], and the Boyer-Moore-Horspool algorithm [22].

Shift-And and Shift-Or. These algorithms simulate an $(m + 1)$-state non-deterministic automaton that recognizes P starting from every position of T. For a window $T[i-m+1..i]$ in T, the j-th state of the automaton $(0 \le j \le m)$ is active if and only if $P[1..j] = T[i - j + 1..i]$. These algorithms represent the automaton as a bit vector and update the active states using bit-parallelism. Their running time is $O(mn/w + n)$, achieving linear time on the size of the text for small patterns. We describe in the full version [15] two UW-RAM algorithms for Shift-And that illustrate different techniques, noting that the UW-RAM implementation of Shift-Or is analogous. We obtain the following theorem:

Theorem 4. *Given a text T of length n and a pattern P of length m, we can find the occ occurrences of P in T in the* UW-RAM *in time $O(nm/w^2 + n/w + occ)$.*

Boyer-Moore-Horspool. BMH [22] keeps a sliding window of length m over the text T and searches backwards in the window for matching suffixes of both the window and the pattern. The worst case running time of BMH is $O(nm)$ (when the entire window is checked for all window positions) but on average the window can be shifted by more than one character, making the running time $O(n)$ [4]. In the UW-RAM, we can take advantage of the wide word to make several character comparisons in parallel, thus achieving a w-fold speedup over the worst case behaviour of BMH. Full details are described in [15].

Theorem 5. *Given T of length n and P of length m over an alphabet of size σ, we can find the occurrences of P in T with a* UW-RAM *implementation of BMH in $O(mn \log \sigma / w^2 + 1)$ time in the worst-case and $O(n)$ time on average.*

6   Conclusions

We introduced the Ultra-Wide Word architecture and model and showed that several classes of algorithms can be readily implemented in this model to achieve a speedup of $\Omega(\log n)$ over traditional word-RAM algorithms. The examples we describe already show the potential of this model to enable parallel implementations of existing algorithms with speedups comparable to those of multi-core computations. We believe that this architecture could also serve to simplify many existing word-RAM algorithms that in practice do not perform well due to large constant factors. We conjecture as well that this model will lead to new efficient algorithms and data structures that can sidestep existing lower bounds.

References

1. AMD: AMD FirePro W9100 Workstation Graphics. http://www.amd.com/Documents/FirePro_W9100_Data_Sheet.pdf. Acessed 20 Nov 2014
2. Andersson, A., Thorup, M.: Dynamic ordered sets with exponential search trees. J. ACM **54**(3), 13 (2007)
3. Baeza-Yates, R., Gonnet, G.H.: A new approach to text searching. Commun. ACM **35**(10), 74–82 (1992)
4. Baeza-Yates, R.A., Régnier, M.: Average running time of the Boyer-Moore-Horspool algorithm. Theoret. Comput. Sci. **92**(1), 19–31 (1992)
5. Beame, P., Fich, F.: Optimal bounds for the predecessor problem and related problems. J. Comput. Syst. Sci. **65**, 2002 (2002)
6. Bellman, R.: Dynamic Programming, 1st edn. Princeton University Press, Princeton (1957)

7. Bose, P., Chen, E.Y., He, M., Maheshwari, A., Morin, P.: Succinct geometric indexes supporting point location queries. In: Proceedings of SODA, pp. 635–644 (2009)
8. Brodnik, A.: Searching in Constant Time and Minimum Space. Ph.D. thesis, University of Waterloo (1995), also available as Technical Report CS-95-41
9. Brodnik, A., Carlsson, S., Fredman, M.L., Karlsson, J., Munro, J.I.: Worst case constant time priority queue. J. Syst. Softw. **78**(3), 249–256 (2005)
10. Brodnik, A., Karlsson, J., Munro, J., Nilsson, A.: An O(1) solution to the prefix sum problem on a specialized memory architecture. In: Navarro, G., Bertossi, L., Kohayakawa, Y. (eds.) TCS 2006. IFIP, vol. 209, pp. 103–114. Springer, Boston (2006)
11. Chan, T.M.: Point location in o(log n) time, Voronoi diagrams in o(n log n) time, and other transdichotomous results in computational geometry. In: Proceedings of FOCS, pp. 333–344 (2006)
12. Cormen, T.H., Leiserson, C.E., Rivest, R.L., Stein, C.: Introduction to Algorithms, 2nd edn. The MIT Press, Cambridge (2001)
13. Crochemore, M., Iliopoulos, C.S., Pinzon, Y.J., Reid, J.F.: A fast and practical bit-vector algorithm for the longest common subsequence problem. Inf. Process. Lett. **80**(6), 279–285 (2001)
14. Fisher, J.A.: Very long instruction word architectures and the ELI-512. SIGARCH Comput. Archit. News **11**, 140–150 (1983)
15. Frazan, A., López-Ortiz, A., Nicholson, P.K., Salinger, A.: Algorithms in the Ultra-Wide Word Model (2014). http://arxiv.org/pdf/1411.7359v2
16. Fredman, M., Saks, M.: The cell probe complexity of dynamic data structures. In: Proceedings of STOC, pp. 345–354 (1989)
17. Fredman, M.L.: The complexity of maintaining an array and computing its partial sums. J. ACM **29**(1), 250–260 (1982)
18. GeForce: GeForce GTX 285 Specifications. http://www.geforce.com/hardware/desktop-gpus/geforce-gtx-285/specifications. Accessed 20 Nov 2014
19. Hagerup, T.: Sorting and searching on the word RAM. In: Meinel, C., Morvan, M. (eds.) STACS 1998. LNCS, vol. 1373, pp. 366–398. Springer, Heidelberg (1998)
20. Hampapuram, H., Fredman, M.L.: Optimal biweighted binary trees and the complexity of maintaining partial sums. SIAM J. Comput. **28**(1), 1–9 (1998)
21. Han, Y.: Deterministic sorting in O(nlog logn) time and linear space. J. Algorithms **50**, 96–105 (2004)
22. Horspool, R.N.: Practical fast searching in strings. Softw. Pract. Exp. **10**(6), 501–506 (1980)
23. Masek, W.J., Paterson, M.: A faster algorithm computing string edit distances. J. Comput. Syst. Sci. **20**(1), 18–31 (1980)
24. Munro, J.: Tables. In: Chandru, V., Vinay, V. (eds.) FSTTCS 1996. LNCS, vol. 1180, pp. 37–42. Springer, Heidelberg (1996)
25. Pisinger, D.: Dynamic programming on the word RAM. Algorithmica **35**, 128–145 (2003)
26. Russell, R.M.: The CRAY-1 computer system. Comm. ACM **21**(1), 63–72 (1978)
27. Thorup, M.: Combinatorial power in multimedia processors. SIGARCH Comput. Archit. News **31**(4), 5–11 (2003)
28. Wu, S., Manber, U.: Fast text searching: allowing errors. Commun. ACM **35**(10), 83–91 (1992)

Uniformity of Point Samples in Metric Spaces Using Gap Ratio

Arijit Bishnu[1]($\boxtimes$), Sameer Desai[1], Arijit Ghosh[2],
Mayank Goswami[2], and Subhabrata Paul[1]

[1] ACM Unit, Indian Statistical Institute, Kolkata, India
arijit@isical.ac.in
[2] MPI for Informatics, Saarbrücken, Germany

Abstract. Teramoto et al. [22] defined a new measure called the *gap ratio* that measures the uniformity of a finite point set sampled from $\mathcal{S}$, a bounded subset of $\mathbb{R}^2$. We attempt to generalize the definition of this measure over all metric spaces. We solve optimization related questions about selecting uniform point samples from metric spaces; the uniformity is measured using gap ratio. We give lower bounds for specific metric spaces, prove hardness and approximation hardness results. We also give a general approximation algorithm framework giving different approximation ratios for different metric spaces and give a $(1 + \epsilon)$-approximation algorithm for a set of points in a Euclidean space.

Keywords: Discrepancy · Metric space · Hardness · Approximation

1 Introduction

Generating uniformly distributed points over a specific domain has applications in digital halftoning; see [1,22,24] and the references therein, numerical integration [10,17], computer graphics [10], etc. Meshing also requires uniform distribution of points over a region of interest [5]. There are different measures of uniformity of points that we discuss below.

One such notion is the discrepancy [10,17] of a point set. For a formalization of this notion, an interested reader is referred to [10,17]. Let $|P| = n$ and vol(B) denote the area of B. The expected number of points that would lie inside B if P is distributed uniformly and independently at random is $n \cdot \text{vol}(B)$. Let $D(P, B)$ denote the deviation of P from uniform distribution inside a particular B, i.e. $D(P, B) = n \cdot \text{vol}(B) - |P \cap B|$. Let $\mathcal{R}$ denote the set of all shapes similar to B. The quantity $D(P, \mathcal{R}) = \sup_{R \in \mathcal{R}} |D(P, R)|$ is the *discrepancy* of P for shapes similar to B. The function $D(n, \mathcal{R}) = \inf_{P \subset S \,\&\, |P|=n} D(P, \mathcal{R})$ captures the notion of the least possible discrepancy of any point set sized n. To compute uniformity using the above measure, the quantity $D(n, \mathcal{R})$ is to be computed for all possible scales and positions of B.

Another notion of uniformity has been captured by the idea of maximizing the minimum distance among points inside $\mathcal{S}$. This is equivalent to packing equal

---

A more comprehensive version of this paper is available at http://arxiv.org/abs/1411.7819.

R. Jain et al. (Eds.): TAMC 2015, LNCS 9076, pp. 347–358, 2015.
DOI: 10.1007/978-3-319-17142-5_30

radius circles inside S [11,18–20]. Packing equal radius circles has remained a difficult problem [16]. This measure does not take into effect large empty areas inside S.

One can observe that both of the above measures are hard to compute. Motivated by problems in digital halftoning, Teramoto et al. [22] defined a new measure of uniformity called the *gap ratio* that measures uniformity in $\mathbb{R}^2$. The basic notion of this uniformity measure is a ratio between the maximum and minimum gaps among points. The minimum gap is the distance between the closest pair of points of P. The maximum gap is the radius of the maximum empty circles among points in P and is linked to the Voronoi diagram [7] of P.

Definition of Gap Ratio. Teramoto et al. [22], who introduced the problem motivated by combinatorial approaches and applications in digital halftoning [1,3,4,21], were interested in the online version of the gap ratio problem. We generalise their definition as follows.

Definition 1. *Let $(\mathcal{M}, \delta)$ be a metric space and P be a set of k points sampled from $\mathcal{M}$. Define the minimum gap as $r_P := min_{p,q \in P,\, p \neq q}\delta(p,q)/2$. The maximum gap brings into play the interrelation between the metric space $\mathcal{M}$ and $P(\subset \mathcal{M})$, the set sampled from $\mathcal{M}$, and is defined as $R_P := \sup_{q \in \mathcal{M}} \delta(q, P)$, where $\delta(q, P) := min_{p \in P}\delta(q, p)$. The gap ratio for the point set P is defined as $GR_P := R_P/r_P$. In the rest of the paper, we would mostly not use the subscript P.*

Gap ratio need not be greater than 1. See the example in [8]. In a geometric sense, the maximum gap is analogous to the covering radius of P, and the minimum gap is analogous to the packing radius of P. In a uniformly distributed point set, we expect the covering to be thin and the packing to be tight. Thus the gap ratio can be a good measure of estimating uniformity of point samples.

The space $\mathcal{M}$, as in Definition 1 can be both continuous and discrete. Using this generalized definition, we can pose the following combinatorial optimization question.

Definition 2 (The gap ratio problem). *Given a metric space $(\mathcal{M}, \delta)$, an integer k and a parameter g, find a set $P \subset \mathcal{M}$ such that $|P| = k$ and $GR_P \leqslant g$.*

Asano [2] in his work opened this area of research, where he asked discrepancy like questions in a discrete setting. Asano opined that the discrete version of this discrepancy-like problem will make it amenable to ask combinatorial optimization related questions. We initiate this line of study in this paper for different metric spaces. As we would go back and forth between different metric spaces, we summarize the results of the paper in the following table.

Previous Results. Teramoto et al. [22] proved a lower bound of $2^{\lfloor k/2 \rfloor/(\lfloor k/2 \rfloor+1)}$ for the gap ratio in the one dimensional case where k points are inserted in the interval $[0, 1]$ and also proposed a linear time algorithm to achieve the same. They got a gap ratio of 2 in 2-dimension using ideas of Voronoi insertion where the new point was inserted in the centre of a maximum empty circle [7]. They also proposed a local search based heuristic for the problem and provided experimental results in support.

Metric space		Lower bounds	Hardness	Approximation
General		None	Yes	2-approx. hard
Discrete	Graph (connected)	$\frac{2}{3}$	Yes	Approx. factor: 3; $\frac{3}{2}$-approx. hard
	Euclidean	-	-	$(1 + \epsilon)$-algorithm
Continuous	Path-connected	1	Yes	Approx. factor: 2
	Unit square in $\mathbb{R}^2$	$\frac{2}{\sqrt{3}} - o(1)$	-	Approx. factor: $\sqrt{3} + o(1)$

Asano [2] discretized the problem and showed a gap ratio of at most 2 where k integral points are inserted in the interval $[0, n]$ where n is also a positive integer and $0 < k < n$. He also showed that such a point sequence may not always exist, but a tight upper bound on the length of the sequence for given values of k and n can be proved.

Zhang et al. [24] focused on the discrete version of the problem and proposed an insertion strategy that achieved a gap ratio of at most $2\sqrt{2}$ in a bounded two dimensional grid. They also showed that no online algorithm can achieve a gap ratio strictly less than 2.5 for a 3×3 grid.

In Sects. 2 and 3, we deal with continuous and discrete metric spaces respectively, where we give lower bounds, hardness, and approximation results. We show a general approximation hardness result in Sect. 4.

2   Continuous Metric Spaces

2.1   Lower Bounds

Here we study the lower bounds for the gap ratio in continuous metric spaces. We first point out that there does not exist a general lower bound on gap ratio. See Example 2 in [8], where we consider two disjoint balls as our metric space. However, if the space is path connected we can fix a general lower bound.

Lemma 3. *The lower bound of gap ratio is 1 when $\mathcal{M}$ path connected.*

For the proof of the above lemma, see Lemma 3 of [8].

Next we consider the metric space, $[0, 1]^2 \subset \mathbb{R}^2$ as in Teramoto et al.'s problem [22]. To prove the lower bound on gap ratio, we would want to increase r and reduce R, as much as possible. To this end we need the definition of packing and covering densities, which can be found in [15,23].

Lemma 4. *The lower bound for gap ratio is $\left(\frac{2}{\sqrt{3}} - o(1) \right)$, when $\mathcal{M} = [0, 1]^2$.*

Proof. Let $2r$ be the minimum pairwise distance between the point of P. Consider a circle of radius r around each point of P. This forms a packing of k circles of radius r in a square of side length $(1 + 2r)$. Suppose the density of such a packing is d_1. Now, we can tile the plane with such squares packed with circles. Thus we have a packing of the plane of density d_1. It is known that the density of the densest packing of equal circles in a plane is $\pi/\sqrt{12}$ [15]. Then obviously $d_1 \leqslant \pi/\sqrt{12}$ as we have packed the plane with density d_1. Hence, $d_1 = k\pi r^2/(1 + 2r)^2 \leqslant \pi/\sqrt{12}$. Consequently we have, $r \leqslant \left(\sqrt{k\sqrt{12}} - 2 \right)^{-1}$.

On the other hand, let $R = \sup_{x \in \mathcal{M}} \delta(x, P)$. Clearly, circles of radius R around each point of P cover $\mathcal{M}$. Suppose the density of such a covering is D_1. Now, we can tile the plane with this unit square. Thus we have a covering of the plane with density D_1. It is known that the density of the thinnest covering of the plane by equal circle is $2\pi/\sqrt{27}$ [15]. Then obviously $D_1 \geqslant 2\pi/\sqrt{27}$ as we have covered the plane with density D_1. Thus we have, $D_1 = k\pi R^2/1 \geqslant 2\pi/\sqrt{27}$, giving us $R \geqslant \sqrt{2}/\sqrt{k\sqrt{27}}$. Hence, the gap ratio is $\frac{R}{r} \geqslant \left(\sqrt{k\sqrt{12}} - 2\right)\sqrt{2}/\sqrt{k\sqrt{27}} = \frac{2}{\sqrt{3}} - o(1)$. $\qquad\square$

Teramoto et al. [22] had obtained a gap ratio of 2 in the online version, whereas, the lower bound for the problem is asymptotically 1.1547.

2.2   Hardness

General NP-Hardness. In this section, we show that the gap ratio problem is hard for a continuous metric space. To show this hardness, we reduce from the problem of system of distant representatives in unit disks [12]. We first define the problem.

Definition 5 ($S(q, l)$-*DR*)**.** *[12] Given a parameter $q > 0$ and a family $\mathcal{F} = \{F_i | i \in I, F_i \subseteq X\}$ of subsets of X, a mapping $f : I \to X$ is called a System of q-Distant Representatives (shortly an Sq-DR) if (i) $f(i) \in F_i$ for all $i \in I$ and (ii) distance between $f(i)$ and $f(j)$ is at least q, for $i, j \in I$ and $i \neq j$. When the family $\mathcal{F}$ is a set of unit diameter disks with centres that are at least l distance apart, we denote the mapping by $S(q, l)$-DR.*

Fiala et al. proved that $S(1, l)$-*DR* is NP-hard [12]. For the general version $S(q, l)$-*DR*, we give a proof sketch using Fiala et al.'s technique. Note that for $q \leqslant l$, the centres of the disks suffice as our representatives. So assume that $q > l$. We restate a generalised version of their result below, see the proof of Theorem 10 in [8].

Theorem 6. *$S(q, l)$-DR is NP-hard for $q > l$ on the Euclidean plane.*

Next we show that the above holds even for a constrained version of the problem.

Lemma 7. *$S(q, l)$-DR-1 is NP-complete for $q > l$, where $S(q, l)$-DR-1 denotes $S(q, l)$-DR with one representative point constrained to lie on the boundary of one of the disks.*

Proof. Clearly, a solution to $S(q, l)$-*DR*-1 is a solution to $S(q, l)$-*DR*. Conversely, a solution of $S(q, l)$-*DR* can be translated until one point hits the boundary to obtain a solution to $S(q, l)$-*DR*-1.

It is easy to see that $S(q, l)$-*DR*-1 is in NP, as any claimed solution can be checked by using a voronoi diagram in polynomial time. Hence, it is NP-complete for $q > l$. $\qquad\square$

We now use the above result to prove the hardness of the gap ratio problem.

Theorem 8. *Let $\mathcal{M}$ be a continuous metric space and $q > 2$. It is NP-hard to find a finite set $P \subset \mathcal{M}$ of cardinality k such that $GR_P \leqslant \frac{2}{q}$.*

Proof. We show that if there is a polynomial algorithm to find a finite set $P \subset \mathcal{M}$ of cardinality k such that the gap ratio of P is at most $\frac{2}{q}$ for some $q > 2$, then there is also a polynomial algorithm for $S(q, l)$-*DR*-1.

Consider an instance of $S(q, l)$-*DR*-1, a family $\mathcal{F} = \{F_1, F_2, \ldots, F_k\}$ of k disks of unit diameter such that their centres are at least distance l apart, where $q > l > 2$ (even with this restriction the proof of Theorem 6 goes through).

We run the algorithm for the gap ratio problem k times, each time on a separate instance. The instance for the i-th iteration would have the disks $\{F_j | j \neq i\}$ and a circle of unit diameter with its centre being the same as the centre of F_i. The following claim, whose proof follows later completes the proof.

Claim 9. *If a single iteration of the above process results "yes", then we have a solution to the $S(q, l)$-DR-1 instance.*

Since $S(q, l)$-*DR*-1 is NP-hard, the gap ratio problem must also be NP-hard. □

Proof of Claim 9. Suppose that the gap ratio of a given point set is at most $\frac{2}{q}$ for the ith instance. If it so happens that two points are within the same disk, then $r \leqslant \frac{1}{2}$. Thus for the gap ratio to fall below $\frac{2}{q}$ we need $R \leqslant \frac{2r}{q} \leqslant 1/q < 1$. But considering the number of points that we are choosing, we must have an empty disk, which would contain a point x such that $R \geqslant d(P, x) \geqslant l - \frac{1}{2} > 1$, giving us a contradiction. Thus we have that each disk contains exactly one point from P. Since, $l > 2$ and F_i is a circle, $R = 1$. Thus, we get $r = \frac{1}{GR} \geqslant \frac{q}{2}$, making the closest pair to be at least a distance q apart.   □

Path Connected Spaces. Next, we show that it is NP-hard to find k points in a path connected space such that $GR = 1$. To prove this, we start by proving that in a path connected space it is NP-Hard to find k points such that $R = r = \frac{3}{2}$ by reducing from the efficient dominating set problem. Later we extend the result for all positive real values of r.

Theorem 10. *It is NP-hard to find a set P of k points in a path connected space $\mathcal{M}$ such that $R_P = r_P = \frac{3}{2}$.*

Proof. Let us consider an instance of the efficient domination problem, an undirected graph $G(V, E)$, and a parameter k. From this graph we form a metric space $(\mathcal{M}, \delta)$ as follows. In $\mathcal{M}$, each edge of E corresponds to a unit length path. We place at each vertex of V an ϵ-path, where $0 < \epsilon < \frac{1}{4}$, which is merely an ϵ long curve protruding from the vertex as shown in Fig. 1a. The vertices merely become points on a path formed by consecutive edges as shown

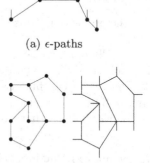

(a) ϵ-paths

(b) The graph and the metric space. The open ended lines are the ϵ-paths

Fig. 1. The reduction for path connected spaces.

in Fig. 1b. If there are edge-crossings, we do not consider the crossing to be an intersection but rather consider it as an embedding in $\mathbb{R}^3$. This ensures that different paths only intersect at vertices of the graph (this makes sure that there is direct correspondence between the path lengths in the graph and the path lengths of the metric space). The distance, δ, between two points in this space is defined by the length of the shortest curve joining the two points.

We show that finding a set P of k points in $\mathcal{M}$ such that $R_P = r_P = \frac{3}{2}$ is equivalent to finding an efficient dominating set of size k in G, using a series of claims.

Claim 11. *Suppose $D \subset V$ is an efficient dominating set in G. Then we have a set $P \subset \mathcal{M}$ with $|D| = |P|$ such that $R_P = r_P = \frac{3}{2}$.*

Conversely, given a set P' of k points in $\mathcal{M}$ such that $R_{P'} = r_{P'} = \frac{3}{2}$, we want to find an efficient dominating set in G. If $P' \subset V$, then we are done as P' is an efficient dominating set in G (refer to Claim 21). Otherwise, if $P' \not\subset V$, then from P' we construct another set $P \subset V$ such that $R_P = r_P = \frac{3}{2}$. We form P by appropriately moving points of P' to the points corresponding to V.

Claim 12. $P' \subset V$ *or* $P' \cap V = \emptyset$.

By Claim 12, if $P' \not\subset V$, then $P' \cap V = \emptyset$. Note that in this case P' cannot have midpoints of the graph edges as between any two midpoints at distance 3 from each other, there is a vertex with an ϵ-path which is distance $\frac{3}{2}$ from both points. Thus the other end of this ϵ-path must be at a distance $\frac{3}{2} + \epsilon$ from both points contradicting the fact that $R_{P'} = \frac{3}{2}$. Thus each point in P' must have a closest vertex. We form the set P by moving each point of P' to its closest vertex.

Claim 13. $R_P = r_P = \frac{3}{2}$.

For the proofs of Claims 11, 12 and 13 refer to the proofs of Claims 8, 16 and 17 of [8].

By Claim 13, without loss of generality, we can assume that the sampled set is a subset of V. Using ideas we present in the proof of Claim 21, it is easy to see that, if we can find a set P of k points in $\mathcal{M}$ such that $R_P = r_P = \frac{3}{2}$, then we can find an efficient dominating set of k vertices in G.

Hence, it is NP-hard to find a set P of k points in a path connected space such that $R_P = r_P = \frac{3}{2}$. $\qquad\square$

In the above reduction, taking the edge lengths to be $\frac{2x}{3}$ instead of 1 and $\frac{2x\epsilon}{3}$-paths instead of ϵ-paths we have that it is NP-hard to find a set of k points in a path connected space such that $R_P = r_P = \frac{3}{2} \times \frac{2x}{3} = x$. Since this can be done for any positive x, the following theorem follows as a corollary to Theorem 10.

Theorem 14. *It is NP-hard to find a set of k points in a path connected space such that gap ratio is 1.*

2.3 Approximation Algorithms

In this section we show that Gonzalez's [14] *farthest point insertion method* (with a slightly tweaked initiation) for k-centre clustering gives a constant factor approximation for gap ratio. We will call it Algorithm 1. The following is an outline of the algorithm.

Let $(\mathcal{M}, \delta)$ be a metric space of n points and k be the number of points to be sampled. The first two points chosen are a pair of farthest points. Let S_i denote the set of first i points. Then, $S_{i+1} = S_i \cup \{q_{i+1}\}$ for $2 < i \leqslant k$, where $\delta(q_{i+1}, S_i) = \sup_{q \in \mathcal{M}} \delta(q, S_i)$.

We now analyse the algorithm. Without loss of generality, let $P = \{p_1, \ldots, p_k\}$ be the set with optimal gap ratio, and let $GR = \alpha$.

Lemma 15. *In Algorithm 1, $R_{S_i} \leqslant R_{S_{i-1}}$ for each $i \in \{2, \ldots, k\}$ and the gap ratio GR_{S_i} is at most 2 after each iteration.*

For the proof of the above lemma, see Lemma 23 of [8]. The main theorem of this section is as follows.

Theorem 16. *Farthest point insertion gives the following approximation guarantees: (i) if $\alpha \geqslant 1$, then the approximation ratio is $\frac{2}{\alpha} \leqslant 2$, (ii) if $\frac{2}{3} \leqslant \alpha < 1$, the approximation ratio is $\frac{2}{\alpha} \leqslant 3$, and (iii) if $\alpha < \frac{2}{3}$, the approximation ratio is $\frac{4}{2-\alpha} < 3$.*

Proof. Case (i) and (ii) follow directly from Lemma 15. We deal with Case (iii). Let us define closed balls centred at p_i's as follows: $B_i = \{x \in \mathcal{M} : \delta(p_i, x) \leqslant r_P\}$ and $B_i' = \{x \in \mathcal{M} : \delta(p_i, x) \leqslant \alpha r_P\}$. We need the following claim.

Claim 17. *For all $i \in \{2, \ldots, k\}$, $2r_{S_i} \geqslant (2 - \alpha)r_P$.*

Proof. Note that B_j''s cover whole of P. The case of $i = 2$ follows from the fact that $2r_{S_2} = diam(\mathcal{M})$. Assume the result is true for some $i \geqslant 2$. We will show it is true for S_{i+1}, if $i \leqslant k-1$, by contradiction. Suppose q_{i+1} falls into a ball B_j' that contains q_t, for some $t \leqslant i$. This would imply $2r_{S_{i+1}} \leqslant \delta(q_t, q_{i+1}) \leqslant 2\alpha r_P$. Note that as $\alpha < 2/3$, we have $2\alpha r_P < (2-\alpha)r_P$. But since, $i \leqslant k-1$, there exists $p_{t'}$ such that $B_{t'}'$ is empty. That implies we could have selected $p_{t'}$ instead of q_{i+1} to get $2r_{S_{i+1}} = \min\{2r_{S_i}, \delta(p_{t'}, S_i)\} \geqslant (2-\alpha)r_P$. Note that last inequality follows from the fact that $2r_{S_i} \geqslant (2-\alpha)r_P$ (by induction) and $\delta(p_{t'}, S_i) \geqslant (2-\alpha)r_P$.

Now that we know q_{i+1} falls into a separate ball B_j', it is easy to see that $2r_{S_{i+1}} \geqslant \min\{2r_{S_i}, \delta(p_j, S_i)\} \geqslant (2-\alpha)r_P$. □

From the proof of Claim 17, we have for all $j \in \{1, \ldots, k\}$, $|B_j' \cap S_k| = 1$. Thus we have $R_{S_k} \leqslant 2\alpha r_P$, since B_j' cover $\mathcal{M}$. Combining this with the fact that $2r_{S_k} \geqslant (2-\alpha)r_P$ (Claim 17), we have $GR_{S_k} \leqslant \frac{4\alpha}{2-\alpha}$ and consequently $\frac{GR_{S_k}}{GR_P} \leqslant \frac{4}{2-\alpha} < 3$. □

From the results in Sect. 2.1, we have the following corollary to Theorem 16.

Corollary 18. *The approximation algorithm gives an approximation ratio of (i) 2 when the metric space is continuous, compact and path connected, and (ii) $\rho(k)$, when the metric space is restricted to a unit square in the Euclidean plane, where $\rho(k) = \frac{\sqrt[4]{27}\sqrt{k}}{\sqrt[4]{3}\sqrt{k}-\sqrt{2}} = \sqrt{3} + o(1)$.*

3 Discrete Metric Space

3.1 Graph

Lower Bounds. Here we study the lower bounds for the gap ratio problem in discrete metric spaces. Again we point out that there does not exist a general lower bound for gap ratio, in discrete spaces as well. See Example 1 in [8] for details.

Next we study the lower bound of gap ratio on a metric space $\mathcal{M}$ which is the vertex set V of an undirected connected graph $G = (V, E)$. The distance between a pair of vertices is the length of the shortest path between them.

Lemma 19. *Gap ratio has a lower bound of $\frac{2}{3}$ when the metric space $\mathcal{M}$ is a connected undirected graph. The bound is achieved only when $R = 1$ and $r = \frac{3}{2}$.*

Proof. Suppose a set of vertices $P \subset \mathcal{M}$ is sampled. Let a closest pair of vertices in P be distance q apart. Thus $r = \frac{q}{2}$. Now between these two vertices, there is a path of $q-1$ vertices in $\mathcal{M}\backslash P$. Among these $q-1$ vertices, the vertex farthest from P is at a distance $\lfloor \frac{q}{2} \rfloor$ from P. Thus $R \geqslant \lfloor \frac{q}{2} \rfloor$ and $GR = \frac{R}{r} \geqslant \frac{2}{q} \lfloor \frac{q}{2} \rfloor$. Note that, when $q = 1$, clearly we have a gap ratio greater or equal to 2. Now, we analyse this expression for even and odd values of q. If q is even, $GR \geqslant \frac{2}{q} \lfloor \frac{q}{2} \rfloor = \frac{2}{q}\frac{q}{2} = 1$ and if q is odd and $q \geqslant 3$, $GR \geqslant \frac{q-1}{q}$. Since, this function is monotonically increasing, $GR \geqslant \frac{2}{3}$, and the equality only occurs for $q = 3$.

Thus, the gap ratio $GR = \frac{2}{3}$ implies $q = 3$, which means $r = \frac{3}{2}$. Therefore, $R = GR \times r = 1$. Hence, $GR = \frac{2}{3}$ only when $R = 1$ and $r = \frac{3}{2}$.   □

Hardness. In this section, we show that the problem of finding minimum gap ratio is NP-complete even for graph metric space. To this end, we need the concept of a variation of domination problem, called efficient domination problem. A subset $D \subseteq V$ is called an *efficient dominating set* of $G = (V, E)$ if $|N_G[v] \cap D| = 1$ for every $v \in V$, where $N_G[v] = \{v\} \cup \{x | vx \in E\}$. An efficient dominating set is also known as *independent perfect dominating set* [6]. Given a graph $G = (V, E)$ and a positive integer k, the efficient domination problem is to find an efficient dominating set of cardinality at most k. The efficient domination problem is known to be NP-complete [9].

Theorem 20. *In graph metric space, gap ratio problem is NP-complete.*

Proof. First note that, the gap ratio problem in graph metric space is in NP. To prove the hardness, we use a reduction from efficient domination problem, to the gap ratio problem. Given an instance of efficient domination problem $G = (V, E)$ and k, set $\mathcal{M} = V$ as the metric space and the shortest path distance between two vertices as the metric δ.

Claim 21. *$G = (V, E)$ has an efficient dominating set of cardinality k if and only if there exists a sampled set P of k points (vertices) whose gap ratio is 2/3.*

See Claim 8 of [8] for the proof of this claim. Thus the gap ratio problem is NP-complete for graph metric space.   □

Approximation Hardness. Here we use the hardness of path connected space from Sect. 2.2 to show that the gap-ratio problem is APX-hard on the graph metric.

Theorem 22. *In an unweighted graph, it is NP-hard to approximate the gap ratio better than a factor of $\frac{3}{2}$.*

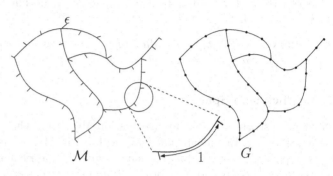

Fig. 2. Illustration of the reduction

Proof. In Sect. 2.2, we reduced the problem of finding a set of k points in a graph such that the gap ratio is $\frac{2}{3}$ to the problem of finding a set of k points in a path-connected space such that the gap ratio is 1. We use this hardness of gap ratio being 1 on instances similar to the one created in the reduction to prove $\frac{3}{2}$ approximation hardness on graphs.

Our starting instance is a space formed by joining integer length curves at their ends (so that points that divide these curves into unit length curves form a connected graph with the unit length curves as edges). Also for some $0 < \epsilon < \frac{1}{4}$ we join curves of length ϵ (at one end) at points such that the integer length curves are divided into unit length curves. Let us call this path connected space $\mathcal{M}$. Note that $\mathcal{M}$ is similar to the path connected space formed in Sect. 2.2, but, the general shape of the space may vary. The reduction is illustrated in Fig. 2. The metric on this space is defined by the length of the shortest path between pairs of points. We form the graph $G = (V, E)$ by putting vertices at the place where the ϵ-length curves are joined to the integer length curves. The ϵ-length protrusions are discarded and the unit length curves between the vertices form the edge set.

Claim 23. *There exists a polynomial time algorithm to find $P \subset \mathcal{M}$ such that $|P| = k$ and $R_P = r_P = \frac{2t+1}{2}$ for some $t \in \{1, 2, ...,\}$ if and only if there exists a polynomial time algorithm to find a set of k vertices in G such that the gap ratio of the set is strictly less than 1.*

See Theorem 22 of [8] for the proof of the above claim.

This gives us that it is NP-hard to find a set with gap ratio less than 1 in graphs, i.e. it is NP-hard to find an algorithm which approximates gap ratio within a factor better than $\frac{3}{2}$.

Note here that if we could have proven Claim 23 for $|P| = k$ and $R_P = r_P = \frac{t}{2}$ for some $t \in \{2, 3, ..., \}$, then we wouldn't need to say strictly less than 1 in the statement.   □

Approximation Algorithm. We start by pointing out that Algorithm 1 and Theorem 16 hold in a discrete metric space as well. Thus, as a consequence of Lemma 19, we have the following corollary to Theorem 16.

Corollary 24. *The approximation algorithm gives an approximation ratio of 3 when the metric space is restricted to graph metric space.*

3.2 Euclidean Space

Next we discuss a $(1 + \epsilon)$- approximation algorithm when the space $\mathcal{M}$ is a set of n points in a Euclidean space. We will call it Algorithm 2.

Suppose, $\mathcal{M}$ is a set of n points in $\mathbb{R}^d$ and the metric δ on $\mathcal{M}$ is the Euclidean metric on $\mathbb{R}^d$. We propose Algorithm 2, and prove that it gives a gap ratio within $(1 + \epsilon)$ factor of the minimum gap ratio, where $\epsilon \in \left(0, \frac{1}{2}\right)$ and $\epsilon_1 := \frac{\epsilon}{(3+2\epsilon)}$. The algorithm is as follows.

Obtain a set $P_1 \subset \mathcal{M}$ of k points by the farthest point method. Create a grid of side-length $\epsilon_2 = \frac{\epsilon_1 R_{P_1}}{2\sqrt{d}}$. Get a set S by choosing 1 point of $\mathcal{M}$ from each grid cell. Of all the $O\left(|S|^k\right)$ subsets of S choose the one with the lowest gap ratio.

For analysing the algorithm we need the following definitions and lemmas.

Define $R_{OPT} := \min_{P \subset \mathcal{M}, |P|=k} \max_{q \in \mathcal{M}} \delta(q, P)$ and $r_{OPT} := \max_{P \subset \mathcal{M}, |P|=k}$ $\min_{p,q \in P, p \neq q} \frac{\delta(p,q)}{2}$. We try to bound the time complexity by estimating the number of grid cells needed to cover $\mathcal{M}$.

Lemma 25. *In Algorithm 2, at most $N := O(k\lceil\frac{1}{\epsilon_1}\rceil^d)$ cells cover $\mathcal{M}$.*

Proof. Consider a set (say P_{cov}) of k points in $\mathcal{M}$, such that $R_{OPT} = \max_{q \in \mathcal{M}} \delta(q, P_{cov})$. Now, we know that balls of radius R_{OPT} around the points of P_{cov} cover $\mathcal{M}$. Each of these balls intersect $O(\lceil\frac{2R_{OPT}}{\epsilon_2}\rceil^d) = O(\lceil\frac{1}{\epsilon_1}\rceil^d)$ grid cells. Thus, $N := O(k\lceil\frac{1}{\epsilon_1}\rceil^d)$ cells cover $\mathcal{M}$.   □

The above lemma shows that the brute force calculation of gap ratio over S takes $O\left(N^k \left(k \log k + (n - k)\, k\right)\right)$ time, where $O(k \log k)$ is required to compute r and $O(k(n - k))$ is required to compute R in each iteration; all other steps in Algorithm 2 are polynomial in n and k. Note that the time is not polynomial in k.

We are now ready to prove the main theorem for this section.

Theorem 26. *In Algorithm 2 we have, $GR_P \geqslant (1 + \epsilon) \cdot GR_{OPT}$.*

Proof. Consider the set P^* of k points in $\mathcal{M}$, which gives the minimum gap ratio, α, in $\mathcal{M}$. Let $r := r_{P^*}$. We have $R_{P_1} \leqslant 2R_{OPT}$ from [14]. For each p_i in P^*, there exists a point q_i in S, such that $\delta(q_i, p_i) \leqslant \sqrt{d}\epsilon_2$, because $\sqrt{d}\epsilon_2$ is the diameter of each grid cell. From the definition of ϵ_2, we have $\delta(q_i, p_i) \leqslant \frac{\epsilon_1 R_{P_1}}{2} \leqslant \epsilon_1 R_{OPT} \leqslant \epsilon_1 R_{P^*} = \epsilon_1 \alpha r$. Also note that $\alpha \leqslant 2$, as the farthest point method itself will yield gap ratio at most 2. Thus, we have $\delta(q_i, p_i) \leqslant r$ (as $\epsilon_1 < \frac{1}{2}$), i.e., $i \neq j \implies q_i \neq q_j$. Let $P_2 := \{q_1, q_2, \ldots, q_k\}$ be a set of such k distinct points in S. Let us compute the gap ratio of P_2. Triangle inequality gives us $R_{P_2} \leqslant (1 + \epsilon_1)\alpha r$ and $r_{P_2} \geqslant (1 - \epsilon_1\alpha)r$. Then the gap ratio of P_2 is $\leqslant \frac{(1+\epsilon_1)\alpha}{(1-\epsilon_1\alpha)} \leqslant \frac{(1+\epsilon_1)\alpha}{(1-2\epsilon_1)} = (1 + \epsilon)\alpha$.

Also by definition, the gap ratio of P is less than the gap ratio of P_2. Thus we have that gap ratio of P in S is at most $(1 + \epsilon)\alpha$. □

4 A General Approximation Hardness Result

In this section we show that the gap ratio problem is hard to approximate within a factor of 2 for the general metric space.

Theorem 27. *In a general metric space, it is NP-hard to approximate the gap ratio better than a factor of 2.*

Proof. To show this hardness, we make a reduction from independent dominating set problem, where the dominating set is also independent set. This problem is known to be NP-hard [13].

Let $G = (V, E)$ and k be an instance of independent domination problem. We make a weighted complete graph over V such that all edges present in G have weight 1 and all other edges have weight 2. Now the metric space $\mathcal{M}$ is given by the vertex set of the complete graph and the metric is defined by the edge weights. The result is easy to see from the following claim.

Claim 28. *$G = (V, E)$ has an independent dominating set of cardinality k if and only if there exists a sampled set P in $\mathcal{M}$ of k points with gap ratio 1.*

See Theorem 20 of [8] for the proof of the above claim and other details of the proof. □

Acknowledgements. The authors want to thank Tetsuo Asano and Geevarghese Philip.

References

1. Asano, T.: Computational geometric and combinatorial approaches to digital halftoning. In: CATS, p. 3 (2006)
2. Asano, T.: Online uniformity of integer points on a line. Inf. Process. Lett. **109**(1), 57–60 (2008)

3. Asano, T., Katoh, N., Obokata, K., Tokuyama, T.: Combinatorial and geometric problems related to digital halftoning. In: Asano, T., Klette, R., Ronse, C. (eds.) Geometry, Morphology, and Computational Imaging. LNCS, vol. 2616, pp. 58–71. Springer, Heidelberg (2003)
4. Asano, T., Katoh, N., Obokata, K., Tokuyama, T.: Matrix rounding under the L_p-discrepancy measure and its application to digital halftoning. SIAM J. Comput. **32**(6), 1423–1435 (2003)
5. Asano, T., Teramoto, S.: On-line uniformity of points. In: Book of Abstracts for 8th Hellenic-European Conference on Computer Mathematics and its Applications, Athens, Greece, pp. 21–22 (2007)
6. Bange, D., Barkauskas, A., Host, L., Slater, P.: Generalized domination and efficient domination in graphs. Discrete Math. **159**(13), 1–11 (1996)
7. Berg, M., Cheong, O., Kreveld, M., Overmars, M.: Computational Geometry: Algorithms and Applications. Springer-Verlag TELOS, Santa Clara (2008)
8. Bishnu, A., Desai, S., Ghosh, A., Goswami, M., Paul, S.: Uniformity of point samples in metric spaces using gap ratio. CoRR, abs/1411.7819v1 (2014)
9. Chain-Chin, Y., Lee, R.: The weighted perfect domination problem and its variants. Discrete Appl. Math. **66**(2), 147–160 (1996)
10. Chazelle, B.: The Discrepancy Method - Randomness and Complexity. Cambridge University Press, New York (2001)
11. Collins, C.R., Stephenson, K.: A circle packing algorithm. Comput. Geom. **25**(3), 233–256 (2003)
12. Fiala, J., Kratochvíl, J., Proskurowski, A.: Systems of distant representatives. Discrete Appl. Math. **145**(2), 306–316 (2005)
13. Garey, M.R., Johnson, D.S.: Computers and Intractability; A Guide to the Theory of NP-completeness. W. H. Freeman & Co., New York (1990)
14. Gonzalez, T.F.: Clustering to minimize the maximum intercluster distance. Theor. Comput. Sci. **38**, 293–306 (1985)
15. Kuperberg, W.: An inequality linking packing and covering densities of plane convex bodies. Geom. Dedicata. **23**(1), 59–66 (1987)
16. Locatelli, M., Raber, U.: Packing equal circles in a square: a deterministic global optimization approach. Discrete Appl. Math. **122**(13), 139–166 (2002)
17. Matoušek, J.: Geometric Discrepancy: An Illustrated Guide. Springer, Heidelberg (1999)
18. Nurmela, K.J., Östergård, P.R.J.: Packing up to 50 equal circles in a square. Discrete Comput. Geom. **18**(1), 111–120 (1997)
19. Nurmela, K.J., Östergård, P.R.J.: More optimal packings of equal circles in a square. Discrete Comput. Geom. **22**(3), 439–457 (1999)
20. Nurmela, K.J., Östergård, P.R.J., aus dem Spring, R.: Asymptotic behavior of optimal circle packings in a square. Can. Math. Bull. **42**(3), 380–385 (1999)
21. Sadakane, K., Chebihi, N.T., Tokuyama, T.: Discrepancy-based digital halftoning: automatic evaluation and optimization. In: Asano, T., Klette, R., Ronse, C. (eds.) Geometry, Morphology, and Computational Imaging. LNCS, vol. 2616, pp. 301–319. Springer, Heidelberg (2003)
22. Teramoto, S., Asano, T., Katoh, N., Doerr, B.: Inserting points uniformly at every instance. IEICE Trans. **89D**(8), 2348–2356 (2006)
23. Tóth, G.: New results in the theory of packing and covering. In: Gruber, P., Wills, J. (eds.) Convexity and Its Applications, pp. 318–359. Birkhuser basel, Basel (1983)
24. Zhang, Y., Chang, Z., Chin, F.Y.L., Ting, H.-F., Tsin, Y.H.: Uniformly inserting points on square grid. Inf. Process. Lett. **111**(16), 773–779 (2011)

On Pure Nash Equilibria in Stochastic Games

Ankush Das[1], Shankara Narayanan Krishna[1], Lakshmi Manasa[1]([✉]),
Ashutosh Trivedi[1], and Dominik Wojtczak[2]

[1] Department of Computer Science and Engineering, IIT Bombay, Mumbai, India
lakshmimanasa.g@gmail.com
[2] Department of Computer Science, The University of Liverpool, Liverpool, UK

Abstract. Ummels and Wojtczak initiated the study of finding Nash equilibria in simple stochastic multi-player games satisfying specific bounds. They showed that deciding the existence of pure-strategy Nash equilibria (PURENE) where a fixed player wins almost surely is undecidable for games with 9 players. They also showed that the problem remains undecidable for the finite-strategy Nash equilibrium (FINNE) with 14 players. In this paper we improve their undecidability results by showing that PURENE and FINNE problems remain undecidable for 5 or more players.

Keywords: Stochastic games · Nash equilibrium · Pure strategy · Finite-state strategy

1 Introduction

Stochastic games are well established formalism for analyzing reactive systems under the influence of random events [1]. Such systems are often modeled as games between the system and its environment, where the environment's objective is the complement of the system's objective: the environment is considered hostile. Therefore, research in this area has traditionally focused on two-player games where each play is won by precisely one of the two players, so-called *two-player zero-sum games*. However, often in the practical settings the system may consist of several components with independent objectives, a situation which is naturally modeled by a multi-player game.

In this paper, we study multi-player *stochastic games* [9] played on finite directed graphs whose vertices are either *stochastic* or controlled by one of the players. A play of such a game evolves by moving a token along edges of the graph in the following manner. The game begins in an initial vertex. Whenever the token arrives at a non-stochastic vertex, the player who controls this vertex must move the token to a successor vertex; when the token arrives at a stochastic vertex, a fixed probability distribution determines the successor vertex. In the most general case, a measurable function maps plays to payoffs. In this paper we consider so-called *simple stochastic games*, where the possible payoffs of a single play are either 0 or 1 (i.e. each player either wins or loses a given play) and depend only on the *terminal vertex* of the play, i.e. a vertex which only has a self-loop edge.

© Springer International Publishing Switzerland 2015
R. Jain et al. (Eds.): TAMC 2015, LNCS 9076, pp. 359–371, 2015.
DOI: 10.1007/978-3-319-17142-5_31

However, due to the presence of stochastic vertices, a player's *expected payoff* (i.e. her probability of winning) can be an arbitrary probability.

The most common interpretation of rational behavior in multi-player games is captured by the notion of a *Nash equilibrium* [8]. In a Nash equilibrium, no player can improve her payoff by unilaterally switching to a different strategy. Chatterjee et al. in [3] gave an algorithm for computing a Nash equilibrium in a stochastic multi-player game with ω-regular winning conditions. However—as observed by Ummels and Wojtczak [11]—the algorithm proposed by Chatterjee et al. may compute an equilibrium where all players lose almost surely, even when there exist other equilibria where all players win almost surely. The equilibrium where all players win almost surely is more *optimal* than the one where all players lose almost surely.

Ummels and Wojtczak [11] successfully argue that in practice it is desirable to look for an equilibrium where as many players as possible win almost surely or where it is guaranteed that the expected payoff of the equilibrium falls into a certain interval. They studied the so-called NE problem as a decision problem where, given a k-player game $\mathcal{G}$ with initial vertex v_0 and two thresholds $\bar{x}, \bar{y} \in [0, 1]^{k\,1}$, the goal is to decide whether $(\mathcal{G}, v_0)$ has a Nash equilibrium with expected payoff at least $\bar{x}$ and at most $\bar{y}$. This problem can be considered as a generalization of the *quantitative decision problem* for two-player zero-sum games, which asks whether in such a game player 0 has a strategy that ensures to win the game with a probability that exceeds a given threshold.

There are several variants of the NE problem depending on the type of strategies permitted. On the one hand, strategies may be *randomized* (allowing randomization over actions) or *pure* (not allowing such randomization). On the other hand, one can restrict to strategies that use (unbounded or bounded) finite memory or even to *stationary* ones (strategies that do not use any memory at all). For the quantitative decision problem, this distinction is often not meaningful since in a two-player zero-sum simple stochastic game with ω-regular objectives both players have optimal pure strategies with finite memory. Moreover, in many games even positional (i.e. both pure and stationary) strategies suffice for optimality. However, regarding NE this distinction leads to distinct decision problems with completely different computational complexity [11].

Contributions. Ummels and Wojtczak [11] showed that deciding the existence of pure-strategy Nash equilibria (PURENE) where a fixed player wins almost surely is undecidable for games with 9 players. They also showed that the problem remains undecidable for the finite-strategy Nash equilibrium (FINNE) with 13 players. In this paper we further refine their undecidability results by showing that PURENE and FINNE problems remain undecidable for 5 or more players.

Related Work. Determining the complexity of Nash equilibria has attracted much interest in recent years. In particular, a series of papers culminated in the result that computing a Nash equilibrium of a two-player game in strategic form is complete for the complexity class PPAD [4,7]. More in the spirit of our

---

[1] The ith element of vector $\bar{x}$ corresponds to the payoff of player i.

work, [6] showed that deciding whether there exists a Nash equilibrium in a two-player game in strategic form where player 0 receives payoff at least x and related decision problems are all NP-hard. For non-stochastic infinite games, a qualitative version of the NE problem was studied in [10]. In particular, it was shown that the problem is NP-complete for games with parity winning conditions but in P for games with Büchi winning conditions.

For stochastic games, most results concern the computation of values and optimal strategies in two player case. In the multi-player case, [3] showed that the problem of deciding whether a (concurrent) stochastic game with reachability objectives has a Nash equilibrium in positional strategies with payoff at least $\bar{x}$ is NP-complete.

Ummels and Wojtczak showed in [11] that the NE problem is undecidable if we allow either arbitrary randomized strategies or arbitrary pure strategies. In fact, even the following, presumably simpler, problem was showed undecidable: Given a game $\mathcal{G}$, decide whether there exists a Nash equilibrium (in pure strategies) where player 0 wins almost surely. Moreover, the problem remains undecidable if one restricts to randomized or pure strategies with finite memory. However, it was also shown there that if one restricts to simpler types of strategies like stationary ones, NE becomes decidable [11]. In particular, for positional strategies the problem is NP-complete, and for arbitrary stationary strategies it is NP-hard but contained in PSPACE. Also, the strictly qualitative fragment of NE is decidable. This fragment arises from NE by restricting the two thresholds to be the same binary payoff. Hence, they were only interested in equilibria where each player either wins or loses almost surely. Formally, the task is to decide, given a k-player game $\mathcal{G}$ with initial vertex v_0 and a binary payoff $\bar{x} \in \{0,1\}^k$, whether the game has a Nash equilibrium with expected payoff $\bar{x}$. It was shown there that for simple stochastic games, this problem is P-complete [11].

Ummels and Wojtczak studied, in [12], the computational complexity of Nash equilibria in concurrent games with limit-average objectives. They showed that the existence of a Nash equilibrium in randomized strategies is undecidable (for at least 14 players), while the existence of a Nash equilibrium in pure strategies is decidable, even if a constraint is put on the payoff of the equilibrium. Their undecidability result holds even for a restricted class of concurrent games, where nonzero rewards occur only on terminal states. Moreover, they showed that the constrained existence problem is undecidable not only for concurrent games but for turn-based games with the same restriction on rewards. They also showed undecidability of the existence of an (unconstrained) Nash equilibrium in concurrent games with terminal-reward payoffs. Finally, Bouyer et al. [2] showed undecidability of the existence of constrained Nash equilibrium in a very similar model – players do no observe the actions taken but only the state of the game – with only three players and 0/1-rewards (i.e., reachability objectives).

2   Simple Stochastic Multi-player Games

We study multi-player extension of *simple stochastic game* introduced by Condon [5] as studied by Ummels and Wojtczak [11].

Definition 1 (Simple Stochastic Multi-player Games). A simple stochastic multi-player game *(SSMG) is a tuple* $(\Pi, V, (V_i)_{i\in\Pi}, \Delta, (F_i)_{i\in\Pi})$ *where:*

- $\Pi = \{0, 1, \ldots, k-1\}$ *is a finite set of* players;
- V *is a finite set of* vertices;
- $V_i \subseteq V$ *is the set of vertices controlled by player i such that $V_i \cap V_j = \emptyset$ for every $i \neq j \in \Pi$;*
- $\Delta \subseteq V \times ([0, 1] \cup \{\bot\}) \times V$ *is the transition relation, and*
- $F_i \subseteq V$ *for each $i \in \Pi$.*

We say that a vertex $v \in V$ is *controlled by player i* if $v \in V_i$. A vertex $v \in V$ is called a *stochastic vertex* if $v \notin \bigcup_{i\in\Pi} V_i$, that is, v is not contained in any of the sets V_i. We require that a transition is labeled by a probability iff it originates in a stochastic vertex: If $(v, p, w) \in \Delta$ then $p \in [0, 1]$ if v is a stochastic vertex and $p = \bot$ if $v \in V_i$ for some $i \in \Pi$. Moreover, for each pair of a stochastic vertex v and an arbitrary vertex w, we require that there exists precisely one $p \in [0, 1]$ such that $(v, p, w) \in \Delta$. As usual, for computational purposes we require that all these probabilities are rational.

For a given vertex $v \in V$, the set of all $w \in V$ such that there exists $p \in (0, 1] \cup \{\bot\}$ with $(v, p, w) \in \Delta$ is denoted by $v\Delta$. For technical reasons, it is required that $v\Delta \neq \emptyset$ for all $v \in V$. Moreover, for each stochastic vertex v, the outgoing probabilities must sum up to 1: $\sum_{(p,w):(v,p,w)\in\Delta} p = 1$. Finally, it is required that each vertex v that lies in one of the sets F_i is a *terminal (sink) vertex*: $v\Delta = \{v\}$. So if F is the set of all terminal vertices, then $F_i \subseteq F$ for each $i \in \Pi$.

A *(mixed) strategy of player i in $\mathcal{G}$* is a mapping $\sigma : V^*V_i \to \mathcal{D}(V)$ assigning to each possible *history* $xv \in V^*V_i$ of vertices ending in a vertex controlled by player i a (discrete) probability distribution over V such that $\sigma(xv)(w) > 0$ only if $(v, \bot, w) \in \Delta$. Instead of $\sigma(xv)(w)$, we usually write $\sigma(w \mid xv)$. A *(mixed) strategy profile of $\mathcal{G}$* is a tuple $\bar{\sigma} = (\sigma_i)_{i\in\Pi}$ where σ_i is a strategy of player i in $\mathcal{G}$. Given a strategy profile $\bar{\sigma} = (\sigma_j)_{j\in\Pi}$ and a strategy τ of player i, we denote by $(\bar{\sigma}_{-i}, \tau)$ the strategy profile resulting from $\bar{\sigma}$ by replacing σ_i with τ.

A strategy σ of player i is called *pure* if for each $xv \in V^*V_i$ there exists $w \in v\Delta$ with $\sigma(w \mid xv) = 1$. Note that a pure strategy of player i can be identified with a function $\sigma : V^*V_i \to V$. A strategy profile $\bar{\sigma} = (\sigma_i)_{i\in\Pi}$ is called *pure* if each σ_i is pure. More generally, a pure strategy σ is called *finite-state* if it can be implemented by a finite automaton with output or, equivalently, if the equivalence relation $\sim \subseteq V^* \times V^*$ defined by $x \sim y$ if $\sigma(xz) = \sigma(yz)$ for all $z \in V^*V_i$ has only finitely many equivalence classes. In general, this definition is applicable to mixed strategies as well, but here, we identify finite-state strategies with pure finite-state strategies. Finally, a *finite-state strategy profile* is a profile consisting of finite-state strategies only.

It is sometimes convenient to designate an initial vertex $v_0 \in V$ of the game. We call the tuple $(\mathcal{G}, v_0)$ an *initialized SSMG*. A strategy (strategy profile) of $(\mathcal{G}, v_0)$ is just a strategy (strategy profile) of $\mathcal{G}$. In the following, we will use the abbreviation SSMG also for initialized SSMGs. It should always be clear from the context if the game is initialized or not.

When drawing an SSMG as a graph, we continue to use the conventions of [11]. The initial vertex is marked by an incoming edge that has no source vertex. Vertices that are controlled by a player are depicted as circles, where the player who controls a vertex is given by the label next to it. Stochastic vertices are depicted as diamonds, where the transition probabilities are given by the labels on its outgoing edges. Finally, terminal vertices are generally represented by their associated payoff vector. In fact, we allow arbitrary vectors of rational probabilities as payoffs. This does not increase the power of the model since such a payoff vector can easily be realized by an SSMG consisting of stochastic and terminal vertices only.

Given an SSMG $(\mathcal{G}, v_0)$ and a strategy profile $\bar{\sigma} = (\sigma_i)_{i \in \Pi}$, the *conditional probability of $w \in V$ given the history $xv \in V^*V$* is the number $\sigma_i(w \mid xv)$ if $v \in V_i$ and the unique $p \in [0, 1]$ such that $(v, p, w) \in \Delta$ if v is a stochastic vertex. We abuse notation and denote this probability by $\bar{\sigma}(w \mid xv)$. The probabilities $\bar{\sigma}(w \mid xv)$ induce a probability measure on the space V^ω in the following way: The probability of a basic open set $v_1 \ldots v_k \cdot V^\omega$ is 0 if $v_1 \neq v_0$ and the product of the probabilities $\bar{\sigma}(v_j \mid v_1 \ldots v_{j-1})$ for $j = 2, \ldots, k$ otherwise. It is a classical result of measure theory that this extends to a unique probability measure assigning a probability to every Borel subset of V^ω, which we denote by $\mathrm{Pr}_{v_0}^{\bar{\sigma}}$. For a set $U \subseteq V$, let $\mathrm{Reach}(U) := V^* \cdot U \cdot V^\omega$.

Given a strategy profile $\bar{\sigma}$, a strategy τ of player i is called a *best response to $\bar{\sigma}$* if τ maximizes the expected payoff of player i, i.e. for all strategies τ' of player i we have that $\mathrm{Pr}_{v_0}^{(\bar{\sigma}_{-i}, \tau')}(\mathrm{Reach}(F_i)) \leq \mathrm{Pr}_{v_0}^{(\bar{\sigma}_{-i}, \tau)}(\mathrm{Reach}(F_i))$. A *Nash equilibrium* is a strategy profile $\bar{\sigma} = (\sigma_i)_{i \in \Pi}$ such that each σ_i is a best response to $\bar{\sigma}$. Hence, in a Nash equilibrium no player can improve her payoff by (unilaterally) switching to a different strategy. In this paper we study the following decision problem.

Definition 2 (Decision Problem NE). *Given an initialized simple stochastic multi-player game $(\mathcal{G}, v_0)$ and two thresholds $\bar{x}, \bar{y} \in [0, 1]^\Pi$, decide whether there exists a Nash equilibrium with payoff $\geq \bar{x}$ and $\leq \bar{y}$.*

As usual, for computational purposes we assume that the thresholds $\bar{x}$ and $\bar{y}$ are vectors of rational numbers. The threshold-free variant of the above problem which omits the thresholds just asks about a Nash equilibrium where some distinguished player, say player 0, wins almost surely.

The following is the key result of this paper.

Theorem 1. *The existence of a pure-strategy-Nash equilibrium SSMG where player 0 wins almost surely is undecidable for games with 5 or more players.*

3   Improved Undecidability Result

In this section we construct an SSMG $\mathcal{G}$ for which we show the undecidability of the existence of pure-strategy Nash equilibria of $(\mathcal{G}, v_0)$ where player 0 wins almost surely, whenever $\mathcal{G}$ has 5 or more players. We then explain how this proof can be adapted to show undecidability of

- finite-strategy Nash equilibrium where player 0 wins almost surely whenever $\mathcal{G}$ has 5 or more players.

3.1   Pure-Strategy Equilibria

In this section, we show that the problem PURENE is undecidable by exhibiting a reduction from an undecidable problem about *two-counter machines*. Our construction is inspired by a construction used in [11]. A two-counter machine $\mathcal{M}$ is given by a list of instructions $\iota_1, \ldots, \iota_m$ where each instruction is one of the following:

- "inc(j); goto k"   (increment counter j by 1 and go to instruction k);
- "zero(j) ? goto k: dec(j); goto l"   (if the value of counter j is zero, go to instruction k; otherwise, decrement counter j by one and go to instruction l);
- "halt"   (stop the computation).

Here j ranges over $1, 2$ (the two counters), and $k \neq l$ range over $1, \ldots, m$. A configuration of $\mathcal{M}$ is a triple $C = (i, c_1, c_2) \in \{1, \ldots, m\} \times \mathbb{N} \times \mathbb{N}$, where i denotes the number of the current instruction and c_j denotes the current value of counter j. A configuration C' is the *successor* of configuration C, denoted by $C \vdash C'$, if it results from C by executing instruction ι_i; a configuration $C = (i, c_1, c_2)$ with $\iota_i = $ "halt" has no successor configuration. Finally, the *computation of* $\mathcal{M}$ is the unique maximal sequence $\rho = \rho(0)\rho(1)\ldots$ such that $\rho(0) \vdash \rho(1) \vdash \ldots$ and $\rho(0) = (1, 0, 0)$ (the *initial configuration*). Note that ρ is either infinite, or it ends in a configuration $C = (i, c_1, c_2)$ such that $\iota_i = $ "halt".

The *halting problem* is to decide, given a machine $\mathcal{M}$, whether the computation of $\mathcal{M}$ is finite. It is well-known that two-counter machines are Turing powerful, which makes the halting problem and its dual, the *non-halting problem*, undecidable.

In order to prove Theorem 1, we show that one can compute from a two-counter machine $\mathcal{M}$ an SSMG $(\mathcal{G}, v_0)$ with five players such that the computation of $\mathcal{M}$ is infinite iff $(\mathcal{G}, v_0)$ has a pure Nash equilibrium where player 0 wins almost surely. This establishes a reduction from the non-halting problem to PURENE.

The game $\mathcal{G}$ is played by player 0 and four other players A^t and B^t, indexed by $t \in \{0, 1\}$. Let $\Gamma = \{\text{init}, \text{inc}(j), \text{dec}(j), \text{zero}(j) : j = 1, 2\}$, and let $q_1 = 2, q_2 = 3$ be two primes. If $\mathcal{M}$ has instructions $\iota_1, \ldots, \iota_m$, then for each $i \in \{1, \ldots, m\}$, each $\gamma \in \Gamma$, each $j \in \{1, 2\}$ and each $t \in \{0, 1\}$, the game $\mathcal{G}$ contains the gadgets $S_{i,\gamma}^t$, $I_{i,\gamma}^t$ and $C_{j,\gamma}^t$, which are depicted in Fig. 1. In the figure, squares represent terminal vertices (the edge leading from a terminal vertex to itself being implicit), and the labeling indicates which players win at the respective vertex. Moreover, the dashed edge inside $C_{j,\gamma}^t$ is present iff $\gamma \notin \{\text{init}, \text{zero}(j)\}$. The initial vertex v_0 of $\mathcal{G}$ is the black vertex inside the gadget $S_{1,\text{init}}^0$.

For any pure strategy profile $\bar{\sigma}$ of $\mathcal{G}$ where player 0 wins almost surely, let $x_0 v_0 \prec x_1 v_1 \prec x_2 v_2 \prec \ldots$ ($x_i \in V^*, v \in V, x_0 = \epsilon$) be the (unique) sequence of all consecutive histories such that, for each $n \in \mathbb{N}$, v_n is a black vertex and $\Pr_{v_0}^{\bar{\sigma}}(x_n v_n \cdot V^\omega) > 0$. Additionally, let $\gamma_0, \gamma_1, \ldots$ be the corresponding sequence of instructions, i.e. $\gamma_n = \gamma$ for the unique instruction γ such that v_n lies in one of the gadgets $S_{i,\gamma}^t$ (where $t = n \mod 2$). For each $j \in \{1, 2\}$ and $n \in \mathbb{N}$, we define two conditional probabilities a_n and p_n as follows:

$$a_n := \Pr_{v_0}^{\bar{\sigma}}(\text{Reach}(F_{A^{n \mod 2}}) \mid x_n v_n \cdot V^\omega) \text{ and}$$

$$p_n := \Pr_{v_0}^{\bar{\sigma}}(\text{Reach}(F_{A^{n \mod 2}}) \mid x_n v_n \cdot V^\omega \setminus x_{n+2} v_{n+2} \cdot V^\omega).$$

Finally, for each $j \in \{1,2\}$ and $n \in \mathbb{N}$, we define an ordinal number $c_j^n \leq \omega$ as follows: After the history $x_n v_n$, with probability $\frac{1}{4}$ the play proceeds to the vertex controlled by player 0 in the counter gadget C_{j,γ_n}^t (where $t = n \mod 2$). The number c_j^n is defined to be the maximal number of subsequent visits to the grey vertex inside this gadget (where $c_j^n = \omega$ if, on one path, the grey vertex

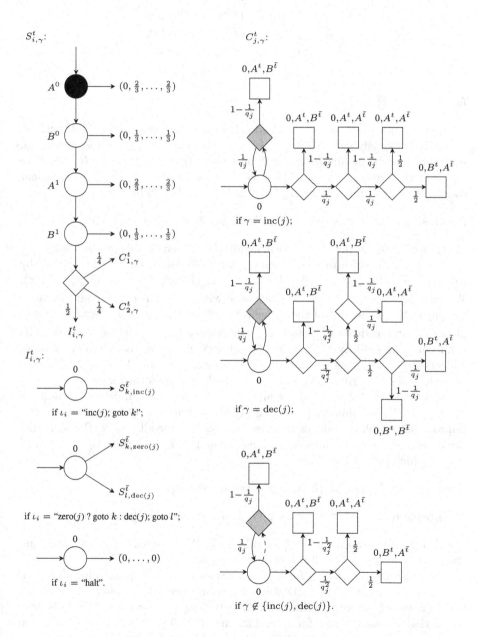

Fig. 1. Simulating a two-counter machine.

is visited infinitely often). Note that, by the construction of $C_{j,\gamma}^t$, it holds that $c_j^n = 0$ if $\gamma_n = \text{zero}(j)$ or $\gamma_n = \text{init}$.

Lemma 1. *Let $\bar\sigma$ be a pure strategy profile of $(\mathcal{G}, v_0)$ where player 0 wins almost surely. Then $\bar\sigma$ is a Nash equilibrium if and only if the following equation holds.*

$$c_j^{n+1} = \begin{cases} 1 + c_j^n & \text{if } \gamma_{n+1} = \text{inc}(j), \\ c_j^n - 1 & \text{if } \gamma_{n+1} = \text{dec}(j), \\ c_j^n = 0 & \text{if } \gamma_{n+1} = \text{zero}(j), \\ c_j^n & \text{otherwise} \end{cases} \tag{1}$$

for all $j \in \{1, 2\}$ and $n \in \mathbb{N}$.

Here $+$ and $-$ denote the usual addition and subtraction of ordinal numbers respectively (satisfying $1 + \omega = \omega - 1 = \omega$). The proof of Lemma 1 goes through several claims. In the following, let $\bar\sigma$ be a pure strategy profile of $(\mathcal{G}, v_0)$ where player 0 wins almost surely. The first claim gives a necessary and sufficient condition on the probabilities a_n for $\bar\sigma$ to be a Nash equilibrium.

Proposition 1. *The profile $\bar\sigma$ is a Nash equilibrium iff $a_n = \frac{2}{3}$ for all $n \in \mathbb{N}$.*

Proof. ($\Rightarrow$) Assume that $\bar\sigma$ is a Nash equilibrium. Clearly, this implies that $a_n \geq \frac{2}{3}$ for all $n \in \mathbb{N}$ since otherwise some player A^t could improve her payoff by leaving one of the gadgets $S_{i,\gamma}^t$. Let $b_n := \Pr_{v_0}^{\bar\sigma}(\text{Reach}(F_{B^{n \bmod 2}}) \mid x_n v_n \cdot V^\omega)$. We have $b_n \geq \frac{1}{3}$ for all $n \in \mathbb{N}$ since otherwise some player B^t could improve her payoff by leaving one of the gadgets $S_{i,\gamma}^t$. Note that at every terminal vertex of the counter gadgets $C_{j,\gamma}^t$ and $C_{j,\gamma}^{\bar t}$ either player A^t or player B^t wins. The conditional probability that, given the history $x_n v_n$, we reach either of those gadgets is $\sum_{k \in \mathbb{Z}} (\frac{1}{2})^k \cdot \frac{1}{2} = 1$ for all $n \in \mathbb{N}$, so we have $a_n = 1 - b_n$ for all $n \in \mathbb{N}$. Since $b_n \geq \frac{1}{3}$, we arrive at $a_n \leq 1 - \frac{1}{3} = \frac{2}{3}$, which proves the claim.

($\Leftarrow$) Assume that $a_n = \frac{2}{3}$ for all $n \in \mathbb{N}$. Clearly, this implies that none of the players A^t can improve her payoff. To show that none of the players B^t can improve her payoff, it suffices to show that $b_n \geq \frac{1}{3}$ for all $n \in \mathbb{N}$. But with the same argumentation as above, we have $b_n = 1 - a_n$ and thus $b_n = \frac{1}{3}$ for all $n \in \mathbb{N}$, which proves the claim. $\square$

The second claim relates the probabilities a_n and p_n.

Proposition 2. *$a_n = \frac{2}{3}$ for all $n \in \mathbb{N}$ if and only if $p_n = \frac{1}{2}$ for all $n \in \mathbb{N}$.*

Proof. ($\Rightarrow$) Assume that $a_n = \frac{2}{3}$ for all $n \in \mathbb{N}$. We have $a_n = p_n + \frac{1}{4} \cdot a_{n+2}$ and therefore $\frac{2}{3} = p_n + \frac{1}{6}$ for all $n \in \mathbb{N}$. Hence, $p_n = \frac{1}{2}$ for all $n \in \mathbb{N}$.

($\Leftarrow$) Assume that $p_n = \frac{1}{2}$ for all $n \in \mathbb{N}$. Since $a_n = p_n + \frac{1}{4} \cdot a_{n+2}$ for all $n \in \mathbb{N}$, the numbers a_n must satisfy the following recurrence: $a_{n+2} = 4a_n - 2$. Since all the numbers a_n are probabilities, we have $0 \leq a_n \leq 1$ for all $n \in \mathbb{N}$. It is easy to see that the only values for a_0 and a_1 such that $0 \leq a_n \leq 1$ for all $n \in \mathbb{N}$ are $a_0 = a_1 = \frac{2}{3}$. But this implies that $a_n = \frac{2}{3}$ for all $n \in \mathbb{N}$. $\square$

Finally, the last claim relates the numbers p_n to Eq. (1).

Proposition 3. $p_n = \frac{1}{2}$ for all $n \in \mathbb{N}$ if and only if Eq. (1) holds for all $n \in \mathbb{N}$.

Proof. Let $n \in \mathbb{N}$, and let $t = n \bmod 2$. The probability p_n can be expressed as the sum of the probability that the play reaches a terminal vertex that is winning for player A^t inside C^t_{j,γ_n} (this probability is denoted as α^j_n) and the probability that the play reaches a terminal vertex winning for player $A^{\bar{t}}$ inside $C^{\bar{t}}_{j,\gamma_{n+1}}$ (denoted as α^j_{n+1}). For counter 1 gadgets, the probability α^1_n of A^t winning in counter gadget C^t_{1,γ_n} is

$$\alpha^1_n = \Sigma_{0 \le i \le c^n_1 - 1} \left(1 - \frac{1}{q_1}\right)\frac{1}{q^i_1} + \frac{1}{q^{c^n_1}_1}\left\{\left(1 - \frac{1}{q^2_1}\right) + \frac{1}{2q^2_1}\right\}$$

$$= 1 - \frac{1}{q^{c^n_1}_1} + \frac{1}{q^{c^n_1}_1}\left\{\left(1 - \frac{1}{q^2_1}\right) + \frac{1}{2q^2_1}\right\}$$

$$= 1 - \frac{1}{q^{c^n_1}_1} + \frac{1}{q^{c^n_1}_1}\left\{1 - \frac{1}{2q^2_1}\right\}$$

$$= 1 - \frac{1}{2q^{c^n_1+2}_1}$$

Suppose $\gamma_{n+1} = \text{inc}(1)$.

Then the probability α^1_{n+1} of $A^{\bar{t}}$ winning in counter gadget $C^{\bar{t}}_{1,\gamma_{n+1}}$ is $\dfrac{1}{q^{c^{n+1}_1}_1} \cdot \dfrac{1}{q_1}$

Similarly, the probabilities α^2_n and α^2_{n+1} corresponding to counter 2 gadgets are as follows:

$$\alpha^2_n = 1 - \frac{1}{2q^{c^n_1+2}_1} \quad \text{and} \quad \alpha^2_{n+1} = \frac{1}{q^{c^{n+1}_2}_2} \cdot \frac{1}{q^2_2}$$

Given, these probabilities, p_n is as follows.

$$p_n = \frac{1}{4}\left[\alpha^1_n + \frac{1}{2}\alpha^1_{n+1}\right] + \frac{1}{4}\left[\alpha^2_n + \frac{1}{2}\alpha^2_{n+1}\right]$$

$$= \frac{1}{4}\left[1 - \frac{1}{2q^{c^n_1+2}_1} + \frac{1}{2q^{c^{n+1}_1+1}_1}\right] + \frac{1}{4}\left[1 - \frac{1}{2q^{c^n_2+2}_2} + \frac{1}{2q^{c^{n+1}_2+2}_2}\right]$$

$$= \frac{1}{2} - \frac{1}{8}\left[\frac{1}{q^{c^n_1+2}_1} - \frac{1}{q^{c^{n+1}_1+1}_1}\right] - \frac{1}{8}\left[\frac{1}{q^{c^n_2+2}_2} - \frac{1}{q^{c^{n+1}_2+2}_2}\right]$$

As q_1 and q_2 are primes, this sum is equal to $\frac{1}{2}$ iff $c^{n+1}_1 = 1 + c^n_1$ and $c^{n+1}_2 = c^n_2$. For γ_{n+1} being any other instruction like decrement, other instructions, the argument is similar. $\qquad\square$

Proof (Proof of Lemma 1). By Proposition 1, the profile $\bar{\sigma}$ is a Nash equilibrium iff $a_n = \frac{2}{3}$ for all $n \in \mathbb{N}$. By Proposition 2, the latter is true if $p_n = \frac{1}{2}$ for all $n \in \mathbb{N}$. Finally, by Proposition 3, this is the case iff Eq. (1) holds for all $j \in \{1, 2\}$ and $n \in \mathbb{N}$. $\square$

To establish the reduction, it remains to show that the computation of $\mathcal{M}$ is infinite iff the game $(\mathcal{G}, v_0)$ has a pure Nash equilibrium where player 0 wins almost surely.

($\Rightarrow$) Assume that the computation $\rho = \rho(0)\rho(1)\dots$ of $\mathcal{M}$ is infinite. We define a pure strategy σ_0 for player 0 as follows: For a history that ends in one of the instruction gadgets $I_{i,\gamma}^t$ after visiting a black vertex exactly n times, player 0 tries to move to the neighboring gadget $S_{k,\gamma'}^{\bar{t}}$ such that $\rho(n)$ refers to instruction number k (which is always possible if $\rho(n-1)$ refers to instruction number i; in any other case, σ_0 might be defined arbitrarily). In particular, if $\rho(n-1)$ refers to instruction $\iota_i = $ "zero(j) ? goto k : dec(j); goto l", then player 0 will move to the gadget $S_{k,\text{zero}(j)}^{\bar{t}}$ if the value of the counter in configuration $\rho(n-1)$ is 0 and to the gadget $S_{l,\text{dec}(j)}^{\bar{t}}$ otherwise. For a history that ends in one of the gadgets $C_{j,\gamma}^t$ after visiting a black vertex exactly n times and a grey vertex exactly m times, player 0 will move to the grey vertex again iff m is strictly less than the value of the counter j in configuration $\rho(n-1)$. So after entering $C_{j,\gamma}^t$, player 0's strategy is to loop through the grey vertex exactly as many times as given by the value of the counter j in configuration $\rho(n-1)$.

Any other player's pure strategy is "moving down at any time". We claim that the resulting strategy profile $\bar{\sigma}$ is a Nash equilibrium of $(\mathcal{G}, v_0)$ where player 0 wins almost surely.

Since, according to her strategy, player 0 follows the computation of $\mathcal{M}$, no vertex inside an instruction gadget $I_{i,\gamma}^t$ where ι_i is the halt instruction is ever reached. Hence, with probability 1 a terminal vertex in one of the counter gadgets is reached. Since player 0 wins at any such vertex, we can conclude that she wins almost surely.

It remains to show that $\bar{\sigma}$ is a Nash equilibrium. By the definition of player 0's strategy σ_0, we have the following for all $n \in \mathbb{N}$: 1. c_j^n is the value of counter j in configuration $\rho(n)$; 2. c_j^{n+1} is the value of counter j in configuration $\rho(n+1)$; 3. γ_{n+1} is the instruction corresponding to the counter update from configuration $\rho(n)$ to $\rho(n+1)$. Hence, Eq. (1) holds, and $\bar{\sigma}$ is a Nash equilibrium by Lemma 1.

($\Leftarrow$) Assume that $\bar{\sigma}$ is a pure Nash equilibrium of $(\mathcal{G}, v_0)$ where player 0 wins almost surely. We define an infinite sequence $\rho = \rho(0)\rho(1)\dots$ of *pseudo configurations* (where the counters may take the value ω) of $\mathcal{M}$ as follows. Let $n \in \mathbb{N}$, and assume that v_n lies inside the gadget S_{i,γ_n}^t (where $t = n \bmod 2$); then $\rho(n) := (i, c_1^n, c_2^n)$.

We claim that ρ is, in fact, the (infinite) computation of $\mathcal{M}$. It suffices to verify the following two properties:

1. $\rho(0) = (1, 0, 0)$;
2. $\rho(n) \vdash \rho(n+1)$ for all $n \in \mathbb{N}$.

Note that we do not have to show explicitly that each $\rho(n)$ is a configuration of $\mathcal{M}$ since this follows easily by induction from 1. and 2. Verifying the first property is easy: v_0 lies inside $S^0_{1,\text{init}}$ (and we are at instruction 1), which is linked to the counter gadgets $C^0_{1,\text{init}}$ and $C^0_{2,\text{init}}$. The edge leading to the grey vertex is missing in these gadgets. Hence, c^0_1 and c^0_2 are both equal to 0.

For the second property, let $\rho(n) = (i, c_1, c_2)$ and $\rho(n+1) = (i', c'_1, c'_2)$. Hence, v_n lies inside $S^t_{i,\gamma}$ and v_{n+1} inside $S^t_{i',\gamma'}$ for suitable γ, γ' and $t = n \mod 2$. We only prove the claim for the case that $\iota_i =$ "zero(2) ? goto k : dec(2); goto l"; the other cases are straightforward. Note that, by the construction of the gadget $I^t_{i,\gamma}$, it must be the case that either $i' = k$ and $\gamma' = \text{zero}(2)$, or $i' = l$ and $\gamma' = \text{dec}(2)$. By Lemma 1, if $\gamma' = \text{zero}(2)$, then $c'_2 = c_2 = 0$ and $c'_1 = c_1$, and if $\gamma' = \text{dec}(2)$, then $c'_2 = c_2 - 1$ and $c'_1 = c_1$. This implies $\rho(n) \vdash \rho(n+1)$: On the one hand, if $c_2 = 0$, then $c'_2 \neq c_2 - 1$, which implies $\gamma' \neq \text{dec}(2)$ and thus $\gamma' = \text{zero}(2)$, $i' = k$ and $c'_2 = c_2 = 0$. On the other hand, if $c_2 > 0$, then $\gamma' \neq \text{zero}(2)$ and thus $\gamma' = \text{dec}(2)$, $i' = l$ and $c'_2 = c_2 - 1$.   □

3.2 Finite-State Equilibria

Theorem 2. *The existence of a finite-strategy-Nash equilibrium SSMG where player 0 wins almost surely is undecidable for games with 5 or more players.*

We now move on to prove Theorem 2. Before showing the undecidability of the existence of FINNE, we first note that FINNE is recursively enumerable: To decide whether an SSMG $(\mathcal{G}, v_0)$ has a finite-state Nash equilibrium with payoff $\geq \bar{x}$ and $\leq \bar{y}$, one can just enumerate all possible finite-state profiles and check for each of them whether the profile is a Nash equilibrium with the desired properties by analyzing the finite Markov chain that is generated by this profile (where one identifies states that correspond to the same vertex and memory state). Hence, to show the undecidability of FINNE, we cannot reduce from the non-halting problem but from the halting problem for two-counter machines (which is recursively enumerable itself).

We now explain how to adapt the proof of Theorem 1 to show the undecidability of FINNE. The construction is similar to the one for proving undecidability of PURENE. Given a two-counter machine $\mathcal{M}$, we modify the SSMG $\mathcal{G}$ constructed in the proof of Theorem 1 by adding another "counter" (sharing the four players from the other two gadgets, but using an additional new prime, say $q_3 = 5$ for checking whether the counter is updated correctly) that has to be incremented in each step. Moreover, additionally to the terminal vertices in the gadgets $C^t_{j,\gamma}$, we let player 0 win at the terminal vertex in each of the gadgets $I_{i,\gamma}$ where $\iota_i =$ "halt". The gadget $\gamma = \text{inc}(j)$ in Fig. 1 is a generic one and when we put $q_j = 5$, it becomes the increment gadget for this new counter. Correctly incrementing this counter comes from Proposition 3 that $p_n = \frac{1}{2}$ iff Eq. (1) is correct. With the extra counter, p_n is the sum of A^t winning in the gadgets of all the three counters. Hence, this will ensure correct updates of all counters.

Let us denote the new game by $\mathcal{G}'$. Now, if $\mathcal{M}$ does not halt, any pure Nash equilibrium of $(\mathcal{G}', v_0)$ where player 0 wins almost surely needs infinite memory:

to win almost surely, player 0 must follow the computation of $\mathcal{M}$ and increment the new counter at each step. On the other hand, if $\mathcal{M}$ halts, then we can easily construct a finite-state Nash equilibrium of $(\mathcal{G}', v_0)$ where player 0 wins almost surely. Hence, $(\mathcal{G}', v_0)$ has a finite-state Nash equilibrium where player 0 wins almost surely iff the machine $\mathcal{M}$ halts.

We shall now compare the above described improved results with their counterparts in [11]. The PURENE undecidability proof in [11] reduced the non-halting problem to a game with 9 players. The game has 4 dedicated players to ensure correctness of each counter - thus using 8 additional players. While we follow their idea of reduction, with the help of primes q_1, q_2 we re-use the 4 players A^t and B^t, $t \in \{0, 1\}$ across the gadgets of both counters. Addtionally, FINNE undecidability proof is achieved by incrementing a third additional counter. While the proof for FINNE in [11] uses 4 new players for the third counter, we use another prime q_3 and re-use the 4 players (A^t and B^t, $t \in \{0, 1\}$) for the third counter.

4 Conclusion

We have showed that PURENE where player 0 wins almost surely is undecidable when the game has 5 or more players. A closely related open problem is PURENE where player 0 wins with probability $p \in [0, 1)$. The decidability of the existence of mixed-strategy NE is an interesting open problem. A further line of work is to explore concurrent moves by all the non-stochastic players, and study the decidability of the existence of various kinds of Nash equilibrium. This concurrent extension of SSMGs is inspired by [12], where the authors consider concurrent moves of all players on finite graphs, with reward vectors attached to the terminal vertices.

References

1. Baier, C., Größer, M., Leucker, M., Bollig, B., Ciesinski, F.: Probabilistic controller synthesis. In: International Conference on Theoretical Computer Science, IFIP TCS 2004, pp. 493–506. Kluwer Academic Publishers (2004)
2. Bouyer, P., Markey, N., Stan, D.: Mixed Nash equilibria in concurrent games. In: FSTTCS (to appear, 2014)
3. Chatterjee, K., Majumdar, R., Jurdziński, M.: On nash equilibria in stochastic games. In: Marcinkowski, J., Tarlecki, A. (eds.) CSL 2004. LNCS, vol. 3210, pp. 26–40. Springer, Heidelberg (2004)
4. Chen, X., Deng, X., Teng, S.-H.: Settling the complexity of computing two-player Nash equilibria. J. ACM 56(3), 1–57 (2009)
5. Condon, A.: On algorithms for simple stochastic games. Adv. Comput. Complex. Theor. 13, 51–73 (1993)
6. Conitzer, V., Sandholm, T.: Complexity results about Nash equilibria. In: Proceedings of the 18th International Joint Conference on Artificial Intelligence, IJCAI 2003, pp. 765–771. Morgan Kaufmann (2003)
7. Daskalakis, C., Goldberg, P.W., Papadimitriou, C.H.: The complexity of computing a Nash equilibrium. SIAM J. Comput. 39(1), 195–259 (2009)

8. Nash, J.F.: Equilibrium points in N-person games. Proc. National Acad. Sci. USA **36**, 48–49 (1950)

9. Neyman, A., Sorin, S. (eds.): Stochastic Games and Applications. NATO Science Series C, vol. 570. Springer, Berlin (2003)

10. Ummels, M.: The complexity of Nash equilibria in infinite multiplayer games. In: Amadio, R.M. (ed.) FOSSACS 2008. LNCS, vol. 4962, pp. 20–34. Springer, Heidelberg (2008)

11. Ummels, M., Wojtczak, D.: The complexity of Nash equilibria in simple stochastic multiplayer games. In: Albers, S., Marchetti-Spaccamela, A., Matias, Y., Nikoletseas, S., Thomas, W. (eds.) ICALP 2009, Part II. LNCS, vol. 5556, pp. 297–308. Springer, Heidelberg (2009)

12. Ummels, M., Wojtczak, D.: The complexity of Nash equilibria in limit-average games. In: Katoen, J.-P., König, B. (eds.) CONCUR 2011. LNCS, vol. 6901, pp. 482–496. Springer, Heidelberg (2011)

Learning, Automata
and Probabilistic Models

Learning from Non-iid Data: Fast Rates for the One-vs-All Multiclass Plug-in Classifiers

Vu Dinh[1], Lam Si Tung Ho[2]($\boxtimes$), Nguyen Viet Cuong[3], Duy Nguyen[4], and Binh T. Nguyen[5]

[1] Department of Mathematics, Purdue University, West Lafayette, USA
vdinh@math.purdue.edu
[2] Department of Biostatistics, University of California, Los Angeles, USA
lamho@ucla.edu
[3] Department of Computer Science, National University of Singapore,
Singapore, Singapore
nvcuong@comp.nus.edu.sg
[4] Department of Statistics, University of Wisconsin-Madison, Madison, USA
dnguyen@stat.wisc.edu
[5] Department of Computer Science, University of Science,
Ho Chi Minh City, Vietnam
ngtbinh@hcmus.edu.vn

Abstract. We prove new fast learning rates for the one-vs-all multiclass plug-in classifiers trained either from exponentially strongly mixing data or from data generated by a converging drifting distribution. These are two typical scenarios where training data are not iid. The learning rates are obtained under a multiclass version of Tsybakov's margin assumption, a type of low-noise assumption, and do not depend on the number of classes. Our results are general and include a previous result for binary-class plug-in classifiers with iid data as a special case. In contrast to previous works for least squares SVMs under the binary-class setting, our results retain the optimal learning rate in the iid case.

1 Introduction

Fast learning of plug-in classifiers from low-noise data has recently gained much attention [1–4]. The first fast/super-fast learning rates[1] for the plug-in classifiers were proven by Audibert and Tsybakov [1] under the Tsybakov's margin assumption [5], which is a type of low-noise condition. Their plug-in classifiers employ the local polynomial estimator to estimate the conditional probability of a label Y given an observation X and use it in the plug-in rule. Subsequently, Kohler and Krzyzak [2] proved the fast learning rate for plug-in classifiers with

---

Vu Dinh and Lam Si Tung Ho—These authors contributed equally to this work. LST Ho was supported by NSF IIS 1251151.

[1] Fast learning rate means the trained classifier converges with rate faster than $n^{-1/2}$, while super-fast learning rate means the trained classifier converges with rate faster than n^{-1}.

R. Jain et al. (Eds.): TAMC 2015, LNCS 9076, pp. 375–387, 2015.
DOI: 10.1007/978-3-319-17142-5_32

a relaxed condition on the density of X and investigated the use of kernel, partitioning, and nearest neighbor estimators instead of the local polynomial estimator. Monnier [3] suggested to use local multi-resolution projections to estimate the conditional probability of Y and proved the super-fast rates of the corresponding plug-in classifier under the same margin assumption. Fast rates for plug-in classifiers were also achieved in the active learning setting [4].

Nevertheless, these previous analyses of plug-in classifiers typically focus on the binary-class setting with iid (independent and identically distributed) data assumption. This is a limitation of the current theory for plug-in classifiers since (1) many classification problems are multiclass in nature and (2) data may also violate the iid data assumption in practice. In this paper, we contribute to the theoretical understandings of plug-in classifiers by proving novel fast learning rates of a multiclass plug-in classifier trained from non-iid data. In particular, we prove that the multiclass plug-in classifier constructed using the *one-vs-all method* can achieve fast learning rates, or even super-fast rates, with the following two types of non-iid training data: data generated from an *exponentially strongly mixing sequence* and data generated from a *converging drifting distribution*. To the best of our knowledge, this is the first result that proves fast learning rates for multiclass classifiers with non-iid data. Moreover, these learning rates do not depend on the number of classes.

Our results assume a multiclass version of Tsybakov's margin assumption. In the multiclass setting, this assumption states that the events in which the most probable label of an example is ambiguous with the second most probable label have small probabilities. This margin assumption was previously considered in the analyses of multiclass empirical risk minimization (ERM) classifiers with iid data [6] and in the context of active learning with cost-sensitive multiclass classifiers [7]. Our results are natural generalizations for both the binary-class and the iid data settings. As special cases of our results, we can obtain fast learning rates for the one-vs-all multiclass plug-in classifiers in the iid data setting and the fast learning rates for the binary-class plug-in classifiers in the non-iid data setting. Our results can also be used to obtain the previous fast learning rates [1] for the binary-class plug-in classifiers in the iid data setting.

In terms of theory, the extension from binary class to multiclass problem is usually not trivial and depends greatly on the choice of the multiclass classifiers. In this paper, our results show that this extension can be achieved with plug-in classifiers and the one-vs-all method. The one-vs-all method is a practical way to construct a multiclass classifier using binary-class classification [8]. This method trains a model for each class by converting multiclass data into binary-class data and then combines them into a multiclass classifier.

Our paper considers two types of non-iid data. Exponentially strongly mixing data is a typical case of identically but not independently distributed data. Fast learning from exponentially strongly mixing data has been previously analyzed for least squares support vector machines (LS-SVMs) [9,10] and ERM classifiers [10]. On the other hand, data generated from a drifting distribution (or drifting concept) is an example of independently but not identically distributed data. Some concept drifting scenarios and learning bounds were previously investigated in [11–14]. In this paper, we consider the scenario where the parameters of the distributions generating the training data converge uniformly to those of the test distribution with some polynomial rate.

We note that even though LS-SVMs can be applied to solve a classification problem with binary data, the previous results for LS-SVMs cannot retain the optimal rate in the iid case [9,10]. In contrast, our results in this paper still retain the optimal learning rate for the Hölder class in the iid case. Besides, the results for drifting concepts can also achieve this optimal rate. Other works that are also related to our paper include the analyses of fast learning rates for binary SVMs and multiclass SVMs with iid data [15,16] and for the Gibbs estimator with ϕ-mixing data [17].

2   Preliminaries

2.1   Settings

Let $\{(X_i, Y_i)\}_{i=1}^n$ be the labeled training data where $X_i \in \mathbb{R}^d$ and $Y_i \in \{1, 2, \ldots, m\}$ for all i. In the data, X_i is an observation and Y_i is the label of X_i. The binary-class case corresponds to $m = 2$, while the multiclass case corresponds to $m > 2$. For now we do not specify how $\{(X_i, Y_i)\}_{i=1}^n$ are generated, but we assume that test data are drawn iid from an unknown distribution $\mathbf{P}$ on $\mathbb{R}^d \times \{1, 2, \ldots, m\}$. In Sects. 4 and 5, we will respectively consider two cases where the training data $\{(X_i, Y_i)\}_{i=1}^n$ are generated from an exponentially strongly mixing sequence with stationary distribution $\mathbf{P}$ and where $\{(X_i, Y_i)\}_{i=1}^n$ are generated from a drifting distribution with the limit distribution $\mathbf{P}$. The case where training data are generated iid from $\mathbf{P}$ is a special case of these settings.

Given the training data, our aim is to find a classification rule $f : \mathbb{R}^d \to \{1, 2, \ldots, m\}$ whose risk is as small as possible. The risk of a classifier f is defined as $R(f) \triangleq \mathbf{P}(Y \neq f(X))$. One minimizer of the above risk is the Bayes classifier $f^*(X) \triangleq \arg\max_j \eta_j(X)$, where $\eta_j(X) \triangleq \mathbf{P}(Y = j|X)$ for all $j \in \{1, 2, \ldots, m\}$. For any classifier $\widehat{f}_n$ trained from the training data, it is common to characterize its accuracy via the excess risk $\mathcal{E}(\widehat{f}_n) \triangleq \mathbf{E}R(\widehat{f}_n) - R(f^*)$, where the expectation is with respect to the randomness of the training data. A small excess risk for $\widehat{f}_n$ is thus desirable as the classifier will perform close to the optimal classifier f^* on average.

For any classifier f, we write $\eta_f(X)$ as an abbreviation for $\eta_{f(X)}(X)$, which is the value of the function $\eta_{f(X)}$ at X. Let $\mathbf{1}_{\{\cdot\}}$ be the indicator function. The following proposition gives a property of the excess risk in the multiclass setting. This proposition will be used to prove the theorems in the subsequent sections.

Proposition 1. *For any classifier $\widehat{f}_n$, we have $\mathcal{E}(\widehat{f}_n) = \mathbf{E}\left[\eta_{f^*}(X) - \eta_{\widehat{f}_n}(X)\right]$, where the expectation is with respect to the randomness of both the training data and the testing example X.*

Proof. $R(\widehat{f}_n) - R(f^*)$

$$= \mathbf{P}(Y \neq \widehat{f}_n(X)) - \mathbf{P}(Y \neq f^*(X)) \quad = \quad \mathbf{P}(Y = f^*(X)) - \mathbf{P}(Y = \widehat{f}_n(X))$$

$$= \mathbf{E}_{X,Y}\left[\mathbf{1}_{\{Y=f^*(X)\}} - \mathbf{1}_{\{Y=\widehat{f}_n(X)\}}\right] = \mathbf{E}_X\left[\mathbf{E}_Y\left[\mathbf{1}_{\{Y=f^*(X)\}} - \mathbf{1}_{\{Y=\widehat{f}_n(X)\}}\big|X\right]\right]$$

$$= \mathbf{E}_X\left[\sum_{j=1}^m \eta_j(X)\left(\mathbf{1}_{\{f^*(X)=j\}} - \mathbf{1}_{\{\widehat{f}_n(X)=j\}}\right)\right] = \mathbf{E}_X\left[\eta_{f^*}(X) - \eta_{\widehat{f}_n}(X)\right].$$

Thus, $\mathcal{E}(\widehat{f}_n) = \mathbf{E}\left[\eta_{f^*}(X) - \eta_{\widehat{f}_n}(X)\right].$ $\qquad\square$

Following the settings for the binary-class case [1], we assume the following Hölder assumption: all the functions η_j's are in the Hölder class $\Sigma(\beta, L, \mathbb{R}^d)$. We also assume that the marginal distribution $\mathbf{P}_X$ of X satisfies the strong density assumption. The definition of Hölder classes and the strong density assumption are briefly introduced below by using the notations in [1].

For $\beta > 0$ and $L > 0$, the Hölder class $\Sigma(\beta, L, \mathbb{R}^d)$ is the set of all functions $g : \mathbb{R}^d \to \mathbb{R}$ that are $\lfloor\beta\rfloor$ times continuously differentiable, and for any $x, x' \in \mathbb{R}^d$, we have $|g(x') - g_x(x')| \leq L\|x - x'\|^\beta$, where $\|\cdot\|$ is the Euclidean norm and g_x is the $\lfloor\beta\rfloor^{th}$-degree Taylor polynomial of g at x. The definition of g_x can be found in Sect. 2 of [1].

Fix $c_0, r_0 > 0$ and $0 < \mu_{\min} < \mu_{\max} < \infty$, and fix a compact set $\mathcal{C} \subset \mathbb{R}^d$. The marginal $\mathbf{P}_X$ satisfies the strong density assumption if it is supported on a compact (c_0, r_0)-regular set $A \subseteq \mathcal{C}$ and its density μ (w.r.t. the Lebesgue measure) satisfies: $\mu_{\min} \leq \mu(x) \leq \mu_{\max}$ for $x \in A$ and $\mu(x) = 0$ otherwise. In this definition, a set A is (c_0, r_0)-regular if $\boldsymbol{\lambda}[A \cap B(x,r)] \geq c_0\boldsymbol{\lambda}[B(x,r)]$ for all $0 < r \leq r_0$ and $x \in A$, where $\boldsymbol{\lambda}$ is the Lebesgue measure and $B(x,r)$ is the Euclidean ball in $\mathbb{R}^d$ with center x and radius r.

2.2   Margin Assumption for Multiclass Setting

As in the binary-class case, fast learning rates for the multiclass plug-in classifier can be obtained under an assumption similar to Tsybakov's margin assumption [5]. In particular, we assume that the conditional probabilities η_j's satisfy the following margin assumption, which is an extension of Tsybakov's margin assumption to the multiclass setting. This is a form of low noise assumption and was also considered in the context of active learning to analyze the learning rate of cost-sensitive multiclass classifiers [7].

Assumption (Margin Assumption). *There exist constants $C_0 > 0$ and $\alpha \geq 0$ such that for all $t > 0$,*

$$\mathbf{P}_X(\eta_{(1)}(X) - \eta_{(2)}(X) \leq t) \leq C_0 t^\alpha$$

where $\eta_{(1)}(X)$ and $\eta_{(2)}(X)$ are the largest and second largest conditional probabilities among all the $\eta_j(X)$'s.

3   The One-vs-All Multiclass Plug-In Classifier

We now introduce the one-vs-all multiclass plug-in classifier which we will analyze in this paper. Let $\widehat{\eta}_n(X) = (\widehat{\eta}_{n,1}(X), \widehat{\eta}_{n,2}(X), \ldots, \widehat{\eta}_{n,m}(X))$ be an m-dimensional function where $\widehat{\eta}_{n,j}$ is a nonparametric estimator of η_j from the training data. The corresponding multiclass plug-in classifier $\widehat{f}_n$ predicts the label of an observation X by

$$\widehat{f}_n(X) = \arg\max_j \widehat{\eta}_{n,j}(X).$$

In this paper, we consider plug-in classifiers where $\widehat{\eta}_{n,j}$'s are estimated using the one-vs-all method and the local polynomial regression function as follows. For each class $j \in \{1, 2, \ldots, m\}$, we first convert the training data $\{(X_i, Y_i)\}_{i=1}^n$ to binary class by considering all (X_i, Y_i)'s such that $Y_i \neq j$ as negative (label 0) and those such that $Y_i = j$ as positive (label 1). Then we construct a local polynomial regression function $\widehat{\eta}_{n,j}^{\mathrm{LP}}(x)$ of order $\lfloor \beta \rfloor$ with some appropriate bandwidth $h > 0$ and kernel K from the new binary-class training data (see Sect. 2 of [1] for the definition of local polynomial regression functions). The estimator $\widehat{\eta}_{n,j}$ can now be defined as

$$\widehat{\eta}_{n,j}(x) \triangleq \begin{cases} 0 & \text{if } \widehat{\eta}_{n,j}^{\mathrm{LP}}(x) \leq 0 \\ \widehat{\eta}_{n,j}^{\mathrm{LP}}(x) & \text{if } 0 < \widehat{\eta}_{n,j}^{\mathrm{LP}}(x) < 1 \\ 1 & \text{if } \widehat{\eta}_{n,j}^{\mathrm{LP}}(x) \geq 1 \end{cases}.$$

In order to prove the fast rates for the multiclass plug-in classifier, the bandwidth h and the kernel K of the local polynomial regression function have to be chosen carefully. Specifically, K has to satisfy the following assumptions, which are similar to those in [1]:

$$\exists c > 0 \text{ such that for all } x \in \mathbb{R}^d, \text{ we have } K(x) \geq c\mathbf{1}_{\{\|x\|\leq c\}},$$

$$\int_{\mathbb{R}^d} K(u)du = 1, \quad \sup_{u \in \mathbb{R}^d} (1+\|u\|^{2\beta})K(u) < \infty, \quad \text{and} \quad \int_{\mathbb{R}^d} (1+\|u\|^{4\beta})K^2(u)du < \infty.$$

Note that Gaussian kernels satisfy these conditions. The conditions for the bandwidth h will be given in Sects. 4 and 5.

4   Fast Learning for Exponentially Strongly Mixing Data

In this section, we consider the case where training data are generated from an exponentially strongly mixing sequence [9,18]. Let $Z_i = (X_i, Y_i)$ for all i. Assume that $\{Z_i\}_{i=1}^{\infty}$ is a stationary sequence of random variables on $\mathbb{R}^d \times \{1, 2, \ldots, m\}$ with stationary distribution $\mathbf{P}$. That is, $\mathbf{P}$ is the marginal distribution of any random variable in the sequence. For all $k \geq 1$, we define the α-mixing coefficients [9]:

$$\alpha(k) \triangleq \sup_{A_1 \in \sigma_1^t, A_2 \in \sigma_{t+k}^{\infty}, t \geq 1} |\mathbf{P}(A_1 \cap A_2) - \mathbf{P}(A_1)\mathbf{P}(A_2)|$$

where σ_a^b is the σ-algebra generated by $\{Z_i\}_{i=a}^b$. The sequence $\{Z_i\}_{i=1}^\infty$ is exponentially strongly mixing if there exist positive constants C_1, C_2 and C_3 such that for every $k \geq 1$, we have

$$\alpha(k) \leq C_1 \exp(-C_2 k^{C_3}). \tag{1}$$

We now state some key lemmas for proving the convergence rate of the multiclass plug-in classifier in this setting. Let $n_e \triangleq \left\lfloor \frac{n}{\lceil \{8n/C_2\}^{1/(C_3+1)} \rceil} \right\rfloor$ be the effective sample size. The following lemma is a direct consequence of Bernstein inequality for an exponentially strongly mixing sequence [18].

Lemma 1. *Let $\{Z_i\}_{i=1}^\infty$ be an exponentially strongly mixing sequence and ϕ be a real-valued Borel measurable function. Denote $W_i = \phi(Z_i)$ for all $i \geq 1$. Assume that $|W_1| \leq C$ almost surely and $\mathbf{E}[W_1] = 0$. Then for all $n \geq 1$ and $\epsilon > 0$, we have*

$$\mathbf{P}^{\otimes n} \left(\left| \frac{1}{n} \sum_{i=1}^n W_i \right| \geq \epsilon \right) \leq 2(1 + 4e^{-2}C_1) \exp\left(-\frac{\epsilon^2 n_e}{2\mathbf{E}|W_1|^2 + 2\epsilon C/3} \right),$$

where $\mathbf{P}^{\otimes n}$ is the joint distribution of $\{Z_i\}_{i=1}^n$ and C_1 is the constant in Eq. (1).

The next lemma is about the convergence rate of the local polynomial regression functions using the one-vs-all method. The proof for this lemma is given in Sect. 7.1.

Lemma 2. *Let β, r_0, and c be the constants in the Hölder assumption, the strong density assumption, and the assumption for the kernel K respectively. Then there exist constants $C_4, C_5, C_6 > 0$ such that for all $\delta > 0$, all bandwidth h satisfying $C_6 h^\beta < \delta$ and $0 < h \leq r_0/c$, all $j \in \{1, 2, \ldots, m\}$ and $n \geq 1$, we have*

$$\mathbf{P}^{\otimes n}(|\widehat{\eta}_{n,j}(x) - \eta_j(x)| \geq \delta) \leq C_4 \exp(-C_5 n_e h^d \delta^2)$$

for almost surely x with respect to $\mathbf{P}_X$, where d is the dimension of the observations (inputs).

Given the above convergence rate of the local polynomial regression functions, Lemma 3 below gives the convergence rate of the excess risk of the one-vs-all multiclass plug-in classifier. The proof for this lemma is given in Sect. 7.2.

Lemma 3. *Let α be the constant in the margin assumption. Assume that there exist $C_4, C_5 > 0$ such that $\mathbf{P}^{\otimes n}(|\widehat{\eta}_{n,j}(x) - \eta_j(x)| \geq \delta) \leq C_4 \exp(-C_5 a_n \delta^2)$ for almost surely x with respect to $\mathbf{P}_X$, and for all $j \in \{1, 2, \ldots, m\}$, $\delta > 0$. Then there exists $C_7 > 0$ such that for all $n \geq 1$,*

$$\mathcal{E}(\widehat{f}_n) = \mathbf{ER}(\widehat{f}_n) - R(f^*) \leq C_7 a_n^{-(1+\alpha)/2}.$$

Using Lemmas 2 and 3, we can obtain the following theorem about the convergence rate of the one-vs-all multiclass plug-in classifier when training data are exponentially strongly mixing. This theorem is a direct consequence of Lemmas 2 and 3 with $h = n_e^{-1/(2\beta+d)}$ and $a_n = n_e^{2\beta/(2\beta+d)}$.

Theorem 1. *Let α and β be the constants in the margin assumption and the Hölder assumption respectively, and let d be the dimension of the observations. Let $\widehat{f}_n$ be the one-vs-all multiclass plug-in classifier with bandwidth $h = n_e^{-1/(2\beta+d)}$ that is trained from an exponentially strongly mixing sequence. Then there exists some constant $C_8 > 0$ such that for all n large enough that satisfies $0 < n_e^{-1/(2\beta+d)} \leq r_0/c$, we have*

$$\mathcal{E}(\widehat{f}_n) = \mathbf{ER}(\widehat{f}_n) - R(f^*) \leq C_8 n_e^{-\beta(1+\alpha)/(2\beta+d)}.$$

The convergence rate in Theorem 1 is expressed in terms of the effective sample size n_e rather than the sample size n since learning with dependent data typically requires more data to achieve the same level of accuracy as learning with independent data (see e.g., [9,19,20]). However, Theorem 1 still implies the fast rate for the one-vs-all multiclass plug-in classifier in terms of the sample size n. Indeed, the rate in the theorem can be rewritten as $O(n^{-\frac{\beta(1+\alpha)}{2\beta+d} \cdot \frac{C_3}{C_3+1}})$, so the fast learning rate is achieved when $2(\alpha - 1/C_3)\beta > (1 + 1/C_3)d$ and the super-fast learning rate is achieved when $(\alpha - 1 - 2/C_3)\beta > d(1 + 1/C_3)$.

5   Fast Learning from a Drifting Concept

In this section, we consider the case where training data are generated from a drifting concept that converges to the test distribution $\mathbf{P}$. Unlike the setting in Sect. 4 where the training data form a stationary sequence of random variables, the setting in this section may include training data that are not stationary. Formally, we assume the training data $\{Z_i\}_{i=1}^n = \{(X_i, Y_i)\}_{i=1}^n$ are generated as follows. The observations X_i are generated iid from the marginal distribution $\mathbf{P}_X$ satisfying the strong density assumption. For each $i \geq 1$, the label Y_i of X_i is generated from a categorical distribution on $\{1, 2, \ldots, m\}$ with parameters $\eta^i(X_i) \triangleq (\eta_1^i(X_i), \eta_2^i(X_i), \ldots, \eta_m^i(X_i))$. That is, the probability of $Y_i = j$ conditioned on X_i is $\eta_j^i(X_i)$, for all $j \in \{1, 2, \ldots, m\}$.

Note that from our setting, the training data are independent but not identically distributed. To prove the convergence rate of the multiclass plug-in classifier, we assume that $\|\eta_j^n - \eta_j\|_\infty \triangleq \sup_{x \in \mathbb{R}^d} |\eta_j^n(x) - \eta_j(x)| = O(n^{-(\beta+d)/(2\beta+d)})$ for all j, i.e., η_j^n converges uniformly to the label distribution η_j of test data with rate $O(n^{-(\beta+d)/(2\beta+d)})$. We now state some useful lemmas for proving our result. The following lemma is a Bernstein inequality for the type of data considered in this section [21].

Lemma 4. *Let $\{W_i\}_{i=1}^n$ be an independent sequence of random variables. For all $i \geq 1$ and $l > 2$, assume $\mathbf{E}W_i = 0$, $\mathbf{E}|W_i|^2 = b_i$, and $\mathbf{E}|W_i|^l \leq b_i H^{l-2} l!/2$ for some constant $H > 0$. Let $B_n \triangleq \sum_{i=1}^n b_i$. Then for all $n \geq 1$ and $\epsilon > 0$, we have*

$$\mathbf{P}^{\otimes n}\left(\left|\sum_{i=1}^n W_i\right| \geq \epsilon\right) \leq 2\exp\left(-\frac{\epsilon^2}{2(B_n + H\epsilon)}\right),$$

where $\mathbf{P}^{\otimes n}$ is the joint distribution of $\{W_i\}_{i=1}^n$.

The next lemma states the convergence rate of the local polynomial regression functions in this setting. The proof for this lemma is given in Sect. 7.3. Note that the constants in this section may be different from those in Sect. 4.

Lemma 5. *Let β, r_0, and c be the constants in the Hölder assumption, the strong density assumption, and the assumption for the kernel K respectively. Let $\widehat{\eta}_{n,j}$ be the estimator of η_j estimated using the local polynomial regression function with $h = n^{-1/(2\beta+d)}$. If $\|\eta_j^n - \eta_j\|_\infty = O(n^{-(\beta+d)/(2\beta+d)})$ for all j, then there exist constants $C_4, C_5, C_6 > 0$ such that for all $\delta > 0$, all n satisfying $C_6 n^{-\beta/(2\beta+d)} < \delta < 1$ and $0 < n^{-1/(2\beta+d)} \leq r_0/c$, and all $j \in \{1, 2, \ldots, m\}$, we have*

$$\mathbf{P}^{\otimes n}(|\widehat{\eta}_{n,j}(x) - \eta_j(x)| \geq \delta) \leq C_4 \exp(-C_5 n^{2\beta/(2\beta+d)} \delta^2)$$

for almost surely x with respect to $\mathbf{P}_X$, where d is the dimension of the observations.

Note that Lemma 3 still holds in this setting. Thus, we can obtain Theorem 2 below about the convergence rate of the one-vs-all multiclass plug-in classifier when training data are generated from a drifting concept converging uniformly to the test distribution. This theorem is a direct consequence of Lemmas 3 and 5 with $a_n = n^{2\beta/(2\beta+d)}$. We note that the convergence rate in Theorem 2 is fast when $\alpha\beta > d/2$ and is super-fast when $(\alpha - 1)\beta > d$.

Theorem 2. *Let α and β be the constants in the margin assumption and the Hölder assumption respectively, and let d be the dimension of the observations. Let $\widehat{f}_n$ be the one-vs-all multiclass plug-in classifier with bandwidth $h = n^{-1/(2\beta+d)}$ that is trained from data generated from a drifting concept converging uniformly to the test distribution. Then there exists some constant $C_8 > 0$ such that for all n large enough that satisfies $0 < n^{-1/(2\beta+d)} \leq r_0/c$, we have*

$$\mathcal{E}(\widehat{f}_n) = \mathbf{E}R(\widehat{f}_n) - R(f^*) \leq C_8 n^{-\beta(1+\alpha)/(2\beta+d)}.$$

6   Remarks

The rates in Theorems 1 and 2 do not depend on the number of classes m. They are both generalizations of the previous result for binary-class plug-in classifiers with iid data [1]. More specifically, $C_3 = +\infty$ in the case of iid data, thus we have $n_e = n$ and the data distribution also satisfies the condition in Theorem 2. Hence, we can obtain the same result as in [1].

Another important remark is that our results for the one-vs-all multiclass plug-in classifiers retain the optimal rate $O(n^{-\beta(1+\alpha)/(2\beta+d)})$ for the Hölder class in the iid case [1] while the previous results in [9,10] for LS-SVMs with smooth kernels do not (see Example 4.3 in [10]). Besides, from Theorem 2, the one-vs-all multiclass plug-in classifiers trained from a drifting concept can also achieve this optimal rate. We note that for LS-SVMs with Gaussian kernels, Hang and Steinwart [10] proved that they can achieve the essentially optimal rate in the iid scenario (see Example 4.4 in [10]). That is, their learning rate is n^ς times of the optimal rate for any $\zeta > 0$. Although this rate is very close to the optimal rate, it is still slower than $\log n$ times of the optimal rate.[2]

7 Technical Proofs

7.1 Proof of Lemma 2

Fix $j \in \{1, \ldots, m\}$. Let $Y_i' \triangleq 1_{\{Y_i = j\}}$ be the binary class of X_i constructed from the class Y_i using the one-vs-all method in Sect. 3. By definition of η_j, note that $\mathbf{P}[Y_i' = 1 | X_i] = \eta_j(X_i)$. Let μ be the density of $\mathbf{P}_X$. We consider the matrix $\mathbf{B} \triangleq (B_{s_1,s_2})_{|s_1|,|s_2| \leq \lfloor \beta \rfloor}$ with the elements $B_{s_1,s_2} \triangleq \int_{\mathbb{R}^d} u^{s_1+s_2} K(u)\mu(x + hu)du$, and the matrix $\widehat{\mathbf{B}} \triangleq (\widehat{B}_{s_1,s_2})_{|s_1|,|s_2| \leq \lfloor \beta \rfloor}$ with the elements $\widehat{B}_{s_1,s_2} \triangleq \frac{1}{nh^d}\sum_{i=1}^{n} (\frac{X_i - x}{h})^{s_1+s_2} K(\frac{X_i - x}{h})$, where s_1, s_2 are multi-indices in $\mathbb{N}^d$ (see Sect. 2 of [1] for details on multi-index). Let $\lambda_\mathbf{B}$ be the smallest eigenvalue of $\mathbf{B}$. Then, there exists a constant c_1 such that $\lambda_\mathbf{B} \geq c_1 > 0$ (see Eq. (6.2) in [1]).

Fix s_1 and s_2. For any $i = 1, 2, \ldots, n$, we define

$$T_i \triangleq \frac{1}{h^d}\left(\frac{X_i - x}{h}\right)^{s_1+s_2} K\left(\frac{X_i - x}{h}\right) - \int_{\mathbb{R}^d} u^{s_1+s_2} K(u)\mu(x + hu)du.$$

It is easy to see that $\mathbf{E}[T_1] = 0$, $|T_1| \leq c_2 h^{-d}$, and $\mathbf{E}|T_1|^2 \leq c_3 h^{-d}$ for some $c_2, c_3 > 0$. By applying Lemma 1, for any $\epsilon > 0$, we have

$$\mathbf{P}^{\otimes n}(|\widehat{B}_{s_1,s_2} - B_{s_1,s_2}| \geq \epsilon) = \mathbf{P}^{\otimes n}\left(\left|\frac{1}{n}\sum_{i=1}^{n} T_i\right| \geq \epsilon\right)$$
$$\leq 2(1 + 4e^{-2}C_1)\exp\left(-\frac{\epsilon^2 n_e h^d}{2c_3 + 2\epsilon c_2/3}\right).$$

Let $\lambda_{\widehat{\mathbf{B}}}$ be the smallest eigenvalue of $\widehat{\mathbf{B}}$. From Eq. (6.1) in [1], we have

$$\lambda_{\widehat{\mathbf{B}}} \geq \lambda_\mathbf{B} - \sum_{|s_1|,|s_2| \leq \lfloor \beta \rfloor} |\widehat{B}_{s_1,s_2} - B_{s_1,s_2}|.$$

Let M be the number of columns of $\widehat{B}$. Then, there exists $c_4 > 0$ such that

$$\mathbf{P}^{\otimes n}(\lambda_{\widehat{\mathbf{B}}} \leq c_1/2) \leq 2(1 + 4e^{-2}C_1)M^2\exp(-c_4 n_e h^d). \tag{2}$$

Let η_j^x be the $\lfloor \beta \rfloor^{th}$-degree Taylor polynomial of η_j at x. Consider the vector $\mathbf{a} \triangleq (a_s)_{|s| \leq \lfloor \beta \rfloor} \in \mathbb{R}^M$ where $a_s \triangleq \frac{1}{nh^d}\sum_{i=1}^{n}[Y_i' - \eta_j^x(X_i)](\frac{X_i - x}{h})^s K(\frac{X_i - x}{h})$. Applying Eq. (6.5) in [1] for $\lambda_{\widehat{\mathbf{B}}} \geq c_1/2$, we have

$$|\widehat{\eta}_{n,j}(x) - \eta_j(x)| \leq |\widehat{\eta}_{n,j}^{\mathrm{LP}}(x) - \eta_j(x)| \leq \lambda_{\widehat{\mathbf{B}}}^{-1} M \max_s |a_s| \leq (2M/c_1) \max_s |a_s|. \tag{3}$$

---

[2] The optimal rates in Example 4.3 and 4.4 of [10] may not necessarily be the same as our optimal rate since Hang and Steinwart considered Sobolev space and Besov space instead of Hölder space.

We also define: $T_i^{(s,1)} \triangleq \dfrac{1}{h^d}[Y_i' - \eta_j(X_i)](\dfrac{X_i - x}{h})^s K(\dfrac{X_i - x}{h})$, and

$$T_i^{(s,2)} \triangleq \frac{1}{h^d}[\eta_j(X_i) - \eta_j^x(X_i)](\frac{X_i - x}{h})^s K(\frac{X_i - x}{h}).$$

Note that $\mathbf{E}[T_1^{(s,1)}] = 0$, $|T_1^{(s,1)}| \leq c_5 h^{-d}$ and $\mathbf{E}|T_1^{(s,1)}|^2 \leq c_6 h^{-d}$ for some $c_5, c_6 > 0$. Similarly, $|T_1^{(s,2)} - \mathbf{E}T_1^{(s,2)}| \leq c_7 h^{\beta - d} + c_8 h^\beta \leq c_9 h^{\beta - d}$ and $\mathbf{E}|T_1^{(s,2)} - \mathbf{E}T_1^{(s,2)}|^2 \leq c_{10} h^{2\beta - d}$, for some $c_7, c_8, c_9, c_{10} > 0$. Thus, by applying Lemma 1 again, for any $\epsilon_1, \epsilon_2 > 0$, we have

$$\mathbf{P}^{\otimes n}\left(\left|\frac{1}{n}\sum_{i=1}^n T_i^{(s,1)}\right| \geq \epsilon_1\right) \leq 2(1 + 4e^{-2}C_1) \exp\left(-\frac{\epsilon_1^2 n_e h^d}{2c_6 + 2c_5 \epsilon_1/3}\right), \text{ and}$$

$$\mathbf{P}^{\otimes n}(|\frac{1}{n}\sum_{i=1}^n (T_i^{(s,2)} - \mathbf{E}T_i^{(s,2)})| \geq \epsilon_2) \leq 2(1 + 4e^{-2}C_1) \exp\left(\frac{-\epsilon_2^2 n_e h^d}{2c_{10}h^{2\beta} + 2c_9 h^\beta \epsilon_2/3}\right).$$

Moreover, $|\mathbf{E}T_1^{(s,2)}| \leq c_8 h^\beta$. By choosing $h^\beta \leq c_1 \delta/(6M c_8)$, there exists $c_{11} > 0$ such that $\mathbf{P}^{\otimes n}\left(|a_s| \geq \dfrac{c_1 \delta}{2M}\right)$

$$\leq \mathbf{P}^{\otimes n}\left(\left|\frac{1}{n}\sum_{i=1}^n T_i^{(s,1)}\right| \geq \frac{c_1\delta}{6M}\right) + \mathbf{P}^{\otimes n}\left(\left|\frac{1}{n}\sum_{i=1}^n (T_i^{(s,2)} - \mathbf{E}T_i^{(s,2)})\right| \geq \frac{c_1\delta}{6M}\right)$$

$$\leq 4(1 + 4e^{-2}C_1)\exp(-c_{11}n_e h^d \delta^2). \tag{4}$$

Let $C_6 = 6M c_8/c_1$. By (2), (3), and (4), there exist $C_4, C_5 > 0$ such that

$$\mathbf{P}^{\otimes n}(|\widehat{\eta}_{n,j}(x) - \eta_j(x)| \geq \delta)$$
$$\leq \mathbf{P}^{\otimes n}(\lambda_{\widehat{\mathbf{B}}} \leq c_1/2) + \mathbf{P}^{\otimes n}(|\widehat{\eta}_{n,j}(x) - \eta_j(x)| \geq \delta, \lambda_{\widehat{\mathbf{B}}} > c_1/2)$$
$$\leq C_4 \exp(-C_5 n_e h^d \delta^2).$$

Note that the constants C_4, C_5, C_6 can be modified so that they are the same for all δ, h, j, and n. Thus, Lemma 2 holds.

7.2   Proof of Lemma 3

Since $\eta_{f^*}(x) - \eta_{\widehat{f}_n}(x) \geq 0$ for all $x \in \mathbb{R}^d$, we denote, for any $\delta > 0$,

$$A_0 \triangleq \{x \in \mathbb{R}^d : \eta_{f^*}(x) - \eta_{\widehat{f}_n}(x) \leq \delta\}, \text{ and}$$

$$A_i \triangleq \{x \in \mathbb{R}^d : 2^{i-1}\delta < \eta_{f^*}(x) - \eta_{\widehat{f}_n}(x) \leq 2^i\delta\}, \text{ for } i \geq 1.$$

By Proposition 1, $\mathbf{E}R(\widehat{f}_n) - R(f^*) = \mathbf{E}[(\eta_{f^*}(X) - \eta_{\widehat{f}_n}(X))\mathbf{1}_{\{\widehat{f}_n(X)\neq f^*(X)\}}]$

$$= \sum_{i=0}^{\infty} \mathbf{E}\left[(\eta_{f^*}(X) - \eta_{\widehat{f}_n}(X))\mathbf{1}_{\{\widehat{f}_n(X)\neq f^*(X)\}}\mathbf{1}_{\{X\in A_i\}}\right]$$

$$\leq \delta\mathbf{P}\left(0 < \eta_{f^*}(X) - \eta_{\widehat{f}_n}(X) \leq \delta\right)$$

$$+ \sum_{i=1}^{\infty} \mathbf{E}\left[(\eta_{f^*}(X) - \eta_{\widehat{f}_n}(X))\mathbf{1}_{\{\widehat{f}_n(X)\neq f^*(X)\}}\mathbf{1}_{\{X\in A_i\}}\right].$$

Let $\widehat{\eta}_{n,\widehat{f}_n}(x)$ denote $\widehat{\eta}_{n,\widehat{f}_n(x)}(x)$. For any x, since $\widehat{\eta}_{n,\widehat{f}_n}(x)$ is the largest among $\widehat{\eta}_{n,j}(x)$'s, we have $\eta_{f^*}(x) - \eta_{\widehat{f}_n}(x) \le |\eta_{f^*}(x) - \widehat{\eta}_{n,f^*}(x)| + |\widehat{\eta}_{n,\widehat{f}_n}(x) - \eta_{\widehat{f}_n}(x)|$. For any $i \ge 1$, we have

$$\mathbf{E}\left[(\eta_{f^*}(X) - \eta_{\widehat{f}_n}(X))\mathbf{1}_{\{\widehat{f}_n(X)\ne f^*(X)\}}\mathbf{1}_{\{X\in A_i\}}\right]$$

$$\le 2^i\delta\,\mathbf{E}\left[\mathbf{1}_{\{|\eta_{f^*}(X)-\widehat{\eta}_{n,f^*}(X)|+|\widehat{\eta}_{n,\widehat{f}_n}(X)-\eta_{\widehat{f}_n}(X)|\ge 2^{i-1}\delta\}}\mathbf{1}_{\{0<\eta_{f^*}(X)-\eta_{\widehat{f}_n}(X)<2^i\delta\}}\right]$$

$$\le 2^i\delta\,\mathbf{E}_X[\mathbf{P}^{\otimes n}(|\eta_{f^*}(X) - \widehat{\eta}_{n,f^*}(X)| + |\widehat{\eta}_{n,\widehat{f}_n}(X) - \eta_{\widehat{f}_n}(X)| \ge 2^{i-1}\delta) \cdot$$

$$\mathbf{1}_{\{0<\eta_{f^*}(X)-\eta_{\widehat{f}_n}(X)<2^i\delta\}}]$$

$$\le c_1 2^i\delta\exp\left(-c_2 a_n(2^{i-2}\delta)^2\right)\,\mathbf{P}_X(0 < \eta_{f^*}(X) - \eta_{\widehat{f}_n}(X) < 2^i\delta),$$

for some $c_1, c_2 > 0$. We have $\mathbf{P}_X(0 < \eta_{f^*}(X) - \eta_{\widehat{f}_n}(X) < \delta) \le \mathbf{P}_X[\eta_{f^*}(X) - \eta_{(2)}(X) < \delta]$, and by the margin assumption, for all $t > 0$, we get $\mathbf{P}_X[\eta_{f^*}(X) - \eta_{(2)}(X) < t] \le C_0 t^\alpha$. Therefore,

$$\mathbf{E}\left[(\eta_{f^*}(X) - \eta_{\widehat{f}_n}(X))\mathbf{1}_{\{\widehat{f}_n(X)\ne f^*(X)\}}\mathbf{1}_{\{X\in A_i\}}\right]$$

$$\le c_1 C_0 2^{i(\alpha+1)}\delta^{\alpha+1}\exp\left(-c_2 a_n(2^{i-2}\delta)^2\right).$$

By choosing $\delta = a_n^{-1/2}$, there exists $C_7 > 0$ that does not depend on n and

$$\mathbf{E}R(\widehat{f}_n) - R(f^*) \le C_0 a_n^{-(\alpha+1)/2} + 2c_1 C_0 a_n^{-(\alpha+1)/2}\sum_{i\ge 1} 2^{i(\alpha+1)/2}\exp(-c_2 2^{2i-4})$$

$$\le C_7 a_n^{-(\alpha+1)/2}.$$

7.3 Proof of Lemma 5

The proof for this lemma is essentially similar to the proof for Lemma 2 in Sect. 7.1, except that we use the Bernstein inequality for iid random variables to bound $\mathbf{P}^{\otimes n}(|\widehat{B}_{s_1,s_2} - B_{s_1,s_2}| \ge \epsilon)$ and thus obtain $\mathbf{P}^{\otimes n}(\lambda_{\widehat{\mathbf{B}}} \le c_1/2) \le 2M^2\exp(-c_4 nh^d)$ as an analogy of Eq. (2) in Sect. 7.1. Besides, Eq. (3) can be obtained in the same way as in Sect. 7.1. To obtain the bound similar to Eq. (4), we define

$$T_i^{(s,1)} \triangleq \frac{1}{h^d}[Y_i' - \eta_j^i(X_i)](\frac{X_i - x}{h})^s K(\frac{X_i - x}{h})$$

$$T_i^{(s,2)} \triangleq \frac{1}{h^d}[\eta_j^i(X_i) - \eta_j(X_i)](\frac{X_i - x}{h})^s K(\frac{X_i - x}{h})$$

$$T_i^{(s,3)} \triangleq \frac{1}{h^d}[\eta_j(X_i) - \eta_j^x(X_i)](\frac{X_i - x}{h})^s K(\frac{X_i - x}{h}).$$

Note that $\mathbf{E}[T_i^{(s,1)}] = 0$, $|T_i^{(s,1)}| \le c_5 h^{-d}$, and $\mathbf{E}|T_i^{(s,1)}|^2 \le c_6 h^{-d}$ for some $c_5, c_6 > 0$. Thus, $\mathbf{E}|T_i^{(s,1)}|^l \le (c_5 h^{-d})^{l-2}\mathbf{E}|T_i^{(s,1)}|^2 \le H_1^{l-2}\mathbf{E}|T_i^{(s,1)}|^2 l!/2$, where $H_1 \triangleq c_5 h^{-d}$ and $l > 2$. Similarly, $|T_i^{(s,2)} - \mathbf{E}T_i^{(s,2)}| \le c_7 h^{-d}$ and $\mathrm{Var}[T_i^{(s,2)}] \le$

$c_8 h^{2-d}$ for some $c_7, c_8 > 0$. Thus, $\mathbf{E}|T_i^{(s,2)} - \mathbf{E}T_i^{(s,2)}|^l \leq H_2^{l-2}\mathrm{Var}[T_i^{(s,2)}]l!/2$, for $H_2 \triangleq c_7 h^{-d}$ and $l > 2$. Furthermore, $|T_i^{(s,3)} - \mathbf{E}T_i^{(s,3)}| \leq c_9 h^{\beta-d}$ and $\mathrm{Var}[T_i^{(s,3)}] \leq c_{10} h^{2\beta-d}$ for some $c_9, c_{10} > 0$. Hence, $\mathbf{E}|T_i^{(s,3)} - \mathbf{E}T_i^{(s,3)}|^l \leq H_3^{l-2}\mathrm{Var}[T_i^{(s,3)}]l!/2$ for $H_3 \triangleq c_9 h^{\beta-d}$ and $l > 2$. Thus, from Lemma 4,

$$\mathbf{P}^{\otimes n}(\frac{1}{n}\sum_{i=1}^{n}|T_i^{(s,1)}| \geq \epsilon_1) \leq 2\exp(-\frac{nh^d\epsilon_1^2}{2(c_6 + c_5\epsilon_1)})$$

$$\mathbf{P}^{\otimes n}(\frac{1}{n}\sum_{i=1}^{n}|T_i^{(s,2)} - \mathbf{E}T_i^{(s,2)}| \geq \epsilon_2) \leq 2\exp(-\frac{nh^d\epsilon_2^2}{2(c_8 h^2 + c_7\epsilon_2)})$$

$$\mathbf{P}^{\otimes n}(\frac{1}{n}\sum_{i=1}^{n}|T_i^{(s,3)} - \mathbf{E}T_i^{(s,3)}| \geq \epsilon_3) \leq 2\exp(-\frac{nh^d\epsilon_3^2}{2(c_{10} h^{2\beta} + c_9 h^\beta\epsilon_3)}),$$

for all $\epsilon_1, \epsilon_2, \epsilon_3 > 0$. Moreover, $\mathbf{E}|T_i^{(s,3)}| \leq c_{11}h^\beta$ for some $c_{11} > 0$, and $\frac{1}{n}\sum_{i=1}^{n}$ $\mathbf{E}|T_i^{(s,2)}| \leq O(h^{-d}\frac{1}{n}\sum_{i=1}^{n}\|\eta_j^i - \eta\|_\infty) \leq O(h^{-d}\frac{1}{n}\sum_{i=1}^{n}i^{-(\beta+d)/(2\beta+d)}) \leq O(h^{-d}\frac{1}{n}(1 + \int_{u=1}^{n}u^{-(\beta+d)/(2\beta+d)}du)) \leq O(h^{-d}n^{-(\beta+d)/(2\beta+d)}) \leq c_{12}h^\beta$ for some $c_{12} > 0$ since $h = n^{-1/(2\beta+d)}$. Thus, we can obtain the new Eq. (4) as $\mathbf{P}^{\otimes n}\left(|a_s| \geq \frac{c_1\delta}{2M}\right) \leq 6\exp(-c_{13}nh^d\delta^2)$ for some $C_6 > 0$ and $c_{13} > 0$. And from the new Eqs. (2), (3), and (4), we can obtain Lemma 5.

References

1. Audibert, J.Y., Tsybakov, A.B.: Fast learning rates for plug-in classifiers. Ann. Stat. **35**(2), 608–633 (2007)
2. Kohler, M., Krzyzak, A.: On the rate of convergence of local averaging plug-in classification rules under a margin condition. IEEE Trans. Inf. Theory **53**(5), 1735–1742 (2007)
3. Monnier, J.B.: Classification via local multi-resolution projections. Electron. J. Stat. **6**, 382–420 (2012)
4. Minsker, S.: Plug-in approach to active learning. J. Mach. Learn. Res. **13**, 67–90 (2012)
5. Tsybakov, A.B.: Optimal aggregation of classifiers in statistical learning. Ann. Stat. **32**, 135–166 (2004)
6. Zhang, T.: Statistical analysis of some multi-category large margin classification methods. J. Mach. Learn. Res. **5**, 1225–1251 (2004)
7. Agarwal, A.: Selective sampling algorithms for cost-sensitive multiclass prediction. In: Proceedings of the International Conference on Machine Learning (2013)
8. Rifkin, R., Klautau, A.: In defense of one-vs-all classification. J. Mach. Learn. Res. **5**, 101–141 (2004)
9. Steinwart, I., Christmann, A.: Fast learning from non-iid observations. In: Bengio, Y., Schuurmans, D., Lafferty, J., Williams, C.K.I., Culotta, A. (eds.) Advances in Neural Information Processing Systems, pp. 1768–1776. MIT Press, Cambridge (2009)
10. Hang, H., Steinwart, I.: Fast learning from alpha-mixing observations. J. Multivar. Anal. **127**, 184–199 (2014)

11. Bartlett, P.L.: Learning with a slowly changing distribution. In: COLT 1992
12. Long, P.M.: The complexity of learning according to two models of a drifting environment. Mach. Learn. **37**(3), 337–354 (1999)
13. Barve, R.D., Long, P.M.: On the complexity of learning from drifting distributions. In: COLT 1996
14. Mohri, M., Muñoz Medina, A.: New analysis and algorithm for learning with drifting distributions. In: Bshouty, N.H., Stoltz, G., Vayatis, N., Zeugmann, T. (eds.) ALT 2012. LNCS (LNAI), vol. 7568, pp. 124–138. Springer, Heidelberg (2012)
15. Steinwart, I., Scovel, C.: Fast rates for support vector machines using gaussian kernels. Ann. Stat. **35**, 575–607 (2007)
16. Shen, X., Wang, L.: Generalization error for multi-class margin classification. Electron. J. Stat. **1**, 307–330 (2007)
17. Pierre, A., Xiaoyin, L., Olivier, W.: Prediction of time series by statistical learning: general losses and fast rates. Depend. Model. **1**, 65–93 (2014)
18. Modha, D.S., Masry, E.: Minimum complexity regression estimation with weakly dependent observations. IEEE Trans. Inf. Theory **42**(6), 2133–2145 (1996)
19. Cuong, N.V., Ho, L.S.T., Dinh, V.: Generalization and robustness of batched weighted average algorithm with v-geometrically ergodic markov data. In: Jain, S., Munos, R., Stephan, F., Zeugmann, T. (eds.) ALT 2013. LNCS (LNAI), vol. 8139, pp. 264–278. Springer, Heidelberg (2013)
20. Ané, C.: Analysis of comparative data with hierarchical autocorrelation. Ann. Appl. Stat. **2**(3), 1078–1102 (2008)
21. Yurinskiĭ, V.: Exponential inequalities for sums of random vectors. J. Multivar. Anal. **6**(4), 473–499 (1976)

Deletion Operations on Deterministic Families of Automata

Joey Eremondi[1], Oscar H. Ibarra[2], and Ian McQuillan[3]([⊠])

[1] Department of Information and Computing Sciences, Utrecht University,
P.O. Box 80.089, 3508 TB Utrecht, The Netherlands
`j.s.eremondi@students.uu.nl`
[2] Department of Computer Science, University of California,
Santa Barbara, CA 93106, USA
`ibarra@cs.ucsb.edu`
[3] Department of Computer Science, University of Saskatchewan,
Saskatoon, SK S7N 5A9, Canada
`mcquillan@cs.usask.ca`

Abstract. Many different deletion operations are investigated applied to languages accepted by one-way and two-way deterministic reversal-bounded multicounter machines as well as finite automata. Operations studied include the prefix, suffix, infix and outfix operations, as well as left and right quotient with languages from different families. It is often expected that language families defined from deterministic machines will not be closed under deletion operations. However, here, it is shown that one-way deterministic reversal-bounded multicounter languages are closed under right quotient with languages from many different language families; even those defined by nondeterministic machines such as the context-free languages, or languages accepted by nondeterministic pushdown machines augmented by any number of reversal-bounded counters. Also, it is shown that when starting with one-way deterministic machines with one counter that makes only one reversal, taking the left quotient with languages from many different language families, again including those defined by nondeterministic machines such as the context-free languages, yields only one-way deterministic reversal-bounded multicounter languages (by increasing the number of counters). However, if there are even just two more reversals on the counter, or a second 1-reversal-bounded counter, taking the left quotient (or even just the suffix operation) yields languages that can neither be accepted by deterministic reversal-bounded multicounter machines, nor by 2-way nondeterministic machines with one reversal-bounded counter. A number of other results with deletion operations are also shown.

The research of O. H. Ibarra was supported, in part, by NSF Grant CCF-1117708. The research of I. McQuillan was supported, in part, by the Natural Sciences and Engineering Research Council of Canada.

© Springer International Publishing Switzerland 2015
R. Jain et al. (Eds.): TAMC 2015, LNCS 9076, pp. 388–399, 2015.
DOI: 10.1007/978-3-319-17142-5_33

1   Introduction

This paper involves the study of various types of deletion operations applied to languages accepted by one-way deterministic reversal-bounded multicounter machines (DCM). These are machines that operate like finite automata with an additional fixed number of counters, where there is a bound on the number of times each counter switches between increasing and decreasing [2,10]. These languages have many decidable properties, such as emptiness, infiniteness, equivalence, inclusion, universe and disjointness [10].

These machines have been studied in a variety of different applications, such as to membrane computing, verification of infinite-state systems and Diophantine equations.

Recently, in [5], a related study was conducted for insertion operations; specifically operations defined by ideals obtained from the prefix, suffix, infix and outfix relations, as well as left and right concatenation with languages from different language families. It was found that languages accepted by one-way deterministic reversal-bounded counter machines with one reversal-bounded counter are closed under right concatenation with Σ^*, but having two 1-reversal-bounded counters and right concatenating Σ^* yields languages outside of DCM and 2DCM(1) (languages accepted by two-way deterministic machines with one counter that is reversal-bounded). It also follows from this analysis that the right input endmarker is necessary for even one-way deterministic reversal-bounded counter machines, when there are at least two counters. Also, concatenating Σ^* to the left of some one-way deterministic 1-reversal-bounded one counter languages yields languages that are neither in DCM nor 2DCM(1). Other recent results on reversal-bounded multicounter languages include a technique to show languages are outside of DCM [3].

Closure properties of some variants of nondeterministic counter machines under deletion operations were studied in [14]. However, in this paper we investigate deterministic machines which were not examined in [14].

2   Preliminaries

The set of non-negative integers is denoted by $\mathbb{N}_0$, and the set of positive integers by $\mathbb{N}$. For $c \in \mathbb{N}_0$, let $\pi(c)$ be 0 if $c = 0$, and 1 otherwise.

We assume knowledge of standard formal language theoretic concepts such as languages, finite automata, determinism, nondeterminism, semilinearity, recursive and recursively enumerable languages [2,9]. Next, we will give some notation used in the paper. The empty word is denoted by λ. If Σ is a finite alphabet, then Σ^* is the set of all words over Σ and $\Sigma^+ = \Sigma^* \setminus \{\lambda\}$. For a word $w \in \Sigma^*$, if $w = a_1 \cdots a_n$ where $a_i \in \Sigma$, $1 \leq i \leq n$, the length of w is denoted by $|w| = n$, and the reversal of w is denoted by $w^R = a_n \cdots a_1$. A language over Σ is any subset of Σ^*. Given a language $L \subseteq \Sigma^*$, the complement of L, $\Sigma^* \setminus L$ is denoted by $\overline{L}$. Given two languages L_1, L_2, the left quotient of L_2 by L_1, $L_1^{-1}L_2 = \{y \mid xy \in L_2, x \in L_1\}$, and the right quotient of L_1 by L_2 is $L_1 L_2^{-1} = \{x \mid xy \in L_1, y \in L_2\}$.

A language L is *word-bounded* or simply *bounded* if $L \subseteq w_1^* \cdots w_k^*$ for some $k \geq 1$ and (not-necessarily distinct) words $w_1, \ldots, w_k$. Further, L is *letter-bounded* if each w_i is a distinct letter. Also, L is *bounded-semilinear* if $L \subseteq w_1^* \cdots w_k^*$ and $Q = \{(i_1, \ldots, i_k) \mid w_1^{i_1} \cdots w_k^{i_k} \in L\}$ is a semilinear set [12].

We now present notation for common word and language operations used throughout the paper.

Definition 1. *For a language $L \subseteq \Sigma^*$, the prefix, suffix, infix and outfix operations are defined by:*

- $\mathrm{pref}(L) = \{w \mid wx \in L, x \in \Sigma^*\}$,
- $\mathrm{suff}(L) = \{w \mid xw \in L, x \in \Sigma^*\}$,
- $\mathrm{inf}(L) = \{w \mid xwy \in L, x, y \in \Sigma^*\}$,
- $\mathrm{outf}(L) = \{xy \mid xwy \in L, w \in \Sigma^*\}$.

Note that $\mathrm{pref}(L) = L(\Sigma^*)^{-1}$ and $\mathrm{suff}(L) = (\Sigma^*)^{-1}L$.

The outfix operation has been generalized to the notion of embedding [13]:

Definition 2. *The m-embedding of a language $L \subseteq \Sigma^*$ is the following set:* $\mathrm{emb}(L, m) = \{w_0 \cdots w_m \mid w_0 x_1 \cdots w_{m-1} x_m w_m \in L,\ w_i \in \Sigma^*, 0 \leq i \leq m,$ $x_j \in \Sigma^*, 1 \leq j \leq m\}$.

Note that $\mathrm{outf}(L) = \mathrm{emb}(L, 1)$.

A *nondeterministic multicounter machine* is a finite automaton augmented by a fixed number of counters. The counters can be increased, decreased, tested for zero, or tested to see if the value is positive. A multicounter machine is *reversal-bounded* if every counter makes a fixed number of changes between increasing and decreasing.

Formally, a *one-way k-counter machine* is a tuple $M = (k, Q, \Sigma, \$, \delta, q_0, F)$, where $Q, \Sigma, \$, q_0, F$ are respectively the finite set of states, the input alphabet, the right input end-marker, the initial state in Q, and the set of final states that is a subset of Q. The transition function δ (defined as in [10] except with only a right end-marker since we only use one-way inputs) is a mapping from $Q \times (\Sigma \cup \{\$\}) \times \{0, 1\}^k$ into $Q \times \{\mathrm{S}, \mathrm{R}\} \times \{-1, 0, +1\}^k$, such that if $\delta(q, a, c_1, \ldots, c_k)$ contains $(p, d, d_1, \ldots, d_k)$ and $c_i = 0$ for some i, then $d_i \geq 0$ to prevent negative values in any counter. The direction of the input tape head movement is given by the symbols S are R for either *stay* or *right* respectively. The machine M is *deterministic* if δ is a function. A *configuration* of M is a $k + 2$-tuple $(q, w\$, c_1, \ldots, c_k)$ for describing the situation where M is in state q, with $w \in \Sigma^*$ still to read as input, and $c_1, \ldots, c_k \in \mathbb{N}_0$ are the contents of the k counters. The derivation relation $\vdash_M$ is defined between configurations, where $(q, aw, c_1, \ldots, c_k) \vdash_M$ $(p, w', c_1 + d_1, \ldots, c_k + d_k)$, if $(p, d, d_1, \ldots, d_k) \in \delta(q, a, \pi(c_1), \ldots, \pi(c_k))$ where $d \in \{\mathrm{S}, \mathrm{R}\}$ and $w' = aw$ if $d = \mathrm{S}$, and $w' = w$ if $d = \mathrm{R}$. Extended derivations are given by $\vdash_M^*$, the reflexive, transitive closure of $\vdash_M$. A word $w \in \Sigma^*$ is accepted by M if $(q_0, w\$, 0, \ldots, 0) \vdash_M^* (q, \$, c_1, \ldots, c_k)$, for some $q \in F$, and $c_1, \ldots, c_k \in \mathbb{N}_0$. The language accepted by M, denoted by $L(M)$, is the set of all words accepted by M. The machine M is *l-reversal bounded* if, in every accepting computation, the count on each counter alternates between increasing and decreasing at most l times.

We denote by $\mathsf{NCM}(k, l)$ the family of languages accepted by one-way non-deterministic l-reversal-bounded k-counter machines. We denote by $\mathsf{DCM}(k, l)$ the family of languages accepted by one-way deterministic l-reversal-bounded k-counter machines. The union of the families of languages are denoted by $\mathsf{NCM} = \bigcup_{k,l \geq 0} \mathsf{NCM}(k, l)$ and $\mathsf{DCM} = \bigcup_{k,l \geq 0} \mathsf{DCM}(k, l)$. We will also sometimes refer to a multicounter machine as being in $\mathsf{NCM}(k, l)$ $(\mathsf{DCM}(k, l))$, if it has k l-reversal bounded counters (and is deterministic).

We denote by REG the family of regular languages, and by NPCM the family of languages accepted by nondeterministic pushdown automata augmented by a fixed number of reversal-bounded counters [10]. We also denote by $\mathsf{2DCM}(1)$ the family of languages accepted by two-way input, deterministic finite automata (both a left and right input tape end-marker are required) augmented by one reversal-bounded counter [11]. A machine of this form is said to be *finite-crossing* if there is a fixed c such that the number of times the boundary between any two adjacent input cells is crossed is at most c [6]. A machine is *finite-turn* if the input head makes at most k turns on the input, for some k. Also, $\mathsf{2NCM}$ is the family of languages accepted by two-way nondeterministic machines with a fixed number of reversal-bounded counters, while $\mathsf{2DPCM}$ is the family of two-way deterministic pushdown machines augmented by a fixed number of reversal-bounded counters.

The next result proved in [12] gives examples of weak and strong machines that are equivalent over word-bounded languages.

Theorem 1. *[12]   The following are equivalent for every word-bounded language L:*

1. *L can be accepted by an NCM.*
2. *L can be accepted by an NPCM.*
3. *L can be accepted by a finite-crossing $\mathsf{2NCM}$.*
4. *L can be accepted by a DCM.*
5. *L can be accepted by a finite-turn $\mathsf{2DCM}(1)$.*
6. *L can be accepted by a finite-crossing $\mathsf{2DPCM}$*
7. *L is bounded-semilinear.*

We also need the following result in [11]:

Theorem 2. *[11]   Let $L \subseteq a^*$ be accepted by a $\mathsf{2NCM}$ (not necessarily finite-crossing). Then L is regular, hence, semilinear.*

3   Closure and Non-closure for Erasing Operations

3.1   Right Quotient for DCM

We begin by showing the closure of DCM under right quotient with any non-deterministic reversal bounded machine, even when augmented with a pushdown store.

Proposition 1. *Let $L_1 \in$ DCM and let $L_2 \in$ NPCM. Then $L_1L_2^{-1} \in$ DCM.*

Proof. Consider a DCM machine $M_1 = (k_1, Q_1, \Sigma, \$, \delta_1, s_0, F_1)$ and NPCM machine M_2 over Σ with k_2 counters where $L(M_1) = L_1$ and $L(M_2) = L_2$. A DCM machine M' will be constructed accepting $L_1L_2^{-1}$.

Let $\Gamma = \{a_1, \ldots, a_{k_1}\}$ be new symbols. For each $q \in Q_1$, let $M_c(q)$ be an interim $k_1 + k_2$ counter (plus a pushdown) NPCM machine over Γ constructed as follows: on input $a_1^{p_1} \cdots a_{k_1}^{p_{k_1}}$, $M_c(q)$ increments the first k_1 counters to $(p_1, \ldots, p_{k_1})$. Then $M_c(q)$ nondeterministically guesses a word $x \in \Sigma^*$ and simulates M_1 on $x\$$ starting from state q and from the counter values of $(p_1, \ldots, p_{k_1})$ using the first k_1 counters, while in parallel, simulating M_2 on x using the next k_2 counters and the pushdown. This is akin to the product automaton construction described in [10] showing NPCM is closed under intersection with NCM. Then $M_c(q)$ accepts if both M_1 and M_2 accept.

Claim. Let $L_c(q) = \{a_1^{p_1} \cdots a_{k_1}^{p_{k_1}} \mid \exists x \in L_2$ such that $(q, x\$, p_1, \ldots, p_{k_1}) \vdash_{M_1}^*$ $(q_f, \$, p'_1, \ldots p'_{k_1}), p'_i \geq 0, 1 \leq i \leq k_1, q_f \in F_1\}$. Then $L(M_c(q)) = L_c(q)$.

Proof. Consider $w = a_1^{p_1} \cdots a_{k_1}^{p_{k_1}} \in L_c(q)$. Then there exists x where $x \in L_2$ and $(q, x\$, p_1, \ldots, p_{k_1}) \vdash_{M_1}^* (q_f^1, \$, p'_1, \ldots p'_{k_1})$, where $q_f^1 \in F_1$. There must then be some final state $q_f^2 \in F_2$ reached when reading $x\$$ in M_2. Then, $M_c(q)$, on input w places $(p_1, \ldots, p_{k_1}, 0, \ldots, 0)$ on the counters and then can nondeterministically guess x letter by letter and simulate x in M_1 from state q on the first k_1 counters and simulate x in M_2 from its initial configuration on the remaining counters and pushdown. Then $M_c(q)$ ends up in state (q_f^1, q_f^2), which is final. Hence, $w \in L(M_c(q))$.

Consider $w = a^{p_1} \cdots a^{p_{k_1}} \in L(M_c(q))$. After adding each p_i to counter i, $M_c(q)$ guesses x and simulates M_1 on the first k_1 counters from q and simulates M_2 on the remaining counters from an initial configuration. It follows that $x \in L_2$, and $(q, x\$, p_1, \ldots, p_{k_1}) \vdash_{M_1}^* (q_f^1, \$, p'_1, \ldots p'_{k_1}), p'_i \geq 0, 1 \leq i \leq k_1, q_f^1 \in F_1$. Hence, $w \in L_c(q)$. □

Since for each $q \in Q_1$, $M_c(q)$ is in NPCM, it accepts a semilinear language [10], and since the accepted language is bounded, it is bounded-semilinear and can therefore be accepted by a DCM-machine by Theorem 1. Let $M'_c(q)$ be this DCM machine, with k' counters, for some k'.

Thus, a final DCM machine M' with $k_1 + k'$ counters is built as follows. In it, M' has k_1 counters used to simulate M_1, and also k' additional counters, used to simulate some $M'_c(q)$, for some $q \in Q_1$. Then, M' reads its input $x\$$, where $x \in \Sigma^*$, while simulating M_1 on the first k_1 counters, either failing, or reaching some configuration $(q, \$, p_1, \ldots, p_{k_1})$, for some $q \in Q_1$, upon first hitting the end-marker $\$$. If it does not fail, we then simulate the DCM-machine $M'_c(q)$ on input $a_1^{p_1} \cdots a_{k_1}^{p_{k_1}}$, but this simulation is done deterministically by subtracting 1 from the first k_1 counters, in order, until each are zero instead of reading input characters, and accepts if $a_1^{p_1} \cdots a_{k_1}^{p_{k_1}} \in L(M'_c(q)) = L_c(q)$. Then M' is deterministic and accepts

$$\{x \mid \text{either } (s_0, x\$, 0, \ldots, 0) \vdash^*_{M_1} (q', a\$, p'_1, \ldots, p'_{k_1}) \vdash_{M_1} (q, \$, p_1, \ldots, p_{k_1}),$$
$$a \in \Sigma, \text{ or } (s_0, x\$, 0, \ldots, 0) = (q, \$, p_1, \ldots, p_{k_1}), \text{ s.t. } a_1^{p_1} \cdots a_{k_1}^{p_{k_1}} \in L_c(q)\}$$
$$= \{x \mid \text{either } (s_0, x\$, 0, \ldots, 0) \vdash^*_{M_1} (q', a\$, p'_1, \ldots, p'_{k_1}) \vdash_{M_1} (q, \$, p_1, \ldots, p_{k_1}),$$
$$a \in \Sigma, \text{ or } (s_0, x\$, 0, \ldots, 0) = (q, \$, p_1, \ldots, p_{k_1}), \text{ where } \exists y \in L_2 \text{ s.t.}$$
$$(q, y\$, p_1, \ldots, p_{k_1}) \vdash^*_{M_1} (q_f, \$, p''_1, \ldots, p''_{k_1}), q_f \in F_1\}$$
$$= \{x \mid xy \in L_1, y \in L_2\}$$
$$= L_1 L_2^{-1}. \qquad \square$$

These immediately show closure for the prefix operation.

Corollary 1. *If $L \in$ DCM, then $\mathrm{pref}(L) \in$ DCM.*

We can modify this construction to show a strong closure result for one-counter languages that does not increase the number of counters.

Proposition 2. *Let $l \in \mathbb{N}$. If $L_1 \in$ DCM$(1, l)$ and $L_2 \in$ NPCM, then $L_1 L_2^{-1} \in$ DCM$(1, l)$.*

Proof. The construction is similar to the one in Proposition 1. However, we note that since the input machine for L_1 has only one counter, $L_c(q)$ is unary (regardless of the number of counters needed for L_2). Thus $L_c(q)$ is unary and semilinear, and Parikh's theorem states that all semilinear languages are letter-equivalent to regular languages [8], and all unary semilinear languages are regular. Thus $L_c(q)$ is regular, and can be accepted by a DFA.

We can then construct M' accepting $L_1 L_2^{-1}$ as in Proposition 1 without requiring any additional counters or counter reversals, by transitioning to the DFA accepting $L_c(q)$ when we reach the end of input at state q. $\qquad \square$

Corollary 2. *Let $l \in \mathbb{N}$. If $L \in$ DCM$(1, l)$, then $\mathrm{pref}(L) \in$ DCM$(1, l)$.*

In fact, this construction can be generalized from NPCM to any class of automata that can be defined using Definition 3. These classes of automata are described in more detail in [7]. We only define it in a way specific to our use in this paper. Only the first two conditions are required for Corollary 3, while the third is required for Corollary 5.

Definition 3. *A family of languages $\mathscr{F}$ is said to be reversal-bounded counter augmentable if*

- *every language in $\mathscr{F}$ is effectively semilinear,*
- *given DCM machine M_1 with k counters, state set Q and final state set F, and $L_2 \in \mathscr{F}$, we can effectively construct, for each $q \in Q$, the following language in $\mathscr{F}$,*

$$\{a_1^{p_1} \cdots a_k^{p_k} \mid \exists x \in L_2 \text{ such that } (q, x\$, p_1, \ldots, p_k) \vdash^*_{M_1} (q_f, \$, p'_1, \ldots p'_k),$$
$$p'_i \geq 0, q_f \in F\},$$

– *given* DCM *machine* M_1 *with* k *counters, state set* Q, *and* $L_2 \in \mathscr{F}$, *we can effectively construct, for each* $q \in Q$, *the following language in* $\mathscr{F}$,

$$\{a_1^{p_1} \cdots a_k^{p_k} \mid \exists x \in L_2 \text{ such that } (q, x, 0, \ldots, 0) \vdash_{M_1}^* (q, \lambda, p_1, \ldots p_k)\}.$$

There are many reversal-bounded counter augmentable families that L_2 could be from in this corollary, such as:

Corollary 3. *Let* $L_1 \in$ DCM *and* $L_2 \in \mathscr{F}$, *a family of languages that is reversal-bounded counter augmentable. Then* $L_1 L_2^{-1} \in$ DCM. *Furthermore, if* $L_1 \in$ DCM$(1, l)$ *for some* $l \in \mathbb{N}$, *then* $L_1 L_2^{-1} \in$ DCM$(1, l)$.

This construction could be applied to several other families of semilinear languages such as:

– MPCA's: one-way machines with k pushdowns where values may only be popped from the first non-empty stack, augmented by a fixed number of reversal-bounded counters [7].
– TCA's: NFA's augmented with a two-way read-write tape, where the number of times the read-write head crosses any tape cell is finitely bounded, again augmented by a fixed number of reversal-bounded counters [7].
– QCA's: NFA's augmented with a queue, where the number of alternations between the non-deletion phase and the non-insertion phase is bounded by a constant [7].
– EPDA's: embedded pushdown automata, modelled around a stack of stacks, introduced in [17]. These accept the languages of tree-adjoining grammars, a semilinear subset of the context-sensitive languages. As was stated in [7], we can augment this model with a fixed number of reversal-bounded counters and still get an effectively semilinear family.

3.2 Right and Left Quotients of Regular Sets

Let $\mathscr{F}$ be any family of languages (which need not be recursively enumerable). It is known that REG is closed under right quotient by languages in $\mathscr{F}$ [9]. However, this closure need not be effective, as it will depend on the properties of $\mathscr{F}$. The following is an interesting observation which connects decidability of the emptiness problem to effectiveness of closure under right quotient:

Proposition 3. *Let* $\mathscr{F}$ *be any family of languages which is effectively closed under intersection with regular sets and whose emptiness problem is decidable. Then* REG *is effectively closed under both left and right quotient by languages in* $\mathscr{F}$.

Proof. We will start with right quotient.

Let $L_1 \in$ REG and L_2 be in $\mathscr{F}$. Let M be a DFA accepting L_1. Let q be a state of M, and $L_q = \{y \mid M$ from initial state q accepts $y\}$. Let $Q' = \{q \mid q$ is a state of $M, L_q \cap L_2 \neq \emptyset\}$. Since $\mathscr{F}$ is effectively closed under intersection with regular sets and has a decidable emptiness problem, Q' is computable.

Then a DFA M' accepting $L_1 L_2^{-1}$ can be obtained by just making Q' the set of accepting states in M.

Next, for left quotient, let L_1 be in $\mathscr{F}$, and L_2 in REG be accepted by a DFA M whose initial state is q_0.

Let $L_q = \{x \mid M \text{ on input } x \text{ ends in state } q\}$. Let $Q' = \{q \mid L_q \cap L_1 \neq \emptyset\}$. Then Q' is computable, since $\mathscr{F}$ is effectively closed under intersection with regular sets and has a decidable emptiness problem.

We then construct an NFA (with λ-transitions) M' to accept $L_1^{-1} L_2$ as follows: M' starting in state q_0 with input y nondeterministically goes to a state q in Q' without reading any input, and then simulates the DFA M.   □

Corollary 4. REG *is effectively closed under left and right quotient by languages in:*

1. *the families of languages accepted by* NPCM *and* 2DCM(1) *machines,*
2. *the family of languages accepted* MPCAs, TCAs, QCAs, *and* EPDAs,
3. *the families of* ET0L *and Indexed languages.*

Proof. These families are closed under intersection with regular sets. They have also a decidable emptiness problem [1,7,16]. The family of ET0L languages and Indexed languages are discussed further in [16] and [1] respectively.   □

3.3 Suffix, Infix and Left Quotient for DCM(1, 1)

In the case of one-counter machines that makes only one counter reversal, it will be shown that a DCM-machine that can accept their suffix and infix languages can always be constructed. However, in some cases, these resulting machines often require more than one counter. Thus, unlike prefix, DCM(1, 1) is not closed under suffix, left quotient, or infix. But, the result is in DCM.

The proof of Lemma 1 is quite lengthy, and due to space constraints is omitted but can be found online in [4]. We will give some intuition for the result here. First, DCM is closed under union and so the second statement of Lemma 1 follows from the first. For the first statement, an intermediate NPCM machine is constructed from L_1 and L that accepts a language L^c. This language contains words of the form qa^i where there exists some word w such that both $w \in L_1$, and also from the initial configuration of M (accepting L), it can read w and reach state q with i on the counter. Then, it is shown that this language is actually a regular language, using the fact that all semilinear unary languages are regular (as $(q)^{-1} L^c$ is unary; see [4] for full details). Then, DCM(1, 1) machines are created for every state q of M. These accept all words w such that $qa^i \in L^c$, and in M, from state q and counter i with w to read as input, M can reach a final state while emptying the counter. The fact that L^c is regular allows these machines to be created.

Lemma 1. *Let* $L \in$ DCM(1, 1), $L_1 \in$ NPCM. *Then* $L_1^{-1} L$ *is the finite union of languages in* DCM(1, 1). *Furthermore, it is in* DCM.

From this, we obtain the following general result (proof also omitted due to space and is found in [4]).

Theorem 3. *Let $L \in$ DCM$(1,1), L_1, L_2 \in$ NPCM. Then both $(L_1^{-1}L)L_2^{-1}$ and $L_1^{-1}(LL_2^{-1})$ are a finite union of languages in DCM$(1,1)$. Furthermore, both languages are in DCM.*

And, as with Corollary 3, this can be generalized to any language families that are reversal-bounded counter augmentable.

Corollary 5. *Let $L \in$ DCM$(1,1), L_1 \in \mathscr{F}_1, L_2 \in \mathscr{F}_2$, where $\mathscr{F}_1$ and $\mathscr{F}_2$ are any families of languages that are reversal-bounded counter augmentable. Then $(L_1^{-1}L)L_2^{-1}$ and $L_1^{-1}(LL_2^{-1})$ are both a finite union of languages in DCM$(1,1)$. Furthermore, both languages are in DCM.*

As a special case, when using the fixed regular language Σ^* for the right and left quotient, we obtain:

Corollary 6. *Let $L \in$ DCM$(1,1)$. Then* suff(L) *and* inf(L) *are both DCM languages.*

It is however necessary that the number of counters increase to accept suff(L) and inf(L), for some $L \in$ DCM$(1,1)$. The result also holds for the outfix operator. The proof is omitted due to space and is found in [4].

Proposition 4. *There exists $L \in$ DCM$(1,1)$ where all of* suff(L), inf(L), outf(L) *are not in* DCM$(1,1)$.

3.4   Non-closure of Suffix, Infix and Outfix with Multiple Counters or Reversals

In [5], a technique was used to show languages are not in DCM and 2DCM(1) simultaneously. The technique uses undecidable properties to show non-closure. As 2DCM(1) machines have two-way input and a reversal-bounded counter, it is difficult to derive "pumping" lemmas for these languages. Furthermore, unlike DCM and NCM machines, 2DCM(1) machines can accept non-semilinear languages. For example, $L_1 = \{a^i b^k \mid i, k \geq 2, i \text{ divides } k\}$ can be accepted by a 2DCM(1) whose counter makes only one reversal. However, $L_2 = \{a^i b^j c^k \mid i, j, k \geq 2, k = ij\}$ cannot be accepted by a 2DCM(1) [11]. This technique from [5] works as follows. The proof uses the fact that there is a recursively enumerable but not recursive language $L_{re} \subseteq \mathbb{N}_0$ that is accepted by a deterministic 2-counter machine [15]. Thus, the machine when started with $n \in \mathbb{N}_0$ in the first counter and zero in the second counter, eventually halts (i.e., accepts $n \in L_{re}$).

Examining the constructions in [15] of the 2-counter machine demonstrates that the counters behave in a regular pattern. Initially one counter has some value d_1 and the other counter is zero. Then, the machine's operation can be divided into phases, where each phase starts with one of the counters equal to some positive integer d_i and the other counter equals 0. During the phase, the positive counter decreases, while the other counter increases. The phase ends with the first counter containing 0 and the other counter containing d_{i+1}. In the next phase, the modes of the counters are interchanged. Thus, a sequence of configurations where the phases are changing will be of the form:

$$(q_1, d_1, 0), (q_2, 0, d_2), (q_3, d_3, 0), (q_4, 0, d_4), (q_5, d_5, 0), (q_6, 0, d_6), \ldots$$

where the q_i's are states, with $q_1 = q_s$ (the initial state), and $d_1, d_2, d_3, \ldots$ are positive integers. The second component of the configuration refers to the value of the first counter, and the third component refers to the value of the second. Also, notice that in going from state q_i in phase i to state q_{i+1} in phase $i + 1$, the 2-counter machine goes through intermediate states.

For each i, there are 5 cases for the value of d_{i+1} in terms of d_i: $d_{i+1} = d_i$, $2d_i, 3d_i, d_i/2, d_i/3$ (the division operation only occurs if the number is divisible by 2 or 3, respectively). The case applied is determined by q_i. Hence, a function h can be defined such that if q_i is the state at the start of phase i, $d_{i+1} = h(q_i)d_i$, where $h(q_i)$ is one of $1, 2, 3, 1/2, 1/3$.

Let T be a 2-counter machine accepting a recursively enumerable language that is not recursive. Assume that $q_1 = q_s$ is the initial state, which is never re-entered, and if T halts, it does so in a unique state q_h. Let Q be the states of T, and 1 be a new symbol.

In what follows, α is any sequence of the form $\#I_1\#I_2\# \cdots \#I_{2m}\#$ (thus we assume that the length is even), where for each i, $1 \le i \le 2m$, $I_i = q1^k$ for some $q \in Q$ and $k \ge 1$, represents a possible configuration of T at the beginning of phase i, where q is the state and k is the value of the first counter (resp., the second) if i is odd (resp., even).

Define L_0 to be the set of all strings α such that

1. $\alpha = \#I_1\#I_2\# \cdots \#I_{2m}\#$;
2. $m \ge 1$;
3. for $1 \le j \le 2m - 1$, $I_j \Rightarrow I_{j+1}$, i.e., if T begins in configuration I_j, then after one phase, T is in configuration I_{j+1} (i.e., I_{j+1} is a valid successor of I_j);

Then, the following was shown in [5].

Lemma 2. L_0 is not in $\mathsf{DCM} \cup \mathsf{2DCM}(1)$.

We will use this language exactly to show taking either the suffix, infix or outfix of a language in $\mathsf{DCM}(1,3)$, $\mathsf{DCM}(2,1)$ or $\mathsf{2DCM}(1)$ can produce languages that are in neither DCM nor $\mathsf{2DCM}(1)$.

Theorem 4. There exists a language L in all of $L \in \mathsf{DCM}(1,3)$, $L \in \mathsf{DCM}(2,1)$, and $L \in \mathsf{2DCM}(1)$ (which makes no turn on the input and 3 reversals on the counter) such that $\mathrm{suff}(L) \notin \mathsf{DCM} \cup \mathsf{2DCM}(1)$, $\mathrm{inf}(L) \notin \mathsf{DCM} \cup \mathsf{2DCM}(1)$, and $\mathrm{outf}(L) \notin \mathsf{DCM} \cup \mathsf{2DCM}(1)$.

Proof. Let L_0 be the language defined above, which is not in $\mathsf{DCM} \cup \mathsf{2DCM}(1)$. Let a, b be new symbols. Clearly, bL_0b is also not in $\mathsf{DCM} \cup \mathsf{2DCM}(1)$. Let $L = \{a^i b \# I_1 \# I_2 \# \cdots \# I_{2m} \# b \mid I_1, \ldots, I_{2m}$ are configurations of the 2-counter machine T, $i \leq 2m - 1, I_{i+1}$ is not a valid successor of I_i}. Clearly L is in DCM$(1,3)$, in DCM$(2,1)$, and in 2DCM(1) (as DCM$(1,3)$ is a subset of 2DCM(1)).

Let L_1 be suff(L). Suppose L_1 is in DCM (resp., 2DCM(1)). Then $L_2 = \overline{L_1}$ is also in DCM (resp., 2DCM(1)).

Let $R = \{b\#I_1\#I_2\cdots\#I_{2m}\#b \mid I_1, \ldots, I_{2m}$ are configurations of $T\}$. Then since R is regular, $L_3 = L_2 \cap R$ is in DCM (resp, 2DCM(1)). We get a contradiction, since $L_3 = bL_0b$.

Non-closure under infix and outfix can be shown similarly.   □

This implies non-closure under left-quotient with regular languages, and this result also extends to the embedding operation, a generalization of outfix.

Corollary 7. *There exists* $L \in$ DCM$(1,3), L \in$ DCM$(2,1), L \in$ 2DCM(1) *(which makes no turn on the input and 3 reversals on the counter), and* $R \in$ REG *such that* $R^{-1}L \notin$ DCM $\cup$ 2DCM(1).

Corollary 8. *Let* $m > 0$. *Then there exists* $L \in$ DCM$(1,3), L \in$ DCM$(2,1)$, $L \in$ 2DCM(1) *(which makes no turn on the input and 3 reversals on the counter) such that* emb$(L,m) \notin$ DCM $\cup$ 2DCM(1).

The results of Theorem 4 and Corollary 7 are optimal for suffix and infix as these operations applied to DCM$(1,1)$ are always in DCM by Corollary 6 (and since DCM$(1,2) =$ DCM$(1,1)$). But whether the outfix and embedding operations applied to DCM$(1,1)$ languages is always in DCM is an open question.

3.5   Closure for Bounded Languages

In this subsection, deletion operations applied to bounded and letter-bounded languages will be examined.

The following is a required straightforward corollary to Theorem 2.

Corollary 9. *Let* $L \subseteq \#a^*\#$ *be accepted by a* 2NCM. *Then* L *is regular.*

Theorem 5. *If* L *is a bounded language accepted by either a finite-crossing* 2NCM, *an* NPCM *or a finite-crossing* 2DPCM, *then all of* pref(L), suff(L), inf(L), outf(L) *can be accepted by a* DCM.

Proof. By Theorem 1, L can always be converted to an NCM. Further, one can construct NCM's accepting pref(L), suff(L), inf(L), outf(L) since one-way NCM is closed under prefix, suffix, infix and outfix. In addition, it is known that applying these operations on bounded languages produce only bounded languages. Thus, by another application of Theorem 1, the result can then be converted to a DCM.   □

The "finite-crossing" requirement in the theorem above is necessary:

Proposition 5. *There exists a letter-bounded language* L *accepted by a* 2DCM(1) *machine which makes only one reversal on the counter such that* suff(L) *(resp., inf(L), outf(L), pref(L)) is not in* DCM $\cup$ 2DCM(1).

Proof. Let $L = \{a^i \# b^j \# \mid i, j \geq 2, j \text{ is divisible by } i\}$. Clearly, L can be accepted by a 2DCM(1) which makes only one reversal on the counter. If $\mathrm{suff}(L)$ is in $\mathsf{DCM} \cup 2\mathsf{DCM}(1)$, then $L' = \mathrm{suff}(L) \cap \# b^+ \#$ would be in $\mathsf{DCM} \cup 2\mathsf{DCM}(1)$. From Corollary 9, we get a contradiction, since L' is not semilinear. The other cases are shown similarly. $\qquad\square$

References

1. Aho, A.V.: Indexed grammars–an extension of context-free grammars. J. ACM **15**(4), 647–671 (1968)
2. Baker, B.S., Book, R.V.: Reversal-bounded multipushdown machines. J. Comput. Syst. Sci. **8**(3), 315–332 (1974)
3. Chiniforooshan, E., Daley, M., Ibarra, O.H., Kari, L., Seki, S.: One-reversal counter machines and multihead automata: Revisited. Theor. Comput. Sci. **454**, 81–87 (2012)
4. Eremondi, J., Ibarra, O., McQuillan, I.: Deletion operations on deterministic families of automata. Technical report 2014–03, University of Saskatchewan (2014). http://www.cs.usask.ca/documents/techreports/2014/TR-2014-03.pdf
5. Eremondi, J., Ibarra, O.H., McQuillan, I.: Insertion operations on deterministic reversal-bounded counter machines. In: Dediu, A.-H., Formenti, E., Martín-Vide, C., Truthe, B. (eds.) LATA 2015. LNCS, vol. 8977, pp. 200–211. Springer, Heidelberg (2015)
6. Gurari, E.M., Ibarra, O.H.: The complexity of decision problems for finite-turn multicounter machines. J. Comput. Syst. Sci. **22**(2), 220–229 (1981)
7. Harju, T., Ibarra, O., Karhumäki, J., Salomaa, A.: Some decision problems concerning semilinearity and commutation. J. Comput. Syst. Sci. **65**(2), 278–294 (2002)
8. Harrison, M.: Introduction to Formal Language Theory. Addison-Wesley Series in Computer Science. Addison-Wesley Publishing Company, Boston (1978)
9. Hopcroft, J.E., Ullman, J.D.: Introduction to Automata Theory, Languages, and Computation. Addison-Wesley, Reading (1979)
10. Ibarra, O.H.: Reversal-bounded multicounter machines and their decision problems. J. ACM **25**(1), 116–133 (1978)
11. Ibarra, O.H., Jiang, T., Tran, N., Wang, H.: New decidability results concerning two-way counter machines. SIAM J. Comput. **23**(1), 123–137 (1995)
12. Ibarra, O.H., Seki, S.: Characterizations of bounded semilinear languages by one-way and two-way deterministic machines. Int. J. Found. Comput. Sci. **23**(6), 1291–1306 (2012)
13. Jürgensen, H., Kari, L., Thierrin, G.: Morphisms preserving densities. Int. J. Comput. Math. **78**, 165–189 (2001)
14. Kari, L., Seki, S.: Schema for parallel insertion and deletion: Revisited. Int. J. Found. Comput. Sci. **22**(07), 1655–1668 (2011)
15. Minsky, M.L.: Recursive unsolvability of post's problem of "tag" and other topics in theory of turing machines. Ann. Math. **74**(3), 437–455 (1961)
16. Rozenberg, G., Salomaa, A.: The Mathematical Theory of L Systems. Academic Press Inc, New York (1980)
17. Vijayashanker, K.: A Study of Tree Adjoining Grammars. Ph.D. thesis, Philadelphia, PA, USA (1987)

ExplicitPRISMSymm: Symmetry Reduction Technique for Explicit Models in PRISM

Reema Patel[1][✉], Kevin Patel[2], and Dhiren Patel[1]

[1] Computer Engineering Department,
NIT Surat, Surat, India
{reema.mtech,dhiren29p}@gmail.com
[2] Department of Computer Science and Engineering,
IIT Bombay, Mumbai, India
kevin.patel@cse.iitb.ac.in

Abstract. Probabilistic model checking of concurrent system involves exhaustive search of the reachable state space associated with the system model. Symmetry reduction is a commonly employed technique that enables model checking of exponentially large models. Most work on symmetry reduction focuses on symbolically represented probabilistic models, which are easy to build and perform reasonably well at property checking. In this work, we rather focus on explicitly represented probabilistic models. We report that explicitly represented models perform well at property checking, but face hurdles in model construction. We present an on-the-fly symmetry reduction technique for explicitly represented models. It significantly reduces build time and thus explicit model representation as an efficient alternative to symbolic model representation.

Keywords: On-the-fly symmetry reduction · Probabilistic model checking · Explicit state representation

1 Introduction

Probabilistic model checking is used to quantitatively analyze the properties of stochastic systems [10]. Properties such as "an adversary will compromise the security protocol with negligible probability" or "message will be delivered with the probability 0.5" can be analyzed with it.

Generally these techniques involve modeling a system as Discrete Time Markov Chains (DTMCs), Markov Decision Process (MDPs), etc. [15]. These models are graph based models, where the vertices represent states, and edges represent transitions among these states. The analysis is then performed using various iterative numerical computation methods e.g. value iteration for MDP.

Real world models tend to be complex, often comprising large number of states and transitions. For instance, a model of concurrent system executing n identical components, each of which can be in one of k possible conditions, have k^n upper bound for reachable states. Such state space explosion invariantly renders analysis through certain algorithms intractable. One of the best techniques

© Springer International Publishing Switzerland 2015
R. Jain et al. (Eds.): TAMC 2015, LNCS 9076, pp. 400–412, 2015.
DOI: 10.1007/978-3-319-17142-5_34

that can be employed to overcome this issue, especially in the case of concurrent systems, is symmetry reduction.

Concurrent systems frequently contain symmetrical components in form of identical processes. For example, consider a concurrent system of n processes, where n_i processes are in i^{th} condition ($1 \leq i \leq k$) then the number of symmetrical states is $\left(\frac{n!}{\Pi_i^k n_i!} \right)$. Analyzing all but one of these states is unnecessary. Symmetry reduction exploits this fact and reduces the state space: only one representative from each equivalence class is chosen. The chosen representative shall preserve the state labeling and transitions. Key challenge in symmetry reduction is to find out the states that belong to the same equivalence class.

Symmetry reduction techniques work on both symbolic and explicit representation of system models. In symbolic, the reachable state space of system model is stored in a very concise manner with the help of Multi Terminal Binary Decision Diagram (MTBDD) [7]. Whereas in explicit, the complete reachable state space is enumerated in memory. Our proposed symmetry reduction technique works with explicitly represented models. We implement this technique in PRISM probabilistic model checker [10].

PRISM works with both symbolic and explicit model representations. However, the existing symmetry reduction technique in PRISM works only with symbolic representations. The structure used for symbolic representation in PRISM is MTBDD [7]. Even though it performs well in general, in certain cases of probabilistic model checking, symmetry reduction results in explosion of MTBDD size by a factor of up to ten [18]. Thus, a study of techniques that can employ representation mechanisms other than MTBDD is warranted.

Thus, we decided to work with explicit state representation of probabilistic models. Our initial explorations showed significant improvement in time required for property checking when explicit models are used instead of symbolic models. However, except for certain trivial cases, explicit model construction is intractable. Major problem with explicit model construction is state space enumeration.

In this paper, we present an on-the-fly symmetry reduction technique to build a quotient reduced model at the time of exploration of reachable states. Using this technique, we are able to build explicit models of concurrent systems with extensive state space. Our algorithm is applicable to well-known probabilistic models e.g. DTMC, MDP and CTMC.

We evaluated our technique using PRISM's benchmark case studies. We compared the performance of our method with the existing PRISM's built-in symbolic symmetry reduction technique. Experimental results indicate substantial improvement in the performance of probabilistic model checking compared to the existing methods.

The rest of the paper is organized as follows: Sect. 2 presents a review of associated work in the literature; Sect. 3 establishes the mathematical notions behind symmetry reduction of DTMCs, MDPs and CTMCs; Sect. 4 explains an on-the-fly symmetry reduction algorithm for explicitly represented probabilistic models; Sect. 5 provides experimental results, followed by conclusion.

2   Related Work

There has been various symmetry reduction techniques developed for both probabilistic and non-probabilistic model checking. [14, 17, 18] cover the mathematical foundation of symmetry reduction and explain different approaches of symmetry reduction for symbolic and explicit-state representation. Here, we focus on symmetry reduction techniques that are applicable in probabilistic setting.

First symmetry reduction technique *PRISM-symm* for probabilistic setting [13] was applicable to symbolically represented probabilistic models. This technique is based on dynamic symmetry reduction concept [6] that is implemented for non-probabilistic model checking. Author has integrated this technique in the popular probabilistic model checker PRISM [10]. *PRISM-symm* reduces the state-space in order of magnitude for system having a large number of processes. It also reduces the size of the MTBDD by a factor of more than two, but in some cases, size of MTBDD increases by a factor of up to ten. Despite this blow-up in MTBDD size, symmetry reduction is beneficial for probabilistic model checking.

Donaldson and Miller have further extended the generic representative approach [5] to the probabilistic setting. This technique performs the language level symmetry reduction of given system model using GRIP tool [3]. GRIP translates a subset of PRISMs language known as Symmetric Probabilistic Specification Language [4] to a reduced form that can be analyzed using PRISM.

Thesis entitled as "Probabilistic Symmetry Reduction" by Christopher Power [16] has presented a new probabilistic model specification language known as *Probabilistic Symmetric System* (PSS). Author has introduced the extended channel diagram approach to detect component and data symmetries from PSS language. For symmetry reduction, they have presented a novel approach to compute the representative state which is based on constraint satisfaction problem. Author has also build a tool by implementing proposed symmetry detection and reduction techniques. This tool generates the symmetry reduced model of given system and set of PCTL formulas which can be analyzed using PRISM. Though unlike PRISM, this tool is not freely available.

On the fly symmetry reduction techniques have been developed for non-probabilistic model checking [1, 8]. The major benefit of these techniques is that symmetric states are discovered while exploring state space, and the corresponding search subtrees are pruned at that instant (thus the name "on the fly"), thereby constraining state space explosion.

In this paper, we present an on-the-fly symmetry reduction technique for explicitly represented probabilistic models. We also discuss the results of integrating this technique in PRISM.

3   Symmetry in Probabilistic Model Checking

In this section, we review some background material on probabilistic models, automorphisms and quotient models. A probabilistic model is a transition system with the state space X, whose behavior is specified by a transition function on X. The best known probabilistic models are: Discrete-time Markov Chains (DTMCs), Markov Decision Processes (MDPs) and Continuous-time Markov Chains (CTMCs). The following definitions are adapted from [5,13,15].

The system with purely probabilistic behavior can be modeled as DTMC.

Definition 1: DTMC is defined as a tuple $D = (S, s_0, P)$ where

- S is a finite set of states,
- $s_0 \in S$, is the initial state,
- $P : S \times S \to [0,1]$ is the transition probability matrix where $\sum_{s' \in S} P(s, s') = 1$ for all $s \in S$.

The system which exhibits both non-deterministic and probabilistic behavior can be modeled as an MDP.

Definition 2: MDP is an ordered tuple $M = (S, s_0, Steps)$ where

- S and s_0 are as for DTMC,
- $Steps : S \to 2^{Dist(S)}$ is a probability transition function where $Dist(S)$ denotes the set of discrete probability distributions over a set S i.e., the set of functions of the form $\mu : S \to [0,1]$ such that $\sum_{s \in S} \mu(s) = 1$.

For each state $s \in S$, $Steps(s)$ maps s to a finite non-empty subset of $Dist(S)$. At a given state s, a distribution μ is chosen non-deterministically from the elements of $Steps(s)$. Now the next state s' from s is chosen probabilistically according to the selected distribution $\mu \in Steps(s)$.

In DTMC and MDP, the progress of time is modeled by discrete time steps, one for each transition of the model. CTMC on the other hand, allows the modeling of real time.

Definition 3: CTMC is defined by a tuple $C = (S, s_0, R)$ where

- S and s_0 are as for DTMC.
- $R : S \times S \to \mathbb{R}$. This gives the rate $R(s, s')$ at which transition occur between each pair of states s, s'.

3.1   Automorphisms

In concurrent system, all the indistinguishable components preserve the system structure (transition relation) under any interchange of the components. An automorphism is a way of mapping the object to itself while preserving all of its structure.

Let's consider a probabilistic model of a concurrent system executing n components. Let $I = \{1, 2, \ldots, n\}$ represents a set of component identifiers. For some $k \geq 0$, let $L = \{0, 1, 2, \ldots, k\}$ be the set of possible local states of the components. A state $s \in S$ of the system can be represented as $s = (glb, l_1, l_2, \ldots, l_n)$, where glb indicates global variables if any. The set of all permutations of I forms a group under composition of mappings, denoted by $Aut(I)$. Let $\alpha \in Aut(I)$. When applied to a state s, α acts in the following way: $\alpha(s) : (glb^{\alpha}, l_{\alpha(1)}, l_{\alpha(2)}, \ldots, l_{\alpha(n)})$.

In case of DTMC D, if for all $s, t \in S, P(s,t)$ there exists $P'(\alpha(s), \alpha(t))$ such that $P(s,t) = P'(\alpha(s), \alpha(t))$, then α is an automorphism of D. The set of all automorphisms of D forms a group $Aut(D) \leq Aut(I)$ under composition of mappings. Similarly, $Aut(C)$ is a set of all automorphisms of CTMC C under composition of mappings.

In case of MDP M, if for all $s \in S$, for each $\mu \in Steps(s)$, there exists $\mu' \in Steps(\alpha(s))$ such that $\mu(s) = \mu'(\alpha(s))$, then α is an automorphism of M. Let $Aut(M) \leq Aut(I)$ be the set of all automorphisms of M under composition of mappings.

Note that the permutation α preserves the initial state and probabilistic transition relation of a particular model. Now consider a subgroup G of either of $Aut(D)$, $Aut(M)$ or $Aut(C)$. This G induces an orbit relation θ on the corresponding DTMC, MDP or CTMC respectively.

3.2 Quotient Probabilistic Models

Definition 4: The orbit relation for G is the set $\theta : \{(s, \alpha(s)) : s \in S, \alpha \in G\} \subseteq S \times S$. The orbit of s under G is the set $[s] = \{t | (s,t) \in \theta\}$. The elements of the orbit of s under G are said to be symmetric to each other.

The quotient model can be constructed by choosing a representative from each orbit. Suppose that we have a total ordering for S and let $min[s]$ denote the lexicographically smallest element of $[s]$ for any state s. Then for all $s \in S$, $min[s]$ can be chosen as a representative in the quotient state space $\overline{S}$.

Let $\overline{D} = (\overline{S}, \overline{s_0}, \overline{P})$, $\overline{M} = (\overline{S}, \overline{s_0}, \overline{Steps})$, and $\overline{C} = (\overline{S}, \overline{s_0}, \overline{R})$ be the quotient models corresponding to D, M and C respectively.

The common denominator among the different model types is a set of reduced states $\overline{S}$ and the corresponding initial state $\overline{s_0} \in \overline{S}$, which can be computed as follows: $\overline{S} : \{min[s] : s \in S\}$ and $\overline{s_0} = min[s_0]$.

In case of DTMC $\overline{D}$, $\overline{P}(min[s], min[t]) = \sum_{x \in [t]} P(min[s], x)$.

Similarly, for MDP $\overline{M}$, for each $min[s] \in \overline{S}$ and $\mu \in Steps(min[s])$, $\overline{Steps}(min[s])$ contains a distribution $\overline{\mu} \in Dist(\overline{S})$ where, for $min[\hat{s}] \in \overline{S}, \overline{\mu}(min[\hat{s}]) = \sum_{x \in [\hat{s}]} \mu(x)$.

For quotient CTMC $\overline{C}$, $\overline{R}(min[s], min[t]) = \sum_{x \in [t]} R(min[s], x)$.

The quotient model is probabilistic bi-simulation equivalent to the original unreduced model [13]. For the formulas in the temporal logics PCTL [9] which is preserved by symmetry, probabilistic model checking can be performed equivalently on the quotient model rather than the unreduced original model [13].

4 Extending Explicit PRISM with Symmetry Reduction

PRISM [10] is one of the most widely used probabilistic model checker in the literature. Though primarily symbolic, it has evolved over the years, and now offers, among other things, the following components for the different representation approaches:

- Symbolic: Model Builder, Symmetry Reduction Technique, Property Checker.
- Explicit: Model Builder, Property Checker.

We begin this section by justifying the need for exploration of symmetry reduction technique for explicitly represented models. We then present *ExplicitPRISMSymm*: our contribution that extends the functionality of PRISM's explicit engine by providing it with a mechanism for symmetry reduction.

4.1   PRISM's Existing Explicit Technique Vs. Symbolic Technique

A first step of quantitative analysis is to build a probabilistic model of a given system. In PRISM, user can choose whether to use symbolic or explicit representation of probabilistic models. As mentioned earlier, symbolic model uses MTBDD [7] structure for compact representation, whereas explicit model enumerate whole reachable state space.

Consider the "Consensus" case study from PRISM's benchmark suite [2]. Table 1 shows the results of quantitative analysis using both symbolic and explicit model. Here N represents the number of symmetric processes in the system. Next two columns give the time taken to build a symbolic and explicit model. Last two columns shows the time required to evaluate one particular quantitative property using both type of models.

From Table 1, we observe that explicit performs really well at model checking, but takes a lot of time for model construction. For verification of any system, building a model is one time process whereas model checking can be performed more than once. So explicit is a better choice. However, current explicit model construction in PRISM takes a lot of resources (both time and memory). For instance, for the given case study, it was not even possible to build an explicit model with more than 6 number of processes[1]. In fact, this is the primary reason for symbolic representation being preferred over explicit representation in the literature.

4.2   On-the-Fly Quotient Model Construction

We now discuss our primary contribution which addresses the problem of explicit model construction. Explicit model construction involves enumeration of the

Table 1. Comparison of Symbolic and Explicit model

Case Study	N	Model Build Time (in Seconds)		Model Checking Time (in Seconds)	
		Symbolic Model	Explicit Model	Symbolic Model	Explicit Model
Consensus Shared Coin Protocol	2	0.133	0.123	0.218	0.058
	4	0.047	0.704	46.734	2.812
	6	0.214	38.17	848.684	373.77

---

[1] The system configuration is given in experiment section.

entire state space. This invariably includes redundant exploration of symmetrical states in the case of concurrent system. This leads to unnecessary consumption of time and memory, thereby limiting the system's (with a given configuration) capability for further analysis. For instance, as mentioned in previous Sect. 4.1, our testbed was not able to analyze the "consensus" protocol with N greater than 6. However, this redundancy can be eliminated using "on-the-fly" approach, which applies symmetry reduction at the time of state space exploration.

Algorithm 1. Original State Space Exploration

// Initialization
$S := \{s_0\}$;
$T := \emptyset$;
$Explore := \{s_0\}$;

while $Explore \neq \emptyset$ **do**
  remove a state s from $Explore$;
  //compute transitions for states
  for each transition $s \rightarrow s'$ **do**
    if $s' \notin S$ **then**
      insert s' into S and $Explore$;
    end if
    //add transition according to type of probabilistic model
    insert $s \rightarrow s'$ into T;
  end for
end while

Algorithm 2. On-the-fly Quotient Model Construction

// Initialization
$\overline{S} := \{min[s_0]\}$;
$\overline{T} := \emptyset$;
$Explore := \{min[s_0]\}$;

while $Explore \neq \emptyset$ **do**
  remove a state s from $Explore$;
  //compute transitions for states
  for each transition $s \rightarrow \hat{s}$ **do**
    $s' = min[\hat{s}]$;
    if $s' \notin \overline{S}$ **then**
      insert s' into $\overline{S}$ and $Explore$;
    end if
    //add transition according to type of probabilistic model
    insert $s \rightarrow s'$ into $\overline{T}$;
  end for
end while

Algorithm 1 represents the original state space enumeration strategy, whereas Algorithm 2 shows our on-the-fly variant. In Algorithm 2, $\overline{S}$ and $\overline{T}$ is defined to store the reduced states and transitions respectively. $\overline{S}$ contains representative from each equivalence class of states. The representative state that is required for further transition computation is also stored in the set $Explore$. $\overline{T}$ stores the transitions of each representative state stored into $\overline{S}$.

In the beginning, $\overline{S}$ and $Explore$ are initialized with initial state s_0. A state s is extracted from $Explore$, and is explored by computing the number of transitions and successor state for each transition. Next, the representative for each successor state of s is computed (discussed later). If the representative of successor state has not been encountered before, then it is added to $\overline{S}$ and $Explore$. The corresponding transitions of the explored state are added into $\overline{T}$. The next state from the set $Explore$ is then extracted, and the process is repeated. Once the $Explore$ set is empty, the state space would have been fully explored, and $\overline{S}$ and $\overline{T}$ will contain the states and transitions of quotient model.

Algorithm 3. Compute representative of a given state	**Example**

// Initialization
$N :=$ Total number of components;
$s := (glb, l_1, l_2, \ldots, l_n)$; //State representation
$N_{glb} :=$ Number of global variables;
	Input : $s = (6, 2, 3, 1, 4, 8)$;
	$//(glb, l_1, l_2, l_3, l_4, l_5)$
	$N = 5$;
	$N_{glb} = 1$; //global variable: 6

// User Defined Parameters
$N_{bfs} :=$ Number of non-symmetric components before symmetric components;
$N_{afs} :=$ Number of non-symmetric components after symmetric components;
 	
$N_{symm} := N - (N_{bfs} + N_{afs})$; //Number of symmetric components;
$rep := ()$; //Representative state
	$N_{bfs} = 1$; //component $l_1 = 2$
	$N_{afs} = 1$; //component $l_5 = 8$
	
	$N_{symm} = 3$; //Three symmetric components $l_2, l_3, l_4 = (3, 1, 4)$
	$rep = ()$;

//Extract global variables - if any
if $N_{glb} > 0$ then
   $add\_to\_rep(rep, glb)$;
end if
	if $N_{glb} > 0$ then
	   $add\_to\_rep(rep, 6)$ // $rep = (6)$
	end if

//If non-symmetric components before symmetric components
if $N_{bfs} > 0$ then
   //Copy non-symmetric components as it is into representative state
   for $i = 1$ to N_{bfs} do
     $add\_to\_rep(rep, l_i)$;
   end for
end if
	if $N_{bfs} > 0$ then
	   $add\_to\_rep(rep, l_1)$ //$rep = (6, 2)$
	end if

//Sort symmetric components
if $N_{symm} > 0$ then
   $start\_ptr := N_{bfs} + 1$;
   $end\_ptr := start\_ptr + N_{symm} - 1$;
   $sorted\_symm\_components =$
     $sort(l_{start\_ptr}, \ldots, l_{end\_ptr})$;
   $add\_to\_rep(rep, sorted\_symm\_components)$;
end if
	if $N_{symm} > 0$ then
	   $start\_ptr := 2$;
	   $end\_ptr := 4$;
	   $sorted\_symm\_components =$
	     $sort(l_2, \ldots, l_4)$;
	
	   $//sorted\_symm\_components =$
	   $(1, 3, 4)$
	   $add\_to\_rep(rep,$
	   $sorted\_symm\_components)$;
	   // $rep = (6, 2, 1, 3, 4)$
	end if

//If non-symmetric components after symmetric components
if $N_{afs} > 0$ then
   $start\_ptr := N_{bfs} + N_{symm} + 1$
   //Copy non-symmetric components as it is into representative state
   for $i = 1$ to N_{afs} do
     $add\_to\_rep(rep, l_{start\_ptr})$;
     $start\_ptr = start\_ptr + 1$;
   end for
end if
	if $N_{afs} > 0$ then
	   $start\_ptr := 5$
	   $add\_to\_rep(rep, l_5)$;
	   $//rep = (6, 2, 1, 3, 4, 8)$
	end if

//Return representative state return(rep);	return rep; $//rep = (6, 2, 1, 3, 4, 8)$;

4.3   Representative Computation

As far as symmetry reduction strategy is concerned, *representative computation* method should be implementable in efficient way.

Algorithm 3 shows the method to compute the representative of a given state. Let $I = \{1, 2, \ldots, n\}$ be the set of component identifiers. A state $s \in S$ has the representation $s = (glb, l_1, l_2, \ldots, l_n)$, where l_i denotes the local state of component i and glb indicates global variables if any. The usual lexicographical ordering of vectors provides a total ordering on S. From each equivalence class of states, we have chosen lexicographically smaller state as a representative of that equivalence class. Kindly refer to an example given beside to Algorithm 3 for a sample dry run of the algorithm.

Once a state is generated, we check if the state consists of any global variables and components which are not symmetric. In the standard explicit representation of a state, non-symmetric components can be defined either before and/or after symmetrical components. But, all symmetrical components must be represented in consecutive manner. In PRISM, as symmetry reduction parameters, we can specify the number of non-symmetric components that appear before and after the symmetric components. To compute the representative, we sort the symmetrical components. This sorted state is lexicographically smaller and is considered as a representative. In the representative state, global variables and non-symmetrical components will be as it is.

Note that on-the-fly symmetry reduction algorithm bypasses the major limitation of explicit models - the requirement to enumerate all reachable states and transitions. The algorithm computes the representative of a state at the time of exploration itself. So those states, whose representative is already explored, need not be stored. This frees up the storage space. Also, such states are not even explored, thereby saving time. Thus using on-the-fly symmetry reduction, we can build an explicit model with a large number of states (the exact figure depends on the model, the degree of symmetry within, and the hardware configuration). In the next section, we present our experimental results for the two different case studies.

5   Experimental Results

In this section we present experimental results for two different case studies - "the randomized consensus shared coin protocol" from [11] and "IEEE 802.3 CSMA/CD protocol" analyzed in [12]. Both the case studies are PRISM benchmarks and available in PRISM distribution [2]. All experiments were executed on a 2.4 GHz Intel Zeon quad core processor, with 8 GB RAM running Linux (Ubuntu 14.0).

Table 2 shows the comparison between full (original) and symmetry reduced quotient explicit model. For two case studies, N denotes the number of symmetric processes in the system. Columns 3–4 gives the state space size of original and symmetry reduced explicit model. As expected, using symmetry reduction we obtained a large reduction in state space size and it increases with N. Next two

Table 2. Experimental Results of Original and Quotient Explicit Model

Case Study	N	Model Size (States)		Explicit Model Build Time (in Seconds)		Model Checking Time (in Seconds)	
		In Full Model	In Quotient Model	Full Explicit Model	Quotient Explicit Model (Proposed)	Full Explicit Model	Quotient Explicit Model (Proposed)
Consensus Shared Coin Protocol	2	272	154	0.123	0.068	0.058	0.048
	4	22656	2e + 3	0.704	0.562	2.812	0.371
	6	1.2e + 6	1e + 4	38.17	2.134	373.77	3.856
	8	6.1e + 7	4e + 4	memory out	6.379	-	28.483
	10	2.8e + 9	1.3e + 5	memory out	21.568	-	155.864
	12	1.2e + 11	3.3e + 5	memory out	59.619	-	600.353
	14	5.04e + 12	7.4e + 5	memory out	224.492	-	2006.925
	16	2.08e + 14	1.4e + 6	memory out	1637.064	-	5883.866
IEEE CSMA/CD Protocol	4	7.6e + 5	4e + 4	16.72	3.013	71.32	3.639
	5	1.5e + 7	1.8e + 5	memory out	11.32	-	23.285
	6	2.7e + 8	7e + 5	memory out	51.206	-	115.385
	7	4.6e + 9	2.2e + 6	memory out	211.659	-	481.434
	8	7.7e + 10	6.6e + 6	memory out	3571.901	-	3453.867

columns show the time taken for building the each model. Last two columns shows the time required to performing probabilistic model checking on each of the two models. In "Consensus" and "CSMA", full explicit model cannot be built with more than 6 and 4 number of processes respectively. However, using on-the-fly symmetry reduction, we are able to build a quotient model with 16 and 8 number of processes for "Consensus" and "CSMA" respectively. Thus, the on-the-fly symmetry reduction technique enables verification of systems with larger number of processes.

We evaluated our technique's performance against PRISM's built-in symbolic technique. For quantitative evaluation of given property, PRISM has four different engines i.e., MTBDD, Explicit and Hybrid. Detail of these engines is given at [2].

Experiment results for comparison between PRISM's built-in symbolic and our on-the-fly symmetry reduction technique are shown in Table 3. Columns 3–4 give the time required to build the quotient model using PRISM's built-in symbolic and our proposed on-the-fly symmetry reduction technique. Here we can see that time to build a model with explicit is more as compared to symbolic technique.

Table 3. Experimental Results of Quotient Symbolic and Quotient Explicit Model

Case Study	N	Model Build Time (in Seconds)		Model Checking Time (in Seconds)			Total Time (in Seconds)		
		Quotient Symbolic	Quotient Explicit (Proposed)	MTBDD	Hybrid	Explicit	MTBDD	Hybrid	Explicit
Consensus Shared Coin Protocol	2	0.037	0.068	0.17	0.02	0.048	0.207	0.057	**0.116**
	4	0.071	0.562	17.259	0.918	0.371	17.33	0.989	**0.933**
	6	0.251	2.134	276.058	12.829	3.856	276.309	13.068	**5.99**
	8	0.628	6.379	1713.06	90.857	28.483	1713.688	91.46	**34.862**
	10	1.306	21.568	7654.605	392.825	155.864	7655.911	394.108	**177.432**
	12	3.476	59.619	mem-out	1449.134	600.353	-	1452.61	**659.972**
	14	6.648	224.492	mem-out	4455.717	2006.925	-	4462.365	**2231.417**
	16	11.983	1637.064	mem-out	11438.624	5883.866	-	11450.607	**7520.93**
IEEE CSMA/ CD Protocol	4	3.307	3.013	71.432	12.414	3.639	74.739	15.721	**6.652**
	5	16.434	11.32	671.374	64.328	23.285	687.808	80.762	**34.605**
	6	49.99	51.206	4750.193	322.208	115.385	4800.183	372.198	**166.591**
	7	132.508	211.659	mem-out	1724.687	481.434	-	1857.195	**693.093**
	8	376.231	3571.901	mem-out	7093.89	3453.867	-	7470.121	**7025.768**

A comparison of symbolic and explicit in terms of probabilistic model checking is interesting. Time required for performing model checking using MTBDD, Hybrid and Explicit engine is given in columns 5–7. For model checking, we have evaluated first and the first two quantitative properties in the "Consensus" and "CSMA" case studies respectively. Model checking using explicit is much faster than MTBDD and Hybrid engine, despite the fact that symmetry reduced model is same for all. In the last three columns we have given a total time (quotient model build + model checking) required by MTBDD, Hybrid and Explicit engine. The model construction time is same as symbolic model for property evaluation using MTBDD and Hybrid engine.

As is observed, the explicit engine outperforms the other two, when it is used in conjunction with the symmetry reduction technique. One should also note here, that for verification of any system, complete builds are typically one-time process, whereas model checking operations can be performed multiple times. So we can conclude that, with increasing number of model checking operations, the explicit model is bound to perform better.

6   Conclusion

We reported significant time gains achieved in property checking by using explicit representation instead of symbolic representation. Our on-the-fly symmetry reduction technique overcomes the major hurdle of explicit model construction thereby expanding the usage of explicit representation beyond trivial cases and into the realm of highly symmetric concurrent systems. We integrated this technique in

PRISM, thus extended the functionality of PRISM's explicit engine. Our encouraging results support in establishing explicit model representations as a significant representation mechanism in probabilistic model checking.

References

1. Barner, S., Grumberg, O.: Combining symmetry reduction and under-approximation for symbolic model checking. In: Brinksma, E., Larsen, K.G. (eds.) CAV 2002. LNCS, vol. 2404, pp. 93–106. Springer, Heidelberg (2002)
2. Kwiatkowska, M., Norman, G., Parker, D.: PRISM: probabilistic symbolic model checker. In: Field, T., Harrison, P.G., Bradley, J., Harder, U. (eds.) TOOLS 2002. LNCS, vol. 2324, pp. 200–204. Springer, Heidelberg (2002). http://www.prismmodelchecker.org/casestudies/index.php
3. Donaldson, A., Miller, A., Parker, D.: GRIP: generic representatives in PRISM. In: Proceeding of the Fourth International Conference on the Quantitative Evaluation of Systems (QEST 2007), pp. 115–116, September 2007
4. Donaldson, A., Miller, A., Parker, D.: Language-level symmetry reduction for probabilistic model checking. In: Proceeding of the Sixth International Conference on the Quantitative Evaluation of Systems (QEST 2009), pp. 289–298, September 2009
5. Donaldson, A.F., Miller, A.: Symmetry reduction for probabilistic model checking using generic representatives. In: Graf, S., Zhang, W. (eds.) ATVA 2006. LNCS, vol. 4218, pp. 9–23. Springer, Heidelberg (2006)
6. Emerson, E.A., Wahl, T.: Dynamic symmetry reduction. In: Halbwachs, N., Zuck, L.D. (eds.) TACAS 2005. LNCS, vol. 3440, pp. 382–396. Springer, Heidelberg (2005)
7. Fujita, M., McGeer, P.C., Yang, J.C.Y.: Multi-terminal binary decision diagrams: an efficient datastructure for matrix representation. Formal Methods Syst. Des. 10(2–3), 149–169 (1997)
8. Gyuris, V., Prasad Sistla, A.: On-the-fly model checking under fairness that exploits symmetry. In: Grumberg, O. (ed.) Computer Aided Verification. LNCS, vol. 1254, pp. 232–243. Springer, Heidelberg (1997)
9. Hansson, H., Jonsson, B.: A logic for reasoning about time and reliability. Formal Aspects Comput. 6(5), 512–535 (1994)
10. Hinton, A., Kwiatkowska, M., Norman, G., Parker, D.: PRISM: a tool for automatic verification of probabilistic systems. In: Hermanns, H., Palsberg, J. (eds.) TACAS 2006. LNCS, vol. 3920, pp. 441–444. Springer, Heidelberg (2006)
11. Kwiatkowska, M., Norman, G., Segala, R.: Automated verification of a randomized distributed consensus protocol using cadence SMV and PRISM. In: Berry, G., Comon, H., Finkel, A. (eds.) CAV 2001. LNCS, vol. 2102, pp. 194–206. Springer, Heidelberg (2001)
12. Kwiatkowska, M., Norman, G., Sproston, J., Wang, F.: Symbolic model checking for probabilistic timed automata. Inf. Comput. 205(7), 1027–1077 (2007)
13. Kwiatkowska, M., Norman, G., Parker, D.: Symmetry reduction for probabilistic model checking. In: Ball, T., Jones, R.B. (eds.) CAV 2006. LNCS, vol. 4144, pp. 234–248. Springer, Heidelberg (2006)
14. Miller, A., Donaldson, A., Calder, M.: Symmetry in temporal logic model checking. ACM Comput. Surv. 38(3), 8 (2006)

15. Parker, D.: Implementation of symbolic model checking for probabilistic systems. Ph.d. thesis, University of Birmingham (2002)
16. Power, C.: Probabilistic symmetry reduction. Ph.d. thesis, University of Glasgow (2012). http://theses.gla.ac.uk/3493/
17. Sistla, A.: Employing symmetry reductions in model checking. Comput. Lang. Syst. & Struct. **30**(3–4), 99–137 (2004)
18. Wahl, T., Donaldson, A.: Replication and abstraction: symmetry in automated formal verification. Symmetry **2**(2), 799–847 (2010)

Parameterised Complexity

Kernelization Algorithms for Packing Problems Allowing Overlaps

Henning Fernau[1], Alejandro López-Ortiz[2], and Jazmín Romero[2]([✉])

[1] FB 4-Abteilung Informatikwissenschaften,
Universität Trier, Trier, Germany
[2] David R. Cheriton School of Computer Science,
University of Waterloo, Waterloo, Canada
hjromero@uwaterloo.ca

Abstract. We consider the problem of discovering overlapping communities in networks which we model as generalizations of the Set and Graph Packing problems with overlap. As usual for Set Packing problems we seek a collection $\mathcal{S}' \subseteq \mathcal{S}$ consisting of at least k sets subject to certain disjointness restrictions. In the r-Set Packing with t-Membership, each element of $\mathcal{U}$ belongs to at most t sets of $\mathcal{S}'$ while in r-Set Packing with t-Overlap each pair of sets in $\mathcal{S}'$ overlaps in at most t elements. For both problems, each set of $\mathcal{S}$ has at most r elements.

Similarly, both of our graph packing problems seek a collection $\mathcal{K}$ of at least k subgraphs in a graph G each isomorphic to a graph $H \in \mathcal{H}$ where each member of $\mathcal{H}$ has at most r vertices. In $\mathcal{H}$-Packing with t-Membership, each vertex of G belongs to at most t subgraphs of $\mathcal{K}$ while in $\mathcal{H}$-Packing with t-Overlap each pair of subgraphs in $\mathcal{K}$ overlaps in at most t vertices.

Here, we show NP-Completeness results for all of our packing problems. Furthermore, we give a dichotomy result for $\mathcal{H}$-Packing with t- Membership analogous to the Kirkpatrick and Hell [12]. Given this intractability, we reduce r-Set Packing with t-Membership and t-Overlap to problem kernels with $O((r+1)^r k^r)$ and $O(r^r k^{r-t-1})$ elements, respectively. Similarly, we reduce $\mathcal{H}$-Packing with t-Membership and t-Overlap to instances with $O((r+1)^r k^r)$ and $O(r^r k^{r-t-1})$ vertices, respectively. In all cases, k is the input parameter while t and r are constants.

1 Introduction

Networks are commonly used to model complex systems that arise in real life, for example, social and protein-interaction networks. A *community* emerges in a network when two or more entities have common interests, e.g., groups of people or related proteins. Naturally, a given person can have more than one social circle, and a protein can belong to more than one protein complex. Thus, communities can share members [15]. The way a community is modeled is highly dependent on the application being modeled by the network. One flexible approach is to use a family of graphs where each graph would be a community model.

© Springer International Publishing Switzerland 2015
R. Jain et al. (Eds.): TAMC 2015, LNCS 9076, pp. 415–427, 2015.
DOI: 10.1007/978-3-319-17142-5_35

The $\mathcal{H}$-Packing with t-Overlap problem [17] captures the problem of discovering overlapping communities. The goal is to find at least k subgraphs (the communities) in a graph G (the network) where each subgraph is isomorphic to a member of a family of graphs $\mathcal{H}$ (the community models) and each pair of subgraphs overlaps in at most t vertices (the shared members). Such type of overlap can be found in certain clustering algorithms as, e.g., in [2]. Here, we also consider the $\mathcal{H}$-Packing with t-Membership problem to bound the number of communities that a member of a network can belong to (instead of bounding the shared members). In any case, each member of $\mathcal{H}$ has at most r vertices. This type of overlap was also previously studied, for instance, in [5] in the context of graph editing. Our graph packing problems generalize the H-Packing problem which consists in finding within G at least k vertex-disjoint subgraphs isomorphic to a graph H.

Similarly, we also consider overlap for the r-Set Packing problem. Given a collection of sets $\mathcal{S}$ each of size at most r drawn from a universe $\mathcal{U}$, an r-Set Packing with t-Overlap consists of at least k sets from $\mathcal{S}$ such that each pair of sets overlaps in at most t elements. In contrast, in an r-Set Packing with t-Membership, each member of $\mathcal{U}$ is contained in at most t of the k sets. The pair $\mathcal{S}$ and $\mathcal{U}$ can be treated as an hypergraph, where the vertices are the members of $\mathcal{U}$ and the hyper-edges are the members of $\mathcal{S}$. Thus, an r-Set Packing with t-Overlap can be seen as k hyper-edges that pairwise intersect in no more than t vertices while an r-Set Packing with t-Membership can be interpreted as k hyper-edges where every vertex is contained in at most t of them.

Some of our generalized problems are NP-complete. This follows immediately from the NP-completeness of the classical H-Packing and Set Packing problems. Our goal is to design *kernelization algorithms*; that is, algorithms that in polynomial time reduce any instance to a size bounded by $f(k)$ (a *problem kernel*), where f is some arbitrary computable function depending only on a parameter. After that a brute-force search on the kernel gives a solution in $g(k)n^{O(1)}$ running time, where g is some computable function. An algorithm that runs in that time is a *fixed-parameter algorithm*. For all our problems, we consider k as the parameter, while t and r are constants.

Related Results. An FPT-algorithm for the $\mathcal{H}$-Packing with t-Overlap problem was developed in [18]. The running time is $O(r^{rk}k^{(r-t-1)k+2}n^r)$, where $r = |V(H)|$ for an arbitrary graph $H \in \mathcal{H}$, $|\mathcal{H}| = 1$, and $0 \leq t < r$. A $2(rk-r)$ kernel when $\{K_r\} = \mathcal{H}$ and $t = r - 2$ is given in [17].

The smallest kernel for the H-Packing problem has size $O(k^{r-1})$, where H is an arbitrary graph and $r = |V(H)|$ [14]. More kernels results when H is a prescribed graph can be found in [7,9,11,16]. There is an $O(r^{rk}k^{(r-t-1)k+2}n^r)$ algorithm for the Set Packing with t-Overlap [18] while the Set Packing (element-disjoint) has an $O(r^r k^{r-1})$ kernel [1].

Summary of Results. Each member in $\mathcal{S}$ has at most r elements while each $H \in \mathcal{H}$ has at most r vertices. For the problems with t-Membership, $t \geq 1$ and for t-Overlap problems, $0 \leq t \leq r - 1$. In any case, t and r are constants.

In Sect. 3, we show that the $\mathcal{H}$-Packing with t-Membership problem is NP-Complete for all values of $t \geq 1$ when $\mathcal{H} = \{H\}$ and H is an arbitrary connected graph with at least three vertices, but polynomial-time solvable for smaller graphs. Hence, we obtain a dichotomy result for the $\{H\}$-Packing with t-Membership which is analogous to the one of Kirkpatrick and Hell [12]. Moreover, we prove that for any $t \geq 0$, there always exists a connected graph H_t such that the $\mathcal{H}$-Packing problem with t-Overlap is NP-Complete where $\mathcal{H} = \{H_t\}$.

In Sect. 4, we give a polynomial parameter transformation (PPT) from r-Set Packing with t-Membership to an instance of $(r+1)$-Set Packing. With this transformation, we reduce r-Set Packing with t-Membership to a problem kernel with $O((r + 1)^r k^r)$ elements. In addition, we obtain a kernel with $O((r + 1)^r k^r)$ vertices for $\mathcal{H}$-Packing with t-Membership by reducing it to r-Set Packing with t-Membership. PPTs are commonly used to show kernelization lower bounds [3]. To our knowledge, this is the first time that PPTs are used to obtain kernels.

Inspired by [6,14], in Sect. 5, we give a kernelization algorithm that reduces r-Set Packing with t-Overlap to a kernel with $O(r^r k^{r-t-1})$ elements. We also achieve a kernel with $O(r^r k^{r-t-1})$ vertices for $\mathcal{H}$-Packing with t-Overlap.

Due to the lack of space, technical details as well as additional complexity and kernelization results, can be found in [8].

2   Terminology

Let $\mathcal{S}$ be a collection of sets drawn from $\mathcal{U}$. For $\mathcal{S}' \subseteq \mathcal{S}$, $val(\mathcal{S}')$ denotes the union of all members of $\mathcal{S}'$. For $P \subseteq \mathcal{U}$, P is *contained* in $S \in \mathcal{S}$, if $P \subseteq S$. Two sets $S, S' \in \mathcal{S}$ *overlap* in $|S \cap S'|$ elements and they *conflict* if $|S \cap S'| \geq t + 1$.

Our graph problems deal with a family of graphs $\mathcal{H}$ where each $H \in \mathcal{H}$ is an arbitrary graph. Let $r(\mathcal{H}) = \max\{|V(H)| : H \in \mathcal{H}\}$ denote the order of the biggest graph in $\mathcal{H}$. Observe that to have this as a reasonable notion, we always (implicitly) require $\mathcal{H}$ to be a finite set. We simply write r instead of $r(\mathcal{H})$ if $\mathcal{H}$ is clear from the context. A subgraph of G that is isomorphic to some $H \in \mathcal{H}$ is called an $\mathcal{H}$-*subgraph*. We denote as $\mathcal{H}_G$ the set of all $\mathcal{H}$-subgraphs in G.

We next introduce the formal definitions of the t-*Membership problems*. Let $r, t \geq 1$ be fixed, that in actual fact defines a whole family of problems.

The r-Set Packing with t-Membership problem

Input: A collection $\mathcal{S}$ of distinct sets, each of size at most r, drawn from a universe $\mathcal{U}$ of size n, and a non-negative integer k.

Parameter: k

Question: Does $\mathcal{S}$ contain a (k, r, t)-*set membership*, i.e., at least k sets $\mathcal{K} = \{S_1, \ldots, S_k\}$ where each element of $\mathcal{U}$ is in at most t sets of $\mathcal{K}$?

The $\mathcal{H}$-Packing with t-Membership problem.
Input: A graph G, and a non-negative integer k.
Parameter: k
Question: Does G contain a (k, r, t)-$\mathcal{H}$-*membership*, i.e., a set of at least k $\mathcal{H}$-subgraphs $\mathcal{K} = \{H_1, \ldots, H_k\}$, where $V(H_i) \neq V(H_j)$ for $i \neq j$, and every vertex in $V(G)$ is contained in at most t subgraphs of $\mathcal{K}$?

Our t-Overlap problems are defined next. Again, $r \geq 1$ and $t \geq 0$ are fixed.

The r-Set Packing with t-Overlap problem
Instance: A collection $\mathcal{S}$ of distinct sets, each of size at most r, drawn from a universe $\mathcal{U}$ of size n, and a non-negative integer k.
Parameter: k.
Question: Does $\mathcal{S}$ contain a (k, r, t)-set packing, i.e., a collection of at least k sets $\mathcal{K} = \{S_1, \ldots, S_k\}$ where $|S_i \cap S_j| \leq t$, for any pair S_i, S_j with $i \neq j$?

The $\mathcal{H}$-Packing with t-Overlap problem
Input: A graph G, and a non-negative integer k.
Parameter: k
Question: Does G contain a (k, r, t)-$\mathcal{H}$-packing, i.e., a set of at least k $\mathcal{H}$-subgraphs $\mathcal{K} = \{H_1, \ldots, H_k\}$ where $|V(H_i) \cap V(H_j)| \leq t$ for any pair H_i, H_j with $i \neq j$?

Notice that when $t = 1$ and $t = 0$ for t-Membership and t-Overlap, respectively, we are back to the classical r-Set Packing and $\mathcal{H}$-Packing problems.

Sometimes the size l will be dropped from the size of a packing, e.g., an (l, r, t)-set membership can be simply denoted as (r, t)-set membership. An (r, t)-$\mathcal{H}$-membership P of G is *maximal* if every $\mathcal{H}$-subgraph of G that is not in P has at least one vertex v contained in t $\mathcal{H}$-subgraphs of P. Similarly, an (r, t)-set packing $\mathcal{M}$ of $\mathcal{S}' \subseteq \mathcal{S}$ is *maximal* if any set of $\mathcal{S}'$ that is not already in $\mathcal{M}$ conflicts with some set in $\mathcal{M}$. That is, for each set $S \in \mathcal{S}'$ where $S \notin \mathcal{M}$, $|S \cap S'| \geq t + 1$ for some $S' \in \mathcal{M}$.

3   Hardness of Packing Problems Allowing Overlaps

Let us first present one concrete $\mathcal{H}$-Packing with t-Membership problem that is hard for all possible values of t.

Theorem 1. *For all $t \geq 1$, the $\{P_3\}$-Packing with t-Membership problem is NP-complete.*

Proof (Sketch). We show a reduction from $\{P_3\}$-Packing with t-Membership to $\{P_3\}$-Packing with $(t + 1)$-Membership, which proves the claim by induction. Without loss of generality, assume that n is divisible by $t + 1$; otherwise, simply add some isolated vertices to obtain an equivalent instance.

Let (G, k) be an instance of the $\{P_3\}$-Packing with t-Membership problem, where $V = \{v_0, \dots, v_{n-1}\}$. Let $U = \{u_0, \dots, u_{(2n)/(t+1)-1}\}$ be a set of new vertices. Create a graph $G' = (V', E')$ as follows: $V' = V \cup U$ and $E' = E \cup \{v_{i+j}u_{2i/(t+1)}, u_{2i/(t+1)}u_{2i/(t+1)+1} \mid 0 \leq i < n, i \pmod{t+1} \equiv 0, 0 \leq j \leq t\}$.

G has a $(k, 3, t)$-$\{P_3\}$-membership if and only if G' has a $(n+k, 3, t+1)$-$\{P_3\}$-membership, where $n = |V|$. If P is a $(3, t)$-$\{P_3\}$-membership set for G, then $P' = P \cup X$ is a $(3, t+1)$-$\{P_3\}$-membership for G', where $X = \{v_{i+j}u_{2i/(t+1)}u_{2i/(t+1)+1} \mid 0 \leq i < n, i \pmod{t+1} \equiv 0, 0 \leq j \leq t\}$. Clearly, $|P'| = |P| + |X| = |P| + n$.

Conversely, let P' be a $(3, t+1)$-$\{P_3\}$-membership for $G' = (V', E')$ of size at least $n + k$. We gradually modify P' in two separate steps, in both of them we always preserve its size and the property that is a $(3, t+1)$-$\{P_3\}$-membership for G'. First, we can always *alter* P' by replacing some paths to end up with P' containing only two types of paths: paths completely consisting of vertices from G (P_G), and those containing exactly one vertex from G and two from U (P_X). After that, we can gradually alter P' once more to force that every vertex in G is contained in exactly one path of P_X. This implies that $P_G = P' \backslash P_X$ would be a $(3, t)$-$\{P_3\}$-membership G of size $|P_G| = |P'| - |P_X| \geq n + k - |P_X| \geq k$. $\square$

A similar reduction can be constructed for each $\{H\}$-Packing with t-Membership problem for each H containing at least three vertices based on classical results due to Kirkpatrick and Hell [12].

On the positive side, we next show that the $\{P_2\}$-Packing with t-Membership problem can be solved in polynomial time. A $(k, 2, t)$-$\{P_2\}$-membership P of a graph G is a subset of at least k edges of $E(G)$ such that every vertex of G is contained in at most t edges of P. Let us denote as G^* the subgraph of G induced by P, i.e., $G^* = (V_{G^*}, P)$ where $V_{G^*} = V(P)$.

Lemma 1. *The graph G^* has maximum degree t.*

Let $b : V(G) \to \mathbb{N}$ a degree constraint for every vertex. The problem of finding a subgraph G^* of G such that each vertex $v \in V(G^*)$ has degree at most $b(v)$ in G^* and the number of edges in G^* is maximized is known as the *degree-constrained subgraph problem* [19], while G^* is called a *degree-constrained subgraph*. Shiloach [19] constructs a graph G' from a given graph G and shows that G has a degree-constrained subgraph with k edges if and only if G' has a maximum matching of size $|E(G)| + k$.

By Lemma 1, we can find a $(k, 2, t)$-$\{P_2\}$-membership P of G, by solving the degree-constrained subgraph problem with $b(v) = t$, for all $v \in V(G)$. Having a closer look at Shiloach's construction of G', we observe that $|V(G')| = 2|E(G)| + t|V(G)|$ and that $|E(G')| = 2t|E(G)| + t|E(G)| = 3t|E(G)|$. Thus, the maximum matching can be solved in $O(\sqrt{2|E(G)| + t|V(G)|} \, 3t|E(G)|)$ by running Micali and Vazirani's algorithm [13]. Hence, we can state:

Corollary 1. *Let $t \geq 1$. P_2-Packing with t-Membership can be solved in time that is polynomial both in the size of the input graph G and in t.*

We can summarize our comments following Theorem 1 and Corollary 1 by stating the following dichotomy result that is analogous to the classical one due to Kirkpatrick and Hell [12].

Theorem 2. *(Dichotomy Theorem) Let $t \geq 1$. Assuming that P is not equal to NP, then the $\{H\}$-Packing with t-Membership problem can be solved in polynomial time if and only if $|V(H)| \leq 2$.*

On the other hand, the next theorem explains at least that there are NP-hard $\{H\}$-Packing with t-Overlap problems for each level $t \geq 0$.

Theorem 3. *For any $t \geq 0$, there exists a connected graph H_t such that the $\{H_t\}$-Packing with t-Overlap problem is NP-complete.*

4 Packing Problems with Bounded Membership

In this section, we introduce a polynomial parametric transformation (PPT) from the r-Set Packing with t-Membership problem to the $(r + 1)$-Set Packing problem. We obtain a kernel result for r-Set Packing with t-Membership by running a kernelization algorithm on the transformed instance of the $(r + 1)$-Set Packing problem. This compression (or bikernel) result can be turned into a proper kernel result by re-interpreting the $(r + 1)$-Set Packing kernel within the original problem.

4.1 Packing Sets with t-Membership

We create an instance for the $(r + 1)$-Set Packing problem (a universe $\mathcal{U}^T$ and a collection $\mathcal{S}^T$) using an instance of the r-Set Packing with t-Membership problem.

Transformation 1. The universe $\mathcal{U}^T$ equals $(\mathcal{U} \times \{1, \ldots, t\}) \cup \mathcal{S}$.

The collection $\mathcal{S}^T$ contains all subsets of $\mathcal{U}^T$ each with at most $r + 1$ elements of the following form: $\{\{(u_1, j_1), \ldots, (u_i, j_i), \ldots, (u_{r'}, j_{r'}), S\} \mid S \in \mathcal{S}, S = \{u_1, \ldots, u_{r'}\}$, for each $1 \leq j_i \leq t$ and $1 \leq i \leq r'$, where $r' \leq r\}$.

The size of $\mathcal{U}^T$ is bounded by $|\mathcal{U}| \cdot t + |\mathcal{S}| < tn + rn^r = O(n^r)$. Each set in $\mathcal{S}^T$ has size at most $r + 1$. For each $S \in \mathcal{S}$, we can form at most t^r sets with the tr ordered pairs from the elements in S. In this way, for each $S \in \mathcal{S}$ there are t^r sets in $\mathcal{S}^T$, and $|\mathcal{S}^T| = t^r |\mathcal{S}| = O(t^r n^r)$. This leads us to the following result.

Lemma 2. *Transformation 1 can be computed in $O(t^r n^r)$ time.*

Note that the parameter k stays the same in this transformation, and t only influences the running time of the whole construction, as the $(r + 1)$-Set Packing instance will grow if t gets bigger.

Lemma 3. *$\mathcal{S}$ has a (k, r, t)-set membership if and only if $\mathcal{S}^T$ has k disjoint sets (i.e., a $(k, r + 1, 0)$-set packing).*

Then, we run the currently best kernelization algorithm for the $(r + 1)$-Set Packing problem [1]. This algorithm would leave us with a new universe $\mathcal{U}'^T$ with at most $2r!((r + 1)k - 1)^r$ elements, as well with a collection $\mathcal{S}'^T$ of subsets. The following property is borrowed from [1].

Lemma 4. $\mathcal{S}^T$ has a $(k, r+1, 0)$-set packing if and only if $\mathcal{S}'^T$ has a $(k, r+1, 0)$-set packing.

Next, we construct the reduced $\mathcal{U}'$ and $\mathcal{S}'$ which together with k, give the reduced r-Set Packing with t-Membership instance we are looking for. To this end, the reduced universe $\mathcal{U}'$ contains each element u appearing in $\mathcal{U}'^T$. In each set of $\mathcal{S}'^T$, there is an element S that will correspond to a set of $\mathcal{S}'$. By our construction, $\mathcal{U}'$ will have at most $2r!((r+1)k-1)^r$ elements. This reduction property allows us to state:

Theorem 4. The r-Set Packing with t-Membership has a problem kernel with $O((r+1)^r k^r)$ elements from the given universe.

4.2 Packing Graphs with t-Membership

To reduce $\mathcal{H}$-Packing with t-Membership to a kernel, we will transform it to the r-Set Packing with t-Membership. Note that in G there could exist more than one $\mathcal{H}$-subgraph with the same set of vertices (but different set of edges). However, we claim next that only one of those $\mathcal{H}$-subgraphs can be in a solution.

Lemma 5. Let H_i and H_j be a pair of $\mathcal{H}$-subgraphs in G such that $V(H_i) = V(H_j)$ but $E(H_i) \neq E(H_j)$. Any (k, r, t)-$\mathcal{H}$-membership of G that contains H_i does not contain H_j (and vice versa). Furthermore, we can replace H_i by H_j in such a membership.

We denote as $\mathcal{H}_G$ the set of all $\mathcal{H}$-subgraphs in G; thus, $|\mathcal{H}_G| = O(|\mathcal{H}|n^{(r(\mathcal{H}))^2})$. We can find a (k, r, t)-$\mathcal{H}$-membership from G by selecting k $\mathcal{H}$-subgraphs from $\mathcal{H}_G$ such that every vertex of $V(G)$ is contained in at most t of those subgraphs. By Lemma 5, we can apply the following reduction rule to $\mathcal{H}_G$. After applying this rule, $|\mathcal{H}_G| = O(|\mathcal{H}|n^{r(\mathcal{H})})$.

Reduction Rule 1. For any pair of $\mathcal{H}$-subgraphs H_1, H_2 in $\mathcal{H}_G$ such that $V(H_1) = V(H_2)$, we arbitrary select one and remove the other from $\mathcal{H}_G$.

Next, we construct an instance for the r-Set Packing with t-Membership as follows.

Transformation 2. The universe $\mathcal{U}$ equals $V(G)$.
   There is a set in $\mathcal{S}$ for each $\mathcal{H}$-subgraph H in $\mathcal{H}_G$ and $S = V(H)$.
   Furthermore, let $r = r(\mathcal{H})$.

In this way, $|\mathcal{U}| = O(n)$ and $|\mathcal{S}| = |\mathcal{H}_G| = O(|\mathcal{H}|n^{r(\mathcal{H})})$. Each set in $\mathcal{S}$ has size at most $r(\mathcal{H})$.

Lemma 6. G has a (k, r, t)-$\mathcal{H}$-membership if and only if $\mathcal{S}$ has a (k, r, t)-set membership.

We obtain the reduced universe $\mathcal{U}'$ and $\mathcal{S}'$ for the constructed instance of the r-Set Packing with t-Membership as shown in Subsect. 4.1. Then, the reduced graph for the original instance is $G' = G[\mathcal{U}']$.

Lemma 7. *G' has a (k, r, t)-$\mathcal{H}$-membership if and only if S' has a (k, r, t)-set membership.*

This reduction property allows us to state:

Theorem 5. *$\mathcal{H}$-Packing with t-Membership has a problem kernel with $O((r + 1)^r k^r)$ vertices where $r = r(\mathcal{H})$.*

5   Packing Problems with Bounded Overlap

In this section, we develop kernelization algorithms for all our packing with t-Overlap problems.

5.1   Packing Sets with t-Overlap

We assume that each set in S has size at least $t + 1$. Otherwise, we can add the sets with size at most t straight to a (k, r, t)-set packing and decrease the parameter k by the number of those sets. We start with a simple reduction rule.

Reduction Rule 2. Remove any element of $\mathcal{U}$ that is not contained in at least one set of S.

Notice that every pair of sets in S overlaps in at most $r - 1$ elements; otherwise there would be two identical sets. Thus for $t = r - 1$, if $|S| \geq k$ then S is a (k, r, t)-set packing; in the other case, S does not have a solution. Henceforth, we assume that $t \leq r - 2$. We say that a set $S_e \in S$ is *extra* if there is a (k, r, t)-set packing that does not include S_e.

Algorithm 1 reduces r-Set Packing with t-Overlap to a kernel in two steps. First, in Lines 3–9 the goal is to identify extra sets of S, and second in Line 16, *unnecessary* elements are removed from $\mathcal{U}$ by triggering a reduction. Lines 3–9 in Algorithm 1 basically keep reducing a maximal $(r, r - 2)$-set packing $\mathcal{R}$ using Algorithm 2. The need to run possibly more than once Algorithm 2 is to preserve the maximality of $\mathcal{R}$.

The basic idea of Algorithm 2 is that if there is more than a specific number of members of S that pairwise overlap in exactly the same subset of elements $P \subset \mathcal{U}$, then some of those members are extra. For each set S in $\mathcal{R}$, Algorithm 2 considers every subset of elements $P \subsetneq S$ from sizes from t_{Ini} to $t + 1$ in decreasing order (Lines 3–5). Given the construction of $\mathcal{R}$ in Algorithm 1, $t_{Ini} = r - 2$. Note that the size of S should be at least $i + 1$ to run Lines 4–13; otherwise S will be considered at a later iteration of the for-loop of Line 3. For every P, we count the number of sets in $\mathcal{R}$ that contain P (collected in $\mathcal{P}$, Line 6). If that number is greater than a specific threshold (Line 7) then we can remove the ones above the threshold (Lines 8–11). The size of P is determined by the variable i in Line 3, and the threshold $f(i)$ is defined as $f(i) = (r - t)(k - 1)f(i + 1) + 1$ where $f(t_{Ini} + 1)$ is initialized to one (Line 2). The function $f(i)$ basically bounds the number of sets in $\mathcal{R}$ that contains a subset P with i elements. Observe that

---

Algorithm 1. Kernelization Algorithm - Set Packing with t-Overlap

---

Input: An instance $\mathcal{U}, \mathcal{S}$
Output: A reduced instance $\mathcal{U}', \mathcal{S}'$

 1: Apply Reduction Rule 2
 2: $\mathcal{R} = \emptyset$, $\mathcal{E} = \emptyset$
 3: **repeat**
 4:    Greedily add sets from $\mathcal{S} \backslash (\mathcal{R} \cup \mathcal{E})$ to $\mathcal{R}$ such that every pair of sets in $\mathcal{R}$ overlaps in at most $r - 2$ elements (i.e., a maximal $(r, r - 2)$-set packing).
 5:    **if** at least one set was added to $\mathcal{R}$ **then**
 6:       $\mathcal{E}' = $ Algorithm 2$(\mathcal{R})$
 7:       $\mathcal{R} = \mathcal{R} \backslash \mathcal{E}'$, $\mathcal{E} = \mathcal{E} \cup \mathcal{E}'$
 8:    **end if**
 9: **until** no more sets have been added to $\mathcal{R}$
10: Reduce $\mathcal{S} = \mathcal{S} \backslash \mathcal{E}$ and re-apply Reduction Rule 2
11: Compute a maximal (r, t)-set packing $\mathcal{M}$ of $\mathcal{R}$
12: **if** $|\mathcal{M}| \geq k$ **then**
13:    Let $\mathcal{S}'$ be any subset of $\mathcal{M}$ of size k
14:    Return $\mathcal{U}$ and $\mathcal{S}'$
15: **end if**
16: Let $\mathcal{U}'$ and $\mathcal{S}'$ be the reduced universe and collection of sets, respectively, after applying Reduction Rule 3.
17: Return $\mathcal{U}'$ and $\mathcal{S}'$

Algorithm 2. Extra Sets Reduction

Input: A set $\mathcal{R} \subseteq \mathcal{S}$ of sets
Output: A set $\mathcal{E} \subseteq \mathcal{R}$ of sets

 1: $t_{Ini} = \max \{|S_i \cap S_j| : S_i, S_j \in \mathcal{R}, i \neq j\}$
 2: $f(t_{Ini} + 1) = 1$, $\mathcal{E} = \emptyset$
 3: **for** $i = t_{Ini}$ downto $t + 1$ **do**
 4:    **for each** $S \in \mathcal{R}$ such that $|S| > i$ **do**
 5:       **for each** $P \subsetneq S$ where $|P| = i$ **do**
 6:          $\mathcal{P} = \{S' \in \mathcal{R} : S' \supsetneq P\}$
 7:          $f(i) = (r - t)(k - 1)f(i + 1) + 1$
 8:          **if** $|\mathcal{P}| > f(i)$ **then**
 9:             Choose any $\mathcal{P}' \subset \mathcal{P}$ of size $f(i)$
10:             Set $\mathcal{E} \leftarrow \mathcal{E} \cup (\mathcal{P} \backslash \mathcal{P}')$ (extra sets)
11:             $\mathcal{R} \leftarrow \mathcal{R} \backslash (\mathcal{P} \backslash \mathcal{P}')$
12:          **end if**
13:       **end for**
14:    **end for**
15: **end for**
16: Return $\mathcal{E}$

$f(i) = (r-t)(k-1)f(i+1)+1 = \sum_{j=0}^{t_{Ini}-i+1}[(r-t)(k-1)]^j$. Algorithm 2 returns $\mathcal{E}$ which is the set of extra sets in $\mathcal{R}$ (Line 16). The next lemma states that Algorithm 2 correctly reduces the given set $\mathcal{R}$.

Lemma 8. *$\mathcal{R}$ has a (k,r,t)-set packing if and only if $\mathcal{R}\backslash\mathcal{E}$ has a (k,r,t)-set packing.*

Lemma 9. *After running Lines 1–9 of Algorithm 1, $\mathcal{S}$ has a (k,r,t)-set packing if and only if $\mathcal{S}\backslash\mathcal{E}$ has a (k,r,t)-set packing.*

By Lemma 9, we reduce $\mathcal{S}$ by removing $\mathcal{E}$ in Line 10 of Algorithm 1. As a consequence, we re-apply Reduction Rule 2. Then we compute a maximal (r,t)-set packing in $\mathcal{R}$. This maximal solution will help us to determine an upper-bound for the number of sets in $\mathcal{R}$. From now on, we assume that the maximal solution was not a (k,r,t)-set packing and Algorithm 1 continues executing.

An element $u \in \mathcal{U}$ is *extra* if there is a (k,r,t)-set packing $\mathcal{K}$ of $\mathcal{S}$ where $u \notin S$ for each set $S \in \mathcal{K}$. We will identify extra elements in $\mathcal{U}\backslash val(\mathcal{R})$, denoted as O henceforth. Before giving our reduction rule, we give a characterization of O.

Lemma 10. *Let $O = \mathcal{U}\backslash val(\mathcal{R})$. (i) Each element in O is contained in at least one set S. (ii) Only sets in $\mathcal{S}\backslash\mathcal{R}$ contain elements from O. (iii) No pair of different elements in O is contained in the same set S for any $S \in \mathcal{S}\backslash\mathcal{R}$. (iv) Each $S \in \mathcal{S}\backslash\mathcal{R}$ contains one element in O and $r-1$ elements of some set in $\mathcal{R}$.*

We will reduce O by applying similar ideas as in [14]. To this end, we first construct an auxiliary bipartite graph $B = (V_O, V_{\mathcal{S}\backslash\mathcal{R}}, E)$ as follows. There is a vertex v_o in V_O for each element $o \in O$. For each set S in $\mathcal{S}\backslash\mathcal{R}$ and each subset $P \subsetneq S$ where $|P| = r-1$, if there is at least one element $o \in O$ such that $\{o\} \cup P \in \mathcal{S}\backslash\mathcal{R}$, add a vertex v_p to $V_{\mathcal{S}\backslash\mathcal{R}}$. We say that v_o and v_p *correspond* to o and P, respectively. We add an edge (v_o, v_p) to E for each pair v_o, v_p if $\{o\} \cup P \in \mathcal{S}\backslash\mathcal{R}$. Then, we apply the following reduction rule.

Reduction Rule 3. Compute a maximum matching M in B. Let V_O' be the set of unmatched vertices of V_O and O' be the elements of $O \subset \mathcal{U}$ corresponding to those vertices. Likewise, let $\mathcal{S}(O')$ be the sets in $\mathcal{S}\backslash\mathcal{R}$ that contain elements of O'. Reduce to $\mathcal{U}' = \mathcal{U}\backslash O'$ and $\mathcal{S}' = \mathcal{S}\backslash\mathcal{S}(O')$.

Lemma 11. *$\mathcal{S}$ has a (k,r,t)-set packing if and only if $\mathcal{S}'$ has a (k,r,t)-set packing.*

Note that after Reduction Rule 3, each element of $\mathcal{U}'$ is contained in at least one element of $\mathcal{S}'$. The correctness of our kernelization algorithm (Algorithm 1) is given by Lemmas 8–11.

Lemma 12. *Algorithm 1 runs in polynomial time.*

After Reduction Rule 3, we obtained a reduced collection $\mathcal{S}'$ and $\mathcal{U}' = val(\mathcal{S}')$. Thus, we will use $\mathcal{S}'$ to upper-bound $\mathcal{U}'$. The collection $\mathcal{S}'$ is equivalent to $\mathcal{S}' = (\mathcal{S}'\backslash\mathcal{R})\cup\mathcal{R}$. Since $(\mathcal{S}'\backslash\mathcal{R}) \subseteq (\mathcal{S}\backslash\mathcal{R})$, by Lemma 10 each set in $(\mathcal{S}'\backslash\mathcal{R})$ contains one element in O and $r - 1$ elements in a set of $\mathcal{R}$. Thus, $val(\mathcal{S}') = O \cup val(\mathcal{R})$. The elements in O' were removed from O (Reduction Rule 3); therefore, $val(\mathcal{S}') = (O \backslash O') \cup val(\mathcal{R}) = \mathcal{U}'$.

Upper bounds for the size of $\mathcal{R}$ and O are given in Lemmas 13 and 14, respectively.

Lemma 13. *The size of $\mathcal{R}$ is at most $2r^{r-1}k^{r-t-1}$, i.e., $|val(\mathcal{R})| \leq 2r^r k^{r-t-1}$.*

Lemma 14. *After applying Reduction Rule 3, there are at most $2r^r k^{r-t-1}$ elements in $O \backslash O'$.*

By Lemmas 12, 13, and 14, we can hence state.

Theorem 6. *The r-Set Packing with t-Overlap possesses a problem kernel with $O(r^r k^{r-t-1})$ elements from the given universe.*

5.2 Packing Graphs with t-Overlap

We next sketch our methodology to reduce our $\mathcal{H}$-Packing with t-Overlap problems to a kernel (see [8] for details). First, we transform an instance of $\mathcal{H}$-Packing with t-Overlap to an instance of r-Set Packing with t-Overlap using Transformation 2. Next, we reduce this transformed instance to a kernel with Algorithm 1. Finally, we re-interpret that kernel as a kernel for the original (graph) problem. Specifically, Algorithm 1 returns a reduced universe $\mathcal{U}'$ which we use to obtain the reduced graph $G' = G[\mathcal{U}']$. Since $|\mathcal{U}'| = O(r^r k^{r-t-1})$, we can conclude:

Theorem 7. *The $\mathcal{H}$-Packing with t-Overlap possesses a problem kernel with $O(r^r k^{r-t-1})$ vertices, where $r = r(\mathcal{H})$.*

6 Conclusions

We commenced a study on the parameterized algorithmics of packing problems allowing overlap, focusing on kernelization issues. This leads to whole new families of problems, whose problem names contain the constant t and the finite set of objects (mostly graphs) $\mathcal{H}$ from which the constant $r(\mathcal{H})$ can be derived. In addition, we have a natural parameter k. In the extended version of this paper, we also added according results on packings with induced overlapping subgraphs or with subgraphs with overlapping edges and discussed additional disjointness conditions between subgraphs.

We list here some of the open problems in this area. Our kernel bounds are not known to be tight. In particular, we have shown no lower bound results, like [4]. Recently, Giannopoulou et al. introduced the notion of *uniform kernelization* [10] for problem families similar to what we considered. This basically raises the question if polynomial kernel sizes could be proven such that the exponent of the kernel bound does not depend (in our case) on r or on t. Notice that for t-Membership, we somehow came half the way, as we have shown kernel sizes that are uniform with respect to t. In particular, $\{K_3\}$-Packing with t-Membership has a uniform kernelization. This could be interesting on its own, as only few examples of uniform kernelizations are known. Very recent work by Marx and his colleagues has shown renewed interest in dichotomy results like the one that we presented in this paper. Apart from the natural questions mentioned above, the case of admitting an infinite number of graphs in the family $\mathcal{H}$ is another natural path for future research. In both respects, Jansen and Marx [11] have obtained very interesting results on classical graph packing problems.

References

1. Abu-Khzam, F.N.: An improved kernelization algorithm for r-set packing. Inf. Process. Lett. **110**(16), 621–624 (2010)
2. Banerjee, S., Khuller, S.: A clustering scheme for hierarchical control in multi-hop wireless networks. In: Proceedings of 20th Joint Conference of the IEEE Computer and Communications Societies (INFOCOM 2001), vol. 2, pp. 1028–1037. IEEE Society Press (2001)
3. Bodlaender, H.L., Thomassé, S., Yeo, A.: Analysis of data reduction: transformations give evidence for non-existence of polynomial kernels. Technical report. UU-CS-2008-030, Department of Information and Computer Sciences, Utrecht University (2008)
4. Dell, H., Marx, D.: Kernelization of packing problems. In: Rabani, Y. (ed.) Proceedings of the Twenty-Third Annual ACM-SIAM Symposium on Discrete Algorithms, SODA. pp. 68–81. SIAM (2012)
5. Fellows, M., Guo, J., Komusiewicz, C., Niedermeier, R., Uhlmann, J.: Graph-based data clustering with overlaps. Discrete Optim. **8**(1), 2–17 (2011)
6. Fellows, M., Knauer, C., Nishimura, N., Ragde, P., Rosamond, F., Stege, U., Thilikos, D., Whitesides, S.: Faster fixed-parameter tractable algorithms for matching and packing problems. Algorithmica **52**(2), 167–176 (2008)
7. Fellows, M., Heggernes, P., Rosamond, F.A., Sloper, C., Telle, J.A.: Finding k disjoint triangles in an arbitrary graph. In: Hromkovič, J., Nagl, M., Westfechtel, B. (eds.) WG 2004. LNCS, vol. 3353, pp. 235–244. Springer, Heidelberg (2004)
8. Fernau, H., López-Ortiz, A., Romero, J.: Kernelization algorithms for packing problems allowing overlaps (Extended Version) (2014). arXiv:1411.6915
9. Fernau, H., Raible, D.: A parameterized perspective on packing paths of length two. J. Comb. Optim. **18**(4), 319–341 (2009)
10. Giannopoulou, A.C., Jansen, B.M.P., Lokshtanov, D., Saurabh, S.: Uniform kernelization complexity of hitting forbidding minors (2014, unpublished). http://www.win.tue.nl/~bjansen/publications.html
11. Jansen, B.M.P., Marx, D.: Characterizing the easy-to-find subgraphs from the viewpoint of polynomial-time algorithms, kernels, and Turing kernels. CoRR abs/1410.0855 (2014)
12. Kirkpatrick, D., Hell, P.: On the completeness of a generalized matching problem. In: Proceedings of the Tenth Annual ACM Symposium on Theory of Computing (STOC), pp. 240–245 (1978)
13. Micali, S., Vazirani, V.V.: An $O(\sqrt{|V|}|E|)$ algorithm for finding maximum matching in general graphs. In: Proceedings of the 21st Annual Symposium on Foundations of Computer Science, SFCS 1980, pp. 17–27. IEEE Computer Society (1980)

14. Moser, H.: A problem kernelization for graph packing. In: Nielsen, M., Kučera, A., Miltersen, P.B., Palamidessi, C., Tůma, P., Valencia, F. (eds.) SOFSEM 2009. LNCS, vol. 5404, pp. 401–412. Springer, Heidelberg (2009)
15. Palla, G., Derényi, I., Farkas, I., Vicsek, T.: Uncovering the overlapping community structure of complex networks in nature and society. Nature **435**(7043), 814–818 (2005)
16. Prieto, E., Sloper, C.: Looking at the stars. Theoret. Comput. Sci. **351**(3), 437–445 (2006)
17. Romero, J., López-Ortiz, A.: The $\mathcal{G}$-packing with t-overlap problem. In: Pal, S.P., Sadakane, K. (eds.) WALCOM 2014. LNCS, vol. 8344, pp. 114–124. Springer, Heidelberg (2014)
18. Romero, J., López-Ortiz, A.: A parameterized algorithm for packing overlapping subgraphs. In: Hirsch, E.A., Kuznetsov, S.O., Pin, J.É., Vereshchagin, N.K. (eds.) CSR 2014. LNCS, vol. 8476, pp. 325–336. Springer, Heidelberg (2014)
19. Shiloach, Y.: Another look at the degree constrained subgraph problem. Inf. Process. Lett. **12**(2), 89–92 (1981)

Parameterized Complexity of Asynchronous Border Minimization

Robert Ganian[1]($\boxtimes$), Martin Kronegger[1], Andreas Pfandler[1,2],
and Alexandru Popa[3]

[1] Vienna University of Technology, Vienna, Austria
{robert.ganian,martin.kronegger,andreas.pfandler}@tuwien.ac.at
[2] University of Siegen, Siegen, Germany
[3] Nazarbayev University, Astana, Kazakhstan
alexandru.popa@nu.edu.kz

Abstract. Microarrays are research tools used in gene discovery as well as disease and cancer diagnostics. Two prominent but challenging problems related to microarrays are the *Border Minimization Problem* (BMP) and the *Border Minimization Problem with given placement* (P-BMP). In this paper we investigate the parameterized complexity of natural variants of BMP and P-BMP, termed BMP^e and $P\text{-}BMP^e$ respectively, under several natural parameters. We show that BMP^e and $P\text{-}BMP^e$ are in FPT under the following two combinations of parameters: (1) the size of the alphabet (c), the maximum length of a sequence (string) in the input (ℓ) and the number of rows of the microarray (r); and, (2) the size of the alphabet and the size of the border length (o). Furthermore, $P\text{-}BMP^e$ is in FPT when parameterized by c and ℓ. We complement our tractability results with corresponding hardness results.

1 Introduction

DNA and peptide microarrays [3,12] are important research tools used in gene discovery, multi-virus discovery as well as disease and cancer diagnosis. Apart from measuring the amount of gene expression [18], microarrays are an efficient tool for making a qualitative statement about the presence or absence of biological target sequences in a sample. For example, peptide microarrays are used for detecting tumor biomarkers [2,16,19].

A microarray is a plastic or glass slide consisting of thousands of sequences of nucleotides called *probes* that are assigned to one cell in the array. The synthesis process [10] consists of two components: *probe placement* and *probe embedding*. In the probe placement, the goal is to determine an assignment of each probe to a unique cell of the array. If the placement is given one has to create the sequences at their respective cells (probe embedding). This can be achieved with help of the following two operations: It is possible to *mask* a certain set of cells. Furthermore, one can *append* a certain nucleotide to the probes in all those cells which

Supported by the Austrian Science Fund (FWF): P25518-N23 and P26696, and the German Research Foundation (DFG) under grant ER 738/2-1.

R. Jain et al. (Eds.): TAMC 2015, LNCS 9076, pp. 428–440, 2015.
DOI: 10.1007/978-3-319-17142-5_36

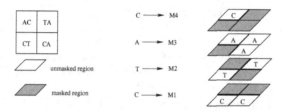

Fig. 1. Asynchronous synthesis of a 2×2 microarray. The deposition sequence $\mathcal{D} =$ CTAC corresponds to four masks $\mathcal{M}_1, \mathcal{M}_2, \mathcal{M}_3$, and $\mathcal{M}_4$. The masked regions are shaded and the border between the masked and unmasked regions is represented by bold lines.

are currently unmasked. Essentially, the nucleotides are represented as characters and the probes as strings. In probe embedding we want to find a common supersequence of all probes, called the *deposition sequence*, and a sequence of 2D arrays describing the masks. The cells of a mask can be either masked (opaque) or unmasked (transparent) allowing the deposition of the nucleotide associated with the mask. For any cell, the concatenation of the nucleotides for which the cell is transparent has to match the probe in that cell of the microarray. See Fig. 1 for an example [15].

Due to diffraction, the cells on the *border* between the masked and the unmasked regions are often subject to unintended illumination [10], and can compromise experimental results. Therefore, unintended illumination should be minimized. The magnitude of unintended illumination can be measured by the *border length* of the masks used, which is the number of borders shared between masked and unmasked regions, e.g., in Fig. 1, the border length of $\mathcal{M}_1, \mathcal{M}_3, \mathcal{M}_4$ is 2 and $\mathcal{M}_2$ is 4 which yields a total border length of 10.

The problem of finding both the placement and the embedding is termed the Border Minimization Problem (BMP). If the placement is given and the task is to find only the embedding, we speak of P-BMP. We refer the reader to Sect. 2 for formal definitions of BMP and P-BMP.

VARIANTS OF BORDER MINIMIZATION. In this paper we consider the exhaustive variants of BMP and P-BMP, termed BMP^e and P-BMP^e respectively. The difference is that in P-BMP^e (and, consequently, in BMP^e) we assume that a mask is always applied exhaustively (we call this the *exhaustive rule*). More precisely, when a mask that synthesizes a character c is applied, the mask has a transparent cell wherever the corresponding sequence begins with the character c.

Without this assumption it is possible to artificially increase the length of the deposition sequence which, as a consequence, also increases the length of the sequence of masks. In most application scenarios this is undesirable, since applying a mask requires an additional cycle of work that causes a waste of material and can also introduce new errors. A second advantage of these exhaustive variants is that they allow the concise description of solutions: a solution to P-BMP^e is fully characterized by the deposition sequence, while for P-BMP it is also necessary to explicitly describe each mask in the sequence. To clarify, we remark

Table 1. Overview of results.

	c or c, r	c, ℓ	c, ℓ, r	c, o
P-BMP^e	paraNP-h (Proposition 2)	FPT (Proposition 4)	FPT (Proposition 4)	FPT (Theorem 4)
BMP^e	paraNP-h (Theorem 1)	open	FPT (Theorem 3)	FPT (Theorem 5)

that an optimal exhaustive solution need not always be an optimal solution for P-BMP (or BMP): there are cases where the border length can increase.

We illustrate the usefulness of the assumption by a simple example. In the P-BMP^e instance $a|b|a$, this assumption indeed helps to reduce the number of masks without increasing the border length. A non-exhaustive optimal solution might work on the left a first, while an exhaustive optimal solution works on both a concurrently. Even though the border length is in both cases 4, the non-exhaustive case could require an additional mask.

OUR RESULTS. Our results are summarized in Table 1. In this paper we investigate the parameterized complexity of the BMP^e and P-BMP^e problems under several natural parameters. First of all, throughout this work we consider the number of available nucleotides c (i.e., the alphabet size) as a parameter. Notice that this assumption does not impose a serious restriction, since in practice the number of available nucleotides is very limited (or even constant). Orthogonal to this assumption we explore the parameterized complexity of the BMP^e and P-BMP^e problem with respect to three natural parameters, i.e., the maximum length of a sequence in the array (ℓ), the maximum border length cost (o), and the maximum number of rows in the array (r). Since errors become more likely as the length of the sequence grows, the length of the constructed probes will be rather limited. Notice that the parameter o models the cost of a solution and hence is also a natural parameter. Finally, with the maximum number of rows r the shape of the array is restricted in the sense that the one dimension does not grow arbitrarily. This is, in particular, interesting because it allows to generalize from the one-dimensional case studied in [17].

More precisely, we show fpt-algorithms for BMP^e and P-BMP^e if we are given either c, ℓ, r or c, o as parameters. We complement these results with parameterized intractability results, i.e., by showing paraNP-hardness. We use a polynomial time reduction from P-BMP^e to BMP^e to build upon the result that P-BMP^e parameterized by c and r is paraNP-hard[1] and obtain hereby paraNP-hardness for BMP^e parameterized by c and r. Notice that with the exception of BMP^e parameterized by c and ℓ, we obtain a full parameterized complexity map of the two considered problems with respect to all additional parameters considered in this paper. We furthermore provide a reduction relating the complexity of BMP^e

---

[1] Although in [17] only NP-hardness is proven for P-BMP, the reduction can also be used to show paraNP-hardness for P-BMP^e when parameterized by c and r.

parameterized by c and ℓ to k-BALANCED PARTITION on grids, a well-studied problem whose parameterized complexity on grids is open (Proposition 3).

The rest of the paper is organized as follows. In Sect. 2 we introduce the problems formally and give preliminaries. Then, in Sect. 3 we show the reduction from P-BMP^e to BMP^e. Section 4 introduces the fpt-algorithms and, finally, in Sect. 5 we present conclusions and open problems.

2   Preliminaries

For $n \in \mathbb{N}$, we use $[n]$ to denote the set $\{1, \ldots, n\}$. For two sequences s_1, s_2, we use $s_1 \cdot s_2$ to mark their concatenation.

The microarray has size $r \times m$, where r is the number of rows and m is the number of columns. The multiset of input sequences (also called *probes*) is denoted by $\mathcal{S} = \{s_1, s_2, \ldots, s_{r \cdot m}\}$ and the input alphabet by Σ. Moreover, let $c = |\Sigma|$. For any sequence s_i, we denote the length of the sequence by ℓ_i and the t-th character of a sequence s_i by $s_i[t]$. We use ℓ for the maximum length of the probes, i.e., $\ell = \max_{i \in [r \cdot m]} \ell_i$. Two cells of the array $v_1 = (x_1, y_1)$ and $v_2 = (x_2, y_2)$ are said to be *neighbors* if $|x_1 - x_2| + |y_1 - y_2| = 1$. For each cell v, we denote the set of neighbors of v by $\mathcal{N}(v)$.

In order to give the formal definition of BMP, we introduce several notions related to the synthesis process.

Definition 1. *A* placement *of the probe sequences is a bijective function φ that maps each probe sequence to a unique cell in the array.*

Definition 2. *A* deposition sequence D *for a set of sequences $\mathcal{S}$ is a sequence of characters which is a common supersequence of all sequences in $\mathcal{S}$.*

Definition 3. *An* embedding *of a sequence s_i into a deposition sequence D is a length-$|D|$ sequence ε_i over alphabet $\Sigma \cup \{-\}$ such that:*

1. *ε_i contains precisely $|s_i|$ characters other than "$-$" occurring at positions $\varepsilon_i[u_1], \varepsilon_i[u_2], \ldots, \varepsilon_i[u_{|s_i|}]$,*
2. *u_1 is the minimum position such that $\varepsilon_i[u_1] = s_i[1]$,*
3. *for $2 \leq j \leq |s_i|$, u_j is the minimum position such that $\varepsilon_i[u_j] = s_i[j]$ and $u_{j-1} < u_j$.*

Informally, ε_i captures how a sequence is built (or, equivalently, deleted) by the deposition sequence; notice that due to the exhaustive rule, the embedding is uniquely determined by the deposition sequence. An *embedding* of a set of probes $\mathcal{S}$ into a deposition sequence D is then denoted by $\varepsilon_D = \{\varepsilon_1, \varepsilon_2, \ldots, \varepsilon_{|\mathcal{S}|}\}$. Note that we will drop the subscript when the associated deposition sequence is clear from the context. The final key notion we need are masks.

Definition 4. *A* mask $\mathcal{M}$ *(for some character c) is a 2D-array such that $\mathcal{M}(i, j)$ is either c or a space "$-$" (here the space means that the character is not deposited into this cell).*

The sequence of masks associated with a deposition sequence D and a placement φ is $\omega = \mathcal{M}_1, \ldots, \mathcal{M}_{|D|}$ where $\mathcal{M}_i(a,b) = \varepsilon_{\varphi^{-1}(a,b)}[i]$ for $i \in [|D|]$. Notice that due to the exhaustive rule, a mask for character c is always maximal with respect to c, i.e., there is no "$-$" in the mask that could be replaced by c. We introduce now the *border length* of a given placement of the probes in the array, which is the value we aim to optimize.

Definition 5. *Let* $\mathrm{border}_D(s_i, s_j)$ *be the Hamming distance between* ε_i *and* ε_j *(with respect to deposition sequence D). The* border length *of a placement φ and a deposition sequence D is then defined as the sum of borders over all pairs of neighboring probe sequences*

$$\mathrm{BL}(\varphi, D) = \sum_{\substack{\forall i,\, j \in \mathbb{N} \,:\, i < j < |\mathcal{S}| \\ \wedge\ \varphi(s_j) \in \mathcal{N}(\varphi(s_i))}} \mathrm{border}_D(s_i, s_j). \tag{1}$$

We can also equivalently define border length in terms of the border length of all the masks.

Definition 6. *For any mask $\mathcal{M}$ of deposition character x, the* border length *of $\mathcal{M}$, denoted by $\mathrm{BL}(\mathcal{M})$, is defined as the number of pairs of neighboring cells (i_1, j_1) and (i_2, j_2) such that $\mathcal{M}(i_1, j_1) = x$ and $\mathcal{M}(i_1, j_1) \neq \mathcal{M}(i_2, j_2)$. For a placement and deposition sequence that corresponds to a sequence of masks $\mathcal{M}_1$, $\mathcal{M}_2, \cdots, \mathcal{M}_{|D|}$, we let*

$$\mathrm{BL}(\varphi, D) = \sum_{h=1}^{|D|} \mathrm{BL}(\mathcal{M}_h) \tag{2}$$

The BMP^e and the $\mathrm{P\text{-}BMP}^e$ problem are defined as follows.

Problem 1. *In the BMP^e problem, we are given $r, m \in \mathbb{N}$ and a multiset of $r \cdot m$ sequences $\mathcal{S}$. The objective is to find a placement φ and a deposition sequence D so that $\mathrm{BL}(\varphi, D)$ is minimized.*

Problem 2. *In the $\mathrm{P\text{-}BMP}^e$ problem, we are given $r, m \in \mathbb{N}$ and a multiset of $r \cdot m$ sequences $\mathcal{S}$ and a placement φ. The objective is to find a deposition sequence D so that $\mathrm{BL}(\varphi, D)$ is minimized.*

For a set $\pi \subseteq \{c, r, \ell, o\}$, we denote by BMP^e_π ($\mathrm{P\text{-}BMP}^e_\pi$) the BMP^e ($\mathrm{P\text{-}BMP}^e$) problem parameterized by π. For a problem BMP^e_π ($\mathrm{P\text{-}BMP}^e_\pi$) where $o \in \pi$, we assume that an upper bound on the border length o is additionally given in the input and only solutions with minimum border length $\leq o$ are admitted.

We conclude this section with some useful observations. A deposition sequence D is called *redundant* if it contains a character $D[i]$ such that $\varepsilon_j[i] = $ "$-$" for each $\varepsilon_j \in \varepsilon$. Note that for any redundant deposition sequence D and any placement φ, it holds that $\mathrm{BL}(\varphi, D) = \mathrm{BL}(\varphi, D')$, where D' is obtained by deleting the redundant character $D[i]$. We say that a deposition sequence D is *good* if it is not redundant.

Observation 1. *Let* (φ, D) *be such that* $\mathrm{BL}(\varphi, D)$ *is minimized for some* $(\mathcal{S}, r, m)$. *If* D *is redundant, then there exists a subsequence* D' *of* D *such that* $\mathrm{BL}(\varphi, D') = \mathrm{BL}(\varphi, D)$ *and* D' *is good.*

As a consequence, when searching for optimal solutions of these problems it suffices to consider only good deposition sequences. Aside from the trivial (quadratic) algorithm for computing the border length for a fixed deposition sequence and placement, we will utilize another algorithm which will in some cases yield better running times:

Proposition 1. *For any given* $(\varphi, D, \mathcal{S}, r, m)$, *there exists an algorithm which computes* $\mathrm{BL}(\varphi, D)$ *in time* $\mathcal{O}(|\mathcal{S}| + p^2 \cdot |D|)$, *where* p *is the number of distinct sequences in* $\mathcal{S}$.

2.1 Parameterized Complexity

Parameterized algorithmics is a promising approach to obtain efficient algorithms for fragments of computationally hard problems. The aim is to find a parameter that describes the structure of the instance such that the combinatorial explosion can be confined to this parameter. In a parameterized complexity analysis the runtime of an algorithm is studied with respect to the input size n and a parameter $k \in \mathbb{N}$ (or a combination of parameters). For a more detailed introduction we refer to the literature [4,9].

Formally, a *parameterized problem* is a subset of $\Sigma^* \times \mathbb{N}$, where Σ is the input alphabet. If a combination of parameters $k_1, \ldots, k_l$ is considered, the second component of an instance (x, k) is given by $k = \sum_{1 \le i \le l} k_i$. The class FPT (*fixed-parameter tractable*) contains all problems that can be decided by an algorithm running in $f(k) \cdot n^{\mathcal{O}(1)}$ time, where f is a computable function and n is the input size. Such algorithms are often called fixed-parameter tractable (fpt).

Let L_1 and L_2 be parameterized problems, with $L_1 \subseteq \Sigma_1^* \times \mathbb{N}$ and $L_2 \subseteq \Sigma_2^* \times \mathbb{N}$. A *parameterized reduction* (or fpt-reduction) from L_1 to L_2 is a mapping $P : \Sigma_1^* \times \mathbb{N} \to \Sigma_2^* \times \mathbb{N}$ such that (1) $(x, k) \in L_1$ iff $P(x, k) \in L_2$; (2) the mapping can be computed by an fpt-algorithm with respect to parameter k; (3) there is a computable function g such that $k' \le g(k)$, where $(x', k') = P(x, k)$.

There is a variety of classes capturing *parameterized intractability*. For our results, we require only the class paraNP [8], which is defined as the class of problems that are solvable by a nondeterministic Turing-machine in fpt-time. We will make use of the characterization of paraNP-hardness given by Flum and Grohe [9], Theorem 2.14: any parameterized problem that remains NP-hard when the parameter is set to some constant is paraNP-hard. Showing paraNP-hardness for a problem rules out the existence of an fpt-algorithm under the usual complexity theoretic assumptions.

3 Hardness

In this section we overview and present new (parameterized) intractability results for BMP^e and $\mathrm{P\text{-}BMP}^e$ with respect to several combinations of parameters.

As our starting point, we notice that the NP-hardness proof for P-BMP of Popa, Wong and Yung [17] can be straightforwardly adapted to P-BMP$^e_{c,r}$.

Proposition 2 (cf. [17, Theorem 1]). *P-BMP$_{c,r}$ is* paraNP-*hard.*

The hardness result for BMP^e relies on a new polynomial-time reduction from P-BMP^e to BMP^e. We believe that this reduction is an interesting result on its own, as it is one of the first results that relates the complexity of these two problems in a general setting. We begin by showcasing a tool for forcibly "separating" any optimal deposition sequence.

Lemma 1. *Let $\mathcal{I} = (\mathcal{S}, r, m)$ be an instance of* BMP^e *such that each $s \in \mathcal{S}$ consists of a prefix $s_{pre} \in \Sigma^*_{pre}$, a fixed separator $sep \in (x^*y^*)^*$ and a suffix $s_{suf} \in \Sigma^*_{suf}$, where $\Sigma_{pre}, \Sigma_{suf}, \{x, y\}$ form a partition of Σ. Let $u \geq 8 \cdot max_{s \in \mathcal{S}}(|s_{pre}|) + 8 \cdot max_{s \in \mathcal{S}}(|s_{suf}|) + 1$. If $sep = (x^{r \cdot m \cdot u} \cdot y^{r \cdot m \cdot u})^{r \cdot m \cdot u}$ then every optimal good deposition sequence has the form $D_{pre} \cdot sep \cdot D_{suf}$ where $D_{pre} \in \Sigma^*_{pre}$ and $D_{suf} \in \Sigma^*_{suf}$.*

Observe that "flipping" the array horizontally or vertically preserves the optimal border length but formally changes the placement φ. The purpose of the following key lemma is to provide a tool to fix the optimal positions of probes in the array; to this end, we will be considering placements which are unique up to these simple symmetries.

Lemma 2. *Let $a, b, x, y \in \Sigma$ and $r, m, t \in \mathbb{N}$. Consider an $r \times m$ array, and probes $\mathcal{S} = \{a^{i \cdot t} \cdot sep \cdot b^{j \cdot t} \mid i \in [r] \text{ and } j \in [m]\}$. Then:*

1. *the unique optimal placement φ_0 (up to simple symmetries) places each probe $a^{i \cdot t} \cdot sep \cdot b^{j \cdot t}$ in cell (i, j),*
2. *the unique optimal good deposition sequence is $D_0 = a^{r \cdot t} \cdot sep \cdot b^{m \cdot t}$, and*
3. *for any placement $\varphi \neq \varphi_0$ (except for symmetries of φ_0) and any deposition sequence D, it holds that $\mathrm{BL}(\varphi, D) \geq \mathrm{BL}(\varphi_0, D_0) + t$.*

With Proposition 2 and Lemma 2, we can proceed to:

Theorem 1. BMP$^e_{c,r}$ *is* paraNP-*hard.*

Theorem 1 and Proposition 2 show that one cannot hope to find an fpt-algorithm for BMP^e or P-BMP^e parameterized by any subset of $\{c, r\}$. These results complete the hardness part of our complexity map for BMP^e or P-BMP^e. For BMP$^e_{c,\ell}$ it remains open whether the problem is fixed parameter tractable. Still, we can relate this problem to k-BALANCED PARTITION, a problem studied well in the literature [1,5,6].

In a k-BALANCED PARTITION instance we are given a graph $G = (V, E)$ with $|V| = n$. The question is to find a partition of the vertices V into k sets $V_1, \ldots, V_k$ such that $|V_i| \leq \lceil \frac{n}{k} \rceil$ for all $1 \leq i \leq k$, and the cut size (i.e., the number of edges $\{x, y\}$ such that $x \in V_i$, $y \in V_j$, and $i \neq j$) is minimized. We remark that, to the best of our knowledge, the parameterized complexity of k-BALANCED PARTITION parameterized by k is open on solid rectangular grids [5]. Below we show that k-BALANCED PARTITION on solid rectangular grids can be reduced to BMP^e and hence BMP^e is at least as hard as k-BALANCED PARTITION.

Proposition 3. *There is a polynomial time reduction from k-BALANCED PAR-TITION on solid rectangular grids to BMP^e.*

4   Fpt-Algorithms

In the following sections we discuss fpt-algorithms for several parameters. The first group focuses on sequences of moderate length and an array whose size is primarily growing in one dimension, i.e., on the parameters c, ℓ, and r. In contrast, the second group parameterizes by c and the maximum admissible border length o.

4.1   Fpt-Algorithm for $P\text{-}BMP^e_{c,\ell}$

Our first algorithm provides a basic introduction to the techniques used later on.

Observation 2. *For any instance $(\mathcal{S}, r, m)$ of $BMP^e_{c,\ell}$, there are at most c^ℓ unique sequences in $\mathcal{S}$.*

Lemma 3. *For any instance $(\mathcal{S}, r, m)$ of $BMP^e_{c,\ell}$ or any instance $(\mathcal{S}, \varphi, r, m)$ of $P\text{-}BMP^e_{c,\ell}$ it holds that $|D| \le c^\ell \cdot \ell$ for any good deposition sequence D.*

At this point we can already prove:

Proposition 4. *$P\text{-}BMP^e_{c,\ell}$ is fixed parameter tractable, and there exists an algorithm for $P\text{-}BMP^e_{c,\ell}$ which runs in time $c^{c^{O(\ell)}} |\mathcal{S}|$.*

4.2   Fpt-Algorithm for $BMP^e_{c,\ell,r}$

We first introduce some notation for our arrays. Given an $r \times m$ array A, a *column* is an $r \times 1$ sub-array of A. A *column placement* into a column of A is a mapping $\varphi : [r] \to \mathcal{S}$ from the cells of A to the multiset of probes.

Observation 3. *For any instance $(\mathcal{S}, r, m)$ of BMP^e, it holds that there are at most $c^{\ell \cdot r}$ distinct column placements.*

Hence for any fixed r and $\mathcal{S}$, we can enumerate all possible column placements as $\varphi_1, \varphi_2, \ldots, \varphi_{c^{\ell \cdot r}}$. Observe that, for any two column placements $\varphi_t, \varphi_{t'}$, it holds that either (i) $t = t'$ and $\varphi_t(x) = \varphi_{t'}(x)$ for all $x \in [r]$, or (ii) $t \neq t'$ and $\varphi_t(x) \neq \varphi_{t'}(x)$ for at least one $x \in [r]$.

Any placement $\varphi : s \in \mathcal{S} \mapsto (a \in \mathbb{N}, b \in \mathbb{N})$ into A can be uniquely decomposed into a sequence of column placements $(\varphi_{i(1)}, \varphi_{i(2)}, \ldots \varphi_{i(m)})$ where $\varphi_{i(x)}(y) = \varphi(x, y)$ and $i : [m] \to [c^{\ell \cdot r}]$. The column placement $\varphi_{i(j)}$ with $j \in [m]$ denotes that the j-th column of A is of placement $i(j)$. Furthermore, since φ is closed under permutation of non-distinct sequences in $\mathcal{S}$, each column placement can be uniquely identified by an r-tuple of sequences from $\mathcal{S}$, formally $\varphi_{i(x)} = (s_1, s_2, \ldots, s_r) \iff \varphi_{i(x)}(y) = s_y$ for all $y \in [r]$.

Next, we prove that when searching for optimal solutions for BMP^e it suffices to restrict ourselves to placements such that identical column placements appear in "consecutive blocks".

Lemma 4. *Let* $(\mathcal{S}, r, m)$ *be an instance of* BMP^e, D *be a deposition sequence and* φ *be a placement which decomposes into* $(\varphi_{i(1)}, \varphi_{i(2)}, \cdots \varphi_{i(m)})$. *Then if there exist* $a, b \in [m]$, $a + 1 < b$, *such that* $\varphi_{i(a)} = \varphi_{i(b)}$ *but* $\varphi_{i(a+1)} \neq \varphi_{i(b)}$, *then* $\mathrm{BL}(\varphi, D) \geq \mathrm{BL}(\varphi', D)$, *where* φ' *decomposes into*

$$(\varphi_{i(1)}, \cdots \varphi_{i(a)}, \varphi_{i(b)}, \varphi_{i(a+1)}, \varphi_{i(a+2)}, \ldots, \varphi_{i(b-1)}, \varphi_{i(b+1)}, \ldots, \varphi_{i(m)}).$$

We say that a placement φ is *consecutive* if it decomposes into column placements $(\varphi_{i(1)}, \varphi_{i(2)}, \cdots \varphi_{i(m)})$ where for each $\varphi_{i(a)}, \varphi_{i(b)}$ such that $\varphi_{i(a)} = \varphi_{i(b)}$ and $a < b$ it holds that $\varphi_{i(a)} = \varphi_{i(c)}$ for all $a < c < b$.

Corollary 1. *For any* BMP^e *instance* $(\mathcal{S}, r, m)$, *there exists an optimal solution* (φ, D) *such that* φ *is consecutive.*

The next algorithm uses an Integer Linear Programming (ILP) subroutine. ILP is a well-known framework for formulating problems and a powerful tool for the development of fpt-algorithms for optimization problems. In following we only give a brief overview of the framework before we present the algorithm.

Definition 7 (p-Variable Integer Linear Programming Optimization). *Let* $A \in \mathbb{Z}^{q \times p}, b \in \mathbb{Z}^{q \times 1}$ *and* $c \in \mathbb{Z}^{1 \times p}$. *The task is to find a vector* $x \in \mathbb{Z}^{p \times 1}$ *which minimizes the objective function* $c \times \bar{x}$ *and satisfies all* q *inequalities given by* A *and* b, *specifically satisfies* $A \cdot \bar{x} \geq b$. *The number of variables* p *is the parameter.*

Lenstra [14] showed that p-ILP, together with its optimization variant p-OPT-ILP (defined above), are in FPT. His running time was subsequently improved by Kannan [13] and Frank and Tardos [11] (see also [7]).

Theorem 2 ([7,11,13,14]). *p-OPT-ILP can be solved using* $\mathcal{O}(p^{2.5p+o(p)} \cdot L)$ *arithmetic operations in space polynomial in* L, L *being the number of bits in the input.*

We are now ready to prove the main theorem of this subsection.

Theorem 3. $\mathrm{BMP}^e_{c,\ell,r}$ *is fixed parameter tractable, and there exists an algorithm for* $\mathrm{BMP}^e_{c,\ell,r}$ *which runs in time* $c^{c^{\mathcal{O}(\ell \cdot r)}} \cdot |\mathcal{S}|$.

Proof (Sketch of Proof). We use Observation 1 and Lemma 3 to branch over the at most $c^{c^{\mathcal{O}(\ell)}}$ good deposition sequences. Then, we use Corollary 1 along with Observation 3 to branch over which column placements appear in φ and the order in which they appear from left to right. In each such branch, we denote the obtained sequence of column placements by the *template* Q_f.

For each such Q_f, we proceed by computing the total cost of the borders between adjacent elements of one specific column placement occurring in Q_f; we refer to these as the "vertical borders" and denote the cost of vertical borders in column placement i as bd_i^{vert}. We also compute the total cost of the "horizontal borders" between different adjacent column placements in Q_f.

Next, we compute the optimal number of times each column placement needs to occur in Q_f so as to minimize the total border length. This step is carried out by using a suitable p-OPT-ILP encoding, where we use one variable for each column placement to denote the number of times it occurs in the consecutive placement. Solving this encoding by Theorem 2 then allows us compute the total optimal cost for any consecutive placement which matches each template Q_f. Finally, we pick the combination of D and Q_f which admits the cheapest border length, and output (φ, D) where φ is computed from Q_f and the optimal multiplicity of each column placement obtained by the p-OPT-ILP encoding. $\square$

4.3   Fpt-Algorithm for P-BMP$^e_{c,o}$

Given an $r \times m$ array, a mask $\mathcal{M}$ is called *trivial* if $\mathcal{M}(i,j) \neq$ "$-$" for all $i \in [r], j \in [m]$. Given a deposition sequence D, we say that a subsequence D' of D is *primal* if it is obtained from D by deleting all characters which are associated with a trivial mask. Notice that the border length of each mask associated with each character in a primal sequence is at least one, and the border length of all trivial masks is 0. For the purpose of providing concise running times, we use n to denote the size of the input.

Observation 4. *For any instance of P-BMP^e and BMP^e, the number of primal sequences is bounded by $\sum_{i=1}^{o} c^i \leq o \cdot c^o$.*

Additionally, since the number of "borders" between distinct probes is bounded from below by the number of distinct probes, we obtain:

Observation 5. *Given a multiset S of probes. For any YES-instance of P-BMP^e and BMP^e over S, the number of distinct probes in S is upper-bounded by $o + 1$.*

Lemma 5. *For any instance of P-BMP^e and BMP^e, any primal sequence D' corresponds to at most one good deposition sequence D. Furthermore, there exists an algorithm which runs in time $\mathcal{O}(o \cdot n)$ and which either computes this D from D' or correctly outputs that no such D exists.*

Theorem 4. *P-BMP$^e_{c,o}$ is fixed-parameter tractable, and there exists an algorithm for P-BMP$^e_{c,o}$ which runs in time $\mathcal{O}(oc^o \cdot (n + o^2))$.*

Proof. This algorithm builds upon Observation 4. We can branch on all primal sequences. For each candidate sequence D' we check whether the primal sequence corresponds to a deposition sequence D via Lemma 5. For each such D, we compute and store $\mathrm{BL}(\varphi, D)$. Finally, a solution with a minimum $\mathrm{BL}(\varphi, D)$ is selected. Observe that an applicable trivial mask can be found in linear time. Along with Observation 5, this yields a total runtime of $\mathcal{O}(oc^o \cdot (n + o^2))$ by Proposition 1 and Lemma 5. $\square$

4.4   Fpt-Algorithm for $\mathrm{BMP}^e_{c,o}$

For a multiset $\mathcal{S}$ and $s \in \mathcal{S}$, we denote by $\mathcal{S}^{-s}$ the set of sequences in $\mathcal{S}$ which are distinct from s. An instance $(\mathcal{S}, r, m, o)$ of $\mathrm{BMP}^e_{c,o}$ is then called s-enveloped if $|\mathcal{S}^{-s}| \leq o^2$.

Lemma 6. *Any instance $(\mathcal{S}, r, m, o)$ of $\mathrm{BMP}^e_{c,o}$ such that $r > o$ and $m > o$ which is not s-enveloped for any $s \in \mathcal{S}$ is a no-instance.*

We now consider two specific subcases of the problem before giving the theorem.

Lemma 7. *There is an algorithm which solves any instance $(\mathcal{S}, r, m, o)$ of $\mathrm{BMP}^e_{c,o}$ such that $m > 2o$ and $r > 2o$ in time $\mathcal{O}(o^3 \cdot c^o \cdot (n + o^2))$.*

Lemma 8. *There is an algorithm which solves any instance $(\mathcal{S}, r, m, o)$ of $\mathrm{BMP}^e_{c,o}$ such that $m > 2o$ and $r \leq 2o$ in time $n \cdot c^{o^{\mathcal{O}(o)}}$.*

Theorem 5. $\mathrm{BMP}^e_{c,o}$ *is fixed parameter tractable, and there exists an algorithm for $\mathrm{BMP}^e_{c,o}$ which runs in time $n \cdot c^{o^{\mathcal{O}(o)}}$.*

Proof. In case $m > 2o$ and $r > 2o$ we use the algorithm described in the proof of Lemma 7. In case $m > 2o$ and $r \leq 2o$ (or, by symmetry, if $m \leq 2o$ and $r > 2o$) we use the algorithm described in the proof of Lemma 8. In case $m \leq 2o$ and $r \leq 2o$ we branch over all of the at most $(4o^2)!$ placements φ, resulting in at most $(4o^2)!$ instances of $\mathrm{P\text{-}BMP}^e_{c,o}$ which can be solved individually in time $\mathcal{O}(oc^o \cdot (n + o^2))$ by Theorem 4.   □

5   Conclusion

In this work we considered the parameterized complexity of BMP^e and $\mathrm{P\text{-}BMP}^e$, two fundamental problems related to the optimal design of microarrays, with respect to combinations of parameters centered around the number of distinct characters c. We presented fpt-algorithms for both BMP^e and $\mathrm{P\text{-}BMP}^e$ if the maximum probe length and the number of rows are viewed as additional parameters (c, ℓ, r); and if the border length is the additional parameter (c, o). In addition, we showed that $\mathrm{P\text{-}BMP}^e$ parameterized by c and ℓ is in FPT. For c, r (and also c alone) we showed paraNP-hardness for both BMP^e and $\mathrm{P\text{-}BMP}^e$. Hence, under the usual complexity theoretic assumptions, one cannot hope to find an fpt-algorithm for these settings.

On our agenda for future work is to settle the question whether there is an fpt-algorithm for BMP^e, parameterized by c, ℓ. Another direction for future research is to study further (structural) parameters for these two problems. Furthermore, in our complexity analysis we plan to consider more sophisticated target functions that take other criteria in addition to the border length into account.

References

1. Andreev, K., Räcke, H.: Balanced graph partitioning. Theor. Comput. Syst. **39**(6), 929–939 (2006)
2. Chatterjee, M., Mohapatra, S., Ionan, A., Bawa, G., Ali-Fehmi, R., Wang, X., Nowak, J., Ye, B., Nahhas, F.A., Lu, K., Witkin, S.S., Fishman, D., Munkarah, A., Morris, R., Levin, N.K., Shirley, N.N., Tromp, G., Abrams, J., Draghici, S., Tainsky, M.A.: Diagnostic markers of ovarian cancer by high-throughput antigen cloning and detection on arrays. Cancer Res. **66**(2), 1181–1190 (2006)
3. Cretich, M., Chiari, M.: Peptide Microarrays Methods and Protocols. Methods in Molecular Biology, vol. 570. Humana Press, New York (2009)
4. Downey, R.G., Fellows, M.R.: Parameterized Complexity. Monographs in Computer Science. Springer, New York (1999)
5. Feldmann, A. E.: Balanced partitions of grids and related graphs. Ph.D. thesis, ETH Zürich (2012)
6. Feldmann, A.E.: Fast balanced partitioning is hard even on grids and trees. Theor. Comput. Sci. **485**, 61–68 (2013)
7. Fellows, M.R., Lokshtanov, D., Misra, N., Rosamond, F.A., Saurabh, S.: Graph layout problems parameterized by vertex cover. In: Hong, S.-H., Nagamochi, H., Fukunaga, T. (eds.) ISAAC 2008. LNCS, vol. 5369, pp. 294–305. Springer, Heidelberg (2008)
8. Flum, J., Grohe, M.: Describing parameterized complexity classes. Inf. Comput. **187**(2), 291–319 (2003)
9. Flum, J., Grohe, M.: Parameterized Complexity Theory. Texts in Theoretical Computer Science. An EATCS Series, vol. XIV. Springer, Berlin (2006)
10. Fodor, S., Read, J.L., Pirrung, M.C., Stryer, L., Lu, A.T., Solas, D.: Light-directed, spatially addressable parallel chemical synthesis. Science **251**(4995), 767–773 (1991)
11. Frank, A., Tardos, É.: An application of simultaneous diophantine approximation in combinatorial optimization. Combinatorica **7**(1), 49–65 (1987)
12. Gerhold, D., Rushmore, T., Caskey, C.T.: DNA chips: promising toys have become powerful tools. Trends Biochem. Sci. **24**(5), 168–173 (1999)
13. Kannan, R.: Minkowski's convex body theorem and integer programming. Math. Oper. Res. **12**(3), 415–440 (1987)
14. Lenstra, H.: Integer programming with a fixed number of variables. Math. Oper. Res. **8**, 538–548 (1983)
15. Li, C.Y., Wong, P.W.H., Xin, Q., Yung, F.C.C.: Approximating border length for DNA microarray synthesis. In: Agrawal, M., Du, D.-Z., Duan, Z., Li, A. (eds.) TAMC 2008. LNCS, vol. 4978, pp. 410–422. Springer, Heidelberg (2008)
16. Melle, C., Ernst, G., Schimmel, B., Bleul, A., Koscielny, S., Wiesner, A., Bogumil, R., Möller, U., Osterloh, D., Halbhuber, K.-J., von Eggeling, F.: A technical triade for proteomic identification and characterization of cancer biomarkers. Cancer Res. **64**(12), 4099–4104 (2004)
17. Popa, A., Wong, P.W.H., Yung, F.C.C.: Hardness and approximation of the asynchronous border minimization problem. In: Agrawal, M., Cooper, S.B., Li, A. (eds.) TAMC 2012. LNCS, vol. 7287, pp. 164–176. Springer, Heidelberg (2012)

18. Slonim, D.K., Tamayo, P., Mesirov, J.P., Golub, T.R., Lander, E.S.: Class prediction and discovery using gene expression data. In: Proceedings of Fourth RECOMB, pp. 263–272 (2000)
19. Welsh, J.B., Sapinoso, L.M., Kern, S.G., Brown, D.A., Liu, T., Bauskin, A.R., Ward, R.L., Hawkins, N.J., Quinn, D.I., Russell, P.J., Sutherland, R.L., Breit, S.N., Moskaluk, C.A., Frierson Jr., H.F., Hampton, G.M.: Large-scale delineation of secreted protein biomarkers overexpressed in cancer tissue and serum. PNAS 100(6), 3410–3415 (2003)

Parametrized Complexity of Length-Bounded Cuts and Multi-cuts

Pavel Dvořák[(⊠)] and Dušan Knop[(⊠)]

Department of Applied Mathematics, Charles University, Prague, Czech Republic
{koblich,knop}@kam.mff.cuni.cz

Abstract. We show that the minimal length-bounded L-cut can be computed in linear time with respect to L and the tree-width of the input graph as parameters. We derive an FPT algorithm for a more general multi-commodity length bounded cut problem when parameterized by the number of terminals also. For the former problem we show a W[1]-hardness result when the parameterization is done by the path-width only (instead of the tree-width).

Keywords: Length bounded cuts · Parameterized algorithms · W[1]-hardness

1 Introduction

The study of network flows and cuts begun in 1960s by the work of Ford and Fulkerson [8]. It has many generalizations and applications now. We are interested in a generalization of cuts related to the flows using only short paths.

Length Bounded Cuts. Let $s, t \in V$ be two distinct vertices of a graph $G = (V, E)$ – we call them source and sink, respectively. We call a subset of edges $F \subseteq E$ of G an L-BOUNDED CUT (or L-CUT for short), if the length of the shortest path between s and t in the graph $(V, E \setminus F)$ is at least $L + 1$. We measure the length of the path by the number of its edges. In particular, we do not require s and t to be in distinct connected components as in the standard cut, instead we do not allow s and t to be close to each other. We call the set F a *minimum L-cut* if it has minimum size among all L-bounded cuts of the graph G.

We state the cut problem formally:

PROBLEM: MINIMUM LENGTH BOUNDED CUT (MLBC)
Instance: graph $G = (V, E)$, vertices s, t and integer $L \in \mathbb{N}$
Goal: find a minimum L-bounded s, t cut $F \subset E$

Length bounded flows were first considered by Adámek and Koubek [1]. They showed that the max-flow min-cut duality cannot hold and also that integral

---

Research was supported by the project SVV-2014-260103.

Dušan Knop— Author supported by the project Kontakt LH12095, project GAUK 1784214 and project CE-ITI P202/12/G061.

© Springer International Publishing Switzerland 2015
R. Jain et al. (Eds.): TAMC 2015, LNCS 9076, pp. 441–452, 2015.
DOI: 10.1007/978-3-319-17142-5_37

capacities do not imply integral flow. Finding a minimum length bounded cut is NP-hard on general graphs for $L \geq 4$ as was shown by Itai et al. [11]. They also found algorithms for finding L-bounded cut with $L = 1, 2, 3$ in polynomial time by reducing it to the usual network cut in an altered graph. The algorithm of Itai et al. [11] uses the fact that paths of length $1, 2$ and 3 are edge disjoint from longer paths, while this does not hold for length at least 4.

Baier et al. [2] studied linear programming relaxation and approximation of MLBC together with inapproximability results for length bounded cuts. They also showed instances of the MLBC having $O(L)$ integrality gap for their linear programming approach, which are series-parallel graphs and thus have constant bounded tree-width. First parametrized complexity study of this and similar topics was made by Golovach and Thilikos [9] who studied parametrization by paths-length and the size of the solution for cuts. They also proved hardness results – finding disjoint paths in graphs of bounded tree-width is a W[1]-hard problem.

The MLBC problem has its applications in the network design and in the telecommunications. Huygens et al. [10] use a MLBC as a subroutine in the design of 2-edge-connected networks with cycles at most L long. The MLBC problem is called *hop constrained* in the telecommunications and the number L is so called number of hops. The main interest is in the constant number of hops, see for example the article of Dahl and Gouveia [4].

Note that the standard use of the Courcelle theorem [3] gives for each fixed L a linear time algorithm for the decision version of the problem. But there is no apparent way of changing these algorithms into a single linear time algorithm. Moreover there is a nontrivial dependency between the formula (and thus the parameter L) and the running time of the algorithm given by Courcelle theorem.

Our Contribution. Our main contribution is an algorithm for the MLBC problem, its consequences and an algorithm for a more general multi-terminal version problem.

Theorem 1. *Let G be a graph of tree-width k. Let s and t be two distinct vertices of G. Then for any $L \in \mathbb{N}$ an minimum L-cut between s and t can be found in time $O((L^{k^2})^2 \cdot 2^{k^2} \cdot n)$.*

Corollary 1. *Let G be a graph, k be the size of a vertex cover of G and s and t be two distinct vertices of G. Then for any $L \in \mathbb{N}$ an minimum L-cut between s and t can be found in time $f(k)n$, where f is a computable function.*

Corollary 2. *Let $G = (V, E)$ be a graph of tree-width k, $s \neq t \in V$ and $L \in \mathbb{N}$. There exists a computable function $f : \mathbb{N} \to \mathbb{N}$, such that a minimum L-cut between s and t can be found in time $O(n^{f(k)})$, where $n = |V|$.*

Theorem 1 gives us that the MLBC problem is fixed parameter tractable (FPT) when parametrized by the length of paths and the tree-width and that it belongs to XP when parametrized by the tree-width only (and thus solvable in polynomial time for graph classes with constant bounded tree-width).

We want to mention that our techniques apply also for more general version of the MLBC problem.

Length-Bounded multi-cut. We consider a generalized problem, where instead of only two terminals, we are given a set of terminals. For every pair of terminals, we are given a constraint—a lower bound on the length of the shortest path between these terminals. More formally:

Let $S = \{s_1, \ldots, s_k\} \subset V$ be a subset of vertices of the graph $G = (V, E)$ and let $a : S \times S \to \mathbb{N}$ be a mapping. We call a subset of edges $F \subseteq E$ of G an **a**-*bounded multi-cut* if length of the shortest path between s_i and s_j in the graph $(V, E \setminus F)$ is at least $a(s_i, s_j)$ long for every $i \neq j$. Again if F has smallest possible size, we call it *minimum* **a**-*bounded* $\{s_1, \ldots, s_k\}$-*multi-cut*. We call the vertices $s_1, \ldots, s_k$ *terminals*. Finally, as there are only finitely many values of the mapping a we write $a_{s,t}$ instead of $a(s, t)$, we also write **a** instead of function a. Let $L \geq \max_{s,t \in S} a(s, t)$, we say that the problem is L-*limited*.

PROBLEM: MINIMUM LENGTH BOUNDED MULTI-CUT (MLBMC)
Instance: graph $G = (V, E)$, set $S \subset V$ and $a_{s,t} \in \mathbb{N}$ for all $s, t \in S$,
            satisfying the triangle inequalities
Goal: find a minimum length bounded S multi-cut $F \subset E$

Theorem 2. *Let $G = (V, E)$ be a graph of tree-width k, $S \subseteq V$ with $|S| = t$ and let $p := t + k$. Then for any $L \in \mathbb{N}$ and any L-limited length-constraints **a** on S an minimum **a**-bounded multi-cut can be computed in time $O((L^p)^2 \cdot 2^{p^2} \cdot n)$.*

2   Preliminaries

In this section we recall some standard definitions from graph theory and state what a tree decomposition is. After this we introduce changes of the tree decomposition specific for our algorithm. Finally we give the notion of auxiliary graphs used in proofs of correctness of our algorithm.

We use the notion of tree decomposition of the graph:

Definition 1. *Let $G = (V, E)$ be graph. We say that $T = (\mathcal{W}, F), B : \mathcal{W} \to 2^V$ is a tree decomposition of the graph G, T is a tree and the following holds:*

1. *for each $v \in V$ there exists a $X \in \mathcal{W}$ such that $v \in B(X)$,*
2. *for each $e \in E$ there exists a $X \in \mathcal{W}$ such that $e \subseteq B(X)$,*
3. *for each $v \in V$ the graph $T[X_v]$ is connected, where $X_v = \{X \in \mathcal{W} : v \in B(X)\}$.*

We call the elements of the set $\mathcal{W}$ the nodes, and the elements of the set F the decomposition edges.

We define a width of a tree decomposition $(\mathcal{W}, F)$ as $\max_{X \in \mathcal{W}} |X| - 1$ and the *tree-width* $\mathrm{tw}(G)$ of a graph G as the minimum width of a tree decomposition of the graph G. Moreover if the decomposition is a path we speak about *path-width* of G, which we denote as $\mathrm{pw}(G)$. For abbreviation, we use same denoting for node X and also for the set $B(X)$—we denote both of them by X. The meaning of the definition with function B is that there can be two nodes $X, Y \in \mathcal{W}$, such that $B(X) = B(Y)$.

Nice Tree Decomposition [12]. For algorithmic purposes it is common to define a *nice tree decomposition* of the graph. We root the decomposition tree in an arbitrary node. We naturally orient the decomposition edge towards the root and for an oriented decomposition edge (Y, X) from Y to X we call X the *parent* of Y and Y a *child* of X.

We also adjust a tree decomposition such that for each decomposition-edge (X, Y) it holds that $||X| - |Y|| \leq 1$ (i.e. it joins nodes that differ in at most one vertex). The in-degree of each node is at most 2 and if the in-degree of the node Y is 2 then for its children X_1, X_2 holds that $X_1 = X_2 = Y$ (i.e. they represent the same vertex set).

We classify the nodes of a nice decomposition into four classes—namely *introduce nodes*, *forget nodes*, *join nodes* and *leaf nodes*. We call the node X an introduce node of the vertex x, if it has a single child Y and $X \setminus Y = \{x\}$. We call the node X a forget node of the vertex x, if it has a single child Y and $Y \setminus X = \{x\}$. If the node X has two children, we call it a join node (of nodes Y_1 and Y_2). Finally we call a node X a leaf node, if it has no child.

Proposition 1 *[12]. Given a tree decomposition of a graph G with n vertices that has width k and $O(n)$ nodes, we can find a nice tree decomposition of G that also has width k and $O(n)$ nodes in time $O(n)$.*

So far we have described a standard nice tree decomposition. Now we change the introduce nodes. Let Y be an introduce node and X its parent. We add another two copies Y_1, Y_2 of Y to the decomposition. We remove decomposition edge (Y, X) and add three decomposition edges $(Y, Y_1), (Y_2, Y_1)$ and (Y_1, X). Note that by these further modifications we preserve linear number of nodes in the decomposition.

Auxiliary Subgraphs. Recall that for each edge there is at least one node containing that particular edge. We choose for each edge a node, moreover it is possible to choose a leaf in our decomposition. Note that after our modification of the decomposition for each edge e there is at least one leaf X of the decomposition satisfying $e \subseteq X$. To see this, suppose this is not the case and that some edge, say e, must be placed into a non-leaf node Y. We may suppose that Y is an introduce node (for join or forget node choose its descendant). But in our construction any introduce node has a sibling Y_2 that is a leaf in the decomposition tree.

Thus we choose an arbitrary leaf satisfying the condition and say that the edge e *belongs* to the leaf X. By this process we have chosen set $E(X) \subset E(G)$ for each leaf node X, we further use the notion of *auxiliary graph* G_X. For a leaf node X we set graph $G_X = (X, E(X))$. For a non-leaf node X we set the graph $G_X = (V, E)$, where $V = X \cup \bigcup_{Y \text{ child of } X} V(G_Y)$ and $E = \bigcup_{Y \text{ child of } X} E(G_Y)$.

3   Minimal Length Bounded Multi-cuts

In this section we give a more detailed study of the length constraints for the length-bounded multi-cut and the triangle inequalities. From this we derive Lemma 1 for merging solutions for edge-disjoint graphs.

The Triangle Inequalities. Note that the solution for MLBMC problem has to satisfy the triangle inequalities with respect to its instance. This means that for any three terminals $s, t, u \in S$ and the distance function $dist$ it holds that $dist(s, u) + dist(u, t) \geq dist(s, t) \geq a_{s,t}$. Thus we can restrict instances of MLBMC problem only to those satisfying these triangle inequalities.

Definition 2 (Length Constraints). *Let $G = (V, E)$ be a graph, $S \subset V$ and let $k = |S|$. We call a vector $\mathbf{a} = (a_{s_1,s_2}, \ldots, a_{s_{k-1},s_k})$ a length constraint if for every $s, t, u \in S$ it holds that $a_{s,u} + a_{u,t} \geq a_{s,t}$.*

For our approach it is important to see the structure of the solution on a graph composed from two edge disjoint graphs.

Lemma 1. *Let $G_1 = (V_1, E_1), G_2 = (V_2, E_2)$ be edge disjoint graphs. Then for the graph $G = G_1 \cup G_2$ and $S = V_1 \cap V_2$ and an arbitrary length constraints $\mathbf{a} \in \mathbb{N}^{\binom{|S|}{2}}$ it holds that the minimum length bounded $(S, \mathbf{a})$ multi-cut F for G is the (disjoint) union of the $(S, \mathbf{a})$ multi-cuts F_1 and F_2 for G_1 and G_2.*

4 Restricted Bounded Multi-cut

In this section we present our approach to the L-bounded cut for the graphs of bounded tree-width. We present our algorithm together with some remarks on our results.

Recall that we use dynamic programming techniques on a tree decomposition of input graph. First we want to root the decomposition in a node containing both source and sink of the L-cut problem. This can be achieved by adding both source and sink to all nodes on the unique path in the decomposition tree between any node containing the source and any node containing the sink. Note that this may add at most 2 to the width of the decomposition.

Length Vectors and Tables. As it was mentioned in the previous section, we solve the L-cut by reducing it to simple instances of generalized MLBMC problem. We begin with a mapping $a : S \times S \to \mathbb{N}$ with meaning $a(s, t) = l_{s,t}$. For simplicity we represent the mapping a by a vector, calling it a *length vector* $\mathbf{a}$ and relax it for a node X $\mathbf{a} = (a_{x_1,x_2}, \ldots, a_{x_{k-1},x_k}) \in \mathbb{N}^{\binom{k}{2}}$, where $k = |X|$ and $X = \{x_1, \ldots, x_k\}$. We reduce the problem to $\mathbf{a}$-bounded multi-cut for k terminals, where $k = tw(G) + 2$ and the additional two is for changing the decomposition. Let us introduce a relation on length vectors $\mathbf{a}, \mathbf{b} \in \mathbb{N}^{\binom{k}{2}}$ on X of the same size. We write $\mathbf{a} \preceq \mathbf{b}$, if $a_{x_i,x_j} \leq b_{x_i,x_j}$ for all $1 \leq i < j \leq |X|$.

Let the set of vertices $X = \{x_1, \ldots, x_k\}$, $\mathbf{a}$ be a length vector, let $I \subset [k]$ and let $Y = \{x_i \in X : i \in I\}$. By $\mathbf{a}|_Y$ we denote the length vector $\mathbf{a}$ containing a_{x_i,x_j} if and only if both $i \in I$ and $j \in I$ (in an appropriate order) – in this case we say $\mathbf{a}|_Y$ is $\mathbf{a}$ *contracted* on the set Y.

Recall that for each node X we have defined the auxiliary graph G_X (see Sect. 2 for definition). With a node X we associate the table Tab_X. The table entry $\mathbf{a} = (a_{1,2}, \ldots, a_{k-1,k})$ of Tab_X (denoting $Tab_X[\mathbf{a}]$) for node $X = \{x_1, \ldots, x_k\}$ contains the size of the $\mathbf{a}$-bounded multi-cut for the set X in the graph G_X. Note that for two length vectors $\mathbf{a} \preceq \mathbf{b}$ it holds that $Tab_X[\mathbf{a}] \leq Tab_X[\mathbf{b}]$.

4.1   Node Lemmas

The leaf nodes are the only nodes bearing some edges. We use an exhaustive search procedure for building tables for these nodes. For this we need to compute the lengths of the shortest paths between all the vertices of the leaf node, for which we use the well known procedure due to Floyd and Warshall [6,13]:

Proposition 2. *Let G be a graph with nonnegative length $f : G(E) \to \mathbb{N}$. It is possible to compute the table of lengths of the shortest paths between any pair $u, v \in V(G)$ with respect to f in time $O(|V(G)|^3)$.*

Lemma 2 (Leaf Nodes). *For all L-limited length vectors and a leaf node X the table Tab_X of sizes of minimum length-bounded multi-cuts can be computed in time $O(L^{k^2} \cdot 2^{k^2} \cdot k^3)$, where $k = |X|$.*

We now use Lemma 1 to prove time complexity of finding a dynamic programming table for join nodes from the table of its children.

Lemma 3 (Join Nodes). *Let X be a join node with children Y and Z, let L be the limit on length vectors components and let $k = |X|$. Then the table Tab_X can be computed in time $O(L^{k^2})$ from the table Tab_Y and Tab_Z.*

Proof. Recall that graphs G_Y and G_Z are edge disjoint and that we store sizes of **a**-bounded multi-cuts. Note also that $X = V(G_Y) \cap V(G_Z)$ and so we can apply Lemma 1 and set $Tab_X[\mathbf{a}] := Tab_Y[\mathbf{a}] + Tab_Z[\mathbf{a}]$, for each **a** satisfying the triangle inequalities.

As there are $O(L^{k^2})$ entries in the table Tab_X we have the complexity we wanted to prove.   □

As the forget node expects only forgetting a vertex and thus forgetting part of the table of the child. This is the optimizing part of our algorithm.

Lemma 4 (Forget Nodes). *Let X be a forget node, Y its child, let L be the limit on length vectors components and let $k = |X|$. Then the table Tab_X can be computed in time $O((L^{k^2})^2)$ from the table Tab_Y.*

Proof. Fix one length vector **a** and compute the set $\mathcal{A}(\mathbf{a})$ of all Y-augmented length vectors. Formally $\mathbf{b} \in \mathcal{A}(\mathbf{a})$ if **b** is a length vector satisfying the triangle inequalities for Y and $\mathbf{b}|_X = \mathbf{a}$. After this we set $Tab_X[\mathbf{a}] := \min_{\mathbf{b} \in \mathcal{A}(\mathbf{a})} Tab_Y[\mathbf{b}]$.

There are at most L^{k^2} of Y-augmented length vectors for each **a** and this gives the claimed time.   □

Also the introduce node (as the counter part for forget node) only adds coordinates to the table of its child. It does no computation as there are no edges it can decide about – these nodes now only add isolated vertex in the graph.

Lemma 5 (Introduce Nodes). *Let X be an introduce node, Y its child, let L be the limit on length vectors components and let $k = |X|$. Then the table Tab_X can be computed in time $O(L^k)$ from the table Tab_Y.*

Proof. Let x be the vertex with the property $x \in X \setminus Y$. The key property is that the vertex x is an isolated vertex in G_X and thus we can set $Tab_X[\mathbf{a}] := Tab_Y[\mathbf{a}|_Y]$, because x is arbitrarily far from any vertex in G_Y, especially from the set Y. □

4.2 Proofs of Theorems

We use Lemmas 2, 3, 4 and 5 to prove the theorem about computing L-bounded (s, t)-cut in graph of bounded tree-width. For this, note that we can use $k = O(tw(G))$ and put it into all the Lemmas as it is an upper bound on the size of any node in the decomposition (the O notation is because $+1$ in the definition and possible $+2$ for producing of node with both s and t in it).

We compute all the L-bounded length-constraint that satisfy the triangle inequalities in advance. This takes additional time $O(L^{k^2} \cdot k^3)$ which can be upper-bounded by $O((L^{k^2})^2 \cdot 2^k)$ for $k \geq 2$ and $L \geq 2$ and so this does not make the overall time complexity worse.

Proof of Theorem 1. As there are $O(n)$ nodes in nice tree decomposition (by Proposition 1) and as we can upper-bound time needed to compute any type of node by $O((L^{k^2})^2 \cdot 2^{k^2})$, we have complexity proposed in state of the Theorem 1. □

Let us now point out that the value of the parameter L can be upper-bounded by the number of vertices n of the input graph G (in fact by $n^{1-\varepsilon}$ as it is proved in [2]).

We now want to sum-up the key ideas leading to Theorem 1. First, it is the use of dynamic program for computing all options of cuts for bounded number of possible choices and for second it is the idea of creating a node that includes both the source and the sink while not harming the tree-width too much. On the other hand, we can use this idea to solve also the generalized version of the problem – length-bounded multi-cut – with the additional parameter the number of terminals. It is easy to see that in this setting that again it is possible to achieve node containing every terminal and thus this yields following Theorem 3. □

Theorem 3. *Let $G = (V, E)$ be a graph of tree-width k, $S \subseteq V$ with $|S| = t$ and let $p := t + k$. Then for any $L \in \mathbb{N}$ and any L-limited length-constraints $\mathbf{b}$ satisfying the triangle inequalities on S an minimum $\mathbf{b}$-bounded multi-cut can be computed in time $O((L^p)^2 \cdot 2^{p^2} \cdot n)$.*

5 Hardness of the L-bounded Cut

In this section we prove MLBC parametrized by path-width is W[1]-hard [7] by FPT-reduction from k-MULTICOLOR CLIQUE.

PROBLEM: k-MULTICOLOR CLIQUE
Instance: k-partite graph $G = (V_1 \dot\cup V_2 \dot\cup \ldots \dot\cup V_k, E)$, where V_i is
            independent set for every i and they are pairwise disjoint
Parameter: k
Goal: find a clique of size k

Denoting. In this section, sets $V_1, \ldots, V_k$ are always partites of k-partite graph G. We denote edges between V_i and V_j by E_{ij}. The problem is W[1]-hard even if every independent set V_i has same size and the number of edges between every V_i and V_j is same. In whole Sect. 5 we denote the size of arbitrary V_i by N and size of arbitrary E_{ij} by M. For FPT-reduction from k-MULTICOLOR CLIQUE to MLBC we need:

1. Create a MLBC instance $G' = (V', E'), s, t, L$ from k-MULTICOLOR CLIQUE instance $G = (V_1 \,\dot\cup\, V_2 \,\dot\cup\, \ldots \,\dot\cup\, V_k, E)$ where size of G' is polynomial from the size of G.
2. Prove that G contains k-clique if and only if G' contains L-bounded cut of size $f(k, N, M)$ where f is polynomial.
3. Prove that path-width of H is smaller than $g(k)$ where g is computable function.

Our ideas were inspired by work of Michael Dom et al. [5]. They proved W[1] hardness of CAPACITATED VERTEX COVER and CAPACITATED DOMINATING SET parametrized tree width of input graph. We remarked that their reduction also proves W[1] hardness of these problems parametrized by path-width.

5.1 Basic Gadget

In k-MULTICOLOR CLIQUE problem we need select exactly one vertex from each independent set V_i and exactly one edge from each E_{ij}. And we have to make certain that if $e \in E_{ij}$ is the selected edge and $u \in V_i, v \in V_j$ are the selected vertices then $e = \{u, v\}$. The idea of the reduction is to have a basic gadget for every vertex and edge. We connect gadgets g_v for every v in V_i into a path P_i. The path P_i is cut in the gadget g_v if and only if the vertex $v \in V_i$ is selected into clique. The same idea will be used for selecting the edges.

Definition 3. *Let $h, Q \in \mathbb{N}$. Butte $B(s', t', h, Q)$ is graph which contains h paths of length 2 and Q paths of length $h + 2$ between s' and t'. The short paths (of length 2) are called shortcuts, the long paths are called ridgeways and the parameter h is called height.*

In our reduction all buttes will have the same parameter Q (it will be computed later). Let $B(s', t', h, Q)$ be a butte. We denote by $s(B), t(B), h(B), Q(B)$ the parameters of butte B s', t', h and Q, respectively. We state easy but important observation about butte path-width:

Observation 4. *Path-width of arbitrary butte B is at most 3.*

Let $B(s', t', h, Q)$ be a butte. Let P_{uv} be a shortest path between u and v, which enters into B in s' and leaves it in t'. The important properties of the butte B are:

1. By removing one edge from all h shortcuts of butte B, we extend path P_{uv} by h. If we cut all shortcuts of butte B we say the butte B is ridged.
2. Let size of the cut is bounded by $K \in \mathbb{N}$ and we can remove edges only from B. If we increase Q to be bigger than K then P_{uv} cannot be cut by removing edges from B (only extended by ridging the butte B).

5.2 Butte Path

In this section we define how we connect buttes into a path, which we call highland. The main idea is to have highland for every pair $(i, j), i \neq j \in [k]$. In highland for (i, j), there are buttes for every vertex $v \in V_i$ and every edge $e \in E_{i,j}$. We connect vertex buttes and edge buttes into a path. Then we set the butte heights and limit the size of the cut in such way that:

1. Exactly one vertex butte and exactly one edge butte have to be ridged.
2. If buttes for vertex v is ridged, then only buttes for edges incident with v can be ridged.

Formal description of highland is in the following definition.

Definition 4. *Highland $H(X, Y, s, t)$ is a graph containing 2 vertices s and t and $Z = X + Y$ buttes $B_1, \ldots, B_Z$ where:*

1. $s = s(B_1), t = t(B_Z)$ and $t(B_i) = s(B_{i+1})$ for every $1 \leq i < Z$.
2. $h(B_i) = X^2 + i$ for $1 \leq i \leq X$.
3. $h(B_i) \in \{X^4, \ldots, X^4 + X - 1\}$ for $X + 1 \leq i \leq Z$.
4. $Q(B_i) = X^4 + X^2$ for every i.

Let $H(X, Y, s, t)$ be a highland. We call buttes $B_1, \ldots, B_X$ from H low and buttes $B_{X+1}, \ldots, B_{X+Y}$ high (low buttes will be used for vertices and high buttes for edges). The vertex $t(B_X) = s(B_{X+1})$, where low and high buttes meet, is called the center of highland H. Note that there can be more buttes with the same height among high buttes and they are not ordered by height as the low buttes.

Proposition 3. *Let $H(X, Y, s, t)$ be a highland. Let $L = 2(X + Y) + X^4 + X^2 + X - 1$. Let C be the L-cut of size $X^4 + X^2 + X$, which cut all path of length L and shorter between s and t, then:*

1. *The cut C contains only edges obtained by ridging the exactly two buttes B_i, B_j, such that B_i is low and B_j is high.*
2. *Let B_i be the ridged low butte and B_j be the ridged high butte. Then, $h(B_j) = X^4 + X - i$.*

5.3 Reduction

In this section we present our reduction. Let $G = (V_1 \,\dot\cup\, V_2 \,\dot\cup\, \ldots \,\dot\cup\, V_k, E)$ be the input for k-MULTICOLOR CLIQUE. As we stated in the last section, the main idea is to have low butte B_v for every vertex v and high butte B_e for every edge e. Vertex v and edge e is selected into the k-clique if and only if butte B_v and butte B_e are ridged. From G we construct MLBC input G', s, t, L:

1. For every $1 \leq i, j \leq k, i \neq j$ we create highland $H^{i,j}(N, M, s, t)$ of buttes $B_1^{i,j}, \ldots, B_{N+M}^{i,j}$.
2. Let $V_i = \{v_1, \ldots, v_N\}$. Vertex $v_\ell \in V_i$ is represented by low butte $B_\ell^{i,j}$ of the highland $H^{i,j}$ for every $j \neq i$. Thus, we have $k - 1$ copies of buttes (in different highlands) for every vertex. Hence, we need to be certain that only buttes representing the same vertex are ridged. Note that buttes representing the same vertex have the same height and the same distance from the vertex s.

3. Let $E_{ij} = \{e_1, \ldots, e_M\}, i < j$. Edge $e_\ell = \{u, v\} \in E_{ij}(u \in V_i, v \in V_j)$ is represented by high butte $B_{N+\ell}^{i,j}$ of the highland $H^{i,j}$ and by high butte $B_{N+\ell}^{j,i}$ of the highland $H^{j,i}$. Note that two buttes represented the same edge has same distance from the vertex s. Let h_i, h_j be the heights of buttes representing vertices u and v, respectively. We set the height of high buttes:
   (a) $h(B_{N+\ell}^{i,j}) = N^4 + N - h_i$
   (b) $h(B_{N+\ell}^{j,i}) = N^4 + N - h_j$
4. We add edge $\{t(B_\ell^{i,j}), t(B_\ell^{i,j+1})\}$ for every $1 \leq i \leq k, 1 \leq j < k, i \neq j$ and $1 \leq \ell < N$.
5. We add paths of length $N - 1$ connected $t(B_\ell^{i,j})$ and $t(B_\ell^{j,i})$ for every $1 \leq i, j \leq k, i \neq j$ and $1 \leq \ell < M$.
6. $L = 2(N + M) + N^4 + N^2 + N - 1$

We call paths between highlands in Items 4 and 5 the valley paths.

Observation 5. *Graph G' has polynomial size from graph G.*

Theorem 6. *If graph G has a clique of size k then (G', s, t) has an L-cut of size $k(k-1)(N^4 + N^2 + N)$.*

Proof. Let G has a k-clique $\{v_1, \ldots, v_k\}$ where $v_i \in V_i$ for every i and $e_{ij} = \{v_i, v_j\} \in E_{ij}$. For every i we ridge all $k - 1$ buttes representing the vertex v_i in G'. And for every $i < j$ we ridge both buttes representing the edge e_{ij}.

We claim that set of removed edges from ridged buttes forms the L-cut. Let $H^{i,j}$ be an arbitrary highland. There is no st-path shorter than L in $H^{i,j}$. Let $h(B_v) = N^2 + \ell$ where B_v is arbitrary butte representing the vertex v_i. By construction of G', the high butte representing the edge e_{ij} in $H^{i,j}$ has height $N^4 + N - \ell$. Thus, ridged buttes in $H^{i,j}$ extend the shortest st-path by $N^4 + N^2 + N$ and it has length $2(M + N) + N^4 + N^2 + N$. Buttes representing the vertex v_i have same height. Thus, path through the low buttes of highlands using some valley path is always longer than path going through low buttes of only one highland. Therefore, it is useless to use valley paths among low buttes for the shortest st-path.

Other situation is among high buttes because buttes representing the same edge have different heights. Butte B_v representing vertex v_i extend the shortest path at least by $N^2 + 1$. Butte B_e representing edge $e_{i,j}$ extend the shortest at least by N^4. However, if $h(B_v) + h(B_e) < N^4 + N^2 + N$ then B_v and B_e have to be in different highlands. Therefore, the st-path going through B_v and B_e has to use a valley path between high buttes, which has length $N - 1$. And such st-path has length at least $2(N + M) + N^4 + N^2 + N$.

We remove $N^4 + N^2 + N$ edges from each highland and there are $k(k - 1)$ highlands in G'. Therefore, G' has L-cut of the size $k(k - 1)(N^4 + N^2 + N)$. □

Theorem 7. *If (G', s, t) has an L-cut of size $k(k - 1)(N^4 + N^2 + N)$ then G has a clique of size k.*

Proof. Let C be an L-cut of G'. Every shortest st-path going through every highland has to be extended by $N^4 + N^2 + N$. By Proposition 3 (Item 1), exactly one low butte and exactly one high butte of each highland has to be ridged. We remove $(N^4 + N^2 + N)$ from every highland in G'. Therefore, there can be only edges from ridged buttes in C.

For fixed i, highlands $H^{i,j}$ are the highlands which low buttes represent vertices from V_i. We claim that ridged low buttes of $H^{i,1}, \ldots, H^{i,k}$ represent the same vertex. Suppose for contradiction, there exists two low ridged buttes B_ℓ of $H^{i,\ell}$ and B_m of $H^{i,m}$ which represent different vertex from V_i. Without loss of generality $H^{i,\ell}$ and $H^{i,m}$ are next to each other (i.e. $|\ell - m| = 1$) and distance from s to $s(B_\ell)$ is smaller than distance from s to $s(B_m)$. Let B'_ℓ be a butte of $H^{i,m}$ such that it has same distance from s as butte B_ℓ. The path s–$t(B'_\ell)$–$t(B_\ell)$–t does not go through any ridged low butte. Therefore, this path is shorter than L, which is contradiction. We can use the same argument to show that there are not two high ridged buttes of highland $H^{i,j}$ and $H^{j,i}$ which represent different edges from E_{ij}.

We put into the k-clique $K \subset V(G)$ the vertex $v_i \in V_i$ if and only if arbitrary butte representing the vertex v_i is ridged. We proved in the previous paragraph that exactly one vertex from V_i can be put into the clique K. Let $e_{ij} \in E_{ij}$ is an edge represented by ridged buttes. We claim that $v_i \in e_{ij}$. Let $B \in H^{i,j}$ be a butte representing v_i with height $N^2 + \ell$. Then by Proposition 3 (Item 2), butte $B' \in H^{i,j}$ of height $N^4 + N - \ell$ has to be ridged. By construction of G', only buttes representing edges incident with v_i have such height. Therefore, chosen edges are incident with chosen vertices and they form the k-clique of the graph G.   □

Observation 8. *Graph G' has path-width in $O(k^2)$.*

The following theorem is corollary of Observations 5 and 8 and Theorems 6 and 7.

Theorem 9. MINIMAL LENGTH BOUNDED CUT *parametrized by path-width is* W[1]*-hard.*

6   Conclusions

There is another standard generalization of the length bounded cut problem – where we add to each edge also its length. If this length is integral it is possible to extend and use our techniques (we only subdivide edges longer than 1 – this doesn't raise the tree-width of the graph on the input). On the other hand, if we allow fractional numbers, it is uncertain how to deal with such a generalization.

Acknowledgments. Authors thank to Jiří Fiala, Petr Kolman and Lukáš Folwarczný for fruitful discussions about the problem. We would like to mention that part of this research was done during Summer REU 2014 at Rutgers University.

References

1. Adámek, J., Koubek, V.: Remarks on flows in network with short paths. Commentationes Mathematicae Universitatis Carolinae **12**, 661–667 (1971)
2. Baier, G., Erlebach, T., Hall, A., Köhler, E., Kolman, P., Pangrác, O., Schilling, H., Skutella, M.: Length-bounded cuts and flows. ACM Trans. Algorithms **7**, 4:1–4:27 (2010)
3. Courcelle, B.: Graph rewriting: an algebraic and logic approach. In: van Leeuwen, J. (ed.) Handbook of Theoretical Computer Science, pp. 194–242. Elsevier, Amsterdam (1990)
4. Dahl, G., Gouveia, L.: On the directed hop-constrained shortest path problem. Oper. Res. Lett. **32**, 15–22 (2004)
5. Dom, M., Lokshtanov, D., Saurabh, S., Villanger, Y.: Capacitated domination and covering: a parameterized perspective. In: Grohe, M., Niedermeier, R. (eds.) IWPEC 2008. LNCS, vol. 5018, pp. 78–90. Springer, Heidelberg (2008)
6. Floyd, R.W.: Algorithm 97: shortest path. Commun. ACM **5**, 345 (1962)
7. Flum, J., Grohe, M.: Parameterized Complexity Theory. Texts in Theoretical Computer Science. An EATCS Series. Springer-Verlag New York Inc., Secaucus (2006)
8. Ford, L.R., Fulkerson, D.R.: Maximal flow through a network. Can. J. Math. **8**, 399–404 (1956)
9. Golovach, P.A., Thilikos, D.M.: Paths of bounded length and their cuts: parameterized complexity and algorithms. Discrete Optim. **8**, 72–86 (2011)
10. Huygens, D., Labbé, M., Mahjoub, A.R., Pesneau, P.: The two-edge connected hop-constrained network design problem: valid inequalities and branch-and-cut. Networks **49**, 116–133 (2007)
11. Itai, A., Perl, Y., Shiloach, Y.: The complexity of finding maximum disjoint paths with length constraints. Networks **12**, 277–286 (1982)
12. Kloks, T. (ed.): Treewidth, Computations and Approximations. LNCS, vol. 842. Springer, Heidelberg (1994)
13. Warshall, S.: A theorem on boolean matrices. J. ACM **9**, 11–12 (1962)

Algorithms and Hardness for Signed Domination

Jin-Yong Lin[1] and Sheung-Hung Poon[2]([✉])

[1] Department of Computer Science, National Tsing Hua University,
Hsinchu, Taiwan, R.O.C.
yongdottw@hotmail.com
[2] School of Computing and Informatics, Institut Teknologi Brunei,
Gadong, Brunei Darussalam
sheung.hung.poon@gmail.com

Abstract. A signed dominating function for a graph $G = (V, E)$ is a function $f: V \rightarrow \{+1, -1\}$ such that for all $v \in V$, the sum of the function values over the closed neighborhood of v is at least one. The weight $w(f(V))$ of signed dominating function f for vertex set V is the sum of $f(v)$ for $v \in V$. The *signed domination number* γ_s of G is the minimum weight of a signed dominating function for G. The *signed domination (SD) problem* asks for a signed dominating function which contributes the signed domination number. First we show that the SD problem is W[2]-hard. Next we show that the SD problem on graphs of maximum degree six is APX-hard. Then we present constant-factor approximation algorithms for the SD problem on subcubic graphs, graphs of maximum degree four, and graphs of maximum degree five, respectively. In addition, we present an alternative and more direct proof for the NP-completeness of the SD problem on subcubic planar bipartite graphs. Lastly, we obtain an $O^*(5.1957^k)$-time FPT-algorithm for the SD problem on subcubic graphs G, where k is the signed domination number of G.

1 Introduction

A *signed dominating function* for a graph $G = (V, E)$ is a function $f : V \rightarrow \{+1, -1\}$ such that $f(N_G[v]) \geq 1$ for all $v \in V$, where $N_G[v]$ denotes the *closed neighborhood* of a vertex v in G, that is, the union of v and all the adjacent vertices of v. The *closed neighborhood* $N_G(U)$ of a subset of vertices U in G is the union of $N_G[v]$ of all vertices v in U. The *weight* $w(f(V))$ of a signed dominating function f for vertex set V is the sum of $f(v)$ for $v \in V$. The *signed domination number* γ_s of G is the minimum weight of a signed dominating function for G. The *signed domination (SD) problem* asks for a signed dominating function which contributes the signed domination number. By assigning value $+1$ or -1 to each vertex in graph G, we can model G as real-world networks in sociology, electronics and operation research [5]. For example, in electronics, value $+1$ or -1 for vertex v may represent a positive or negative spin of electron v; and in social network, value $+1$ or -1 for vertex v may represent an agreeing or disagreeing vote of voter v. In such model of signed domination, although the total negative votes might be more than the total positive votes, we can still

© Springer International Publishing Switzerland 2015
R. Jain et al. (Eds.): TAMC 2015, LNCS 9076, pp. 453–464, 2015.
DOI: 10.1007/978-3-319-17142-5_38

assure that all local groups of voters (represented by closed neighborhoods) have more positive votes than negative votes.

A graph is *planar* if it can be embedded in the plane (drawn with points for vertices and curves for edges) without any edge crossing. A graph is *bipartite* if its vertex set can be partitioned into two independent sets. A graph is *subcubic* if the degrees of its vertices are all at most three. A graph is *grid* if it is an induced subgraph of a grid. Note that a grid graph is a planar bipartite graph.

Next, we describe the related work on the SD problem. First, Hattingh et al. [5] showed that the SD problem is NP-complete for chordal and bipartite graphs, respectively. Thus the SD problem for general graphs is NP-complete. Damaschke [2] then showed that the SD problem is NP-complete for subcubic planar graphs. Later, Lee [6] showed that the SD problem for planar bipartite graphs is NP-complete. Furthermore, Zheng et al. [7] showed that the SD problem for subcubic grid graphs is NP-complete.

A problem is called *fixed-parameter tractable (FPT)* with respect to a parameter k if there exists a solution running in $f(k) \cdot n^{O(1)}$ time, where f is a function of the solution size k which is independent of n, and the corresponding algorithm which contributes such a solution is called an *FPT-algorithm*. Zheng et al. [7] showed that the SD problem has a kernel of size $(k^2 + k)/2$ for general graphs, where the parameter k is the number of $+1$ values assigned to vertices of the given graph. Thus, there is an FPT-algorithm for SD problem on general graphs in such a parameter k [3]. Note that $FPT = W[0] \subseteq W[1] \subseteq W[2] \subseteq \ldots \subseteq W[t]$. However, for the specific parameter k in our paper, which is the weight of the signed dominating function for a given graph, we show that the SD problem on general graphs is W[2]-hard. Furthermore, for subcubic graphs, we are able to obtain an FPT-algorithm for our parameter.

2   W[2]-hardness for SD Problem

A dominating set of a graph $G = (V, E)$ is a vertex set $D \subseteq V$ such that for each vertex $v \in V$, there exists at least one vertex in $N_G[v]$ belonging to set D. In other words, we can label V with $\{0, +1\}$, such that the closed neighborhood of each vertex is positive. The *domination number* $\gamma(G)$ is the cardinality of the minimum dominating set of G. *Dominating set problem* asks for a dominating set which contributes the domination number. Downey et al. [3] showed that domination problem on general connected graphs is W[2]-complete. In this section, we show that the SD problem on general connected graphs is $W[2]$-hard.

In our reduction, we need to make use of a graph X, which was first introduced by Hattingh [5]. The graph X is defined as in Fig. 1(a). Faria [4] found that if we label the vertices of X as shown in Fig. 1(b) such that the weight of such a signed domination function is -1. In the following lemma, we proceed to show that -1 is in fact the minimum weight of a legitimate signed domination function the graph X can have.

Lemma 1. *The weight of any signed domination function for the graph X is greater than or equal to -1.*

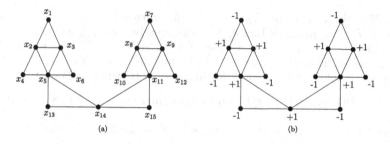

Fig. 1. (a) Graph X. (b) Graph X with weight -1.

Proof. Let f be a singed dominating function for graph X. First, we focus on vertex x_5. Since x_5 has 7 vertices in its closed neighborhood $N[x_5]$, at least four of the seven vertices $x_2, x_3, x_4, x_5, x_6, x_{13}, x_{14}$ of $N[x_5]$ are labeled with $+1$ by function f. Similarly, for vertex x_{11}, at least four of the seven vertices $x_8, x_9, x_{10}, x_{11}, x_{12}, x_{14}, x_{15}$ of $N[x_{11}]$ are labeled with $+1$ by function f. As there is only one common vertex x_{14} in $N[x_5]$ and $N[x_{11}]$, we obtain that at least 7 vertices in graph X are labeled with $+1$. Then at most 8 vertices in X are labeled with -1. Hence, the weight of function f will be greater than or equal to -1. □

Then we are prepared to show the W[2]-hardness of the SD problem.

Theorem 1. *The SD problem on general connected graphs is W[2]-hard.*

Proof. We describe our reduction from dominating set problem as follows. Given a graph $G = (V, E)$, we construct a new graph G' as described below. Initially we set G' to be G. Then we add each vertex v of G' a set of $deg_G(v) + 1$ paths of length two, where $deg_G(v)$ is the degree of vertex v in G, and we let U be the set of newly added vertices on the paths of length two. Note that $|U| = 4m + 2n$, where n is number of vertices of G and m is number of edges of G. Next, we add $4m + n$ copies of graph X into graph G' and denote them by $X_1, X_2, \ldots, X_{4m+n}$, respectively. $Y = X_1 \cup \ldots \cup X_{4m+n}$. Then we take any previously added length-two path, say ρ. Let p be the degree-1 vertex of ρ, and q the neighbor of p on ρ. Now, we join the vertex q with the corresponding vertex x_2 of each copy $X_i(i = 1, \ldots, 4m + n)$ of graph X. This completes the construction of the graph G'. Let V' be the set of vertices in G'. As G is a connected graph, graph G' is also a connected graph by our construction. We then show that the graph G has a dominating set of size at most k if and only if there is a signed dominating function with weight at most $2k$ in G'.

Assume that there is a dominating set D of size k in G. Then we will show that there is a signed dominating function f with weight at most $2k$ in G'. We construct a function f, which labels the vertices in U and vertices in D with value $+1$, and label the corresponding vertices $x_2, x_3, x_5, x_8, x_9, x_{11}, x_{14}$ for each $X_i(i = 1, \ldots, 4m + n)$ with value $+1$. Then we label the other vertices in G' with value -1. Now, we claim that f is a signed dominating function of G' with weight at most $2k$. Clearly, the weight of the closed neighborhood of any vertex in U or in Y is positive. Then we consider vertices in $V' \setminus (U \cup V(Y))$. For any vertex v in $V' \setminus (U \cup V(Y))$, there is at least one vertex in $N_G[v] \cap D$. Thus more than half of the vertices in $N_{G'}[x]$ are labeled with +1 in function f. Thus the weight of $N'_G[x]$ is positive. Hence, f is a legitimate signed dominating function of G'. Now, we construct two subsets of V', $P(V')$ and $M(V')$, where $P(V')$ contains the vertices labeled with +1 in V' for function f, and $M(V')$ contains the vertices labeled with -1 in V' for function f. Hence, we obtain that

$$w(f(V')) = |P(V')| - |M(V')| \le k + (4m+2n) + 7(4m+n) - (n-k) - 8(4m+n) = 2k.$$

Conversely, we assume that f is a signed dominating function of G' with weight at most $2k$. We collect the vertices in $V' \setminus (U \cup V(Y))$ labeled with +1 as a set D. Then we claim that D is a dominating set of G of size at most k. We only need to show that the vertices in $V \setminus D$ are dominated by vertices in D. Let v be a vertex in $V \setminus D$. As v lies in $V \setminus D$, we have $f(x) = -1$. As f is a signed dominating function and $deg_{G'}(v) = 2deg_G(v) + 1$, at least $deg_G(v) + 2$ vertices are labeled with +1 by function f. Thus at least one vertex u in $N_{G'}[v] \setminus (U \cup V(Y))$ is labeled with +1. This implies that $u \in D$ such that $u \in N_G[v]$. Thus, D is a dominating set of G. Now, we compute the size of set D. We use the same notations $P(V')$ and $M(V')$ as defined above. We are going to bound the number of vertices in $P(V) = P(V') \setminus (U \cup V(Y))$, which is equal to the size of D. First, we have

$$w(f(V')) = w(f(U)) + w(f(V(Y))) + w(f(V' \setminus (U \cup V(Y)))) \le 2k.$$

As $w(f(V(Y))) \ge -(4m+n)$, it implies that:

$$(4m+2n) - (4m+n) + w(f(V' \setminus (U \cup V(Y)))) \le 2k,$$

$$w(f(V' \setminus (U \cup V(Y)))) \le 2k - n.$$

Consequently,

$$w(f(V' \setminus (U \cup V(Y)))) = |P(V)| - |M(V)| = 2|P(V)| - |V| = 2|P(V)| - n \le 2k - n.$$

Thus we obtain that $|D| = |P(V)| \le k$. This completes the proof.   □

3   APX-hardness for Graphs of Maximum Degree Six

Alimonti and Kann [1] show that the domination problem is APX-hard on sub-cubic graphs. For the signed domination problem, we manage to show that the SD problem for graphs of maximum degree six is APX-hard in the following theorem, whose proof is omitted due to lack of space. Whether the SD-problem is APX-hard for graphs of maximum degree lower than six is open.

Theorem 2. *The SD problem on graphs of maximum degree six is APX-hard.*

4   Approximation Algorithm for Small Degree Graphs

In this section, we present constant-factor approximation algorithms for the SD problem on subcubic graphs, graphs of maximum degree four, and graphs of maximum degree five, respectively.

4.1 3-approximation for Subcubic Graphs

Theorem 3. *There is a linear-time 3-approximation algorithm for the SD problem on subcubic graphs.*

Proof. Let $G = (V, E)$, and let f^* be the optimal signed domination function of G. We use the 3-element subsets as defined in the proof of Lemma 3. Then by the same reasoning as the proof of Lemma 3, we know that for each such subset, one of its elements is labeled with value -1. Moreover, other vertices in V must be labeled with $+1$. Thus we have $w(f^*(V)) \geq \dfrac{|V|}{3}$. By labeling all the vertices in V with $+1$, we can obtain a signed domination function with weight $|V|$. Hence we have a 3-approximation for the SD problem on subcubic graphs.   □

4.2 13-approximation Graphs with Maximum Degree Four

In our approximation algorithm for graphs of maximum degree four, we use similar terminologies as Damaschke [2]. Let P be the set of vertices with label value $+1$, and M be the set of vertices with label value -1. Let D_i be the set of vertices of degree i. Let P_i denote the set of vertices labeled with $+1$, each of which has exactly i neighbors labeled with -1. Similarly, M_i denotes the set of vertices labeled with -1, each of which has exactly i neighbors labeled with $+1$. Furthermore, we let p, m, p_i and m_i denote the cardinalities of sets P, M, P_i and M_i, respectively. Moreover, we use $N[\cdot]$ to represent $N_G[\cdot]$ for simplicity. Damaschke [2] showed the following lemma.

Lemma 2 (Damaschke [2]). *Any signed dominating function in a graph of maximum degree five satisfies $p - m = p_0 + p_1/2 + m_3/2 + m_4 + 3m_5/2$.*

As all the terms on the right hand side of the equality are non-negative, this lemma implies that the weight of a signed domination function of a graph of maximum degree five is greater than or equal to zero. Moreover, Damaschke [2] also showed that it can be checked in linear time whether $\gamma_s = 0$ for a graph of maximum degree five. Thus in following theorems, namely Theorems 4 and 5, the designed algorithms are executed only when $\gamma_s > 0$ after such a linear-time checking procedure has been executed. Since γ_s may not be bounded in graphs of maximum degree greater than five, γ_s can be a negative value with arbitrary large magnitude in graphs of degree greater than five. Hence, such graphs cannot be approximated.

Theorem 4. *There is a linear-time 13-approximation algorithm for the SD problem on graphs of maximum degree four.*

Proof. Let $G = (V, E)$ be a graph with maximum degree at most four and $\gamma_s > 0$. As the maximum vertex degree in G is four, we have $m_5 = 0$. By substituting this into the equation in Lemma 2, we thus have:

$$\gamma_s = p - m = p_0 + p_1/2 + m_3/2 + m_4.$$

Now we bound $p_2 - m_2$. By considering the number of edges which incident to vertices labeled with $+1$ and vertices labeled with -1, respectively, we have $p_1 + 2p_2 = 2m_2 + 3m_3 + 4m_4$. Thus

$$p_2 - m_2 = 3m_3/2 + 2m_4 - p_1/2.$$

Let R be the set of vertices in P_2, which have neighbors in $P \cap D_2$. Then we have that $|R| \leq 2p_0 + p_1$. Note that vertices in P_2 also lie in D_4. Now let S be the set of vertices in D_4, which at most two neighbors in D_2. It is clear that $P_2 \subseteq (R \cup S)$, and $(P_2 \setminus S) \subseteq R$. Thus we have $|P_2 \setminus S| \leq 2p_0 + p_1$.

Next, at most p_1 vertices of M_2 may have neighbors not in P_2. In other words, there are at least $m_2 - p_1$ vertices which have both neighbors in P_2, and we have seen above that at most $2p_0 + p_1$ of them lie in $P_2 \setminus S$. Then at most $4p_0 + 2p_1$ vertices of M_2 may have neighbors in $P_2 \setminus S$. Thus we have that at least $m_2 - 4p_0 - 3p_1$ vertices of M_2 have both neighbors in $S \cap P_2$. We further let T be the set of vertices in D_2 having both neighbors in S. Then we have

$$|T| \geq m_2 - 4p_0 - 3p_1.$$

Now we construct a function f, which labels vertices in T with value -1, and labels other vertices with $+1$. Note that the set T can be constructed in linear time. Then we claim that the function f is a signed dominating function for G. Let $N[T]$ be the union of vertices of $N[v]$ for all vertices v in T. We only need to consider the vertices in $N[T]$ to verify whether $w(f(N[v])) \geq 1$ for any $v \in N[T]$. First, we consider the vertex $v \in T$. Since v is of degree 2 and connects to vertices in S which were labeled with value $+1$, we then have $w(f(N[v])) = 1$. Then we consider a vertex u in $N[T] \setminus T$. Since u belongs to the set S, u is a degree 4 vertex and is adjacent to at most two neighbors in D_2. This implies that there are at most two vertices labeled with -1 in $N[u]$. Thus we have $w(f(V)) \geq 1$. Hence, function f is a signed dominating function for G.

Hence, we can show that the signed dominating function f is a 13-approximation of the optimal signed dominating function.

$$\begin{aligned}
w(f(V)) &= (n - |T|) - |T| \\
&\leq p_0 + p_1 + p_2 + m_2 + m_3 + m_4 - 2|T| \\
&\leq 9p_0 + 7p_1 + p_2 - m_2 + m_3 + m_4 \\
&= 9p_0 + 13p_1/2 + 5m_3/2 + 3m_4 \\
&\leq 13\gamma_s.
\end{aligned}$$

$\square$

4.3   17-approximation for Graphs with Maximum Degree Five

Using the same strategy as in Theorem 4, we can obtain the following theorem, whose proof is omitted due to lack of space.

Theorem 5. *There is a linear-time 17-approximation algorithm for the SD problem on graphs of maximum degree five.*

5   NP-completeness for Subcubic Planar Bipartite Graphs

Recently, it has been shown that the SD problem for subcubic grid graphs is NP-complete [7]. In this section, we present an alternative and more direct NP-completeness proof for the SD problem on subcubic planar bipartite graphs in the following theorem, whose proof is omitted due to lack of space.

Theorem 6. *The SD problem on subcubic planar bipartite graphs is NP-complete.*

6   FPT-algorithm for Subcubic Graphs

Since the SD problem for subcubic graphs is NP-complete, in this section we consider FPT-algorithm for subcubic graphs. Damoschke [2] showed that the SD problem has a naïve FPT-algorithm run in $O(4^k \times 2^{2k}) = O(16^k)$ for subcubic graphs.

In the following lemma, we present a kernel of size $3k$ for the SD problem for subcubic graphs, where the parameter k is the weight of the signed dominating function.

Lemma 3. *There is a kernel of size $3k$ for the SD problem on subcubic graphs, where k is the weight of the signed dominating function.*

Proof. Let $G = (V, E)$ be a subcubic graph. For any signed dominating function of G with weight k, we claim that $|V| \leq 3k$. For each vertex v labeled with -1, it is clear that v has distance at least three to any other vertex labeled with -1. In other words, we obtain a 3-element subset of V contains v with label -1 and its two neighbors, which must be labeled with $+1$. Moreover, for any pair of such subsets, their intersection is empty. Hence, we can obtain a kernel of size $3k$ as an upper bound for the SD problem on subcubic graphs.   □

First we consider a brute-force algorithm by assigning each vertex $+1$ and -1. Hence we have an $O^*(2^n)$-time algorithm. By Lemma 3, we have a kernel of size $3k$. Thus the brute-force algorithm can be preformed in $O^*(2^{3k}) = O^*(8^k)$ time. In the following, we present an improved $O^*(5.1957^k)$-time FPT-algorithm which solves the SD problem on subcubic graphs.

In the following theorem, we first state the detailed steps of our algorithm and its correctness proof. Then we analyze its time complexity.

Theorem 7. *The SD problem of subcubic graphs G can be solved in $O^*(5.1957^k)$ time, where k is the signed dominating function number of G.*

Proof (Sketch). Due to Lemma 3, we only need to consider the given subcubic graph G with kernel size of $3k$ vertices. Furthermore, since disconnected components of a graph can be handled separately, we assume that the given graph G is connected in the following context.

The details of our algorithm are as follows. In our algorithm, we grow a potential optimal signed dominating set D incrementally, where the *signed dominating set* D is a set of vertices in V labeled with values $+1$ or -1. The *label* of a vertex is the value of vertex assigned by a specific signed dominating function. A vertex is called *labeled* is if it has been assigned a value; otherwise, it is called *unlabeled*. The weight of the closed neighborhood $N[v]$ of a labeled vertex v is called *valid* (resp. *invalid*) if all vertices in $N[v]$ are labeled, and the sum of weights of vertices in $N[v]$ is positive (resp. non-positive). In the process, we maintain a list L of unlabeled vertices which are the neighboring vertices of the currently labeled vertices in D. Initially, L is set to contain one degree-3 vertex of the input graph G, and $D = \emptyset$. During each iteration of our algorithm, we select an arbitrary unlabeled vertex y from list L as the focus vertex, and we assume that x is a labeled vertex adjacent to y in $D \cap N[y]$. We set $\Delta = (N[x] \cup N[y]) \setminus D$. Then our algorithm makes execution branches on all different ways of assigning values to vertices in Δ of new labeled vertices in $(N[x] \cup N[y]) \setminus D$, by performing detailed case analysis from *Case 1.1* to *Case 4.2*, which is provided in the following context. We remark that for technical reasons, in the analysis of *Subcases 1.6, 2.4, 3.3* and *4.2*, we need to set $\Delta = (N[x] \cup N[N[y]]) \setminus D$ instead. In the case analysis, our algorithm makes subsequent recursions only on those feasible ways of value assignment in the corresponding subcases. For each of such subsequent recursions, the vertices of Δ, which have been labeled, are added into D, resulting in a larger labeled dominating set $D + \Delta$. Then for each vertex v in $D + \Delta$, if all vertices in $N[v]$ are labeled, we then check whether the weight of $N[v]$ is valid. If we reach any weight of closed neighborhood of a vertex is invalid, then the current execution branch is aborted; otherwise, we proceed to update D and L for next round of execution. We update D by setting $D = D + \Delta$, and then we update list L accordingly by visiting the neighboring vertices of the neighbors of vertices in Δ. More precisely, the vertices in Δ are removed from L, and the unlabeled vertices in the neighborhoods of vertices in Δ are added into L. It is clear that such an update takes only $O(1)$ time. Then we proceed to the next execution round with the updated D and L as parameters. We repeat such a selection step until all vertices in G are labeled, that is, L becomes empty. Thus we obtain a candidate signed dominating set D.

In the selection process, we enumerate and store all possible candidates of D according to the above recursive procedure. When all branches of the selection process finished, we obtain a set of candidate signed dominating sets D for the input graph G. We choose the one with minimum weight among all these candidates. This completes our algorithm.

Case Analysis for Selection Step. In each selection step as mentioned in the overview of our algorithm, we are going to label the vertices in set Δ with values $+1$ or -1, where $\Delta = (N[x] \cup N[y]) \setminus D$ or $\Delta = (N[x] \cup N[N[y]]) \setminus D$, by performing careful case analysis for all following subcases from *Case 1.1* to *Case 4.2*. In the step, we plan to add the newly labeled vertices in Δ into D such that the resulting larger signed dominating set $D = D \cup \Delta$, in which the weight of each vertex v in D is valid if all vertices in $N[v]$ are labeled.

In the initial selection step of our algorithm, the set Δ and list L are empty. We choose a vertex with degree three as y, and the vertex x as defined previously does not exist for current situation. See Fig. 2. For such a case, since at most one of the four vertices in $N[y]$ can be labeled with -1, we have five choices to label the four vertices. Note that this kind of selection step only executes once in our whole algorithm, and after this initial selection step, the current signed domination set D has already contained the four elements in $N[x]$, that is, $|D| = 4$.

In the subsequent selection steps of our algorithm, we consider the following subcases from *Case 1.1* to *Case 4.2* for different feasible ways of value assignment for the unlabeled vertices in Δ. Then we proceed to call subsequent recursions according to these different feasible ways of value assignment for Δ. First of all, if the degree of y in G is one, then both x and y must be labeled with $+1$. Thus we have one feasible way to label vertex y.

Fig. 2. This figure shows the vertex y and its neighbors.

Therefore, we assume that in the following case analysis, the degrees of both vertices x and y are at least two. Moreover, during the execution of a branch in our algorithm, we need to run recursively until $|D| = |V|$, that is, a candidate signed dominating set D is found. We consider four cases according to different degrees of x and y. These four cases consist of 15 subcases in total, ranging from *Case 1.1* to *Case 4.2*.

For *Case 1* and its six subcases, we give the details in the following. Due to lack of space, the analysis for *Cases 2* to *4* is omitted.

Case 1. The degrees of both x and y are three. Due to the initial selection step, we know that there is another neighbor x_1 of x lying in D. Let x_2 be the third neighbor of x, and y_1 and y_2 be the other two neighbors of y. For this case, since y_1 and y_2 are symmetric, we need to consider six subcases depending on the content of Δ, the set of unlabeled vertices in the neighborhood of x and y. See Figs. 3(a) to (f). We will analyze these six subcases one by one in the following.

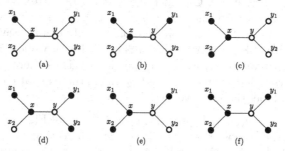

Fig. 3. This figure shows the six subcases of *Case 1*, where the labeled vertices in D are drawn as black disks, and the unlabeled vertices in Δ as circles. (a) *Subcase 1.1*: $\Delta = \{y, x_2, y_1, y_2\}$. (b) *Subcase 1.2*: $\Delta = \{y, x_2, y_2\}$. (c) *Subcase 1.3*: $\Delta = \{y, y_1, y_2\}$. (d) *Subcase 1.4*: $\Delta = \{y, x_2\}$. (e) *Subcase 1.5*: $\Delta = \{y, y_2\}$. (f) *Subcase 1.6*: $\Delta = \{y\}$.

Subcase 1.1. $\Delta = \{y, x_2, y_1, y_2\}$. That is, $\{x, x_1\} \subset D$. See Fig. 3(a).

In order to reduce the number of subcases we need to consider under a specific case, we only need to consider the worst-case scenarios in each of the subcases. For such a purpose, we make the following two assumptions.

(i) When two vertices are connected by an edge, the number of feasible ways of labeling these two vertices will possibly decrease. Thus the worst-case scenario of *Subcase 1.1* is the subgraph as shown in Fig. 3(a) in the way that there is no edge connecting any pair of vertices in $(N[x] \cup N[y]) \setminus \{x, y\}$.

(ii) We observe that a vertex labeled with -1 only have vertices labeled with $+1$ as its neighbors, and a vertex labeled with $+1$ can have vertices labeled with $+1$ or -1 as its neighbors. Thus for a specific subcase, the worst-case scenario for the number of feasible ways to label the vertices in Δ is when the labeled vertices for the specific subcase (for instance, vertices x and y in *Subcase 1.1*) are all assigned value $+1$.

We remark that these two assumptions are usually made for subsequent subcases we considered in this proof. This dramatically shorten the length of the proof required for each subcase.

For *Subcase 1.1*, according to assumption (i), we suppose that there is no edge connecting any pair of vertices in $(N[x] \cup N[y]) \setminus \{x, y\}$, and according to assumption (ii), we suppose that x and x_1 are labeled with $+1$. Consider the labeling of vertex y. There are two ways to label y, say with value -1 or $+1$. First, if y is labeled with value -1, then vertices x_2, y_1 and y_2 have only one feasible way of labeling, say all of them being labeled with value $+1$. Second, if y is labeled with value $+1$, then x_2 can be assigned value $+1$ or -1, and at most one of y_1 and y_2 can be assigned value -1. Thus there can be at most six feasible ways to label vertices x_2, y_1 and y_2. In all, there are at most seven feasible ways in total to label the four vertices in Δ for this subcase.

Subcase 1.2. $\Delta = \{y, x_2, y_2\}$. That is, $\{x, x_1, y_1\} \subset D$. See Fig. 3(b). According to assumption (i), we suppose that there is no edge connecting any pair of vertices in $(N[x] \cup N[y]) \setminus \{x, y\}$, and according to assumption (ii), we suppose that x, x_1 and y_1 are labeled with $+1$. Consider the labeling of vertex y. There are two ways to label y, say with value -1 or $+1$. First, if y is labeled with value -1, then vertices x_2 and y_2 have only one feasible way of labeling, say both of them being labeled with value $+1$. Second, if y is labeled with value $+1$, then x_2 and y_2 can be assigned value $+1$ or -1, independently. Thus there can be at most four feasible ways to label vertices x_2 and y_2. In all, there are at most five feasible ways in total to label the three vertices in Δ for this subcase.

Subcase 1.3. $\Delta = \{y, y_1, y_2\}$. That is, $\{x, x_1, x_2\} \subset D$. See Fig. 3(c). According to assumption (i), we suppose that there is no edge connecting any pair of vertices in $(N[x] \cup N[y]) \setminus \{x, y\}$, and according to assumption (ii), we suppose that x, x_1 and x_2 are labeled with $+1$. Since at most one of y, y_1 and y_2 can be assigned -1, there are at most four feasible ways in total to label the three vertices y, y_1 and y_2 in Δ for this subcase.

Subcase 1.4. $\Delta = \{y, x_2\}$. That is, $\{x, x_1, y_1, y_2\} \subset D$. See Fig. 3(d). According to assumption (i), we suppose that there is no edge connecting any pair of vertices in $(N[x] \cup N[y]) \setminus \{x, y\}$, and according to assumption (ii), we suppose that x, x_1, y_1 and y_2 are labeled with $+1$. Since at most one of x_2 and y can be assigned -1, there are at most three feasible ways in total to label the two vertices y and x_2 in Δ for this subcase.

Subcase 1.5. $\Delta = \{y, y_2\}$. That is, $\{x, x_1, x_2, y_1\} \subset D$. See Fig. 3(e). According to assumption (i), we suppose that there is no edge connecting any pair of vertices in $(N[x] \cup N[y]) \setminus \{x, y\}$, and according to assumption (ii), we suppose that x, x_1, x_2 and y_1 are labeled with $+1$. Since at most one of y and y_2 can be assigned -1, there are at most three feasible ways in total to label the two vertices y and y_2 in Δ for this subcase.

Subcase 1.6. $\Delta = \{y\}$. That is, $\{x, x_1, x_2, y_1, y_2\} \subset D$. See Fig. 3(f). There are two ways to label the only unlabeled vertex y, say with value -1 or $+1$. Thus, in order to achieve a better running time, we consider a larger subgraph. See Fig. 4. We further consider the vertices in neighborhoods of y_1 and y_2. Since $y_1, y_2 \in D$, and after the initial selection step, the subgraph induced by D is always connected and $|D| \geq 4$, there is another labeled neighbor $z_1 \in D$ of y_1, and there is another labeled neighbor $z_2 \in D$ of y_2. There may or may not be third neighbor of y_1, say z_3 if any. There may or may not be third neighbor of y_2, say z_4 if any. According to assumption (i), we suppose that there is no edge connecting any pair of vertices in $(N[x] \cup N[N[y]]) \setminus \{x, y\}$ for this subcase. Let Z be the set containing these new neighbors, z_1, z_2, z_3 (if any) and z_4 (if any), of y_1 and y_2. Note that $|Z| \leq 4$. Now we add the unlabeled vertices in Z into Δ, which originally only contains vertex y. As $z_1, z_2 \in D$, we have that $|Z \cap \Delta| \leq 2$. We consider three situations depending on the value of $|Z \cap \Delta|$ which is either 0, 1, or 2. See Figs. 4(a) to (c), respectively.

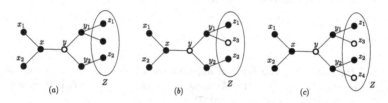

Fig. 4. This figure shows the three situations of *Subcase 1.6.* (a) *Situation 1.6a:* $|Z \cap \Delta| = 0$. Then $\Delta = \{y\}$. (b) *Situation 1.6b:* $|Z \cap \Delta| = 1$. W.l.o.g., $\Delta = \{y, z_3\}$. (c) *Situation 1.6c:* $|Z \cap \Delta| = 2$. Then $\Delta = \{y, z_3, z_4\}$.

Situation 1.6a. $|Z \cap \Delta| = 0$. That is, all vertices in Z are labeled. See Fig. 4(a). Note that we don't use assumption (ii) for this situation. It is clear that if all of the vertices x, x_1, x_2, y_1, y_2 and vertices in Z are labeled with $+1$, then we can label y with -1. Otherwise, vertex y must be labeled with $+1$. Hence, there are only one feasible way to label the only vertex y in Δ for this situation.

Situation 1.6b. $|Z \cap \Delta| = 1$. Let z_3 be the only vertex in $Z \cap \Delta$. Then $\Delta = \{y, z_3\}$. Since y_1 and y_2 are symmetric, without loss of generality, let z_3 be the third neighbor of y_1. See Fig. 4(b). According to assumption (ii), we suppose that x, x_1, x_2, y_1, y_2 and the vertices in $Z \cap D$ are labeled with $+1$. Since at most one of y and z_3 can be assigned -1, there are at most three feasible ways in total to label the two vertices y and z_3 in Δ for this situation.

Situation 1.6c. $|Z \cap \Delta| = 2$. Then one vertex in $Z \cap \Delta$ must be the third neighbor of y_1, say z_3, and the other vertex in $Z \cap \Delta$ must be the third neighbor of

y_2, say z_4. Thus $\Delta = \{y, z_3, z_4\}$. See Fig. 4(c). According to assumption (ii), we suppose that x, x_1, x_2, y_1, y_2 and the vertices in $Z \cap D$ are labeled with $+1$. Consider the labeling of vertex y. There are two ways to label y, say with value -1 or $+1$. First, if y is labeled with value -1, then vertices z_3 and z_4 have only one feasible way of labeling, say both of them being labeled with value $+1$. Second, if y is labeled with value $+1$, then z and y_2 can be assigned value $+1$ or -1, independently. Thus there can be at most four feasible ways to label vertices z_3 and z_4. In all, there are at most five feasible ways in total to label the three vertices in Δ for this subcase.

This completes the case analysis of *Case 1*.

In the above detailed analysis for *Case 1*, the *Subcases 1.4*, *1.5*, *1.6b* take the worst-case running time. Even in the omitted analyses of *Cases 2* to *4*, the subcases with the worst-case running time also have the same time complexity as the aforementioned worst subcases in *Case 1*. Let $T(n)$ be the worst-case running time of the whole algorithm. Then we have the following recurrence relation:

$$T(n) \leq 3T(n-2) + O(1).$$

By solving this recurrence relation, we thus obtain that:

$$T(n) = O^*(1.7320^n) = O^*(1.7320^{3k}) = O^*(5.1957^k)$$

as $n = 3k$ for subcubic graphs by Lemma 3.   □

References

1. Alimonti, P., Kann, V.: Hardness of approximating problems on cubic graphs. In: Bongiovanni, G., Bovet, D.P., Di Battista, G. (eds.) CIAC 1997. LNCS, vol. 1203, pp. 288–298. Springer, Heidelberg (1997)
2. Damaschke, P.: Minus signed dominating function in small-degree graphs. Discrete Appl. Math. **108**, 53–64 (2001)
3. Downey, R.G., Fellows, M.R.: Parameterized Complexity. Springer, Heidelberg (1999)
4. Faria, L., Hon, W.-K., Kloks, T., Liu, H.-H., Wang, T.-M., Wang, Y.-L.: On complexities of minus domination. In: Widmayer, P., Xu, Y., Zhu, B. (eds.) COCOA 2013. LNCS, vol. 8287, pp. 178–189. Springer, Heidelberg (2013)
5. Hattingh, J.H., Henning, M.A., Slater, P.J.: The algorithmic complexity of signed dominating function in graphs. Australas. J. Comb. **12**, 101–112 (1995)
6. Lee, C.-M.: Labelled signed dominating function and its variants, Ph.D. thesis, National Chung Cheng University, Taiwan (2006)
7. Zheng, Y., Wang, J., Feng, Q.: Kernelization and Lower Bounds of the Signed Domination Problem. In: Fellows, M., Tan, X., Zhu, B. (eds.) FAW-AAIM 2013. LNCS, vol. 7924, pp. 261–271. Springer, Heidelberg (2013)

Author Index